INTERNATIONAL UNION OF THEORETICAL
AND APPLIED MECHANICS

BOUNDARY LAYER RESEARCH

SYMPOSIUM FREIBURG/BR. AUGUST 26–29, 1957

EDITED BY

H. GÖRTLER

WITH 206 FIGURES

SPRINGER-VERLAG BERLIN HEIDELBERG GMBH

1958

INTERNATIONALE UNION FÜR THEORETISCHE
UND ANGEWANDTE MECHANIK

GRENZSCHICHTFORSCHUNG

SYMPOSIUM FREIBURG/BR. 26. BIS 29. AUGUST 1957

HERAUSGEGEBEN VON

H. GÖRTLER

MIT 206 ABBILDUNGEN

SPRINGER-VERLAG BERLIN HEIDELBERG GMBH

1958

ISBN 978-3-540-02273-2   ISBN 978-3-642-45885-9 (eBook)
DOI 10.1007/978-3-642-45885-9

Vorwort

Der Beschluß des Generalrats der Internationalen Union für Theoretische und Angewandte Mechanik (IUTAM), ein Symposium über Grenzschichtforschung vorzubereiten, war von dem Wunsch getragen, Wissenschaftlern aus aller Welt, die in eigenen Arbeiten zu den neueren Fortschritten auf diesem wichtigen Gebiet der Strömungsmechanik wesentlich beigetragen haben, zu gründlichen Diskussionen Gelegenheit zu geben. Die grundsätzlichen Probleme der Grenzschichtforschung sollten bei der Aussprache im Vordergrund stehen, die technischen Anwendungen mußten im Hintergrund bleiben.

Da eine echte Diskussion nur in einem relativ kleinen Kreis möglich ist, hat das Wissenschaftliche Komitee den Kreis der Teilnehmer beschränkt, dabei aber nach Möglichkeit Vertreter aus allen an den Fortschritten der Grenzschichtforschung beteiligten Nationen zusammengeführt. Die Teilnehmer wurden in Beratungen des Wissenschaftlichen Komitees ausgewählt und entweder zur Erstattung eines allgemeinen oder eines speziellen Berichtes eingeladen bzw. zur Teilnahme an den Diskussionsveranstaltungen aufgefordert.

Das Ergebnis dieser Tagung legen wir hiermit vor. Im Druck sind alle jene Fragen und Bemerkungen aus den Diskussionen weggelassen worden, die sich durch das Vorliegen des vollen Vortragstextes in diesem Buch von selbst erledigen, bzw. die durch nachträgliche Bearbeitung des Manuskriptes von den Autoren berücksichtigt worden sind.

Wir empfinden es mit Dank als eine Ehrung, daß die erste Veranstaltung der IUTAM in Deutschland die Grenzschichtforschung behandelt und damit das Gedenken an LUDWIG PRANDTL wachhält, den Begründer der Grenzschichtforschung und den Förderer der modernen Strömungsforschung.

Der Herausgeber als Vorsitzender des von der IUTAM eingesetzten Wissenschaftlichen Komitees dankt hiermit allen Förderern des Symposiums, zunächst den Teilnehmern aus aller Welt, die durch ihre Vorträge und die Beteiligung an der Diskussion das Gelingen des Symposiums sicherten; ebenso aber auch den Mitgliedern des Wissenschaftlichen Komitees für ihre Mitwirkung bei der Planung der Veranstaltung. Außer der IUTAM haben auch andere Stellen durch erhebliche Beiträge das Zustandekommen des Symposiums finanziell ermöglicht. Hier gilt mein

besonderer Dank dem Bundesministerium des Innern, dem Kultus-
ministerium des Landes Baden-Württemberg, der Rand Development
Corporation, Cleveland/Ohio, und besonders derem Vizepräsidenten,
Herrn Professor C. J. PEIRCE, ferner aber auch der Albert Ludwigs-
Universität, Freiburg im Breisgau, und der Deutschen Versuchsanstalt
für Luftfahrt.

Mein herzlicher Dank gilt allen meinen Mitarbeitern sowohl des Insti-
tuts für Angewandte Mathematik der Universität Freiburg als auch des
mit dieser Veranstaltung erstmals an die Öffentlichkeit getretenen Insti-
tuts für Angewandte Mathematik und Mechanik der Deutschen Versuchsanstalt für Luftfahrt in Freiburg.

Bei den umfangreichen Korrekturarbeiten zur Fertigstellung des Ban-
des hat mich in dankenswerter Weise mein Mitarbeiter Dr. G. JUNGCLAUS
weitgehend entlastet. Der Springer-Verlag hat meinen Wünschen in
jeder Weise entsprochen und bringt das Werk in der bekannten vor-
züglichen Ausstattung heraus. Es ist mir eine Freude, im Namen aller
Beteiligten hierfür zu danken.

Freiburg/Br., im Juli 1958

H. Görtler

Wissenschaftliches Komitee — Scientific Committee:

H. L. DRYDEN, Washington.

H. GÖRTLER (Vorsitzender — Chairman), Freiburg i. Br.

L. HOWARTH, Bristol.

M. ROY, Paris.

R. TIMMAN, Delft.

Mitarbeiter des örtlichen Tagungsausschusses — Members of the Local Organizing Committee:

P. ČOLAC-ANTIĆ, Frl. S. KAHMANN, Frl. H. KOMPE, K. MAGNUS, H. OSER, V. SALJ-NIKOV, H. WITTING

aus dem Institut für Angewandte Mathematik der Albert-Ludwigs-Universität Freiburg i. Br.;

Frau E. AHMED, T. GEIS, G. HÄMMERLIN, G. HELKE, K. KIRCHGÄSSNER, Frl. D. LICHTBLAU, E. WRAGE

aus dem Institut für Angewandte Mathematik und Mechanik der Deutschen Versuchsanstalt für Luftfahrt, Freiburg i. Br.

Die Teilnehmer am Symposium

Verzeichnis der Teilnehmer — List of Participants

Argentinien — Argentine:
J. E. DE KRASINSKI.

Deutschland — Germany:
E. FÖRTHMANN, T. GEIS, K. GERSTEN, H. GÖRTLER, G. HÄMMERLIN, G. JUNG-
CLAUS, H. J. KAEPPELER, G. LUDWIG, K. MAGNUS, K. NICKEL, A. W. QUICK,
F. W. RIEGELS, N. SCHOLZ, K. SCHRÖDER, R. SCHWARZENBERGER, O. TIETJENS,
W. VAN NES, A. WALZ, O. WEHRMANN, J. WEISSINGER, K. WIEGHARDT, R.
WILLE, H. WITTING, F. X. WORTMANN, W. WUEST.

England — England:
G. E. GADD, M. B. GLAUERT, W. MANGLER, E. C. MASKELL, W. J. STERN,
K. STEWARTSON, G. I. TAYLOR, G. TEMPLE, B. THWAITES, A. A. TOWNSEND.

Frankreich — France:
E. A. EICHELBRENNER, J. KAMPÉ DE FÉRIET, D. MASSIGNON, R. MICHEL, M. ROY.

Holland — Netherlands:
H. BERGH, J. L. VAN INGEN, W. T. KOITER, R. TIMMAN, J. A. ZAAT.

Indien — India:
B. R. SETH.

Israel — Israel:
J. B. POPPER, M. REINER.

Italien — Italy:
C. FERRARI, E. PISTOLESI.

Japan — Japan:
I. TANI, T. TATSUMI.

Jugoslawien — Yugoslavia:
P. ČOLAC-ANTIĆ, V. SALJNIKOV, K. VORONJEC.

Kanada — Canada:
G. N. PATTERSON

Norwegen — Norway:
O. BJÖRGUM, L. N. PERSEN.

Polen — Poland:
M. ŁUNC.

Schweden — Sweden:
G. DROUGGE, H. FAXÉN, N. FRÖSSLING, F. K. G. ODQVIST, H. SCHUH.

Schweiz — Switzerland:
J. ACKERET.

Union der Sozialistischen Sowjetrepubliken — Soviet Russia:
A. A. DORODNICYN, A. A. NIKOLSKIJ, H. A. RACHMATULIN.

Vereinigte Staaten von Amerika — United States of America:
H. L. DRYDEN, A. IBERALL, TH. VON KÁRMÁN, G. KUERTI, J. LAUFER, C. C.
LIN, H. MIRELS, F. K. MOORE, S. OSTRACH, C. J. PEIRCE, W. PFENNINGER,
S. A. SCHAAF, M. G. SCHERBERG, G. B. SCHUBAUER, I. R. SCHWARTZ, M. P. TULIN.

Inhaltsverzeichnis — Contents

IV. SITZUNG

Dienstag, den 27. 8. 57, nachmittags

Vorsitzender: A. W. QUICK

Diskussionsveranstaltung zur IV. Sitzung

V. SITZUNG

Mittwoch, den 28. 8. 57, vormittags

Vorsitzender: G. N. PATTERSON

Diskussionsveranstaltung zur V. Sitzung

VI. SITZUNG

Mittwoch, den 28. 8. 57, nachmittags

Vorsitzender: M. ROY

VII. SITZUNG

Donnerstag, den 29. 8. 57, vormittags

Vorsitzender: J. Ackeret

VIII. SITZUNG

Donnerstag, den 29. 8. 57, nachmittags

Vorsitzender: M. Łunc

The turbulent boundary layer

By

A. A. Townsend

Cavendish Laboratory, Cambridge

Summary

Two experimental studies of turbulent boundary layers are described. The first was begun as part of a search for more accurate knowledge of the structure and motion of the large-scale eddy motions that appear to control the rate of entrainment of non-turbulent fluid and which may also play a dominant part in the constant-stress layer. The experimental technique depends on the measurement of mean products of the various components of the fluctuation velocities at two separated points in the flow, and a comparison of the measurements with those to be expected from the presence of hypothetical simple eddy structures. Within the inner, constant-stress layer, the measurements are consistent with the irregular occurrence of nearly two-dimensional jets, directed outwards from the viscous layer and of comparatively long and indeterminate length in the direction of mean flow. In the outer part of the layer the dominant large-scale motion appears to be jets of turbulent fluid which may arise from the release of the anisotropic Reynolds stresses set up by the shearing motion at the free boundary. The connection between these results and the hypothesis of wall similarity ("law of the wall") is discussed.

The second part is concerned with the boundary layer on a flat plate of finite aspect ratio, in particular the flow in the immediate neighbourhood of the free edge parallel to the stream and its effect on the boundary layer as a whole. The measurements include mean velocities, turbulent intensities and local surface friction. For fully turbulent flow at effective Reynolds numbers over one million, well-developed eddies are found with their axes parallel to the edge and one on each side of the plate. These are caused by the cross-flow which must arise if a boundary layer is bounded by a free edge as a consequence of the well-known inequality of the normal Reynolds stresses in a boundary layer. Measurements at lower Reynolds numbers show that the laminar flow near the edge becomes unstable at a lower Reynolds number than the rest of the layer and may become turbulent before the remainder of the flow is in any way unstable. The interaction between the two parts of the flow has some very interesting features and it is hoped that these experiments may help in the interpretation of recent measurements of "turbulent" skin friction on flat plates at very low Reynolds numbers.

1. Introduction

In this paper I propose to describe two separate pieces of work on boundary layers, one concerned with the fundamental problem of the nature of the fluid motion in turbulent flow and the other with the effects of a free edge on the flat-plate boundary layer. The first part is an outline of the results of a comprehensive survey of the double velocity correlation function in a turbulent boundary layer, carried out with the purpose of determining the structure of the largest-scale components of the turbulent motion. Almost everyone has his own private idea of a typical "eddy" and of the nature of the complex motion that forms a turbulent flow, but an attempt to give these ideas quantitative form soon leads to difficulty and argument. Our group in the Cavendish Laboratory has been occupied with this problem for some time and progress has been made with the identification and description of the largest eddies. The importance of these eddies depends on the hypothesis that they exercise control over the transfer of energy from the mean flow to the turbulent motion (Townsend 1956), and they are also particularly easy to study since they determine the behaviour of the correlation function for large separations. All of this may seem to confirm the popular opinion that we insist on our work having no practical significance, but knowledge of this kind helps to show in what way the wall stress determines the mean motion next the wall and lead to a better understanding of the basis and limitations of wall similarity.

The second part describes some preliminary measurements in the boundary layer on a flat plate of finite aspect ratio, particularly in the neighbourhood of the free edge. It has been shown (Townsend 1954) that the inequality of the Reynolds normal stresses within the boundary layer must cause a cross-flow, at least in the neighbourhood of a free edge, and the nature and extent of this flow is of considerable interest when it comes to correcting measurements of skin-friction for finite aspect ratio. The nature of the transition to turbulent flow has also been studied.

All the experimental work described below was done either by Dr. H. L. Grant (correlation measurements in the boundary layer) or by Mr. J. W. Elder (boundary layer near a free edge) and will be reported by them separately and in more detail. The author is greatly indebted to them for allowing him to use this material, and his contribution is limited to report and comment on their work.

2. Inference of eddy structure from correlation measurements

It is well known that the general form of the velocity correlation function is determined by the larger-scale, energy-containing components of the turbulent motion, the small-scale components producing only fine detail at small separations of the points of observation. This is due in part to the lesser energy of the small-scale motions but also to their lack of correlation over large distances. So, if there is present a group of eddies appreciably larger in size than any of the remaining eddies, the correlation function for large separations will be very nearly the correlation function for the velocity field of the large eddies alone. Many, though not all, of the correlation components that have been measured in turbulent shear flows suggest strongly that a moderately distinct group of eddies is always present (TOWNSEND 1956), and there seems to be a physical meaning in proceeding on the assumption that a turbulent shear flow may be considered in three parts, a mean flow, a set of large eddies and an "ordinary" turbulent motion made up of a wide and continuous range of eddy sizes.

The usual definition of the velocity correlation function (strictly the covariance) is

$$R_{ij}'\left(\boldsymbol{x} \; ; \; \boldsymbol{r}\right) = \overline{u_i\left(\boldsymbol{x}\right) u_j\left(\boldsymbol{x} + \boldsymbol{r}\right)} \qquad (2.1)$$

where u_i is the component of the velocity fluctuation along the $0\,x_i$ axis at the vector position $\boldsymbol{x}$, and u_j is the component along the $0\,x_j$ axis at the position $\boldsymbol{x} + \boldsymbol{r}$. It is convenient, both experimentally and otherwise, to use non-dimensional quantities, and this may be done by defining

$$R_{ij}\left(\boldsymbol{x} \; ; \; \boldsymbol{r}\right) = \frac{\overline{u_i\left(\boldsymbol{x}\right) u_j\left(\boldsymbol{x} + \boldsymbol{r}\right)}}{\overline{u_i{}^2\left(\boldsymbol{x}\right)}} \qquad (2.2)$$

(repeated suffices do *not* imply summation). This definition corresponds with the experimental arrangement in which one anemometer wire is fixed with respect to the flow and the other is moved and has more significance in inhomogeneous flows than the ordinary correlation coefficient. The measurements have been confined to those components of the correlation function for which $i = j$, and, with few exceptions, for which the vector separation $\boldsymbol{r}$ is parallel to one of the co-ordinate axes.

Inference of the structure of a system of eddies from the velocity correlation function is neither a simple nor a unique process, and depends on the prejudices of the operator. With experience, methods may be developed for the rejection of classes of eddy structure, but to make a single choice from the possible structures remaining demands strong prejudices. The most important of these is a belief that the correlation function at large separations is determined by a homogeneous group of large eddies of simple structure. The component eddy structures are assumed to occur with centres randomly distributed in directions of effective homogeneity of the flow, for a boundary layer over planes parallel to the wall. Conformity of a particular eddy structure with the observations can be tested by computing the correlation function for such a random arrangement of the selected eddies, but this may be a laborious process and is not recommended as a first step. The general character of the basic eddy structure is more easily inferred by relating the form of each component of the correlation function to the form of the average velocity profile of the velocity component concerned for a transit in the direction of separation through the eddy. Thus, a correlation function that is positive and decreases monotonically to zero with increase of separation (type A) is produced by the passage of eddy structures with pulse-like velocity profiles. One that becomes negative and then tends monotonically to zero (type B) corresponds to profiles resembling a single cycle of a sine-wave. And so on, Simple spherical vortices lead to correlation components of types A or B. depending on the orientation of the plane of circulation, and vortex rings may lead to types A, B, or C. A compact way of representing the correlation material is to use a pattern of the type letters for the various components in the array

$$\begin{pmatrix} R_{11}\,(r, 0, 0) & R_{11}\,(0, r, 0) & R_{11}\,(0, 0, r) \\ R_{22}\,(r, 0, 0) & R_{22}\,(0, r, 0) & R_{22}\,(0, 0, r) \\ R_{33}\,(r, 0, 0) & R_{33}\,(0, r, 0) & R_{33}\,(0, 0, r) \end{pmatrix}$$

Turbulence composed of simple long eddies with axes along $0x$ and circulation in planes equally inclined to the $y0z$ and $x0z$ planes has a pattern $\begin{pmatrix} A\,A\,B \\ A\,A\,B \\ A\,B\,A \end{pmatrix}$ while isotropic turbulence made up of simple eddies randomly orientated has one $\begin{pmatrix} A\,B\,B \\ B\,A\,B \\ B\,B\,A \end{pmatrix}$. In this notation, the motion in the outer part of the boundary layer has a pattern $\begin{pmatrix} A\,A\,\,B \\ B\,A\,\,B? \\ B\,B?\,A \end{pmatrix}$, and in the constant stress layer $\begin{pmatrix} A\,A\,\,B \\ B\,A\,\,A \\ A\,B?\,B \end{pmatrix}$.

3. Experimental Details

The measurements were made in the boundary layer on the wall of a small wind-tunnel of rectangular section, approximately 61 cm. by 15 cm. but with a slow expansion of the section to maintain a negligible longitudinal pressure gradient. The free stream velocity was about 700 cm./sec.$^{-1}$ and the total thickness of the layer at the position of measurement was about 5.5 cm. A more easily determined thickness is the distance from the wall at which the difference of the mean velocity from the free stream velocity equals the friction velocity. This thickness, δ_0, was 3.8 cm. The observed profiles of mean velocity have been compared with other measurements on flat plates by plotting in the form $[U_1 - U(y)]/\tau_0^{1/2}$ vs. y/δ_0 (U_1 is the free stream velocity and τ_0 is the [kinematic] wall stress), and are in good agreement. The hot-wire assemblies used to measure the three components of the fluctuation velocity were

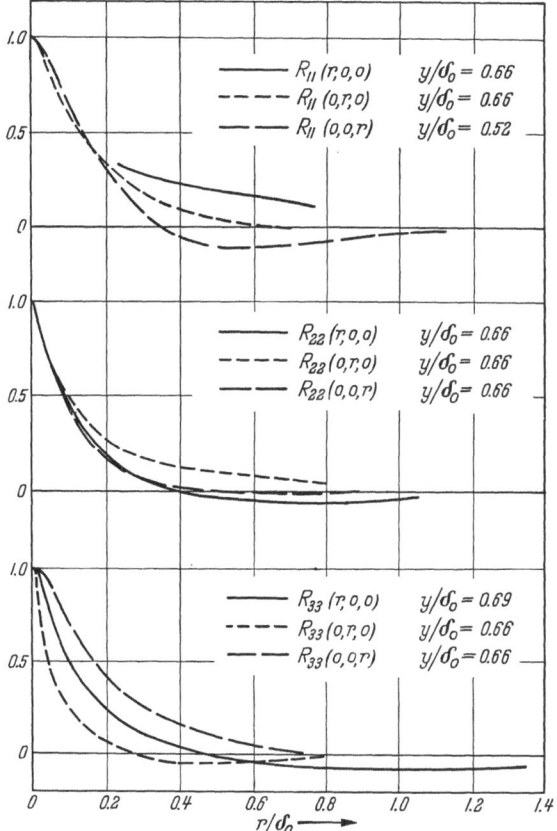

Fig. 1. Correlations in the outer part of a boundary layer

either single wires between 0.5 and 1 mm. long for the measurement of the 0x component or X-form wires of similar dimensions for the measurement of transverse velocity components.

The measurements were made with one wire at a fixed distance from the plate and the other at various positions displaced with respect to the fixed wire parallel to the three co-ordinate axes, which are conventionally, 0x parallel to the free stream, 0y at right angles to the surface and 0z in the plane of the plate and at right angles to the mean stream. The results considered here refer broadly to two different positions of the fixed wire, one set with the fixed wire at or near $y/\delta_0 = 0.70$

and the other for $y/\delta_0 = 0.04$. In each set the $0y$ displacements are outwards, and so the first set refers to the outer layer (intermittency of flow
becomes appreciable near $y/\delta_0 = 0.9$) and the second to motion in the
constant stress layer.

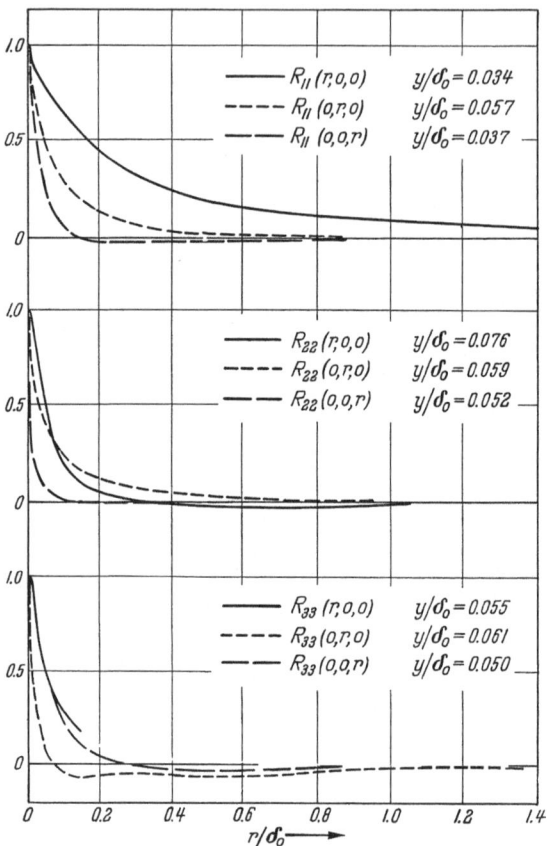

Fig. 2. Correlations in the constant-stress region of a boundary layer

4. The outer layer

A short examination of the correlation components in fig. 1 is sufficient
to show that simple roller eddies of the kind postulated by the author
(TOWNSEND 1956) cannot lead to correlations of these forms. After considering many others types of eddy, the conclusion has been reached that
the dominant form of large-eddy motion are motions which may be described as jets of turbulent fluid emerging from the interior of the flow
into the relatively undisturbed flow outside. The mean velocity gradient
causes the jet to follow a curved path and the flow pattern may resemble

that sketched in fig. 3. Partial confirmation of the existence of these jets has been obtained by measuring the correlation $R_{22}\,(r_1, r_2, 0)$, which shows considerable asymmetry corresponding to the jet deflection, but the main reason for suspecting their presence was the outcome of similar work on the large eddies in a two-dimensional wake (GRANT 1957). In this flow, clear evidence was found of the existence of these jets, very homogeneous in size and often appearing in evenly spaced groups of four to six. Their structure was studied in detail, by measuring correlation components and by visual observations of the diffusion of dye filaments introduced into the flow. GRANT suggests that the jets arise by instability of the bounding surface between the turbulent and the non-turbulent fluid, the cause of the instability being the presence of anisotropic Reynolds stresses within the turbulent fluid. However this may be, their presence in the wake is well-established and their presence in the wake-like part of the boundary layer is probable, although they do not appear to be as regular and homogeneous in the boundary

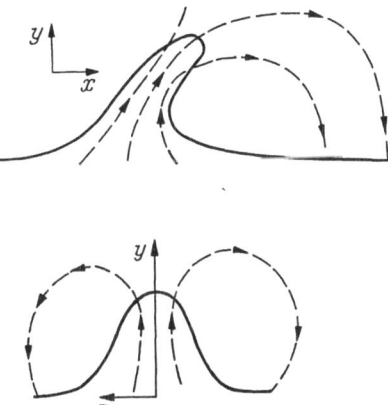

Fig. 3. Structure of a mixing jet in the outer part of a boundary layer. (The full line represents the edge of the jet and is nearly the boundary of the turbulent fluid. The dashed lines are streamlines relative to the mean velocity at the base of the jet.)

layer as they are in the wake. It is not unreasonable to suppose that this is a consequence of the wide range of turbulent intensities and scales to be found in a wall flow as compared with the near homogeneity of wakes and jets.

A remarkable feature of the observations is the close similarity of the component $R_{11}\,(r, 0, 0)$ at all distances from the wall, a feature that might have been inferred from any of the published measurements of the spectrum function of the longitudinal component of the motion. This similarity seems to imply a close correlation between movements at different depths in the layer and this is true if the points of observation lie along a certain curved line in the x0y plane. The course of this line, which is nearly the mean path of the postulated jets, can be inferred from the extremely interesting measurements of space-time correlations by FAVRE, GAVIGLIO & DUMAS (1957), which include, in effect, a complete survey of the $R_{11}\,(r_1, r_2, 0)$ correlation. The correlation is found to be a maximum along curved lines for which r_1/r_2 is of order -1, i.e. roughly parallel to the direction of maximum positive rate of strain. It is to be hoped that

these experimenters will soon extend their measurements to the normal
(0y) component of the turbulent motion, the correlations of which would
be more informative about these eddy structures.

5. The constant stress layer

The measurements taken with the fixed wire deep within the constant-
stress layer (fig. 2) show large differences in scale and character of the
correlation components. The most remarkable feature is the wide range
of scales, from the relatively enormous scale of the R_{11} $(r, 0, 0)$ component
to the scale of the R_{22} $(0, 0, r)$ component which is comparable with
the distance from the wall of the wires. At first sight, these results
are in total disagreement with the corollary of the well-established "law
of the wall", that the turbulent motion responsible for the Reynolds
shear stress must have an effective scale proportional to distance from the
wall. To reconcile these measurements in the constant stress layer with
this scale variation, we must distinguish between the total turbulent
motion and the part responsible for the Reynolds shear stress, and
between the effective scale, which is broadly the scale determining the
rate of energy transfer to the dissipating range of eddies, and the scales
of the motion in particular directions. The differences between the ratios
of turbulent intensities to the wall stress as measured in pipes, channels
and boundary layers (LAUFER 1955, LAUFER 1951, SCHUBAUER 1954)
show that the total motion depends to some extent on the particular flow,
although the universality of the relation between mean velocity and wall
stress implies that the part of the motion responsible for the shear stress
is essentially the same in all these flows. This motion might be called the
universal motion to indicate that it behaves in the way required be wall
similarity, while the remainder of the motion, the "irrelevant" motion,
is characteristic of the particular flow and does not interact with the
universal motion or contribute to the Reynolds shear stress. Comparison
of intensities measured in different flows suggests that the irrelevant
motion is nearly confined to planes parallel to the wall and it may be
regarded as slight surging and changes in direction of the mean flow.

If we now consider the correlation measurements and insist that the
scales of the universal motion should be of order the distance of the fixed
wire from the wall, i.e. $0.04\,\delta_0$, there is immediate difficulty with the
components with separations parallel to the mean flow. None of these
show signs of consisting of two components, one for the universal motion
of scale near $0.04\,\delta_0$ and one for the irrelevant motion of larger scale.
In fact, they are remarkably smooth and we are forced to the conclusion
that scales in the stream direction for all parts of the motion are large
compared with distance from the wall and that these scales are irrelevant

in the sense used above. The large eddies of the universal motion apparently have no definite length in the direction of mean flow.

Turning to the more significant components with separations parallel to $0y$ and $0z$, we notice that $R_{33}(0, 0, r)$, which is a "longitudinal" correlation, takes negative values which indicates either a convergence or divergence of flow in the $0z$ direction. This suggests that the motion may be a two-dimensional jet, originating at the edge of the viscous layer, with its surrounding induced flow (fig. 4), and this hypothesis is in good qualitative agreement with the form of the remaining correlation components. The jet is naturally passing through a velocity gradient transverse to its direction of motion, and turbulent exchange will keep the longitudinal velocity within the jet consistently less than the mean velocity and the velocity outside consistently greater. It may be observed that, by representing the motion in the constant-stress layer as a series of two-dimensional turbulent jets aligned with the mean flow, a consistent account may be obtained of the variations of scale and intensity with distance from the wall and also of the observed skewness of the distribution of normal velocity fluctuations.

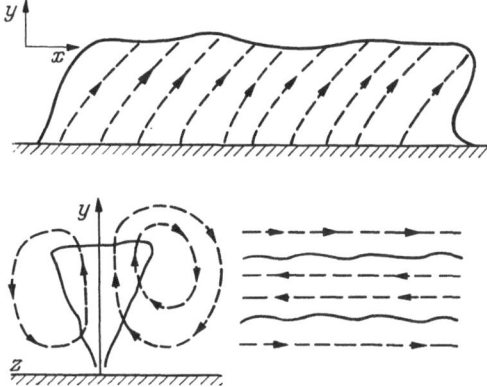

Fig. 4. Structure of the "two-dimensional" jets in the constant-stress layer. (The full line represents the edge of the jet. The dashed lines are streamlines relative to the mean velocity at the base of the jet.)

So far, no mention has been made of the origin of these jets or of the factors determining the irrelevant scale in the direction of flow. This irrelevant scale, which is roughly the mean length of the jets, is nearly identical with the scales of the outer flow, suggesting that disturbances in the outer flow produce jets from the constant-stress layer. Since the motion in the outer flow is of comparatively low intensity, the production of jets can only be through some kind of trigger action and not through direct energy transfer. It is possible that the viscous layer, which is near a condition of critical stability, contains Tollmien-Schlichting waves which will develop most rapidly in local regions of adverse pressure gradient and may break up along lines parallel to the mean stream (for example, see SCHUBAUER 1957). The line of broken waves may then join up to form a nearly two-dimensional jet of length set by the extent of the region of adverse pressure gradient. The large-scale and persistent adverse pressure gradients that could act in this way can only be produced by the motion in the outer part of the layer, and their scale will be much the same as the scale of the velocity fluctuations in this part of the flow.

6. Boundary layer near a free edge: introduction

The work described below is the first stage of a not yet complete investigation, but a number of interesting results have already been obtained that seem to justify a short account. In these experiments, a rectangular steel plate is suspended from the roof of a wind-tunnel with one edge along the centre-line of the tunnel and parallel to the flow in the free stream. The study of the three-dimensional boundary layer near the free edge was begun for two reasons. First, measurements of skin-friction at very high Reynolds numbers are necessarily made using flat plates of low aspect ratio, and the corrections for aspect ratio that have been used depend on measurements of the edge effect at very much lower Reynolds numbers (HUGHES 1954). In an attempt to provide a theory of the edge effect, the author pointed out that the inequality of normal Reynolds stresses in the layer must cause a secondary flow near the free edge, and suggested that the re-entry of the flow into the layer might lead to a general increase in skin friction (TOWNSEND 1954). Some evidence of this had been found by ALLAN & CUTLAND (1952) who made total head traverses in the wake behind flat plates of various aspect ratios. It seemed useful to study the flow near the free edge to find out how much of the total increase in friction is a true edge effect, appearing as additional skin friction located close to the edge, and how much is a distributed edge effect. Second, it appeared likely that the laminar flow near a free edge would become unstable at a lower Reynolds number than would the Blasius profile remote from the edge. If this is true, the flow at the edge may become turbulent while the flow remote from the edge is still stable and laminar. This situation of an active turbulent flow interacting with a stable flow has considerable theoretical interest, and may have some practical interest in the interpretation of friction measurements at low Reynolds numbers.

The results are described in terms of a co-ordinate system in which $0z$ is the leading edge of the plate and $0x$ is the free edge on the working side. The total length of the plate was 100 cm., the breadth 19 cm. and the thickness 2 mm. The leading and trailing edges were tapered to an edge of moderately small radius, while the free edge is square. Measurements of velocity fluctuations were made using conventional techniques of hot-wire anemometry. Laminar skin friction was measured using a hot-wire anemometer to measure the velocity gradients close to the surface, and turbulent skin friction using a surface total head tube. Most of the measurements were made at a single distance from the leading edge and the Reynolds number was varied by changing the wind-speed. It should be mentioned that the flow patterns were substantially symmetrical about the plate, indicating absence of circulation and lift.

7. Distribution of skin friction near a free edge

Some measurements of the distribution of skin friction are shown in fig. 5. The upper set of measurements, taken at a Reynolds number near 10^5 with laminar flow, shows the existence of a large increase in friction towards the edge of the plate. This increase extends for a surprising distance from the free edge. From these results the direct edge effect is found to be

$$\Delta C_f / C_{f0} = 0 \cdot 1\, x/D \quad (7.1)$$

where ΔC_f is the increase in skin friction averaged across the width of the plate, D, and C_{f0} is the skin friction far from the edge. No appreciable secondary flow is present in the laminar flow and the increase in friction arises from the comparative ease of momentum diffusion near the edge.

In contrast with the results for laminar flow, the turbulent skin friction increases relatively little as the edge is approached, the maximum increase being about 30%, and the region of increase is comparatively restricted. From this and similar curves, the direct edge effect is estimated to be

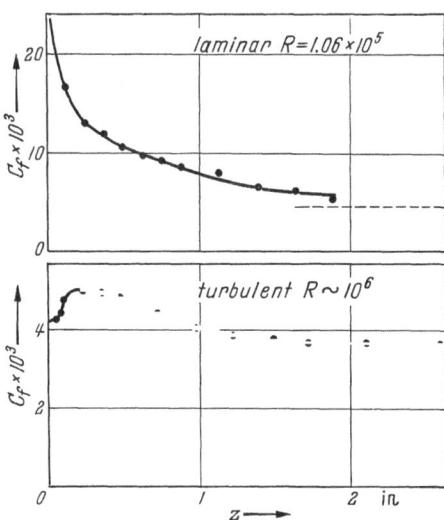

Fig. 5. Distributions of skin friction near a free edge for laminar and turbulent flow

$$\Delta C_f / C_{f0} = 0 \cdot 002\, x/D \quad (7.2)$$

This remarkable difference in the magnitude of the direct edge effect is certainly due to the existence of a secondary flow, the development of which is clearly visible in the measurements of mean velocity distri-

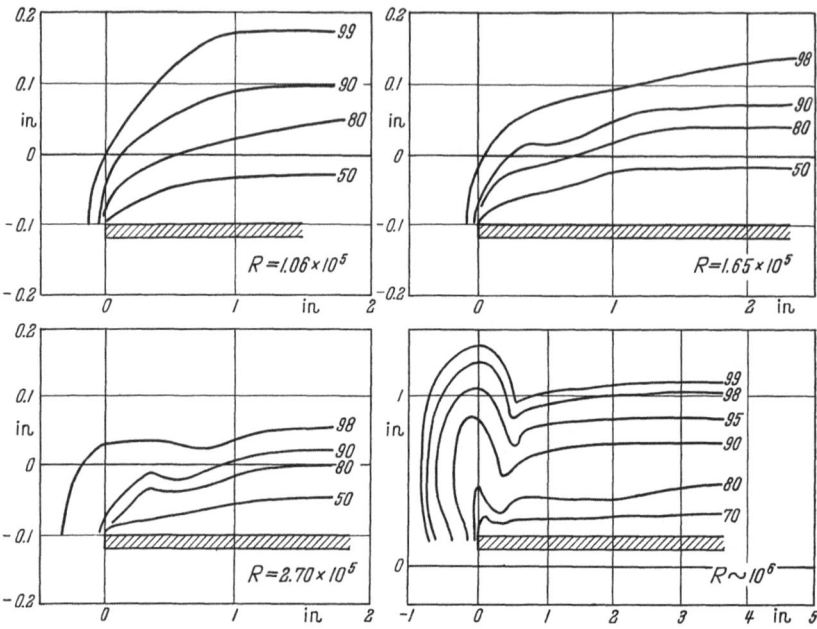

Fig. 6. Distributions of mean velocity in the neighbourhood of a free edge. (The lines of equal velocity have numbers giving the percentage of the free stream velocity.)

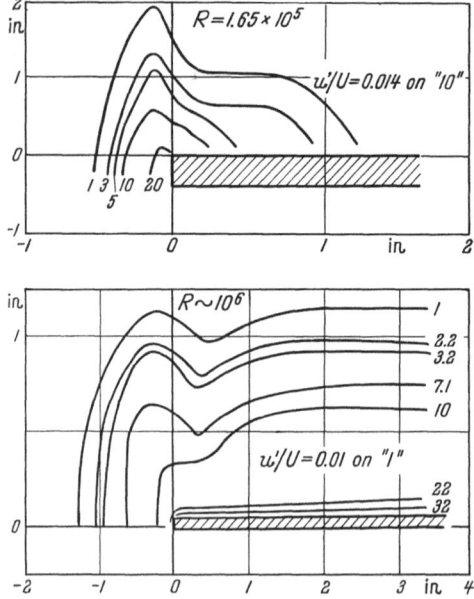

Fig. 7. Contours of equal turbulent intensity near a free edge. (The numbers are proportional to the root-mean-square intensity.)

butions (fig. 6). Comparing the laminar and turbulent flows, it will be
seen that the turbulent contours of equal velocity tend to be rectangular
while the laminar contours are more like confocal parabolas. Contours
of equal turbulent intensity have a similar shape. There is obvious
similarity with the well-known distortion of velocity contours in pipes
of non-circular section, and this is an example of secondary flow of
Prandtl's second kind.

The present measurements are not sufficiently extensive for a proper
comparison with measurements of the effect of aspect ratio on coefficients of total skin friction, but it is certain that the direct edge effect is
considerably less than the total effect. This seems to confirm the hypothesis that the secondary flow issuing from the edge of the boundary
layer re-enters the layer almost uniformly across the entire width of the
plate and leads to a general rise in skin friction. The very large edge
effect for laminar flow is a warning that the inference of edge effects
from measurements at Reynolds numbers below one million may lead to
considerable overestimates of the magnitudes at high Reynolds numbers.

8. The transition from laminar to turbulent flow

The laminar boundary layer on a flat plate owes its considerable
stability to the restraining effect of the solid boundary, and it is natural

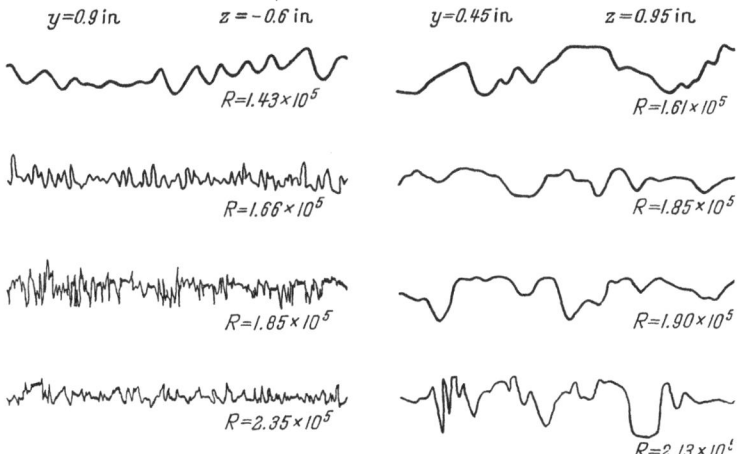

Fig. 8. Oscillograms of velocity fluctuations near a free edge

to expect the layer to be less stable near the free edge where the restraint
is less. The expectation, that transition to turbulent flow occurs at a
considerably lower Reynolds number in the neighbourhood of the edge,
has been confirmed in these experiments as can be seen from the oscillo-

graph records reproduced in fig. 8. The records taken at $z = -0.6$ in.,
i.e. outside the edge, show a first appearance of persistent oscillations at
a Reynolds number near 10^5, and a fully developed turbulent flow at a
Reynolds number of 1.66×10^5. At this Reynolds number, the laminar
flow on a flat plate is not yet unstable to Tollmien-Schlichting waves,
and in the present experiments continuously turbulent flow occured over
the whole surface only at Reynolds numbers above 4.4×10^5.

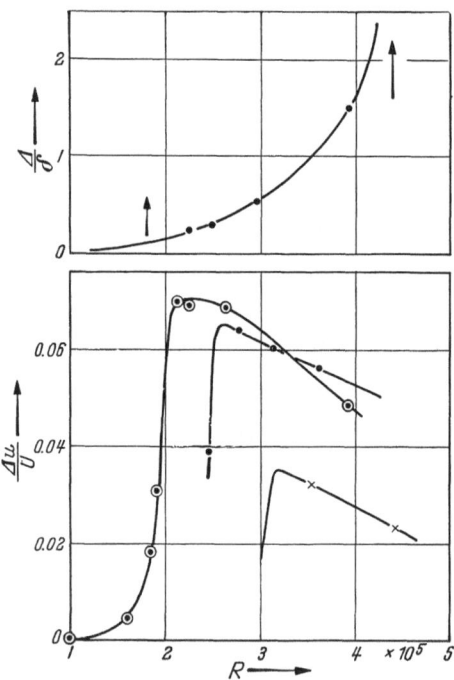

Fig. 9. Variations of the extent and intensity of turbulent "spots" with Reynolds number. ($\varDelta$ is the half-width of the distribution of fluctuation intensity, and $\varDelta u$ is the total range of velocity fluctuation within a turbulent spot.)

Naturally the presence of a turbulent flow at the edge produces disturbances in the neighbouring laminar flow. For Reynolds numbers less
than 2×10^5, these disturbances are irregular but wave-like and may be
stable Tollmien-Schlichting waves excited by the edge flow. For higher
Reynolds numbers, the laminar boundary layer is unstable and the
excitation leads to the production of turbulent spots which increase in
frequency and lateral distribution up to a Reynolds number of 4.4×10^5.
At this Reynolds number, transition is complete.

The turbulent spots observed in' this experiment differ considerably
from the spots observed in free transition on flat plates of large aspect

ratio. There is a strong tendency for the velocity within them to oscillate between two quasi-stable values with comparatively few high-frequency oscillations of the usual kind. As the Reynolds number increases, the range of velocity variation decreases and the similarity to "normal" turbulent spots becomes more apparent (fig. 9). It is probable that these peculiarities are characteristic of the comparatively low Reynolds number of the flow.

References

ALLAN, J. F., and R. S. CUTLAND (1952): Trans. N.E. Coast Inst. Eng. and Ship-builders, p. 53.

FAVRE, A.J., J.J. GAVIGLIO and R. DUMAS (1957): J. Fluid Mechanics, 2, 313.

GRANT, H. L. (1958): J. Fluid Mechanics, 4, (in press).

HUGHES, G. (1954): Trans. Inst. Naval Architects, p. 1.

LAUFER, J. (1951): N.A.C.A. Report no. 1033.

LAUFER, J. (1955): N.A.C.A. Report no. 1174.

SCHUBAUER, G. B. (1954): J. Appl. Phys. 25, 188.

SCHUBAUER, G. B. (1957): IUTAM Symposium on Boundary Layer Research, Freiburg.

TOWNSEND, A. A. (1954): European Shipbuilding, 3, 86.

TOWNSEND, A. A. (1956): Structure of Turbulent Shear Flow. Cambridge University Press.

Aus der Diskussion

R. WILLE (Berlin): Herr TOWNSEND hat gezeigt, daß große Wirbel in turbulenten Grenzschichten von Bedeutung sind. Da dies auch für die freie Grenzschicht eines Strahls gilt, hat Herr O. WEHRMANN eine besondere Studie über Hitzdrahtsignale, die Wirbeln entsprechen, gemacht. Die Ergebnisse sind in [1] enthalten. Einige Anwendungen werden in dem Beitrag von O. WEHRMANN und R. WILLE zur VIII. Sitzung dieses Symposiums mitgeteilt werden. (Vgl. S. 387.)

Literatur:

[1] WEHRMANN, O.: Hitzdrahtmessungen in einer aufgespaltenen Kármánschen Wirbelstrasse. DVL-Bericht Nr. 43 (1957).

On the possibility of a mathematical theory of shear-flow turbulence[1]

By

Oddvar Björgum

Matematisk Institutt, Universitetet i Bergen

In this note the following assumptions are introduced:

I. That Navier-Stokes equations may be applied to the study of shear-flow turbulence.

II. That steady-state turbulent flows exist.

Assumption I furnishes us with dynamical laws which are to be valid "microscopically" in turbulent flows. The problem is then to invent mathematical processes which yield the corresponding "macroscopic" laws.

In the case of a boundary layer the main problem may be said to be the determination of the mean-velocity distribution. To this end it is not necessary to endeavour to derive a theory valid for all classes of fluctuations. If the consideration of a special class of fluctuations yields the correct answer one should be satisfied.

Because of this fact I have tentatively considered only two-dimensional fluctuations, although the treatment might be easily extended to three-dimensional fluctuations. However, the mathematics would then be much more complicated.

From a theoretical point of view the flow between two parallel planes is simpler than a boundary layer. I shall therefore consider the former case although the same considerations may be applied to boundary layers when the usual approximations are introduced.

Let the mean velocity $U(y)$ be directed in the x direction and let the flow extend to infinity in both x directions. The process of averaging I shall define by integration over x and time t, i.e. the average $\bar{f}$ of a quantity $f(x, y, t)$ shall be defined by

---

[1] In the paper read at the Symposium with the title: "On the foundation of a mathematical theory of shear-flow turbulence" there was an error which was kindly pointed out by Prof. C. C. Lin and by Prof. Kampé de Fériet. The present note therefore differs from the paper read at the Symposium.

$$\bar{f} = \lim_{X, T \to \infty} \frac{1}{4XT} \int_{-X}^{X} dx \int_{-T}^{T} dt\, f(x, y, t) \,. \tag{1}$$

In order that the stream function $\psi(x, y, t)$ of the fluctuations shall be a stationary random function with respect to x and t we must have

$$\psi(x, y, t) = \int_{\alpha}\int_{\beta} e^{i(\alpha x - \beta t)}\, dS(y, \alpha, \beta)$$

$$= \int_{\alpha}\int_{\beta} \varphi(y, \alpha, \beta)\, e^{i(\alpha x - \beta t)}\, dA(\alpha, \beta) \tag{2}$$

where, for convenience, I have put

$$dS(y, \alpha, \beta) = \varphi(y, \alpha, \beta)\, dA(\alpha, \beta) \tag{3}$$

where $\varphi(y, \alpha, \beta)$ may be normalized in some way. The integration in (2) has to be extended over all real values of α and β. The integrals have to be stochastic Fourier-Stieltjes integrals.

In order that ψ shall be real we must have

$$\varphi(y, -\alpha, -\beta) = \varphi^*(y, \alpha, \beta),\ dA(-\alpha, -\beta) = dA^*(\alpha, \beta) \tag{4}$$

and in order that ψ shall not contribute to the mean motion we must have

$$dA(0,0) = 0 \,. \tag{5}$$

For the problem considered, the vorticity equation is

$$\frac{\partial \nabla^2 \psi}{\partial t} + U \frac{\partial \nabla^2 \psi}{\partial x} - U'' \frac{\partial \psi}{\partial x} + \frac{\partial(\nabla^2 \psi, \psi)}{\partial(x, y)} = \frac{1}{R}(U''' + \nabla^4 \psi) \tag{6}$$

where R denotes Reynolds number of the flow considered. Averaging this equation we obtain

$$U''' = R \frac{\overline{\partial(\nabla^2 \psi, \psi)}}{\partial(x, y)} \,. \tag{7}$$

Adding (6) and (7) we obtain

$$\nabla^4 \psi - R \frac{\partial \nabla^2 \psi}{\partial t} - RU \frac{\partial \nabla^2 \psi}{\partial x} + RU'' \frac{\partial \psi}{\partial x} =$$

$$= R \frac{\partial(\nabla^2 \psi, \psi)}{\partial(x, y)} - R \frac{\overline{\partial(\nabla^2 \psi, \psi)}}{\partial(x, y)} \,. \tag{8}$$

Making use of the representation (2) Eq. (8) reads

$$L\varphi(y, \alpha_0, \beta_0)\, dA(\alpha_0, \beta_0) =$$

$$= R \int_{\alpha}\int_{\beta} Q\big(\varphi(y, \alpha, \beta), \varphi(y, \alpha_0 - \alpha, \beta_0 - \beta)\big)\, dA(\alpha, \beta)\, dA(\alpha_0 - \alpha, \beta_0 - \beta) \tag{9}$$

where α_0 and β_0 cannot vanish simultaneously and where the linear operator L is defined by

$$L\varphi(y, \alpha, \beta) = \tag{10}$$

$$= \varphi'''' - \big\{2\alpha^2 + i\alpha RU - iR\beta\big\}\varphi'' + \big\{\alpha^4 + i\alpha^3 RU - i\alpha^2 R\beta + i\alpha RU''\big\}\varphi$$

and the quadratic operator Q is defined by

$$Q\left(\varphi\left(y, \alpha, \beta\right), \varphi\left(y, \alpha_0 - \alpha, \beta_0 - \beta\right)\right) =$$
$$= \alpha\left\{\varphi''\left(\alpha, \beta\right)\varphi'\left(\alpha_0 - \alpha, \beta_0 - \beta\right) - \varphi\left(\alpha, \beta\right)\varphi'''\left(\alpha_0 - \alpha, \beta_0 - \beta\right) + \quad (11)\right.$$
$$\left. + \alpha_0\left(\alpha_0 - 2\alpha\right)\varphi\left(\alpha, \beta\right)\varphi'\left(\alpha_0 - \alpha, \beta_0 - \beta\right)\right\}.$$

As a first approximation we may neglect the non-linear terms of (9). We then obtain

$$L\,\varphi\left(y, \alpha, \beta\right) = 0 \tag{12}$$

for all values of α and β for which $d\,A\left(\alpha, \beta\right)$ does not vanish. Eq. (12) is the stability equation of Orr and Sommerfeld. The boundary conditions are that φ and φ' have to vanish at the boundaries. This furnishes us with a determinantal equation. The problem is then to determine $U\left(y\right)$ such that β becomes real for all values of α in question. The investigations of MALKUS [2] indicate that this problem is not determinate. Indeed, in applying LIN's results [1] which were obtained by the use of asymptotic series, MALKUS assumed in addition a maximum rate of dissipation of potential energy into heat. The question is then if other approximations applied to Orr-Sommerfeld's equation and/or a process of successive approximations in which the non-linear terms of (9) are taken into account can make the problem determinate. If in addition it may be proved that such a process converges we should have a mathematical theory of shear-flow turbulence.

References

[1] LIN, C. C. (1945): Quart. Appl. Math., **III**, 117—142, 218—234, 277—301.
[2] MALKUS, W. V. R. (1956): Journ. Fluid Mech., **1**, 521—539.

Aus der Diskussion

Kritische Bemerkungen der Herren C. C. LIN und KAMPÉ DE FÉRIET zum Vortrag von Herrn BJØRGUM sind bereits in der vorstehenden revidierten Fassung berücksichtigt worden. Die Wiedergabe dieser Ausführungen erübrigt sich daher. Die revidierte Fassung wurde den Herren C. C. LIN und KAMPÉ DE FÉRIET vom Herausgeber vorgelegt. Sie erklärten sich mit dieser Fassung einverstanden.

H. J. KAEPPELER (Stuttgart): If I understand correctly, the author of the present paper assumes the validity of the Navier-Stokes equations for the bulk of the fluid flow and/or more or less small parts of it. I wish to state my belief that a disturbance or fluctuation theory should rather start from a Liouville type equation proceeding via the basic transport equation. Dr. BJØRGUM proceeds from a macroscopic equation (Navier-Stokes) by a quasistochastic method to obtain final macroscopic expressions. I rather feel that a consequent stochastic theory of disturbances or fluctuations should not start with a macroscopic equation which itself is an integration of microscopic behavior to obtain *average* values. Under certain circumstances, an integration of microscopic behavior including fluctuations or disturbances may easily lead to equations different from those by Navier-Stokes. It should be kept in mind that the consequent development in stochastic descriptions starts with the Liouville equation via the generalized transport equation and only then through possibly several stages of integration to the final macroscopic equations.

C. C. LIN (Cambridge, Mass.): A theory based on the superposition of linear oscillations was developed by HEISENBERG (1924), but obviously non-linear effects must be considered to yield a valid theory of turbulence. Regarding the applicability of Navier-Stokes equations, I would like to comment in support of Professor BJÖRGUM. When the scale of motion is comparable with that of mean free path, it is indeed often desirable to approach the problem from the molecular point of view. However, in the case of ordinary turbulent flow through channels, the scale of motion involved is very much larger than the molecular scale, and there is no question that the Navier-Stokes equations should apply. Furthermore, in all the cases where the deductions of the theory of turbulence based on Navier-Stokes equations have been checked by experimental measurements, there is no evidence that the basic equations should be modified.

Diskussionsveranstaltung zur I. Sitzung

Beitrag zum Thema

Turbulente Grenzschichten

Von **K. Wieghardt**, Institut für Schiffbau der Universität Hamburg

J. C. COOKE [1] hat die laminare Grenzschicht an längsangeströmten allgemeinen Zylindern berechnet und fand, daß der Reibungswiderstand eines elliptischen Zylinders kleiner ist als der eines Kreiszylinders mit demselben Umfang. Ähnliches scheint auch für turbulente Reibungsschichten zu gelten.

Im Windkanal wurde der Impulsverlust längs eines Zylinders (Durchmesser $2\,r_0 = 150$ mm, Länge 3,5 m) und eines quadratischen Prismas (Kantenabstand $a = 2\,r_\square = 150$ mm, Länge 3,5 m) gemessen. Der Zylinder hatte eine Halbkugel als Kopf; den Prismakopf bildete dementsprechend die Durchdringungsfläche der beiden Halbzylinder über gegenüberliegenden Kanten der Stirnfläche des Prismas. Der turbulente Umschlag erfolgte dicht beim Kopfende ($x = 0$); der gemessene Impulsverlust enthält also auch den laminaren Widerstand des Kopfes (beim Zylinder: $W_{\text{Kopf}} = \dfrac{\varrho}{2}\,U^2\,\pi\,r_0^2\,3{,}48\,(Ur_0/\nu)^{-1/2}$, mit ϱ = Dichte, U = Anblasgeschwindigkeit, ν = kinematische Zähigkeit).

Der auf den Staudruck und die Oberfläche bezogene Widerstand c_{F0} bzw. $c_{F\square}$ für Zylinder und Prisma hängt nicht nur von der Re-Zahl der Gesamtlänge $U\left(\dfrac{\pi}{2}r + x\right)\!/\nu$ ab, sondern auch noch vom Verhältnis Grenzschichtdicke zu Körperradius; für einen allgemeinen Vergleich wären daher Meßreihen an verschieden großen Körpern nötig. Für den Spezialfall zweier Körper mit gleichem „Radius" $r_0 = r_\square$ ist nach Abb. 1 der Flächenwiderstand des Zylinders 8 bis 11% größer als der des Prismas. Ebenso verhalten sich die auf die Stirnfläche bezogenen Widerstandsbeiwerte, da $c_W = 2\,c_F\,(1 + x/r)$. Auf die $^2/_3$-Potenz des Volumens bezogen

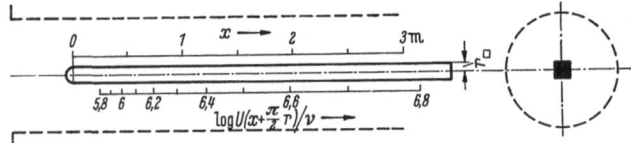

Prisma im Meßkäfig

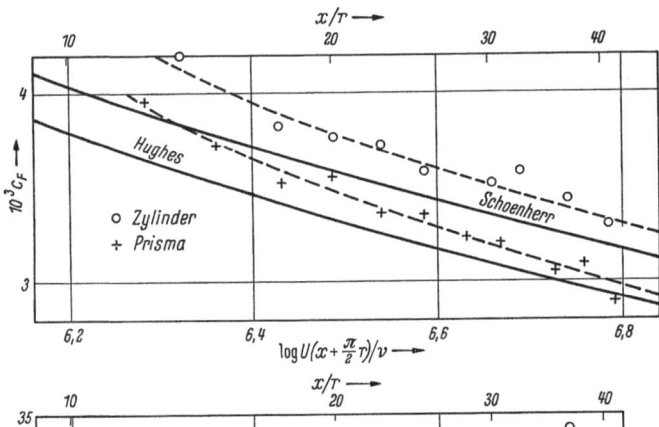

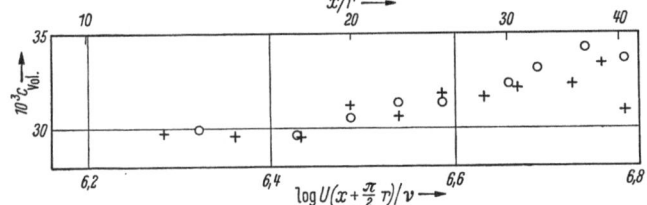

Abb. 1. Widerstandsbeiwerte von
Zylinder und quadratischem Prisma
verglichen mit dem der Platte

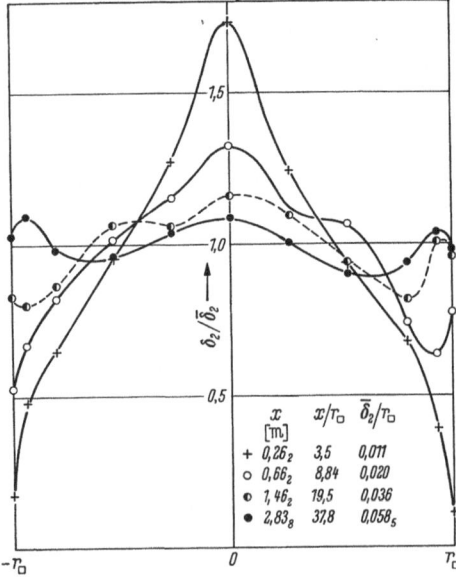

Abb. 2. Verteilung der Impulsverlustdicke über der Seitenfläche des Prismas
bei verschiedenen Rücklagen

wird in diesem Fall $c_{\text{Vol.}_0} \approx c_{\text{Vol.}_\square}$, da für $x/r \gg 1$ gilt $c_{F_0}/c_{F_\square} = (4/\pi)^{1/3} (r_0/r_\square)^{1/3} c_{\text{Vol.}_0}/$ $c_{\text{Vol.}_\square}$, mit $(4/\pi)^{1/3} = 1{,}083$; also wird bei $r_0 = r_\square$ und $c_{F_0}/c_{F_\square} = 1{,}08$ bis $1{,}11$ $c_{\text{Vol.}_0} \approx c_{\text{Vol.}_\square}$.

Wie sich längs des Prismas die über dem Querschnitt zunächst sehr ungleichförmige Reibungsschicht ausgleicht, zeigt Abb. 2, wo die relativen Schwankungen der Impulsverlustdicke δ_2 über dem jeweiligen Querschnitt aufgetragen sind.

Nach Abb. 1 unterscheidet sich der Flächenwiderstand am Prisma kaum von dem der idealen ebenen Platte, der zwischen der Schoenherr- und der Hughes-Linie liegen dürfte. Jedenfalls erhöht die Querkrümmung der Strömung am Zylinder den Reibungswiderstand gegenüber dem an der Platte weit mehr, als es die Kanten des Prismas tun.

Weitere Ergebnisse werden in der Diplomarbeit von W. KLEUTERS beschrieben werden.

Literatur:

[1] COOKE, J. C.: The Flow along Cylinders. Quart. Journ. Mech. and Appl. Math. X, 312 (1957).

Grenzschichten in geraden und gekrümmten Diffusoren

Von

J. Ackeret

Aerodynamisches Institut der Eidg. Technischen Hochschule, Zürich

Ich möchte über einige Betrachtungen über die Wirkungsweise der *Diffusoren* berichten und über *Experimente*, die in Zürich von meinen Mitarbeitern (vor allem H. SPRENGER) ausgeführt worden sind.

Wenn man versucht, die Grenzschichttheorie etwas schematisch anzuwenden, so gelangt man zu einer Schwierigkeit, die darin besteht, daß im Gegensatz zum Falle der frei-umströmten Körper der *Druckverlauf* nicht einfach aufgeprägt, sondern vom *GS-Verlauf* selber abhängig ist. Nur so ist es verständlich, daß der Diffusor-Erweiterungswinkel so wichtig ist. Vom Potentialstandpunkt aus wäre er ja gleichgültig, da der Druckverlauf bei einer divergenten Strömung, die etwa einer punkt- oder linienförmig angenommenen Quelle entspricht, stets derselbe ist. Offenbar spielen die Platzverhältnisse, also die *pauschale Kontinuitätsgleichung*, eine wichtige Rolle. — Zulässige Winkel sind (total) $8 \div 12°$ im Maximum; darüber hinaus tritt meist einseitige Ablösung auf. Wir wollen vor allem solche Diffusoren betrachten, bei denen man auch in der Nähe des Austritts noch von *Wandgrenzschichten* sprechen darf (wenn auch von sehr dicken). Das ist immer der Fall, wenn die Eintrittsgrenzschichtdicke klein und die flächenmäßige Erweiterung nicht zu groß ist, z.B. $2 \div 3$. Wir betrachten ferner nur turbulente Grenzschichten.

Für das, was ich zeigen möchte, spielt es keine wesentliche Rolle, daß eine vollbefriedigende Theorie turbulenter Grenzschichten bei Druckanstieg noch nicht besteht. Es genügt etwa die Methode von GRUSCHWITZ heranzuziehen. — Wir wollen zunächst vom *ebenen* Fall sprechen.

Mit den üblichen Bezeichnungen (SCHLICHTING) lautet die GS-Gleichung:

$$\frac{d\vartheta}{dx} + \frac{\vartheta}{U} \cdot \frac{dU}{dx} (H+2) = \frac{\tau_0}{\varrho U^2}.$$

Wir nehmen nun die *pauschale Kontinuitätsgleichung*

$$U(h - 2\delta^*) = \text{const.} = Q; \quad (h = \text{Kanalhöhe}$$
$$h_e \quad \text{am Eintritt})$$

hinzu und erhalten:

$$\frac{d\vartheta}{dx} = \frac{\frac{\tau_0}{\varrho Q^2}(h - 2H\vartheta)^3 + \vartheta(H+2)\left\{\frac{dh}{dx} - 2\vartheta\frac{dH}{dx}\right\}}{h + 2H\vartheta(H+1)}.$$

Wir führen ferner den Gruschwitz-Parameter ein:

$$\eta = 1 - \frac{u^2\vartheta}{U^2}$$

mit der zugehörigen Bestimmungsgleichung:

$$\vartheta\frac{d(U^2 \cdot \eta)}{dx} + A(U^2 \cdot \eta) = BU^2$$

$$A = 0{,}00894 \qquad\qquad B = 0{,}00461,$$

ferner die empirische Abhängigkeit (nach PRETSCH):

$$\eta = 1 - \left[\frac{H-1}{H(H+1)}\right]^{H-1}.$$

Für die Wandschubspannung benützen wir den alten Ansatz:

$$\frac{\tau_0}{\varrho U^2} = \frac{\zeta}{\left(\frac{U \cdot \vartheta}{\nu}\right)^{1/4}}.$$

Nun läßt sich die modifizierte Gleichung numerisch schrittweise integrieren, wobei keine Wiederholungen nötig sind, wie etwa im Falle, wo man die pauschale Kontinuität nachträglich als *Korrektur* einführt — ein Verfahren, das schlecht oder gar nicht konvergiert.

Das wichtigste Ergebnis ist eine überraschend *starke Abhängigkeit* des Grenzschichtdickenverlaufes δ^* von der Anfangsgrenzschichtdicke $\delta_1^* = \delta_e^*$, d.h. von den Einlaufbedingungen in den Diffusor. Man kann den Betrag der Druckumsetzung im Diffusor durch einen Umsetzungsgrad η_d ausdrücken, wobei die Definition noch verschieden angenommen werden kann. Wir wollen die volle kinetische Energie am Eintritt als im Prinzip ausnützbar ansehen und als günstigste Austrittsverteilung der Geschwindigkeit eine konstante u_2 annehmen. Dann ist η_d gegeben durch (1 Eintritt; 2 Austritt):

$$\eta_d = \frac{Q(p_2 - p_1)}{(\varrho/2)\int_1 u^3\,dy - (\varrho/2)Q u_2{}^2}.$$

Die Abweichung von $\eta_d = 1$ drückt nun nicht nur irreversible Vorgänge aus, sondern auch die Tatsache, daß am Austritt viel mehr kinetische Energie noch vorhanden ist, als u_2 entsprechen würde. Die *Verdrängung* der Strömung ist ein wesentlicher Grund für die fehlende Umsetzung und erklärt weitgehend die starke Abhängigkeit von η_d von den *Eintrittsbedingungen*. Vielfach ist es in der Praxis so, daß die kinetische Austrittsenergie für den weiteren Prozeß verloren ist. Verschiedene Beobachtungen haben dort gezeigt, wie empfindlich Diffusoren auf Eintrittsverhältnisse reagieren können und wie sorgfältig deshalb die Montage erfolgen muß. Über $\delta_e{}^*/h_e$ aufgetragen (Abb. 1), sieht man den großen Abfall. Freilich ist hier noch eine Bemerkung zu machen. GRUSCHWITZ hat als Ablösekriterium aus seinen Versuchen $\eta > 0,8$ entnommen. In

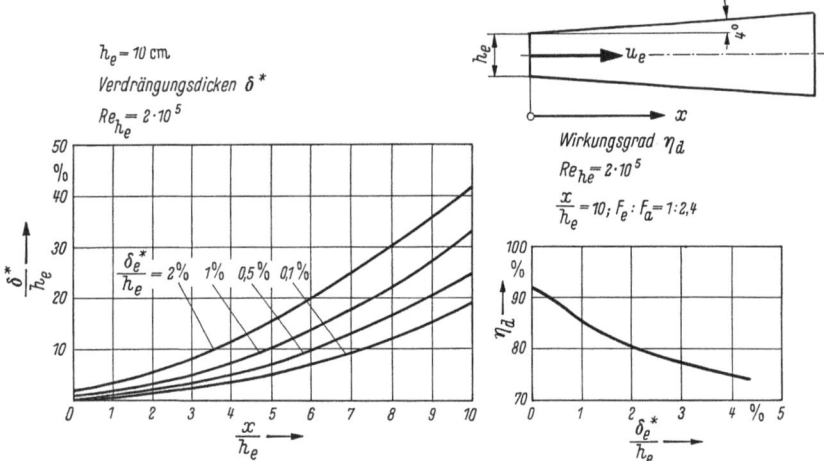

Abb. 1. Berechneter Einfluß der Eintrittsverdrängungsdicke $\delta_e{}^*$ auf Grenzschichtverlauf und Druckumsetzungsgrad η_d beim ebenen Diffusor

einem engen Kanal kommt nun aber auch hier die pauschale Kontinuität in Betracht. Die Strömung kann nicht einfach ablösen, wenn dafür kein Platz da ist. Eine Stagnation der dicken Grenzschichten würde ja sogleich zu einem sehr verringerten Druckanstieg bzw. sogar Druckabfall führen. Man kommt deshalb bei mäßigen Erweiterungswinkeln über $\eta = 0,8$ hinaus. Bei großen Winkeln tritt die Versperrung weit weniger in Erscheinung; deshalb tritt Ablösung deutlich auf. Solche Diffusoren arbeiten sehr unregelmäßig und geben geringen Druckumsatz.

Daß die Eintrittsverhältnisse so entscheidend sind, ist experimentell schon lange bekannt, und zwar in der Weise, daß man bemerkte, daß vorgeschaltete gerade Rohre η_d sehr wesentlich reduzieren. — Es wurde sogar ein Gesetz im alten Rom erlassen. Die Anschlüsse an die öffentliche Wasserleitung durften nur durch Rohre erfolgen, die eine längere Strecke (50 Fuß) *konstante Weite* (gleich derjenigen der geeichten Entnahmemündungen) hatten. Erst hinterher durften sie erweitert werden (FRONTINUS [*1*]). Damit wurde verhindert, daß jemand durch Erweiterung unmittelbar hinter der Entnahmestelle sich rechtswidrig einen größeren Anteil verschaffen konnte. EYTELWEIN [2] hat um 1800 den Einfluß vorgeschalteter Rohre experimentell untersucht und dabei gefunden, daß die Diffusorwirkung (ausgedrückt durch die Ausflußzahl) mit wachsender vorgeschalteter Rohrlänge ständig abnimmt. — Neuere Messungen von SPRENGER [3], PETERS [4] und WINTERNITZ [5] zeigen dies noch deutlicher (bessere Einläufe als bei EYTELWEIN). Auch hier kommt die Verschlechterung deutlich zum Ausdruck (Abb. 2).

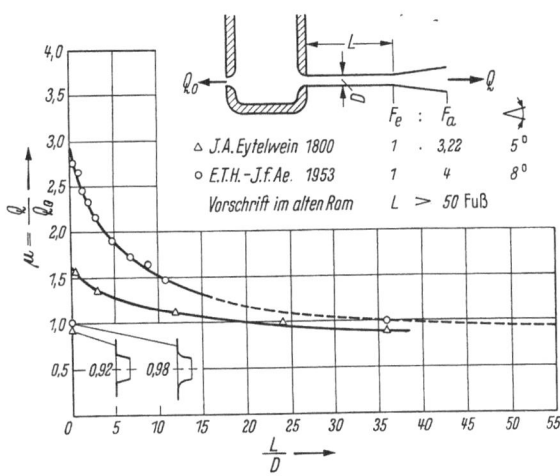

Abb. 2. Ausflußmengen bei verschieden großen Vorschaltlängen. Der Ausfluß wird durch die Diffusorwirkung vergrößert

Die Reynoldszahl kann indirekt einen relativ großen Einfluß ausüben. Wenn man für nicht zu lange Zuläufe noch im Anlaufgebiet ist, macht sich mit Re eine ziemlich starke Änderung von δ_e^* bemerkbar, die in der erwähnten Weise auf die Druckumsetzung im Diffusor einwirkt.

Abb. 3 zeigt die Versuchsanordnung mit geradem Diffusor. Die verschiedenen Beruhigungssiebe sind angegeben, ebenso ein Filter, um Staub u. a. zurückzuhalten (Staub hat vor allem einen störenden Einfluß auf die Druckentnahmestellen, besonders im konkaven Teil der gekrümmten Diffusoren). Die Geschwindigkeiten waren noch klein genug, um Kompressibilitätseffekte auszuschließen.

Abb. 4 zeigt u. a., wie groß die *Empfindlichkeit* der Diffusoren auf Einlaufstörungen ist. An der Gesamtdruckverteilung am Austritt ist noch sichtbar, daß beim Übergang des viereckigen Zulaufrohres zum achteckigen Zwischenstück und dann zum relativ weiten runden Sammelraum *Verluste* auftreten, die gewissermaßen verstärkt zum Vorschein kommen. Links sieht man für zwei vorgeschaltete Rohre die Druckverläufe (Unterschied in der Asymptote) und die lokal gerechneten Wirkungsgrade der Druckumsetzung. Im zweiten Fall ist die dickere Grenzschicht zunächst ohne größeren Einfluß, gibt aber nachher, d.h. weiter stromabwärts, um so größere Abfälle.

Der untersuchte gerade 8°-Kreisdiffusor ist in Abb. 5 gezeichnet, wo-

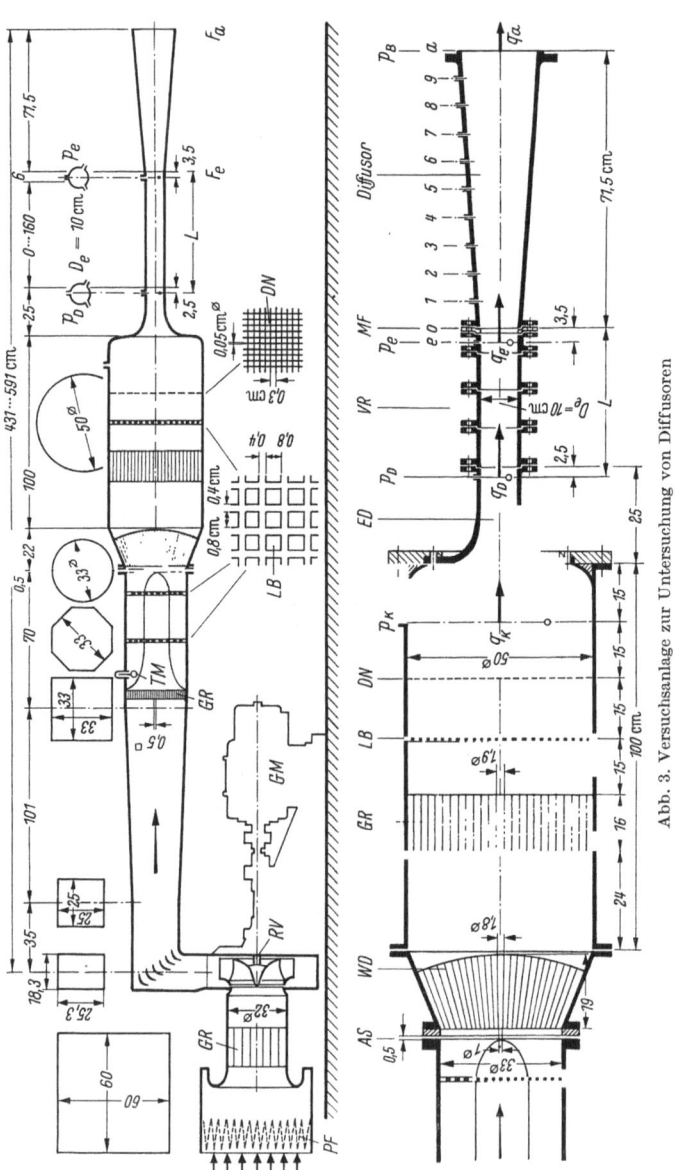

Abb. 3. Versuchsanlage zur Untersuchung von Diffusoren

bei die gemessenen Geschwindigkeitsprofile und die Druckverteilungen für gleiche Durchflußmengen, aber zwei verschiedene Eintrittsverdrängungsdicken ($2\,\delta_e{}^*/D_e = 0,61\%$ bzw. $2,33\%$) aufgetragen wurden. Das starke Anwachsen der Verdrängungsdicke besonders im zweiten Fall

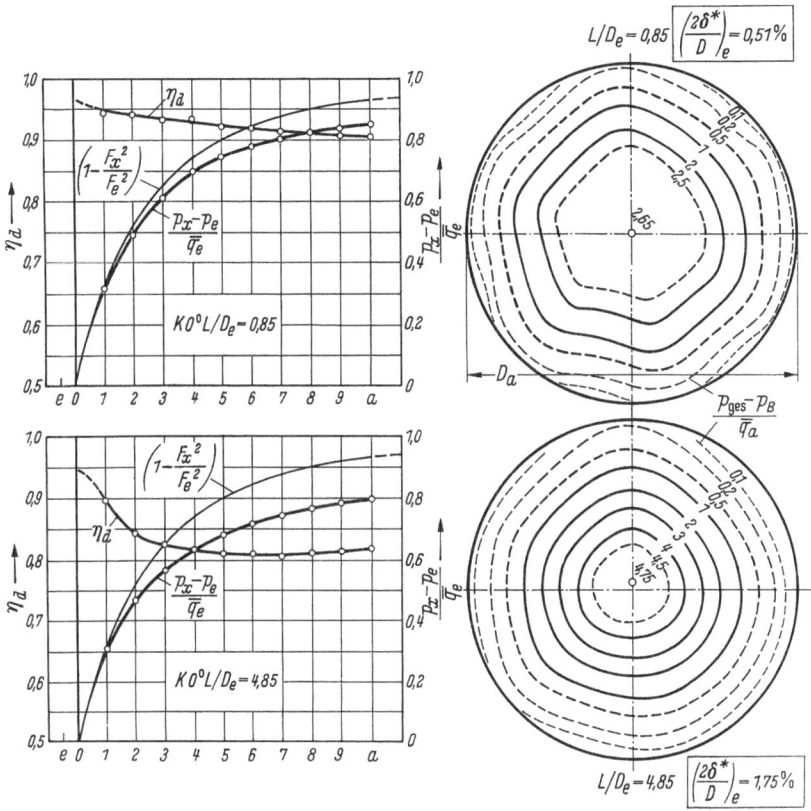

Abb. 4. Druckanstiege im geraden Diffusor (örtliche η_d) bei verschiedenen Vorschaltlängen. In der Staudruckverteilung am Austritt (rechts) sind die kleinen Eintrittsstörungen noch zu erkennen

gibt die Erklärung für die verminderten Umsetzungsgrade, die nur $0,898$ bzw. $0,776$ betragen.

Noch deutlicher ergibt sich der Verdrängungseffekt aus Abb. 6, wo die Gesamtdrücke am Austritt rechts eingetragen sind für verschieden lange Eintrittsrohre (L/D_e) und damit auch verschieden große Verdrängungsdicken. Da das Gesamtdruckrohr bei der registrierenden Anzeige langsam über den Querschnitt verschoben wurde, ergab sich gleichzeitig auch ein grobes Bild der turbulenten Schwankungen, die schließlich bis zur Rohrachse reichen. Die Umsetzung nimmt stetig ab. — Angaben über Diffusor-

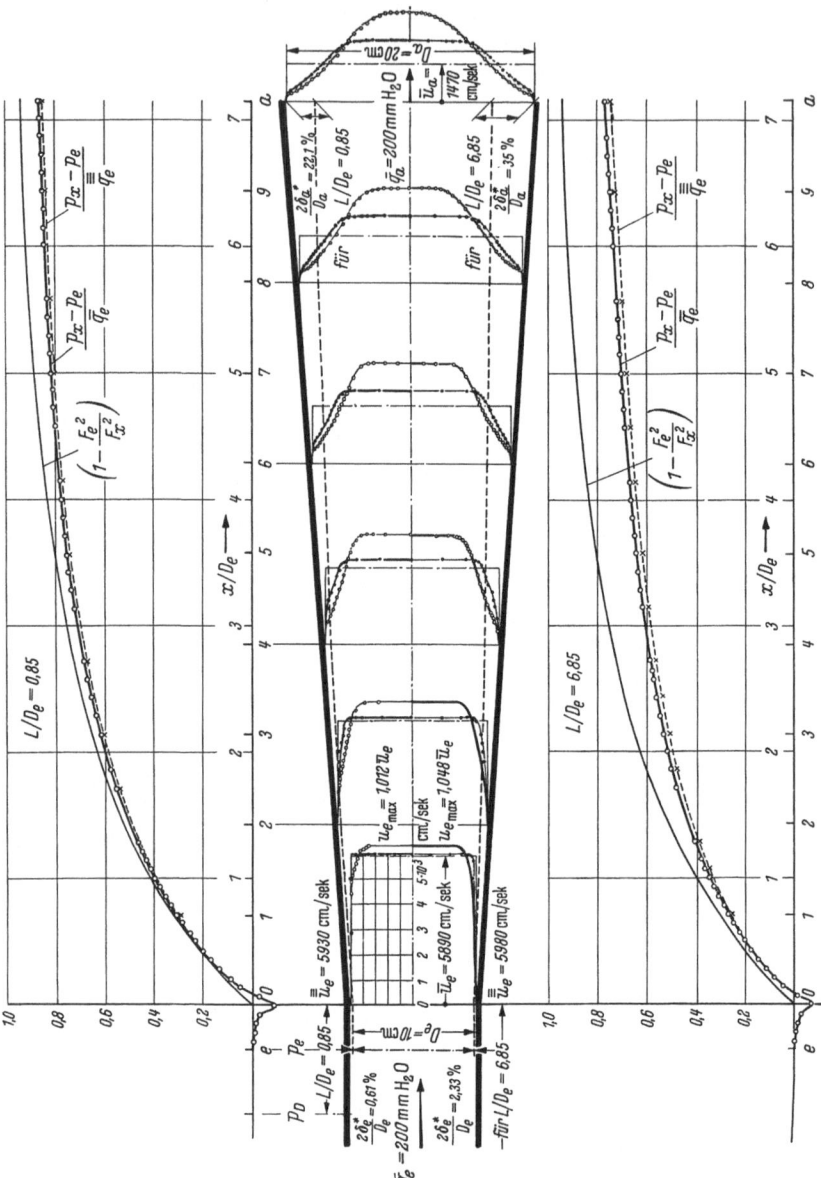

Abb. 5. Geschwindigkeits- und Druckverteilungen am geraden 8°-Diffusor bei gleichen Durchflußmengen, aber verschiedenen Verdrängungsdicken δe^* am Eintritt. Bei den Druckverteilungen ist
sowohl der Staudruck der mittleren Geschwindigkeit ($\overline{q}$) als auch der Staudruck, der der gemittelten

kinetischen Eintrittsenergie entspricht ($\overline{\overline{q}}$), als Bezugsdruck genommen. Am Austritt ist noch ein
Geschwindigkeits-„Plateau" vorhanden. Das gemessene Wachstum der Verdrängungsdicke längs der
Wand ist gestrichelt angegeben

„Wirkungsgrade" sind also so lange fast wertlos, als nicht auch die Eintrittsverhältnisse genau definiert sind.

Für den *Kegeldiffusor* sind in Abb. 7 rechts unten die gemessenen

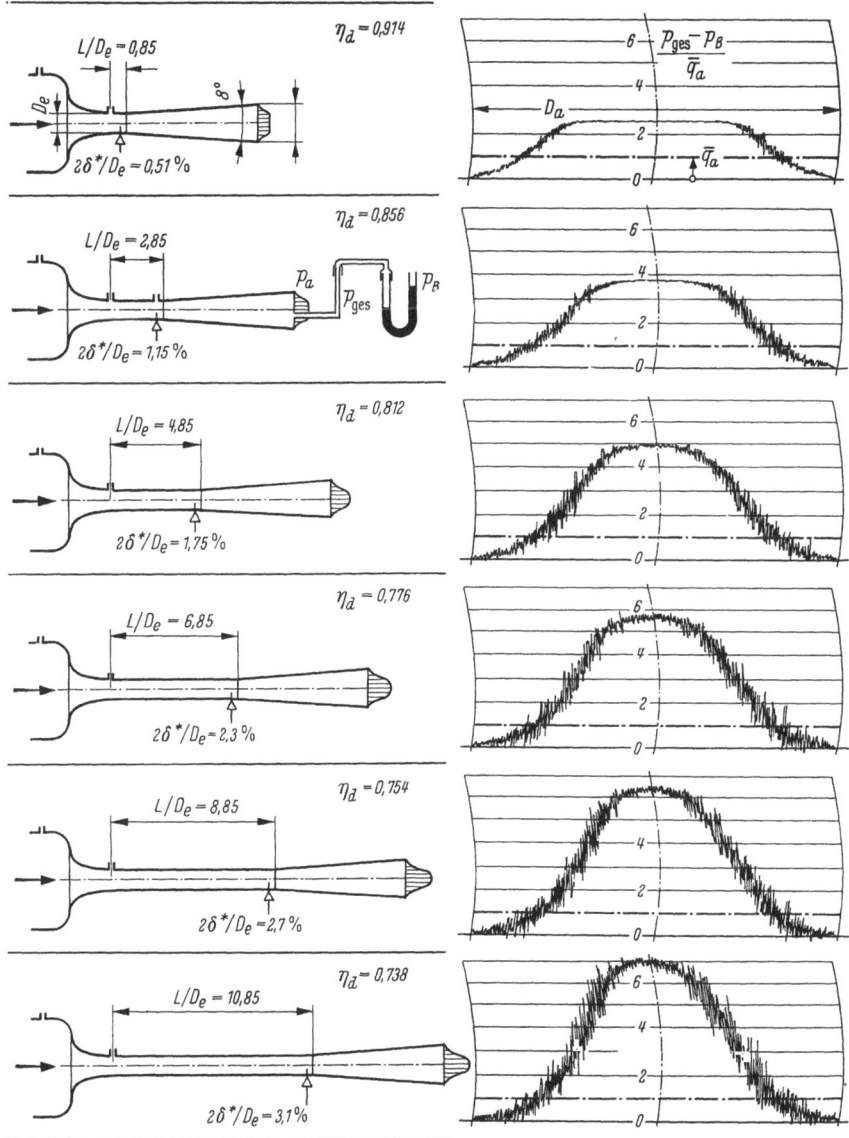

Abb. 6. Druckumsetzungsgrad η_d und Austrittsstaudruckverteilungen bei verschiedenen Vorschaltlängen bzw. Eintrittsverdrängungsdicken. Die Turbulenz am Austritt ist dadurch näherungsweise sichtbar geworden, daß das Staurohr langsam über den Querschnitt verschoben wurde (etwa 1 mm je Sekunde)

η_d-Werte über $2\delta_e^*/D_e$ eingetragen. Dabei sind sowohl Punkte verwendet worden, bei denen die Verdrängungsdicke durch Längen- als auch (bei gegebener Länge des Vorschaltrohres) durch Änderung der Reynoldszahl variiert wurde. Es ergibt sich ein ziemlich gut definierter Verlauf, und man kann versuchen, ihn nach Gruschwitz zu berechnen. Dabei zeigt es sich allerdings, daß man mit den früher erwähnten Konstanten A und B nicht ohne weiteres zum Ziel gelangt. Vielmehr muß eine (andernorts

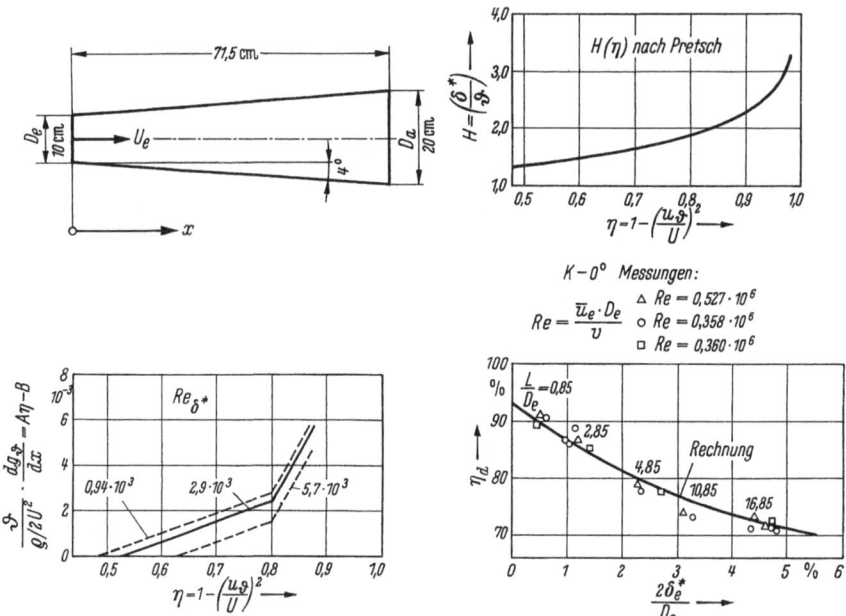

Abb. 7. Kegeldiffusor 8°. Die η_d-Werte ordnen sich nach den mit Vorschaltlänge bzw. Reynoldszahl variierten Eintrittsverdrängungsdicken (rechts unten). Um die Rechnung nach Gruschwitz den Ergebnissen anzupassen, müssen die Konstanten der zweiten Gruschwitz-Gleichung geändert werden (links unten)

schon vermutete) Abhängigkeit von B mit der Reynoldszahl (Abb. 7 links unten) und etwa von $\eta = 0{,}8$ an eine Änderung der Steigung A angenommen werden. Wie weit dies tiefere Gründe hat, bleibt noch abzuklären; doch erscheint eine Vergrößerung des turbulenten Austausches an dieser Stelle nicht unwahrscheinlich.

Wir haben weiterhin Diffusoren untersucht, die in systematischer Weise gekrümmte (Kreisbogen-) Achsen haben und deren Querschnittsformen vom runden Eintritt (100 mm $\varnothing$) mit allmählichem Übergang zu elliptischen Austrittsöffnungen führen (Abb. 8). Dabei wurde darauf gesehen, daß die Querschnittsfläche $F(s)$ den gleichen Verlauf hat. Es sind sowohl die großen Achsen hochkant (in der Krümmungsebene) als

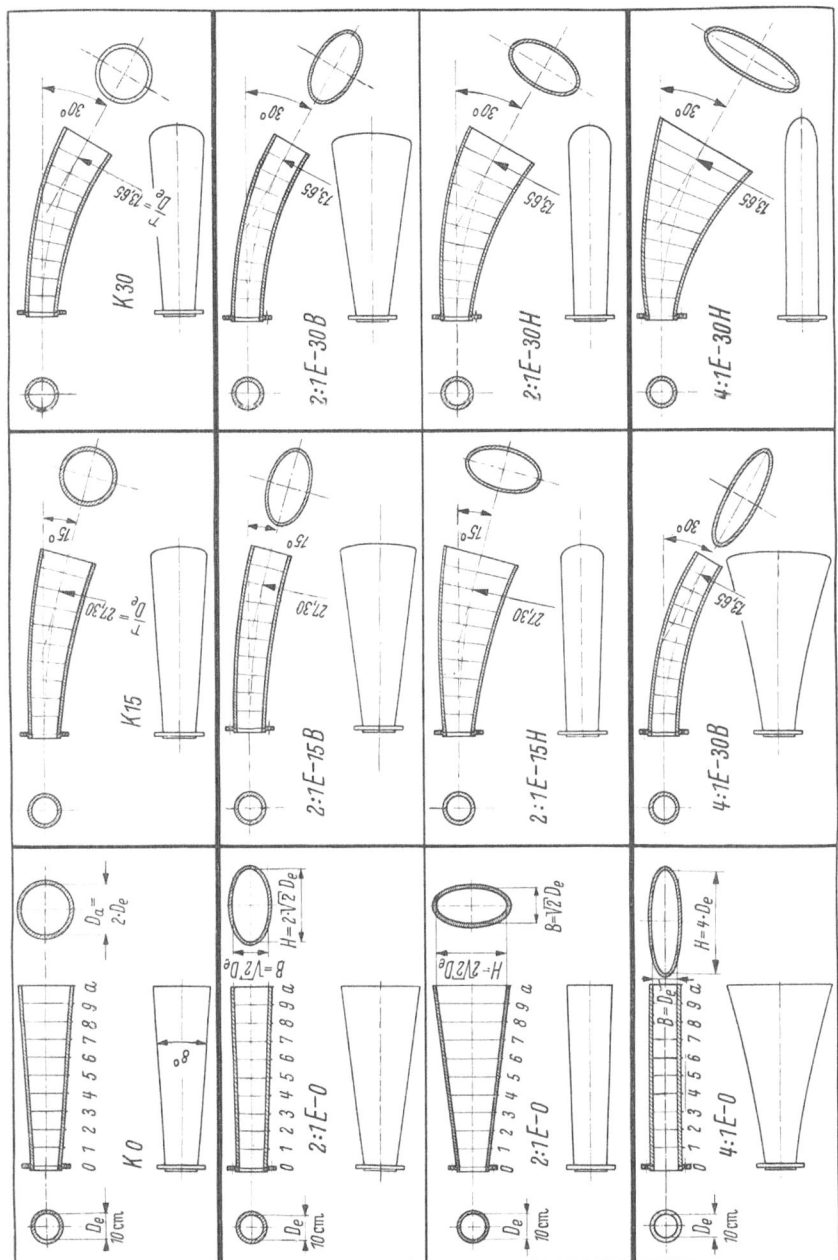

Abb. 8. Die untersuchten Diffusoren. — Verschiedene Achsenverhältnisse der elliptischen Austritts-
querschnitte und verschiedene Krümmungsradien der Mittellinien. Hochkant- (*H*) und Flachkant-
(*B*)-Stellungen

auch flachkant (kleine Achse in der Krümmungsebene) gestellt worden
(*B* bzw. *H*). Ebenso wurden Vorsatzrohre zur Veränderung von δ_e^* ein-
geschaltet und soweit als möglich die Reynoldszahl durch Geschwindig-
keitsänderung variiert.

Die Ellipsen-Achsenverhältnisse waren:

$1:1$ Kreis K

$2:1$ Ellipse E   breit B

                        hoch H

$4:1$ Ellipse E   breit B

                        hoch H.

Die Ablenkungen, auf der Rohrachse gemessen, waren: $0°$, $15°$, $30°$.
Die Ablenkwinkel wurden absichtlich nicht sehr groß gewählt, um Verhältnisse zu schaffen, die gegebenenfalls theoretisch leichter zu erfassen
sind. Die einfache Grenzschichtrechnung mit über dem ganzen Umfang
gleicher Grenzschichtdicke kann hier natürlich nicht in Betracht kommen — es sei denn für den Anfang der Erweiterung, der immerhin schon
einen großen Teil des gesamten Druckanstieges ergibt. Bei Krümmern
treten Sekundärbewegungen auf, die die Grenzschicht im konvexen Teil
anhäufen und deshalb zu Ablösungen führen können.

Es ist nun interessant zu bemerken, daß trotzdem auch hier ein starker
Einfluß der Eintrittsverdrängungsdicke δ_e^* auf die Druckumsetzung
sichtbar ist (Abb. 9). Besonders ungünstig ist die Hochkantstellung der
flachen Ellipse (4:1 E, 30° H), bei der schon bei etwas mehr als 1% Verdrängungsdicke die Umsetzung auf 50% sinkt. Die Flachkantstellung
hingegen (4:1 E, 30° B) liegt sogar etwas höher als die ungekrümmte
Form (4:1 E, 0°) — in Übereinstimmung übrigens mit den Erfahrungen
des Wasserturbinenbaues.

Abb. 10 zeigt noch Einzelheiten für eine feste Verdrängungsdicke von
0,35%. Aufgetragen sind die Austrittsverteilungen:

$$\frac{p_{ges} - p_a}{\bar{q}_a},$$

die mit dem Staudruck der mittleren Austrittsgeschwindigkeit $\bar{q}_a$ dimensionslos gemacht wurden. Schon bei K 15 treten Ablösungen auf, die bei
2:1 E, 30° H sehr ausgeprägt sind. Die Turbulenz, die wiederum an-
deutungsweise zum Vorschein kommt, ist besonders stark beim flachen
ungekrümmten Ellipsendiffusor 4:1. — Die in Abb. 10 rechts unten auf-
geführten Hochkantellipsen zeigen den sehr bemerkenswerten Einfluß
eines Eintritts mit *Drall*, der hier sehr günstig wirkt und noch etwas
näher untersucht wurde.

In Abb. 11 sind die verschiedenen Krümmer mit 30° Ablenkung mit kleinen Hilfsflügeln ausgerüstet, die nach Art der bekannten Vortex-Generatoren im Gegen- oder Gleichsinn auf der konvexen, also ablöse-gefährdeten Krümmerseite wirken können. Es zeigt sich, daß die gleich-sinnige, also drallerzeugende Wirkung, günstiger ist und daß selbst

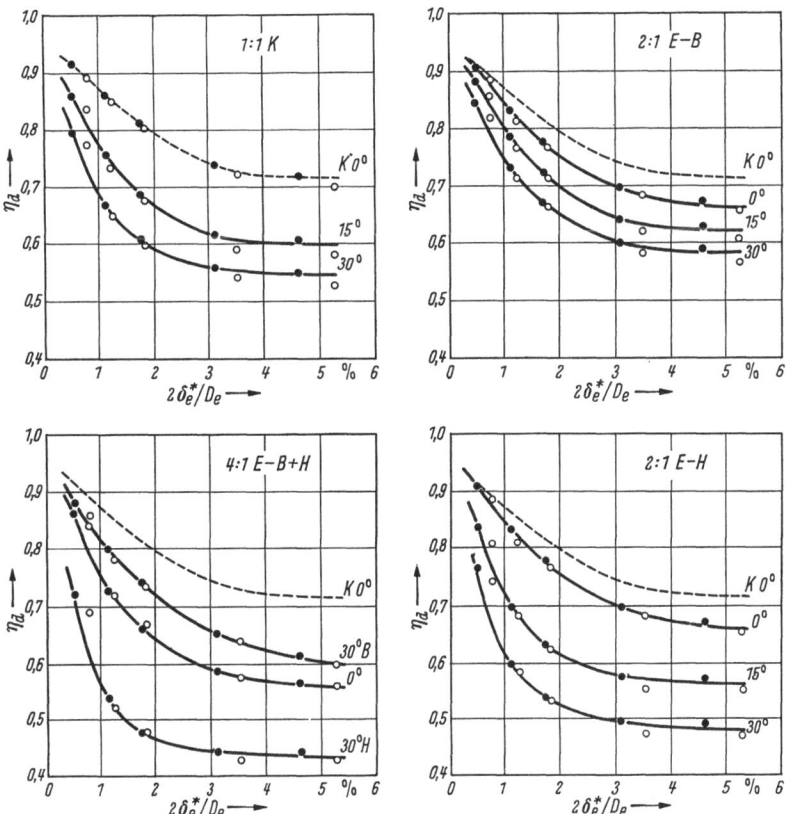

Abb. 9. Umsetzungsgrade η_d der Diffusoren bei verschiedenen Eintrittsverdrängungsdicken

der sonst sehr schlechte Krümmer (4:1 E, 30° H) erheblich verbessert werden kann (im Druckumsetzungsgrad η_d ist der Eigenverlust der Hilfsflügel eingeschlossen).

Betrachtet man die Ergebnisse dieser ja unter begrenzt variierten Be-dingungen angelegten Untersuchungen, so muß man sagen, daß die Diffusoren strömungstechnische Gebilde sind, die wir auch heute noch nur wenig genau kennen. Ihre große praktische Bedeutung recht-fertigt vermehrte Anstrengungen in theoretischer und experimenteller Hinsicht.

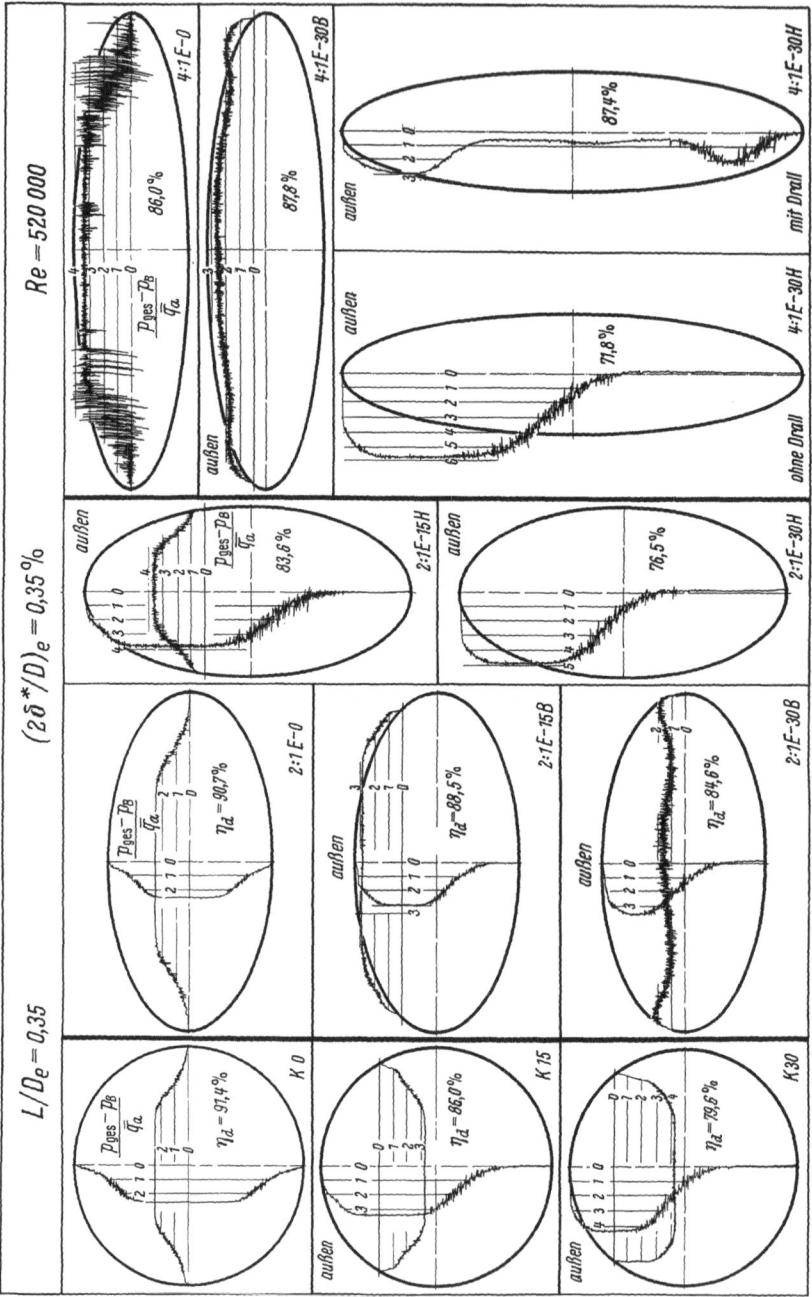

Abb. 10. Verteilungen der Austrittsstaudrücke für $2\,\delta e^{*}/De = 0{,}35\%$ und verschiedene Austritts-querschnitte. Ablösungen und Turbulenz sichtbar. Rechts unten: Einwirkung eines Eintrittsdralls (siehe Abb. 11)

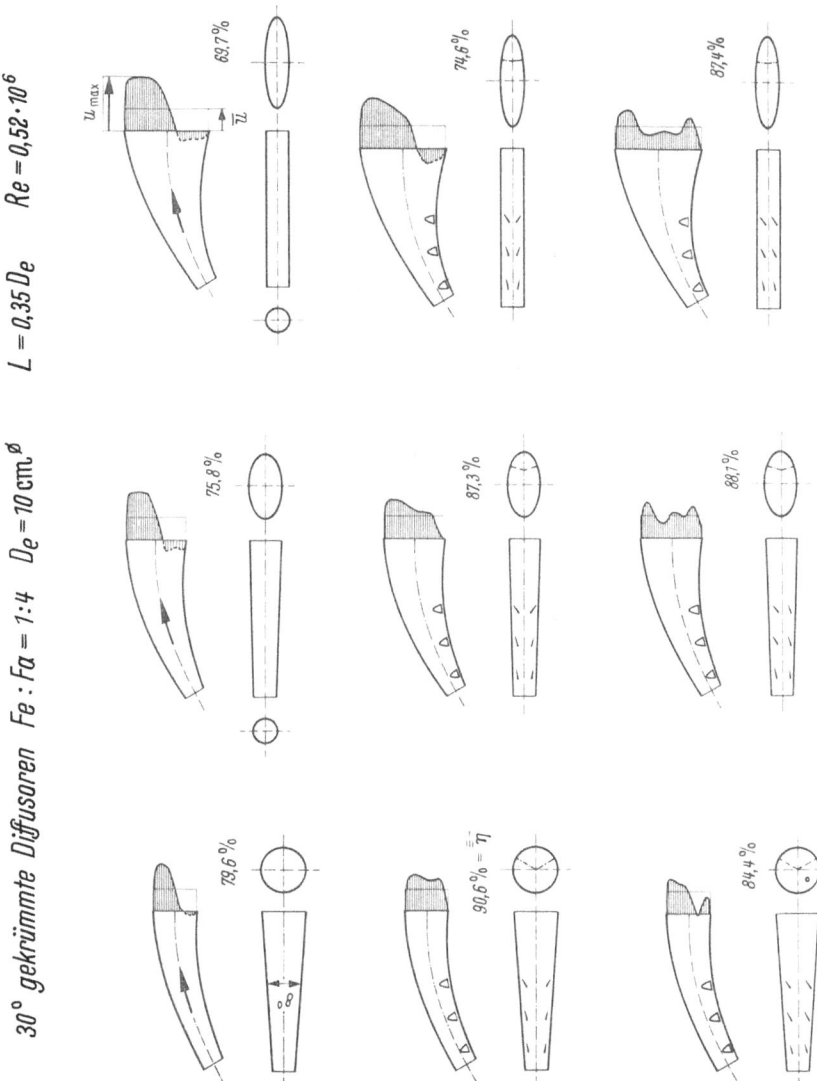

Abb. 11. Einfluß von Hilfsflügeln (gleich- und gegengerichtet) auf die Druckumsetzung in gekrümmten Diffusoren

Anhang

Bei gekrümmten Rohren ist auf der konkaven Seite infolge der Zentrifugalkräfte eine gegen die Wand gerichtete Hauptströmung, die die Grenzschicht zur Seite drängt, dünner macht und damit erhöhte Reibung verursacht. Um einen möglichst einfachen Fall einer solchen Verdrängung

zu verwirklichen, wurde der folgende Versuch ausgeführt (Abb. 12):
Eine um ihre Achse rotierende Trommel (Umfangsgeschwindigkeit w)
wird durch eine Ringdüse radial angeströmt. Die der Drehung entgegen-
wirkende Schubspannung in Umfangsrichtung kann im laminaren Fall

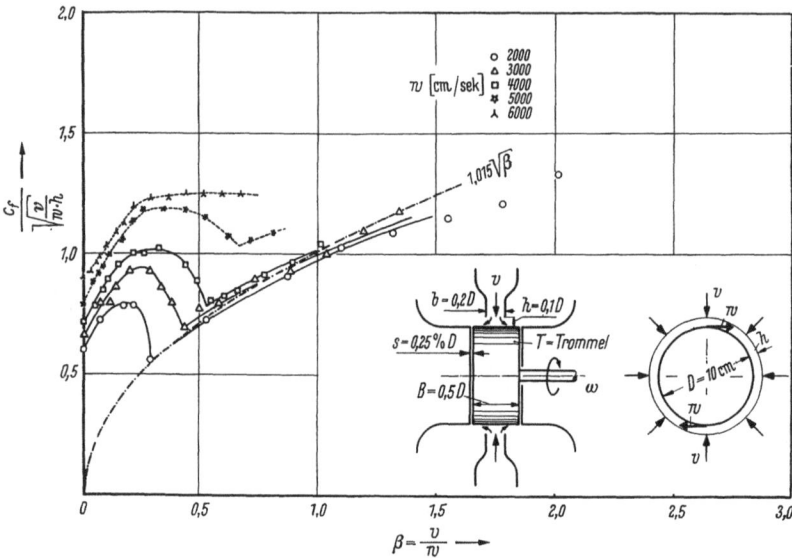

Abb. 12. Reibung an einer rotierenden radial angeblasenen Trommel. Theoretischer Wert für laminare Staupunktströmung 1,015 $\sqrt{\beta}$ strichpunktiert. Übergang zur Turbulenz bei kleinen β

einer Staupunktsströmung mit überlagerter Wandbewegung relativ einfach gerechnet werden und führt zum Resultat:

$$c_f = \frac{\tau}{(\varrho/2)\,w^2} = 1,15 \cdot \frac{\sqrt{a\,v}}{w},$$

wo a die Staupunktsströmung beschreibt ($u = ax$, $v = -ay$). In unserem
Falle ist nur in der Nähe der Trommel-Mittelebene eine genaue Staupunktsströmung vorhanden. Immerhin kann die Freistrahlströmung, die
in Wirklichkeit vorhanden ist, näherungsweise als ebene Strömung gerechnet werden und es ergibt sich

$$a = 0,775\ v/h,$$

wo v die Ausströmgeschwindigkeit aus der Düse und h der Abstand der
Düsenmündung von der Trommeloberfläche bedeuten. Führen wir noch
das Verhältnis $\beta = v/w$ ein, so folgt schließlich:

$$\frac{c_f}{\sqrt{\dfrac{v}{w \cdot h}}} = 1,015 \sqrt{\beta}\,.$$

Die so berechnete Parabel schließt sich nun den Meßpunkten recht gut an und zeigt, daß bei größeren Anblasgeschwindigkeiten laminare Strömung herrscht. Bei kleineren β tritt ein scharfer Umschlag in Turbulenz auf. Er scheint verwandt mit der TAYLOR-GÖRTLERschen Unstabilität und auch numerisch in der richtigen Größenordnung. — Weitere Untersuchungen mit verbesserter Apparatur sind zur Zeit im Gange.

Literatur

[1] HERSCHEL, CL: The two books on the water supply of the City of Rome of SEXTUS JULIUS FRONTINUS (etwa 100 n.Chr.). Boston 1899, P. 205.

[2] EYTELWEIN, J. A.: Handbuch der Mechanik fester Körper und der Hydraulik. Berlin 1800, § 96—99.

[3] SPRENGER, H.: Messungen an Diffusoren. VDI Berichte — 3, 109—110 (1955). Experimentelle Untersuchungen über den Einfluß der Eintrittsgrenzschicht auf die Druckumsetzung in geraden und gekrümmten Diffusoren (Vortragsbericht). ZAMP 1956, S. 372—374.

[4] PETERS, H.: Energie-Umsetzung in Querschnittserweiterungen bei verschiedenen Zulaufbedingungen. Ingenieur-Archiv 1931, S. 92—107.

[5] WINTERNITZ, F. A. L., und W. J. RAMSAY: Effects of inlet boundary layer on pressure recovery energy conversion and losses in conical diffusors. Journal of the Roy. Aer. Soc. 1957, p. 116—124.

Aus der Diskussion

H. GÖRTLER (Freiburg i. Br.): Von dem Herrn Vorsitzenden aufgefordert, mich speziell zu der zuletzt von Herrn ACKERET behandelten Strömung und ihrer Stabilität zu äußern, kann ich nur sagen, daß wir die Stabilität der ebenen Staupunktströmung gegenüber longitudinalen Wirbeln untersucht haben [1, 2] und dabei gefunden haben, daß eine Instabilität gegenüber solchen Störungen möglich ist. Dort handelt es sich jedoch um eine ruhende Wand, während im Falle der von Herrn ACKERET untersuchten Strömung der Zylinder rotiert. Die Wirkung der Zentrifugalkräfte dürfte einen erheblichen Einfluß auf die Stabilität haben. Es ist denkbar, daß sich unsere frühere Untersuchung auf den von Herrn ACKERET untersuchten Fall erweitern läßt.

Ich möchte die Gelegenheit benutzen, auf eine Erscheinung hinzuweisen, die zeigt, daß die Rotation einer Flüssigkeit von erheblichem Einfluß auf das Verhalten kleiner Störungen sein kann. Wenn man eine starr um eine Achse rotierende Flüssigkeit erzwungenen, zeitlich periodischen Störungen unterwirft (etwa durch Eintauchen eines harmonisch schwingenden Körpers) und nach der Ausbreitung solcher Störungen vom Störzentrum aus durch die Flüssigkeit fragt, so ergeben die für kleine Störungen aufgestellten Differentialgleichungen eine überraschende Antwort. Bei Vernachlässigung der inneren Reibung der Flüssigkeit erhält man für die ortsabhängigen Amplituden der erzwungenen Schwingungen eine lineare partielle Differentialgleichung zweiter Ordnung, die ihren Typus wechselt, wenn die Störungsfrequenz β einen gewissen kritischen Wert $2\,\omega$, die doppelte Winkelgeschwindigkeit der Rotation, durchschreitet. Für Kreisfrequenzen $\beta > 2\,\omega$ ist die Differentialgleichung elliptisch, also wie in einem nicht rotierenden Medium, und die Störungsamplituden nehmen allseitig mit wachsendem Abstand vom Störungszentrum

38   J. Ackeret

ab, wobei das Amplitudenfeld durch die Rotation lediglich gegenüber dem in einer nicht-rotierenden Flüssigkeit deformiert ist. Für unterkritische Frequenzen $\beta < 2\,\omega$ hingegen ist die Differentialgleichung hyperbolisch; Störungen breiten sich vom Störungszentrum aus längs den reellen Charakteristikenflächen durch die Flüssigkeit fort. Dieses zunächst theoretisch gefundene Ergebnis [3] ist kürzlich experimentell voll bestätigt worden [4, 5]. In einer zähen Flüssigkeit bleiben zwar die Charakteristikenflächen nicht als Unstetigkeitsflächen bestehen, sie bleiben jedoch, wie das Experiment zeigt, als Schichten endlicher aber geringer Dicke erkennbar, durch die hindurch das Geschwindigkeitsfeld sich extrem stark ändert. Mit anderen Worten, in der zähen Flüssigkeit breiten sich bei unterkritischen Frequenzen vom Störungszentrum aus freie Reibungsschichten durch die Flüssigkeit aus. (Auch die Bewegung der Flüssigkeit im einzelnen in den durch die charakteristischen Störungsflächen getrennten Strömungsbereichen zeigt überraschendes Verhalten, wie es auch theoretisch bei strenger Integration der linearisierten Bewegungsgleichungen für die reibungslose Flüssigkeit gefunden wird [6].) Diese Ergebnisse lassen erwarten, daß sich Störungen in einer rotierenden Flüssigkeit, wie sie etwa nach Zufall von außen in die Flüssigkeit hereingetragen werden, ganz verschieden auf die Stabilität der Strömung auswirken werden, je nach dem Frequenzspektrum. Da sich unsere Ergebnisse auf Schwingungen in *starr* rotierenden Flüssigkeiten beziehen, versuchen wir nun zunächst, sie theoretisch auf einfache Fälle nicht starrer Rotation auszudehnen. Grundsätzlich wenigstens erscheint eine sehr weitgehende Verallgemeinerung möglich. Sollte sie auch praktisch gelingen, so müßte das Auftreten freier Reibungsschichten, die sich vom Störungszentrum aus durch die Flüssigkeit ausbreiten, zu neuartigen Einsichten, insbesondere im Hinblick auf den Mechanismus des laminar-turbulenten Umschlags in rotierenden Flüssigkeiten, führen.

Literatur:

[1] Görtler, H.: Dreidimensionale Instabilität der ebenen Staupunktströmung gegenüber wirbelartigen Störungen. „50 Jahre Grenzschichtforschung". Braunschweig: Vieweg, 1955. S. 304—314.
[2] Hämmerlin, G.: Zur Instabilitätstheorie der ebenen Staupunktströmung, ebenda, S. 315—327.
[3] Görtler, H.: Einige Bemerkungen über Strömungen in rotierenden Flüssigkeiten. ZAMM **24**, 210—214 (1944).
[4] Oser, H.: Experimentelle Untersuchung über harmonische Schwingungen in rotierenden Flüssigkeiten. ZAMM (1958). Erscheint in Kürze.
[5] Görtler, H.: On forced oscillations in rotating fluids. Proceedings of the 5th Midwestern Conference on Fluid Mechanics, University of Michigan 1957. Erscheint in Kürze.
[6] Oser, H.: Erzwungene Schwingungen in rotierenden Flüssigkeiten. Archive for Rational Mechanics and Analysis, **1**, 81—96 (1957).

N. Scholz (München): Ich möchte im Zusammenhang mit der Grenzschichtberechnung in Diffusoren auf ein anderes sehr wichtiges technisches Anwendungsgebiet der Theorie der turbulenten Grenzschichten hinweisen, mit dem wir uns in Braunschweig sehr gründlich befaßt haben. Es handelt sich um die Grenzschichten in Schaufelgittern von Strömungsmaschinen. Zur theoretischen Vorausberechnung der Charakteristik und des Wirkungsgrades von Strömungsmaschinen spielt die Kenntnis des Reibungseinflusses auf die Strömung durch Schaufelgitter eine entscheidende Rolle. Für das Einzelprofil ist die theoretische Berechnung der Profilpolare mit Hilfe grenzschichttheoretischer Methoden bereits mit gutem Erfolg durchgeführt worden. In den letzten Jahren konnte eine entsprechende Methode auch für das ebene Schaufelgitter entwickelt werden. Dabei sind im wesentlichen drei Einflüsse der Reibungsschicht auf die Strömung im Schaufelgitter zu unterscheiden:

1. Der durch die Grenzschicht verursachte Energieverlust einschließlich der im Nachlauf eintretenden Mischverluste;

2. Die infolge der Verdrängungswirkung der Grenzschicht verursachte Auftriebsänderung;

3. Die infolge der unterschiedlichen Grenzschichtdicken auf Saug- und Druckseite geänderte Abflußbedingung an der Profilhinterkante.

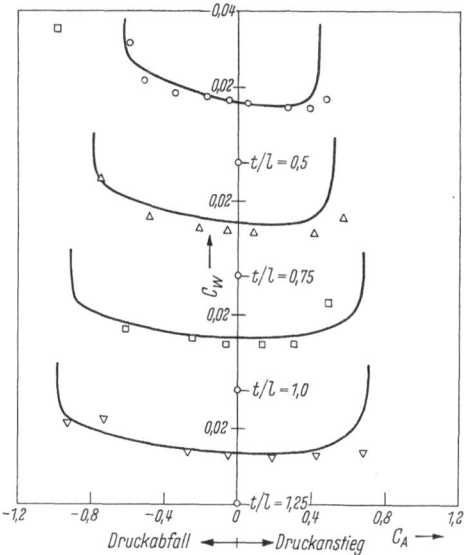

Abb. 1. Einige Profilpolaren c_W (c_A) für ein Schaufelgitter aus Profilen NACA 0010 bei einem Schaufelwinkel von 60° und für verschiedene Teilungsverhältnisse t/l, nach [1]

Umfangreiche parallel durchgeführte theoretische und experimentelle Untersuchungen an Schaufelgittern [1] haben gezeigt, daß theoretische Vorausberechnungen der Profilpolaren im Gitter eine ausgezeichnete Übereinstimmung mit dem Experiment ergeben. Abb. 1 zeigt als Ergebnis einige Profilpolaren $c_W(c_A)$ für ein Schaufelgitter aus Profilen NACA 0010 bei einem Schaufelwinkel von $\beta_s = 60°$ (Winkel zwischen Profilsehne und Gitterfront) und verschiedene Teilungsverhältnisse t/l (t die Schaufelteilung, l die Profiltiefe). Dabei gehören negative c_A-Werte zu einem Gitter mit Druckabfall, positive c_A-Werte zu einem Gitter mit Druckanstieg in Strömungsrichtung. Dementsprechend werden im Druckabfallgebiet höhere c_A-Werte erreicht.

Gewisse Unsicherheiten treten im Gebiet der $c_{A\max}$-Werte auf, bei denen bereits größere Ablösungsgebiete an der Schaufelkontur vorhanden sind. Der Einfluß dieser Totwassergebiete kann nach den heutigen Erkenntnissen quantitativ nur roh abgeschätzt werden. Wegen der technischen Bedeutung ist eine genauere Erfassung dieser turbulenten Totwasserzonen sehr erstrebenswert. Eine weitere noch nicht befriedigend gelöste Frage ist die Vorausbestimmung der Lage des laminar-turbulenten Umschlagpunktes, die auf den Strömungsverlust einen erheblichen Einfluß hat. Bei den Untersuchungen nach [1] wurde die Grenzschicht als voll-turbulent vorausgesetzt und im Experiment durch Turbulenzfäden an der Profil-nase turbulent gemacht.

Literatur:

[1] SPEIDEL, L., und N. SCHOLZ: Untersuchungen über die Strömungsverluste in ebenen Schaufelgittern. (Mitt. aus dem Inst. f. Strömungsmechanik der T.H. Braunschweig, Leiter: Prof. Dr. H. SCHLICHTING.) Erscheint in Kürze als VDI-Forschungsheft (1957).

Grenzschichtströmung in dissoziierendem Stickstoff bei hohen Machzahlen

Von

G. Ludwig und J. Metzdorf

Institut für theoretische Physik der Freien Universität Berlin

(Vorgetragen von G. Ludwig)

Im Institut für theoretische Physik der Freien Universität Berlin wurde das Problem in Angriff genommen, den Wärmeübergang bei hohen Machzahlen in dichter und auch sehr verdünnter Luft zu untersuchen. Als erster Schritt, über den hier nur berichtet werden möge, wurde die Grenzschichtströmung in reinem Stickstoff im Falle des Dissoziationsgleichgewichtes untersucht, um diese Ergebnisse mit dem Falle des Nichtgleichgewichtes zu vergleichen. Weiterhin sollte dieser erste Schritt eine Verbesserung der Arbeit von L. L. MOORE [1] bringen. Eine ähnliche Verbesserung ist in letzter Zeit von Y. J. KUO [2] durchgeführt worden.

Die Grundgleichungen der Strömung seien gleich für das Nichtgleichgewicht angegeben (über doppelt auftretende Indizes ist immer von 1 bis 3 zu summieren):

$$\frac{dn_1}{dt} = - n_1 v_{\sigma|\sigma} + r_1 - d_{1\sigma|\sigma} \tag{1}$$

$$\frac{dn_2}{dt} = - n_2 v_{\sigma|\sigma} + r_2 - d_{2\sigma|\sigma} \tag{2}$$

$$\varrho \frac{dv_\nu}{dt} = - p_{|\nu} + (\mu \, \Pi_{\nu\sigma})_{|\sigma} \tag{3}$$

$$\varrho \frac{du}{dt} = - p \, v_{\sigma|\sigma} + \mu \, v_{\varrho|\sigma} \, \Pi_{\varrho\sigma} - q_{\sigma|\sigma} \tag{4}$$

$$p = (n_1 + n_2) \, k \, T = n \, k \, T \tag{5}$$

mit den Abkürzungen:

$$f_{|\sigma} = \frac{\partial f}{\partial x_\sigma}, \qquad \frac{df}{dt} = \frac{\partial f}{\partial t} + v_\sigma f_{|\sigma} \tag{6}$$

$$\Pi_{\varrho\sigma} = v_{\varrho|\sigma} + v_{\sigma|\varrho} - \frac{2}{3} v_{\tau|\tau} \, \delta_{\sigma\varrho} \tag{7}$$

$$r_1 = (k_1 n_1 + k_2 n_2) (n_2 - k \, n_1{}^2), \qquad 2 \, r_2 = - r_1 \tag{8}$$

$$u = \frac{1}{\varrho} \left[\frac{3}{2} k \, T n_1 + \left(\frac{3}{2} k \, T + \bar{\varepsilon} \right) n_2 + \frac{1}{2} E_D n_1 \right] \tag{9}$$

$$q_\sigma = -\lambda\, T_{/\sigma} + \frac{5}{2}\, k\, T\, (d_{1\sigma} + d_{2\sigma}) + d_{1\sigma}\, \frac{E_D}{2} + d_{2\sigma}\, \bar{\varepsilon} \qquad (10)$$

$$d_{1\sigma} = -\frac{m_2\, n^2}{\varrho}\, D_{12}\, d_{12\sigma} \qquad (11)$$

$$d_{1\sigma} = \frac{m_1\, n^2}{\varrho}\, D_{12}\, d_{12\sigma} \qquad (12)$$

$$d_{12\sigma} = \left(\frac{n_1}{n}\right)_{/\sigma} + \frac{n_1\, n_2\, (m_2 - m_1)}{n\, \varrho}\, (\log\, p)_{/\sigma}\,. \qquad (13)$$

Hierbei bedeuten n_1 die Zahl der Atome N in der Volumeneinheit und n_2 die entsprechende Zahl der Moleküle N_2. $d_{1\sigma}$ ist der Diffusionsstrom der Atome, $d_{2\sigma}$ der der Moleküle. Die Gln. (1) und (2) geben die Änderungen der Atom- und Moleküldichten an, wobei r_1 bzw. r_2 die Zahl der durch Reaktionen pro Zeiteinheit in der Volumeneinheit entstandenen Atome bzw. Moleküle darstellen, so daß nach (8) $r_1 = -2\,r_2$ sein muß. Die Form (8) des Wertes von r_1 ergibt sich anschaulich aus folgender Überlegung:

Atome N können entstehen durch Zusammenstoß zweier Moleküle, wobei das eine dissoziiert, oder durch Zusammenstoß eines Atoms und eines Moleküls. Im ersten Fall erhält man ein Glied der Form $k_2 n_2{}^2$, im zweiten Falle $k_1 n_1 n_2$. Atome können zu einem Molekül rekombinieren, wenn Dreierstöße erfolgen: z.B. wenn drei Atome oder zwei Atome und ein Molekül zusammenstoßen. Das ergibt zwei Glieder der Form $-k_3 n_1{}^3$ und $-k_4 n_1{}^2 n_2$. Die Summe aller dieser vier Glieder muß aber gerade Null werden, wenn chemisches Gleichgewicht eingetreten ist, d.h., wenn nach dem Massenwirkungsgesetz

$$n_2 = K n_1{}^2 \qquad (14)$$

ist (K = Konstante des Massenwirkungsgesetzes). Deshalb müssen $k_3 = k_1 K$ und $k_4 = k_2 K$ sein, so daß sich für r_1 die Form (8) ergibt. (3) ist die übliche Form des Impulssatzes mit μ als Koeffizient der inneren Reibung. (4) hat ebenfalls die übliche Form des Energiesatzes mit u als innerer Energie pro Masseneinheit und q_σ als Wärmestrom. (5) ist die Zustandsgleichung des idealen Gases. Wichtig sind die Formen von u und q_σ, die in (9) bis (13) angegeben sind: In (9) ist $\frac{3}{2}\,kT$ die kinetische Energie eines Atoms bzw. eines Moleküls. Für die Atome tritt hinzu die auf je zwei Atome aufzuteilende Dissoziationsenergie E_D; also $\frac{1}{2}\,E_D$ pro Atom; für die Moleküle die mittlere Anregungsenergie $\bar{\varepsilon}$ der Rotations- und Schwingungsfreiheitsgrade. Der Wärmestromvektor setzt sich zusammen aus einem Anteil, der proportional dem Gradienten der Temperatur ist. Dazu kommt aber der durch Diffusion der Atome und Moleküle bedingte Energietransport. (11), (12), (13) geben die Form der Diffusionsströme an, in die die Diffusionskonstante D_{12} eingeht (m_1 und m_2 sind die Massen von Atom und Molekül).

Der Fall des Dissoziationsgleichgewichtes ist dadurch charakterisiert, daß n_1 und n_2 als Funktionen von Temperatur und Druck durch (14) gegeben sind. Statt (1), (2) ist dann die übliche Kontinuitätsgleichung

$$\frac{d\varrho}{dt} = -\varrho \, v_{\sigma/\sigma} \tag{15}$$

zu benutzen. Das bedeutet aber nicht etwa, daß dieser Fall auftritt, wenn r_1 (und r_2) gleich Null werden. Im Gegenteil, man kann aus (1) rückwärts berechnen, wie groß r_1 wird. Aus (8) ergäbe sich dann, bei bekanntem k_1 und k_2, wie stark $n_2 - K n_1{}^2$ von Null abweicht, d.h. wie groß der Fehler der Voraussetzung des Gleichgewichtes ist. Dies erfordert aber eine theoretische Untersuchung der Wirkungsquerschnitte für Dissoziation bei Stoß von Atomen und Molekülen, um die Konstanten k_1 und k_2 zu berechnen. Über diese langwierigen Untersuchungen sei hier nicht berichtet. Es sollen hier nur die Ergebnisse mitgeteilt werden unter der Voraussetzung des Dissoziationsgleichgewichtes.

Dann sind aber n_1 und n_2 Funktionen der Temperatur und des Druckes. Damit kann (10) zusammengefaßt werden in der Form:

$$q_\sigma = -\tilde{\lambda} \, T_{/\sigma} - \alpha \, p_{/\sigma} \tag{16}$$

mit einer von λ verschiedenen Wärmeleitungskonstanten $\tilde{\lambda}$. In der oben erwähnten Arbeit von L. L. Moore ist λ statt $\tilde{\lambda}$ benutzt worden. Wir wollen im Folgenden bei Benutzung von λ von dem Fall „ohne Diffusion" und bei Benutzung von $\tilde{\lambda}$ von dem Fall „mit Diffusion" sprechen.

Die genauere Bestimmung der Koeffizienten $\mu, \lambda, \tilde{\lambda}$ erfordert die Lösung der Boltzmannschen Stoßgleichungen für den vorliegenden Fall, worüber aber hier nicht näher berichtet sei. Vielmehr sollen hier Ergebnisse mitgeteilt werden, die man schon mit Hilfe von halbempirischen Mischungsformeln und den Koeffizienten für einheitliche Gase erhält:

$$\mu = \beta_1 \mu_1 + \beta_2 \mu_2 ; \qquad \lambda = \beta_1 \lambda_1 + \beta_2 \lambda_2$$

$$\left.\begin{array}{l}
\mu_i = \dfrac{1}{1 + \dfrac{C}{T}} \cdot \dfrac{5}{16\,\sigma_i{}^2} \left(\dfrac{k\,m_i\,T}{\pi}\right)^{1/2} \\[4ex]
\lambda_i = \dfrac{1}{1 + \dfrac{C}{T}} \cdot \dfrac{75}{64\,\sigma_i{}^2} \left(\dfrac{k^3\,T}{m_i\,\pi}\right)^{1/2} ;
\end{array}\right\} \quad i = 1 \,;\, 2 \tag{17}$$

$$\sigma_{12} = \frac{1}{2}\,(\sigma_1 + \sigma_2)$$

$$\beta_1 = \left[1 + \frac{n_2}{n_1}\,\frac{\sigma_{12}{}^2}{\sigma_1{}^2}\,\sqrt{\frac{3}{2}}\,\right]^{-1} ; \qquad \beta_2 = \left[1 + \frac{n_1}{n_2}\,\frac{\sigma_{12}{}^2}{\sigma_2{}^2}\,\sqrt{\frac{3}{2}}\,\right]^{-1}$$

$$D_{12} = \frac{3}{8\,n\,\sigma_{12}{}^2} \left(\frac{k\,T}{2\,\pi}\,\frac{m_1 + m_2}{m_1\,m_2}\right)^{1/2}.$$

In Abb. 1 ist λ (die Wärmeleitfähigkeit ohne Diffusion) und $\tilde{\lambda}$ (die Wärmeleitfähigkeit mit Diffusion) aufgetragen. Man erkennt einen be- trächtlichen Unterschied in dem Gebiet, wo gleichzeitig Moleküle und

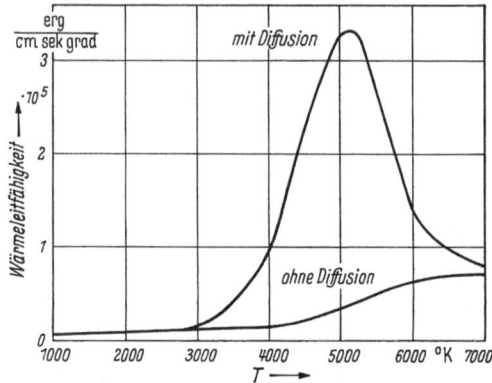

Abb. 1. Die Wärmeleitfähigkeit von dissoziierendem Stickstoff $p = 3 \cdot 10^5$ dyn cm^{-2}

Atome vorhanden sind. In Abb. 2 ist entsprechend die Prandtl-Zahl $Pr = \dfrac{\mu\, c_p}{\lambda}$ aufgetragen. Man sieht, daß man im Gebiet der Dissoziation einen wesentlichen Fehler macht, wenn man die Diffusion nicht berück-

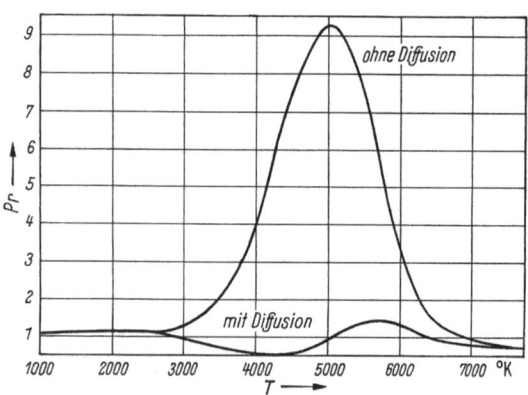

Abb. 2. Die Prandtl-Zahl von dissoziierendem Stickstoff $p = 3 \cdot 10^5$ dyn cm^{-2}

sichtigt. Tatsächlich bleibt die Prandtl-Zahl in der Größenordnung von Eins.

Mit den Gln. (1) bis (4) kann man den üblichen Grenzübergang zu den Grenzschichtgleichungen durchführen. Dabei fällt das Glied $\alpha\, p_{/\sigma}$ in (16) wegen der Grenzschichtvereinfachungen weg. Für den Fall des Dissoziationsgleichgewichtes und der Umströmung einer ebenen Platte kann man mit der Transformation

$$\eta = \frac{1}{\sqrt{x}} \int\limits_0^y \varrho\,(x, y')\,d y' \; ; \tag{18}$$

dieses partielle Differentialgleichungssystem in ein gewöhnliches überführen und es dann numerisch integrieren. Abb. 3 gibt ein Beispiel für die Grenzschicht an einer ebenen Platte für den Fall des Thermometer-

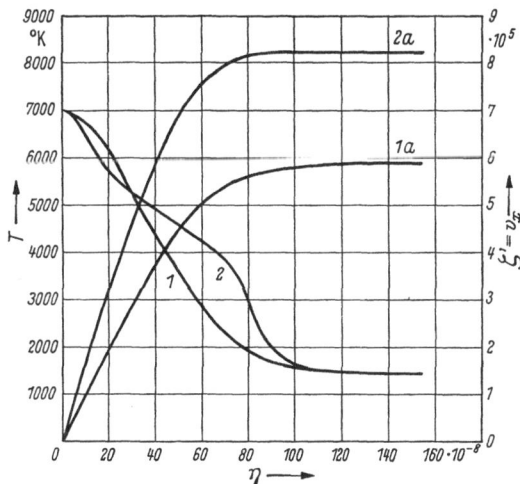

Abb. 3. Die Temperatur- und Geschwindigkeitsverteilung in der Plattengrenzschicht im Falle des Thermometerproblems. Die Geschwindigkeit $\varsigma' = v_x$ ist in cm sec^{-1} gemessen. $p_\infty = 3 \cdot 10^5$ dyn cm^{-2} $T_\infty = 1400°$ K. Der Zusammenhang zwischen η und der BLASIUSschen Variablen τ besteht in

$$\tau = \eta \cdot \frac{1}{\rho_\infty} \sqrt{\frac{V x_\infty}{\nu_\infty}} \; ; \quad \nu = \frac{\mu}{\rho}$$

problems. Es ist der Temperatur- und Geschwindigkeitsverlauf durch die Grenzschicht aufgetragen, (1) für den Fall ohne Diffusion, (2) mit Diffusion. Man sieht, daß bei gleicher Temperatur der Wand bei Berücksichtigung der Diffusion eine wesentlich höhere Außengeschwindigkeit möglich ist als für den Fall ohne Diffusion.

Literatur

[1] MOORE, L. L.: Journ. aeron. Sci. **19**, 505 (1952).
[2] KUO, Y. L.: Journ. aeron. Sci. **24**, 345 (1957).

Aus der Diskussion

H. J. KAEPPELER (Stuttgart): Ich möchte einige Bemerkungen zur Größe der Prandtl-Zahl machen. Die von L. L. MOORE gerechneten Werte ($Pr > 1$) sind von HANSEN [1] sowie auch von mir und KRAUSE [2] und von KRAUSE und KÜB-

LER [3] beanstandet worden. Auch für Luft dürfte die Prandtl-Zahl nicht größer als Eins werden. Die Rechnungen von uns [2] für Gleichgewicht in Luft stimmen überein mit den Rechnungen von Prof. LUDWIG. Der Verlauf der PRANDTL-Zahl war wie in der untenstehenden Abb. 1.

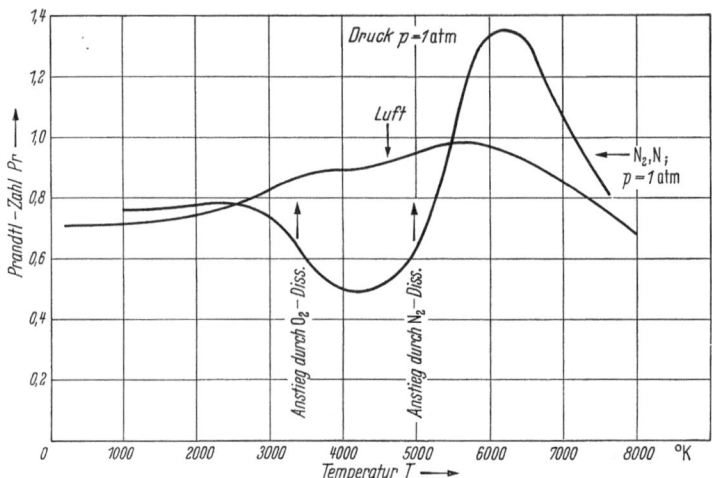

Abb. 1. Prandtl-Zahlen für dissoziierende Luft nach KAEPPELER-KRAUSE [2] und KRAUSE-KÜBLER [3] im Vergleich mit Werten für dissoziierendes N_2, N-Gemisch nach G. LUDWIG

Literatur:

[1] HANSEN, C. F.: Journ. Aeron. Sci. **20**, 789—790 (1953).
[2] KAEPPELER, H. J., und H. G. L. KRAUSE: Mitt. Forsch. Inst. Physik der Strahlantriebe, Stuttgart, Nr. **1** (1954).
[3] KRAUSE, H. G. L., und M. E. KÜBLER: Mitt. Forsch. Inst. Physik der Strahlantriebe, Stuttgart, Nr. **7** (1956).

M. LUNC (Warszawa): L'équation d'équilibre entre les molécules et les ions, présentée par l'auteur, néglige 1) l'ionisation par chocs des molécules contre la paroi, 2) les collisions non élastiques entre molécules — avec émission des photons — qui prédominent pour les températures envisagées par l'auteur.

G. LUDWIG stimmt zu und erklärt, daß bei höheren Temperaturen Glieder hinzugefügt werden müßten, die den Strahlungseffekten Rechnung tragen.

Synthetic method for boundary layer thickness

By

B. R. Seth

Indian Institute of Technology, Kharagpur

1. Introduction

The classical boundary layer equations are a reduced form of the Navier-Stokes equations in which some terms have been omitted on certain order considerations. EULER's equations are also another simplified form of the Navier-Stokes equations where viscosity is completely ignored. For two-dimensional (or axisymmetrical) motion the stream function ψ satisfies a fourth order non-linear differential equation if the full Navier-Stokes equations are taken into account; a fourth order linear differential equation for slow viscous motions; a third order non-linear equation according to Prandtl's boundary layer assumptions; and a second order linear equation if the fluid is regarded inviscid. The reduction in order of the differential equation also necessitates some sacrifice of the boundary conditions. Thus in inviscid flow only the relative normal velocity on the solid must vanish at the wall, while in viscous flow both tangential and normal velocities must be zero. Linearization of the Navier-Stokes equations, either by regarding the fluid non-viscous or the motion to be so slow that the inertia terms can be completely or partially ignored, has resulted in only limited applications of the theory. The resistance to a solid moving through a real fluid is caused by the viscosity of the fluid, and if the parameter of viscosity is small, the viscous effects are confined only to a thin region surrounding the boundary – the boundary layer region.

The truncated boundary layer equations and the inviscid fluid equations are fundamentally different in character and hence it is not possible to obtain a single solution from them which holds at all points near and away from the boundary. Both these sets of equations are, however, derivable from the Navier-Stokes equations. As no exact solution is known even for a sphere moving though a viscous liquid it follows, as FRIEDRICHS [1] has pointed out, that the boundary layer phenomenon is asymptotic in character. In other words, the actual solution should be the limiting case of an extended field. Synthetic method is an attempt in this direction. A plausible flow pattern is taken and the field equations and the boundary conditions are exactly satisfied by introducing an external constraining force. The vanishing of this force at all points ensures the correctness of the solution. But this is found to vanish exactly only in the case of inviscous or vortex motion, while for a general viscous liquid it can be made to vanish asymptotically. Thus, the actual solution is obtained as a limiting case.

Although reduced in order and complexity the classical boundary layer equations still possess no exact solution in a closed form. Series solutions have been developed by various workers but their convergence has not been proved satisfactorily. In a recent paper GÖRTLER [2] has suggested a more general series solution applicable to all cases of cylinders with rounded or wedge-shaped noses, as also to cylinders with arbitrary outer pressure distribution. This new series includes all the known series solutions [3, 4, 5] and many more which were previously untractable. Regarding convergence, GÖRTLER admits that in the case of rounded noses no improvement is accomplished by the new series over the Blasius solution, and that convergence cannot be maintained far downstream from the front stagnation point, for it must end near the separation in any case.

Synthetic method not only gives a flow field holding at all points near and away from the boundary, but also indicates the existence of a boundary layer theoretically. The boundary layer theory developed by this method seems to be an improvement over the existing classical theory because —

a) The field equations (Navier-Stokes equations) and the boundary conditions are exactly satisfied at all points in the field by introducing an external constraining force.

b) The edge of the boundary layer is an arbitrary line in the fluid such that the viscous effects which are predominant near the boundary die out rapidly as we proceed away from it. For flow past a fixed obstacle large vorticity is present very near the boundary but vanishes asymptotically away from it. The formula for the thickness of the boundary layer should depend on what order of vorticity we regard as neglegible outside the layer, and should not be assigned in the simplifying assumptions.

c) The viscous effects as compared to inertia effects are predominant near the walls but negligibly small not very far from the boundary. Thus, however thin the boundary layer region, the ratio of the viscous to inertia forces is not uniformly equal to unity inside the boundary layer, but it must be a rapidly, though continuously, decreasing function. The ratio, in fact, almost suddenly drops from an infinite value on the boundary to a very small quantity much inside the boundary layer, and is sufficiently small at the outer edge of the boundary layer.

d) The assumptions of the classical boundary layer theory also fix the order of thickness as $R^{-1/2}$. In a recent paper PROUDMAN [6] has taken the order of boundary layer thickness to lie between $R^{-1/3}$ and $R^{-1/4}$. In the present investigation it is shown that the order of thickness must depend on two parameters-one depending on the Reynolds number and the other on the vorticity allowable outside the boundary layer. Thus, in any case, the order is greater than $R^{-1/2}$.

e) It should also be possible to include the case of turbulent boundary layers by this method.

As in many branches of applied mathematics quite a few problems in fluid mechanics have been tackled by the indirect approach. Exact solutions are not available; approximate solutions are not obtained but plausible solutions are tried and fitted. In the synthetic method we start with a plausible flow pattern. For bluff symmetrical bodies the stream function ψ is chosen in the form [7]

$$\psi = y f (\xi),$$

where $\xi = $ const. represents the solid boundary in orthogonal curvilinear coordinates and y-direction is the direction normal to the undisturbed stream. Such a choice, good enough for non-viscous and slow viscous (Stokes' approximation) motion, is not satisfactory for the general viscous motion. For that we should choose a non-symmetrical flow pattern such that ψ is expressible in the form

$$\psi = y [f (\xi) + x \varphi (\xi)],$$

where f and φ are unknown functions to be determined. Corresponding choice can be made for flow past axisymmetrical bodies. KAWAGUTI [8] has chosen a similar flow pattern in discussing the critical Reynolds number for motion past a sphere.

2. Viscous flow along a flat plate

We take the x-axis along the length of the plate. The origin coincides with the leading edge. The equations of motion of incompressible viscous fluid without heat transfer are

$$u \frac{\partial u}{\partial x} + v \frac{\partial u}{\partial y} = X - \frac{1}{\varrho} \frac{\partial p}{\partial x} + v \nabla^2 u,$$

$$u \frac{\partial v}{\partial x} + v \frac{\partial v}{\partial y} = Y - \frac{1}{\varrho} \frac{\partial p}{\partial y} + v \nabla^2 v, \qquad (2.1)$$

together with the equation of continuity

$$\frac{\partial u}{\partial x} + \frac{\partial v}{\partial y} = 0. \qquad (2.2)$$

Eq. (2.2) is satisfied by a stream function ψ such that

$$u = \frac{\partial \psi}{\partial y}, \qquad v = -\frac{\partial \psi}{\partial x}. \qquad (2.3)$$

Substituting in (2.1) and eliminating p by cross-differentiation we get

$$v \nabla^4 \psi + \frac{\partial (\psi, \nabla^2 \psi)}{\partial (x, y)} = \frac{\partial Y}{\partial x} - \frac{\partial X}{\partial y}. \qquad (2.4)$$

Choosing

$$\eta = \frac{y}{\sqrt{lx}}, \qquad \psi = U \sqrt{lx}\, f(\eta) \qquad (2.5)$$

where l is some length measured along the plate, we get

$$\frac{\partial Y}{\partial x} - \frac{\partial X}{\partial y} = \frac{v\,U}{l^{3/2}\,x^{3/2}} \left[f'' + \frac{1}{2}\left(\frac{l}{x}\right)(\eta^2 f'''' + 5\,\eta f''' + 3\,f'') + \right.$$

$$+ \frac{1}{16}\left(\frac{l}{x}\right)^2 (\eta^4 f'''' + 14\,\eta^3 f''' + 45\,\eta^2 f'' + 15\,\eta f' - 15 f) \Big] +$$

$$+ \frac{U^2}{2\sqrt{lx^3}} \left[f f''' + f' f'' + \frac{1}{4}\left(\frac{l}{x}\right)\{\eta^2 f f''' + 3\,\eta\,(\eta f' + f) f'' + \right.$$

$$\left. + 3\,\eta f'^2 - 3 f f'\} \right]. \qquad (2.6)$$

Eq. (2.6) can be satisfied if we choose

$$Y = 0$$

and

$$X = -\frac{U^2}{2x}\left(\frac{2}{R} f''' + f f''\right) - \frac{U^2 l}{2 x^2}\left[\frac{1}{R}\,(\eta^2 f''' + 3\,\eta f'') + \right.$$

$$+ \frac{1}{4}\,(\eta^2 f f'' + \eta^2 f'^2 + \eta f f' - 2 f^2) \Big] -$$

$$- \frac{U^2 l^2}{16 x^3 R}\,(\eta^4 f'''' + 10\,\eta^3 f''' + 15\,\eta^2 f' - 15\,\eta f), \qquad (2.7)$$

where
$$R = \frac{Ul}{\nu}.$$

The force X also vanishes if
$$\eta f' - f = 0 \quad \text{or} \quad f = A \eta, \tag{2.8}$$
which gives the non-viscous solution.

The Blasius equation
$$\frac{2}{R} f''' + f f'' = 0, \tag{2.9}$$
is obtained by equating to zero only the first bracketed expression in the value of X. Blasius' solution is, however, defective in that the velocity in the normal direction does not vanish at infinity. For a general viscous solution holding near and away from the boundary we take
$$f(\eta) = A \eta + B \eta e^{-n\eta}. \tag{2.10}$$

The boundary conditions
$$\begin{aligned} u, v &= 0, &&\text{at } \eta = 0, \\ u &= U, v = 0 &&\text{at } \eta = \infty, \end{aligned} \tag{2.11}$$

determine the values of the constants A and B, but n is still undetermined. The solution then becomes
$$f(\eta) = \eta(1 - e^{-n\eta}). \tag{2.12}$$

If n is sufficiently large, the solution approaches the irrotational value at all points except near the plate. X is small not very far from the boundary. On the plate
$$X = -\frac{\nu U}{lx} f'''(0) = -\frac{\nu U}{lx} \cdot 3 n^2 \sim \frac{n^2}{R}. \tag{2.13}$$

This can also be made small provided we choose
$$R \sim n^{2+\lambda}, \quad \lambda > 0. \tag{2.14}$$

We find that the vorticity $\nabla^2 \psi$ is of order n on the boundary and n is large, so that the vorticity is large near the boundary but dies out rapidly as we proceed away from the boundary, thus indicating the existence of a thin region surrounding the boundary which we may define as the boundary layer region. The thickness of this region will depend on the order of vorticity allowable outside the region. If $\eta = \delta$ represents the edge of the boundary layer, the vorticity outside this region must be vanishingly small. This is possible if $n\delta$ is large while δ is small. We should therefore have
$$\delta \sim n^{-k} \quad 0 < k < 1. \tag{2.15}$$

Hence we get

$$\delta \sim R^{-k/(2+\lambda)} . \tag{2.16}$$

The momentum thickness is given by

$$\Theta = \frac{5\sqrt{lx}}{4\,n} \sim \sqrt{x}\, R^{-\frac{1}{2+\lambda}} . \tag{2.17}$$

The ratio of the viscous to inertia forces becomes

$$\frac{V}{I} = \frac{\mu\,\nabla^2 u}{\varrho\left(u\,\dfrac{\partial u}{\partial x} + v\,\dfrac{\partial u}{\partial y}\right)} = \tag{2.18}$$

$$= \frac{1}{R}\,\frac{\left[2\,n^2\,(3 - n\,\eta) - \dfrac{l}{2\,x}\,n\,\eta\,(n^2\,\eta^2 - 6\,n\,\eta + 6)\right]}{n\,\eta\,(-2 + n\,\eta)\,(1 - e^{-n\,\eta})} .$$

Now $n\,\eta = 0$ on the boundary and becomes large of order n^{1-k} at the outer edge of the boundary layer. The ratio (V/I) therefore falls from an infinite value at the boundary to a quantity of order $n^{k-1-\lambda}$, which is small. Thus the viscous forces are predominant near the boundary, but become negligible as compared to the inertia forces at the outer edge of the layer.

The results of the classical theory are obtained by taking

$$\lambda = 0, \qquad k = 1 , \tag{2.19}$$

which give

$$\frac{V}{I} \sim 0\,(1) \text{ and } \delta \sim \frac{1}{\sqrt{R}} ; \tag{2.20}$$

but (2.19) do not hold since $\lambda = 0$, violates X being vanishingly small throughout the field, and $k = 1$ makes the vorticity large at the outer edge of the boundary layer.

If

$$k = \frac{1}{2} , \qquad \lambda = \frac{1}{2} , \tag{2.21}$$

we get

$$\delta \sim R^{-\frac{1}{5}} , \tag{2.22}$$

which is a formula for the turbulent case.

The drag on the plate in the x-direction is given by

$$D = 8\,n\,\mu\,U = \frac{8\,n\,U^2\,l\,\varrho}{R} , \tag{2.23}$$

so that the drag coefficient is

$$c_D \sim R^{-(1+\lambda)/(2+\lambda)} .$$

4*

3. Viscous flow past a circular cylinder

Let the origin of the xy-plane be at the centre of a circular cross-section of the cylinder, the axis of x in the direction of the undisturbed stream, and the axis of y perpendicular to it.

We choose the stream function ψ in the form

$$\psi = y\,[f\,(r) + r\,\varphi\,(r)]. \tag{3.1}$$

The equations of motion are

$$\frac{\partial}{\partial x}\left(\frac{u^2+v^2}{2}+\frac{p}{\varrho}\right) = X + \nu\,\nabla^2 u - v\left(\frac{\partial u}{\partial y}-\frac{\partial v}{\partial x}\right), \tag{3.2}$$

$$\frac{\partial}{\partial y}\left(\frac{u^2+v^2}{2}+\frac{p}{\varrho}\right) = Y + \nu\,\nabla^2 v + u\left(\frac{\partial u}{\partial y}-\frac{\partial v}{\partial x}\right). \tag{3.3}$$

Substituting (3.1) into (3.2) and (3.3) we get

$$\frac{\partial}{\partial x}\left(\frac{u^2+v^2}{2}+\frac{p}{\varrho}\right)-\nu\left[F'+x\,\Phi'-\frac{x^2}{r}\frac{r\,F'-F}{r^2}-\frac{x^3}{r}\frac{r\,\Phi'-\Phi}{r}\right]+$$

$$-\left[r\,F\,\varphi+x(r\,\varphi\,\Phi+F\,F')+x^2\left(\Phi\,f'+F\,\varphi'-\frac{F\,\varphi}{r}\right)-\right. \tag{3.4}$$

$$\left.+x^3\left(\Phi\,\frac{r\,\varphi'-\varphi}{r}-\frac{F\,f'}{r^2}\right)-\frac{x^4}{r^2}\,(\Phi\,f'+F\,\varphi')-\frac{x^5}{r^2}\,\Phi\,\varphi'\right]-X=0$$

and

$$\frac{\partial}{\partial y}\left(\frac{u^2+v^2}{2}+\frac{p}{\varrho}\right)+\nu\,\frac{y}{r}\left(\Phi+x\,\frac{r\,F'-F}{r^2}+x^2\,\frac{r\,\Phi'-\Phi}{r^2}\right)-$$

$$-\frac{y}{r}\left[F\,(r\,\varphi'+f)+x\,\{\Phi\,(r\,f'+f)+F\,(r\,\varphi'+\varphi)\}+\right. \tag{3.5}$$

$$\left.+x^2\left\{\Phi\,(r\,\varphi'+\varphi)-\frac{F\,f'}{r}\right\}-\frac{x^3}{r}\,(F\,\varphi'+\Phi\,f')-\frac{x^4}{r}\,\Phi\,\varphi'\right]-Y=0$$

where

$$F = r\,f''+3\,f', \qquad \Phi = r\,\varphi''+5\,\varphi'. \tag{3.6}$$

We can now choose the extraneous forces X and Y so that the eqs. (3.4), (3.5) and (3.6) are integrable. Thus if we take $Y = 0$, and

$$X=-\nu\left[F'+\frac{F}{r}+x\left(\Phi'+\frac{3\,\Phi}{r}\right)\right]+\left[\{\Phi\,r\,f-\textstyle\int r\,(f\,\Phi'+\Phi\,F')\,d\,r\}\,+\right.$$

$$+x\left\{r\,\varphi\,\Phi+\frac{f\,F}{r}-2\textstyle\int\left(r\,\varphi\,\Phi+\frac{F\,f'}{r}\right)d\,r\right\}+x^2\left\{\frac{2\,F\,\varphi+\Phi\,f}{r}-\right. \tag{3.7}$$

$$\left.-3\textstyle\int\frac{F\,\varphi'+\Phi\,f'}{r}\,d\,r\right\}+2\,x^3\left\{\frac{\Phi\,\varphi}{r}-2\textstyle\int\frac{\Phi\,\varphi'}{r}\,d\,r\right\}\right],$$

the integral of the equations becomes

$$\frac{u^2 + v^2}{2} + \frac{p}{\varrho} - \left[\int F\,(r\,f' + f)\,d\,r + x \int \{ F\,(r\,\varphi' + \varphi) + \varPhi\,(r\,f' + f) \}\,d\,r + \right.$$

$$+ x^2 \int \left\{ -\frac{F\,f'}{r} + \varPhi\,(r\,\varphi' + \varphi) \right\}\,d\,r - x^3 \int \frac{F\,\varphi' + \varPhi\,f'}{r}\,d\,r - \tag{3.8}$$

$$\left. - x^4 \int \frac{\varPhi\,\varphi'}{r}\,d\,r \right] + v \left[\int \varPhi\,d\,r + \frac{x\,F}{r} + \frac{x^2\,\varPhi}{r} \right] = a \text{ constant.}$$

The force X vanishes exactly if $F = \varPhi = 0$ and this gives the perfect fluid solution.

The non-viscous terms in the value of X in (3.7) vanish if we choose

$$F = r\,f'' + 3\,f' = -\,m\,r\,f,$$
$$\varPhi = r\,\varphi'' + 5\,\varphi = -\,m\,r\,\varphi, \tag{3.9}$$

m being an arbitrary constant. For $v = 0$, the solutions of (3.9) give the rotational solution corresponding [9] to $(\nabla^2 + k^2)\,\psi = 0$. These cannot be made to satisfy the viscous boundary conditions. Hence these are of no use for the viscous problem.

The coefficient of v in (3.7) vanishes if

$$F = \frac{A}{r},$$
$$\varPhi = \frac{B}{r^3}. \tag{3.10}$$

This corresponds to Stokes' slow viscous solution with the well-known [10] defect of the boundary conditions at infinity.

We may take away some non-viscous terms from X to include them in Y. Thus if we take

$$X = -v \left[F' + \frac{F}{r} + x \left(\varPhi' + \frac{3\,\varPhi}{r} \right) \right] - \left[r\,F\,\varphi + x \left(r\,\varphi\,\varPhi + \frac{F\,f}{r} \right) + \right.$$

$$+ x^2 \left(\varPhi\,f' + F\,\varphi' - \frac{F\,\varphi}{r} \right) + x^3 \left(\varPhi\,\varphi' - \frac{\varPhi\,\varphi}{r} - \frac{F\,f'}{r^2} \right) - \tag{3.11}$$

$$\left. - \frac{x^4}{r^2}\,(\varPhi\,f' + F\,\varphi') - \frac{x^5}{r^2}\,\varPhi\,\varphi' \right],$$

and

$$Y = \frac{x\,y}{r} \left[\varPhi\,(r\,f' + f) + F\,(r\,\varphi' + \varphi) + x \left\{ \varPhi\,(r\,\varphi' + \varphi) - \frac{F\,f'}{r} \right\} - \right.$$

$$\left. - \frac{x^2}{r}\,(F\,\varphi' + \varPhi\,f') - \frac{x^3}{r}\,\varPhi\,\varphi' \right], \tag{3.12}$$

all non-viscous terms in X, Y vanish over $r = a$.

With this choice of the constraining forces the solution of eqs. (3.4), (3.5) and (3.6) is

$$\frac{u^2 + v^2}{2} + \frac{p}{\varrho} + v\left[\int \Phi \, dr + x\,\frac{F}{r} + x^2\,\frac{\Phi}{r}\right] -$$

$$- \int F\,(r\,f' + f)\,dr = \text{constant.}$$

(3.13)

Putting $F = A\,r^{-n}$, $\Phi = B\,r^{-s} + C\,r^{-t}$, we find that the forces X and Y and the vorticity ω vanish asymptotically as we proceed away from the boundary. If we suppose the constants n, s, and t to be large quantities of the same order of magnitude, a supposition which is physically justifiable, the vanishing of X, Y, and ω can be attained not very far from the boundary. The force Y vanishes exactly on the boundary, but X and ω do not. On the boundary

$$Y = 0, \quad X \sim n^2/R, \quad \omega \sim n.$$

(3.14)

In this case, again, the force X can be made negligibly small everywhere in the field if we take

$$R \sim n^{2+\lambda}$$

If $r = a + \delta$ represents the edge of the boundary layer, at least near the front stagnation point, we have at $r = a + \delta$, the vorticity given by

$$\omega \sim n\,e^{-\dfrac{n\,\delta}{a}}.$$

(3.15)

We have to make a choice similar to (2.15) in order to set the vorticity small outside the edge $r = a + \delta$ and hence we get, once again, the thickness of the boundary layer as

$$\frac{\delta}{a} \sim R^{-k/(2+\lambda)}.$$

(3.16)

Comparing viscous to inertia forces at any point we have

$$\frac{V}{I} = v\left[F' + x\,\Phi' - \frac{x^2}{r}\,\frac{r\,F' - F}{r^2} - \frac{x^3}{r}\cdot\frac{r\,\Phi' - \Phi}{r^2}\right]\Big/\Big[(f + x\,\varphi)$$

$$\left(x\,\varphi' + \varphi + \frac{x\,f'}{r}\right) + \frac{y^2}{r}\left\{(f' + x\,\varphi')\left(r\,\varphi' - 2\,\varphi - \frac{2\,x\,f'}{r}\right) - 3\,x^2\,\frac{\varphi'}{r}\right.$$

(3.17)

$$(f' + x\,\varphi')\left(\frac{r\,f'' - f'}{r^2} + x\,\frac{r\,\varphi'' - \varphi'}{r^2}\right) - \frac{y^4}{r^2}\,\varphi\left(\frac{r\,f'' - f'}{r^2} + x\,\frac{r\,\varphi'' - \varphi'}{r^2}\right)\Big],$$

from which we have

$$\frac{V}{I} \sim \frac{n\,\omega}{R} \sim \frac{1}{n^\lambda\,e^{n^{1-k}}}, \qquad \lambda > 0, \quad 0 < k < 1.$$

(3.18)

Thus, the viscous forces are predominant in a thin region surrounding the boundary—the boundary layer region—and become negligible as compared to the inertia forces outside this region.

For the classical case $k = 1$, $\lambda = 0$ and the viscous and inertia forces become the same in the boundary layer, where thickness now varies as $R^{-1/2}$.

The drag coefficient of the cylinder is found to be

$$C_D \sim \frac{n}{R} \sim R^{-(1+\lambda)/(2+\lambda)} . \tag{3.19}$$

4. Viscous flow past a sphere

Let the centre of the sphere be the origin of the cylindrical coordinates $(x,\ y,\ \theta)$ where the x-axis is taken in the direction of the undisturbed stream, y is the perpendicular distance from x-axis and θ is the azimuthal angle. Since the component of velocity in the direction of θ increasing is zero and all quantities are independent of θ, the equations of motion are

$$\frac{\partial}{\partial x}\left(\frac{u^2+v^2}{2}+\frac{p}{\varrho}\right) = X - v\left(\frac{\partial u}{\partial y}-\frac{\partial v}{\partial x}\right) + \nu\left(\frac{\partial^2 u}{\partial x^2}+\frac{\partial^2 u}{\partial y^2}+\frac{1}{y}\frac{\partial u}{\partial y}\right),$$
$$\frac{\partial}{\partial y}\left(\frac{u^2+v^2}{2}+\frac{p}{\varrho}\right) = Y + u\left(\frac{\partial u}{\partial y}-\frac{\partial v}{\partial x}\right) + \nu\left(\frac{\partial^2 v}{\partial x^2}+\frac{\partial^2 v}{\partial y^2}+\frac{1}{y}\frac{\partial v}{\partial y}-\frac{v}{y^2}\right). \tag{4.1}$$

The equation of continuity is

$$\frac{\partial u}{\partial x}+\frac{\partial v}{\partial y}+\frac{v}{y}=0 . \tag{4.2}$$

This last equation is satisfied by a Stokes' stream function ψ so that

$$u = \frac{1}{y}\frac{\partial \psi}{\partial y} , \qquad v = -\frac{1}{y}\frac{\partial \psi}{\partial x} .$$

We now choose a non-symmetrical form of ψ, viz

$$\psi = y^2\left[f(r)+x\,\varphi(r)\right], \qquad \text{where } r^2 = (x^2+y^2) . \tag{4.3}$$

This gives

$$u = r f' + 2f + x\left(r f' + 2\varphi\right) - \frac{x^2 f'}{r} - x^3\frac{\varphi'}{r} ,$$
$$v = -\frac{y}{r}\left[r\varphi + x f' + x^2 \varphi'\right] , \tag{4.4}$$

and

$$\omega = \frac{\partial u}{\partial y}-\frac{\partial v}{\partial x} = \frac{y}{r}(F + x\,\Phi) \tag{4.5}$$

where

$$F = r f'' + 4f' , \qquad \Phi = r\varphi'' + 6\varphi' . \tag{4.6}$$

Substituting in equations (4.1) we get

$$\frac{\partial}{\partial x}\left(\frac{p}{\varrho}+\frac{u^2+v^2}{2}\right)=X+\frac{y^2}{r^2}(F+x\,\Phi)(r\,\varphi+x\,f'+x^2\,\varphi')+$$
$$+\nu\left[\frac{r\,F'+F}{r}+x\,\frac{r\,\Phi'+\Phi}{r}-x^2\,\frac{r\,F'-F}{r^3}-x^3\,\frac{r\,\Phi'-\Phi}{r^3}\right], \tag{4.7}$$

$$\frac{\partial}{\partial y}\left(\frac{p}{\varrho}+\frac{u^2+v^2}{2}\right)=Y+\frac{y}{r}(F+x\,\Phi)\left[(r\,f'+2\,f)+x\,(r\,\varphi'+2\,\varphi)-\right.$$
$$-x^2\,\frac{f'}{r}-x^3\,\frac{\varphi'}{r}\Big]-\frac{\nu\,y}{r}\left[\Phi+x\,\frac{r\,F'-F}{r^2}+x^2\,\frac{r\,\Phi'-\Phi}{r^2}\right]. \tag{4.8}$$

These equations can be integrated if we choose

$$X=-\nu\left[\frac{r\,F'+2\,F}{r}+x\,\frac{r\,\Phi'+2\,\Phi}{r}\right]-\left[r\,F\,\varphi+x\left(r\,\Phi\,\varphi-\frac{2\,F\,f}{r}\right)+\right.$$
$$+x^2\left(F\,\frac{r\,\varphi'-\varphi}{r}+\Phi\,f'\right)+x^3\left(\Phi\,\frac{r\,\varphi'-\varphi}{r}-\frac{F\,f'}{r^2}\right)-$$
$$-\frac{x^4}{r^2}(F\,\varphi'+\Phi\,f')-\frac{x^5}{r^2}\,\Phi\,\varphi'\right], \tag{4.9}$$

$$Y=-\frac{x\,y}{r}\left[\Phi\,(r\,f'+2\,f)+F\,(r\,\varphi'+2\,\varphi)+x\left\{\Phi\,(r\,\varphi'+2\,\varphi)-\frac{F\,f'}{r}\right\}-\right.$$
$$-\frac{x^2}{r}(\Phi\,f'+F\,\varphi')-\frac{x^3}{r}\,\Phi\,\varphi'\right],$$

and the solution, then, becomes

$$\frac{p}{\varrho}+\frac{u^2+v^2}{2}-\int F\,(r\,f'+2\,f)\,dr+\nu\left[\int\Phi\,dr+\frac{x\,F}{r}+\frac{x^2\,\Phi}{r}\right]=\text{const.} \tag{4.10}$$

Again, the vanishing of F and Φ gives the perfect fluid solution. If, however, we keep X only in the direction of x-axis, we get

$$Y=0,$$

$$X=[\int\{F\,(r\,\varphi'+2\,\varphi)+\Phi\,(r\,f'+2\,f)\}\,dr-r\,F\,\varphi]+$$
$$+x\left[\frac{2\,F\,f}{r}+2\int\left\{\Phi\,(r\,\varphi'+2\,\varphi)-\frac{F\,f'}{r}\right\}dr-r\,\Phi\,\varphi\right]+$$
$$+x^2\left[\frac{3\,F\,\varphi+2\,\Phi\,f}{r}-3\int\frac{F\,\varphi'+\Phi\,f'}{r}\,dr\right]+ \tag{4.11}$$
$$+x^3\left[\frac{3\,\Phi\,\varphi}{r}-4\int\frac{\Phi\,\varphi'}{r}\,dr\right]-\nu\left[\frac{r\,F'+2\,F}{r}+x\,\frac{r\,\Phi'+4\,\Phi}{r}\right].$$

This leads to the case of Hill's spherical vortex for the symmetrical case $\varphi=0$. Stokes' slow viscous motion is obtained by taking

$$r F' + 2 F = 0,$$

and
$$r \Phi' + 4 \Phi = 0, \tag{4.12}$$

and satisfying the boundary conditions

$$
\begin{array}{lll}
u = U, & v = 0 & \text{at } r = \infty \\
u = 0, & v = 0 & \text{at } r = a
\end{array} \tag{4.13}
$$

or

$$
\begin{array}{lll}
f = f' = \varphi = \varphi' = 0 & \text{at} & r = a, \\
f = U, f' = \varphi = \varphi' = 0 & \text{at} & r = \infty
\end{array} \tag{4.14}
$$

The solution thus obtained is

$$f = u \left[1 - \frac{3}{2} \left(\frac{a}{r} \right) + \frac{1}{2} \left(\frac{a}{r} \right)^3 \right], \qquad \varphi = 0 . \tag{4.15}$$

In this solution the external forces do not vanish but these become negligible if U^2/a is small, i.e. if the motion is slow. Thus we see that for slow viscous motion φ vanishes automatically, and the motion becomes symmetrical.

For a general viscous solution we take

$$F = \frac{A}{r^n}, \qquad \Phi = \frac{B}{r^s} + \frac{C}{r^t} . \tag{4.16}$$

Satisfying the boundary conditions (4.14) we get

$$f = U \left[1 - \frac{n-1}{n-4} \left(\frac{a}{r} \right)^3 + \frac{3}{n-4} \left(\frac{a}{r} \right)^{n-1} \right],$$

$$\varphi = \frac{C}{(t-1)(t-6)(s-6) a^{t-1}} \left[(s-6) \left(\frac{a}{r} \right)^{t-1} - (t-6) \left(\frac{a}{r} \right)^{s-1} - \right. \tag{4.17}$$

$$\left. - (s-t) \left(\frac{a}{r} \right)^5 \right].$$

Proceeding as in the case of a circular cylinder we again get

$$\frac{\delta}{a} \sim n^{-k}, \qquad 0 < k < 1, \tag{4.18}$$

and
$$R \sim n^{2+\lambda} \qquad \lambda > 0,$$

so that
$$\frac{\delta}{a} \sim R^{-k/(2+\lambda)} . \tag{4.19}$$

The drag suffered by the sphere is given by

$$D = 6 \pi \mu U a (n - 1) \tag{4.20}$$

and
$$C_D \sim \frac{n}{R} \sim R^{-(1+\lambda)/(2+\lambda)} .$$

References

[1] FRIEDRICHS, K. O.: Asymptotic phenomena in mathematical physics. Bul. Amer. Math. Soc. **61**, 485—504 (1955) No. 6.

[2] GÖRTLER, H.: A new series for the calculation of steady laminar boundary layer flows. Jour. of Maths. and Mech. **6**, 1—67 (1957) No. 1.

[3] HOWARTH, L.: Steady laminar flow in the boundary layer near the surface of a cylinder in a stream. Aero. Res. Comm., R.M.No.1632 (1935).

[4] BLASIUS, H.: Zeit. Math. Physik. **56**, 1—37 (1908).

[5] TANI, I.: On the solution of laminar boundary layer equations. Jour. Phys. Soc. Japan, **4**, 149—154 (1949).

[6] PROUDMAN, J.: Jour. Fluid Mech. Nov. 56, p. 505.

[7] WADHWA, Y. D.: Axisymmetrical viscous flow, unpublished thesis (see also ZAMM, Jan. 55, p. 69).

[8] KAWAGUTI, M.: Critical Reynolds number for the flow past a sphere, Jour. Phys. Soc., Japan **10**, 694—699 (1955) No. 8.

[9] LAMB, H.: Hydrodynamics. 6th Edition, p. 244.

[10] —: Hydrodynamics. 6th Edition, p. 614.

Aus der Diskussion

G. TEMPLE (Oxford): The method employed by Dr. SETH to determine the accuracy of his approximate solutions is to calculate the residual body forces needed to produce the field of flow assumed. This seems to be a development of the method employed by OSEEN for similar problems and of the method now regularly employed in relaxation calculations.

The great difficulty seems to be to prescribe an appropriate upper bound to the body forces which will ensure a prescribed degree of approximation in the velocity field.

On rotating laminar boundary layers

By

K. Stewartson

Department of Mathematics, University of Bristol[1]

1. The rotating sphere

The theory of the laminar boundary layer on a sphere of radius a rotating, without translation, about a diameter with angular velocity Ω was first considered by HOWARTH [6]. He showed that a consistent solution can be obtained on the assumption that there are two boundary layers, originating at opposite ends of the diameter of rotation and converging towards each other at the equator. Near the poles each boundary layer is the same as that on a disc rotating about its axis in a fluid otherwise at rest. In this boundary layer, first discussed by VON KÁRMÁN [8], fluid is drawn towards the disc and also moves radially outwards from the axis. As a result of the curvature of the sphere the boundary layers on the sphere must be modified as they spread outwards from the poles and by applying the momentum integral HOWARTH was able to extend them as far as the equatorial plane. According to his solution the boundary layers at the equator were infinitely thick with zero longitudinal components of velocity but his view was that this would be unlikely in the exact solution. In any case the boundary layers from the two poles must meet at the equator and the parabolic nature of their governing equations made them inadequate to describe the flow in its vicinity.

His argument has been criticized by NIGAM [11] who wrote down a slightly different form for the series expansion of the velocity components and inferred that although there is an inflow to the sphere near the poles it was balanced by an outflow near the equator so that when the boundary layers met each had lost all its fluid. This view can however be refuted as follows.

Let us take spherical polar coordinates R, θ, φ with R measured radially outwards from the centre of the sphere, θ measured from the axis of rotation and φ the azimuth angle. Let (w, u, v) respectively be the components of the velocity in the directions of (R, θ, φ) increasing. Then [6] the boundary layer equations are

[1] Present adress: Department of Mathematics, the Durham Colleges in the University of Durham.

$$\frac{u}{a}\frac{\partial u}{\partial \theta} + w\frac{\partial u}{\partial R} - \frac{v^2}{a}\cot \theta = v\frac{\partial^2 u}{\partial R^2}\,, \tag{1.1}$$

$$\frac{u}{a}\frac{\partial v}{\partial \theta} + w\frac{\partial v}{\partial R} + \frac{uv}{a}\cot \theta = v\frac{\partial^2 v}{\partial R^2}\,, \tag{1.2}$$

and

$$\frac{1}{a}\frac{\partial u}{\partial \theta} + \frac{\partial w}{\partial R} + \frac{u}{a}\cot \theta = 0\,. \tag{1.3}$$

The boundary conditions are that

$$u = w = 0,\, v = \Omega\, a \sin \theta \text{ at } R = a \text{ and } u \to v \to 0 \text{ as } R \to \infty. \tag{1.4}$$

From (1.3) we may introduce a stream function ψ so that

$$\frac{\partial \psi}{\partial R} = au \sin \theta\,, \qquad \frac{\partial \psi}{\partial \theta} = -a^2\, w \sin \theta \tag{1.5}$$

and on multiplying (1.1) by $a\psi \sin \theta$ we have, on integrating with respect to R,

$$\frac{\partial}{\partial \theta}\int_a^\infty \psi u^2 \sin \theta\, dR = \int_a^\infty \psi v^2 \cos \theta\, dR. \tag{1.6}$$

Hence

$$\int_0^{\pi/2} d\theta \int_a^\infty \psi v^2 \cos \theta\, dR = \left\{\int_a^\infty \psi u^2 dR\right\}_{\theta = \pi/2}. \tag{1.7}$$

Now since the boundary layer for u is being accelerated by the centrifugal term $(v^2/a)\cot \theta$, it may be proved that ψ is never negative, as follows. Considering the boundary layer which starts at $\theta = 0$, we have initially $u/\theta > 0$ in $a < R < \infty$. As θ increases $\partial u/\partial R$ at $R = a$ cannot vanish before $u = o$ at some greater value of R because $\partial^2 u/\partial R^2 < 0$ at $R = a$. Further at the point in $R > a$ where u vanishes first $\partial u/\partial R$ must also vanish and $\partial^2 u/\partial R^2 > 0$. However from (1.1) when $u = \partial u/\partial R = 0$, $\partial^2 u/\partial R^2 < 0$ which is a contradiction. Hence u and therefore ψ is never negative, implying that the left-hand side of (1.7) is positive. At $\theta = \pi/2$ therefore the right-hand side of (1.7) is non-zero [its value from HOWARTH'S theory is $0.0025\ \Omega^2 v a$] the boundary layers there cannot be empty of fluid, and a collision is unavoidable.

After the collision the fluid originally in the boundary layers will move outwards along the equatorial plane ultimately taking up the velocity distribution of SQUIRE's radial jet [15]. Since the thicknesses of the shear layers both before and after the collision are $0\ (v^{1/2})$ it is reasonable to expect that the region in which the boundary layer assumptions break down is within a distance $0\ (v^{1/2})$ of the equator. The flow in the neighbourhood of $R = a$, $\theta = \pi/2$ may be found by writing

$$\psi = a^2\, (\Omega v)^{1/2}\, \Psi\, (\eta,\, \xi)\,, \qquad v = a\,\Omega\, V\, (\eta,\, \xi)$$

$$\eta = (R - a)\left(\frac{\Omega}{v}\right)^{1/2}\,, \qquad \xi = (\theta - \pi/2)\left(\frac{\Omega}{v}\right)^{1/2} \tag{1.8}$$

and substituting in the full equations of motion. On proceeding to the limit $\nu = 0$ these reduce to

$$V = F(\Psi) \quad \text{and} \quad \frac{\partial^2 \Psi}{\partial \xi^2} + \frac{\partial^2 \Psi}{\partial \eta^2} = G(\Psi) \tag{1.9}$$

where F and G are functions of Ψ only. The boundary conditions are that V and Ψ are prescribed as $\xi \to \pm \infty$ for fixed η, being given by the values of ψ and v from the boundary layer solution at $\theta = \pi/2 \pm 0$. Further grad $\Psi \to 0$ as $\xi^2 + \eta^2 \to \infty$ and $\Psi = 0$ at $\eta = 0$.

Since the order of the governing equation has been decreased by two, two boundary conditions, that $\partial \Psi/\partial \eta = 0$ on $\eta = 0$ and $\partial^2 \Psi/\partial \xi^2 = 0$

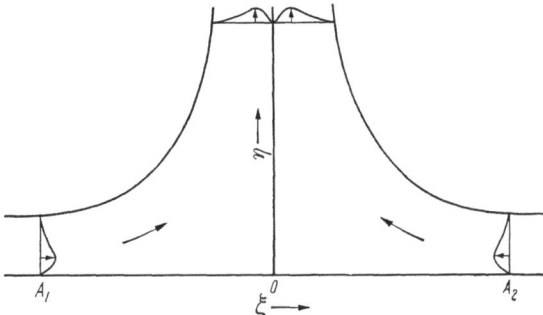

Fig. 1. Flow in the neighbourhood of the equator of a rotating sphere

on $\xi = 0$, must be abandoned. According to the solution of (1.9) u must have the form $\Omega\, a\, U_0(\xi)$ on $\eta = 0$ where $U_0{}^2 = 0$ (1) rising from zero as ξ^2 decreases from infinity and falling to zero again as $\xi^2 \to 0$. This must be cancelled by an inner boundary layer, whose thickness is $0\ (\nu^{3/4})$ since a typical length in the direction of $A_1\, 0\, A_2$ (fig. 1) is $0\ (\nu^{1/2})$. Although this inner layer is thin in comparison with the outer inviscid layer its effect cannot be neglected altogether, for since U_0 vanishes both at $\xi = \pm \infty$ and $\xi = 0$ it must separate somewhere between A_1 and 0, and between 0 and A_2. It is believed however that U_0 sgn ξ is a positive quantity so that the separated region is confined to the vicinity of 0 and further that U_0 sgn ξ is not large enough for the separated region seriously to modify the flow described by (1.9).

The discontinuity in the tangential stress across $\xi = 0$ also needs an inner boundary layer of thickness $0\ (\nu^{3/4})$ to cancel it but the induced tangential component of velocity is only $0\ (\nu^{1/4})$ and is negligible in comparison with the velocity in the outer inviscid layer.

If we may neglect the effect of the inner boundary layer the behaviour of Ψ as $\eta \to \infty$ can be found without solving (1.9) completely. For we may expect that $\partial \Psi/\partial \eta \to 0$ as $\eta \to \infty$ for fixed ξ in which case Ψ is the same function of ξ when $\eta \to \infty$, as it is of η when $\xi \to \infty$. Beyond the region of validity of (1.9) the governing equations are those for a radial jet [15] whose initial profile is the final profile as $\eta \to \infty$ according to (1.9) and hence is simply related to the final profile of the boundary layers on the sphere as $\theta \to \pi/2 \pm 0$.

2. The rotating cylindrical half-body

Let us suppose that the half-body consists of an infinitely long circular cylinder of radius a whose axis lies on $\theta = \pi$ and which is closed at one end by a hemispherical boss $R = a$, $|\theta| < \pi/2$. If it rotates about the axis of symmetry then as in the case of the sphere a boundary layer occurs centred on the pole of the hemisphere and spreading out to the equator. There however instead of being deflected, it continues along the surface of the cylinder. Let us define cylindrical polar coordinates (r, φ, z) where r denotes distance from the axis of symmetry, z distance from the plane base of the hemisphere and φ the azimuth angle as before. Further let us extend the definitions of the velocity components so that in $z > 0$ (w, v, u) respectively are the components in the direction of (r, φ, z) increasing. Then the controlling equations in the boundary layer on the cylinder are

$$u \frac{\partial u}{\partial z} + w \frac{\partial u}{\partial r} = v \frac{\partial^2 u}{\partial r^2} \tag{2.1}$$

$$u \frac{\partial v}{\partial z} + w \frac{\partial v}{\partial r} = v \frac{\partial^2 u}{\partial r^2} \tag{2.2}$$

and

$$\frac{\partial u}{\partial z} + \frac{\partial w}{\partial r} = 0$$

to begin with, together with boundary conditions

$$u = w = v - \Omega a = 0 \quad \text{at} \quad r = a \quad z > 0$$

$$u \to v \to 0 \quad \text{as} \quad r \to \infty \tag{2.3}$$

and u, v prescribed at $z = 0$ from the boundary layer on the hemisphere. If $au = \partial \Psi / \partial r$ then

$$\int_a^\infty \psi u^2 \, dr = \alpha \Omega^2 a^4 v \tag{2.4}$$

is constant and equal to its value at $z = 0$ which is prescribed. In fact on making use of Howarth's [6] theory of the rotating sphere $a \approx 0.0025$. After a distance $0(a)$ in which the initial velocity profiles are adjusted we have an example of Glauert's wall jet [4] in which

$$u = \Omega a \left[\frac{10 \alpha a}{z + \beta} \right]^{1/2} g(\eta) g'(\eta), \quad v = \Omega a [1 - g(\eta)] \tag{2.8}$$

where

$$g' = \frac{1}{3}\,(1 - g^3)\,, \qquad g\,(0) = 0\,, \qquad \eta = (r - a)\left[\frac{5\,\alpha\,a^3\,\Omega^2}{32\,\nu^2\,(z + \beta)^3}\right]^{1/4} \qquad (2.9)$$

and β is a constant length. In HOWARTH's theory there was another quantity which tends to a finite limit as $\theta \to \pi/2$ in addition to (2.4) namely

$$\int\limits_a^\infty u\,d\,r \to 0.497\,(\Omega\,\nu)^{1/2}\,. \qquad (2.10)$$

An estimation to the value of β may be made by comparing the right-hand side of (2.10) with the value of the left-hand side when (2.8) is substituted for u and z is set equal to 0. It is found that

$$\beta \approx 5\,a/8\,.$$

As z increases this wall jet will continue to grow and lose its momentum until either the curvature terms in the boundary layer equations, which have been neglected in (2.1) — (2.3), or the pressure gradient across the boundary layer which is caused by the centrifugal acceleration becomes significant. The second is of more importance and can no longer be neglected when

$$\frac{1}{\varrho}\,\frac{\partial\,p}{\partial\,z} = 0\left(\nu\,\frac{\partial^2\,u}{\partial\,r^2}\right) \quad \text{where} \quad \frac{1}{\varrho}\,\frac{\partial\,p}{\partial\,r} = \frac{v^2}{r} \qquad (2.11)$$

$$\text{i. e.} \quad z/a = 0\left[\frac{a^5\,\Omega^2\,a^4}{\nu^2}\right]^{1/7}\,. \qquad (2.12)$$

Thus strictly speaking it only occurs in the double limit $\nu \to 0$, $a \to 0$ keeping $a^{11}\nu^{-2}$ finite. The governing equations now become

$$\frac{\partial\,p}{\partial\,r} = \frac{\varrho\,v^2}{a}\,, \quad u\,\frac{\partial\,u}{\partial\,z} + w\,\frac{\partial\,u}{\partial\,r} = -\frac{1}{\varrho}\,\frac{\partial\,p}{\partial\,z} + \nu\,\frac{\partial^2\,u}{\partial\,r^2}$$

$$u\,\frac{\partial\,v}{\partial\,z} + w\,\frac{\partial\,v}{\partial\,r} = \nu\,\frac{\partial^2\,v}{\partial\,r^2} \quad \text{and} \quad \frac{\partial\,u}{\partial\,z} + \frac{\partial\,w}{\partial\,r} = 0 \qquad (2.13)$$

with the same boundary conditions as in (2.3) except that the initial profile is the wall jet taken at some value of z. As z increases the pressure gradient gradually takes control until eventually the contribution from the boundary layer on the hemisphere may be neglected. The boundary layer on the cylinder tends towards a "similar" solution of (2.13) obtained by writing

$$a\,u = +\frac{\partial\,\psi}{\partial\,r}\,, \quad \zeta = \frac{r - a}{a}\left\{\frac{a^3\,\Omega}{\nu\,(z + \gamma)}\right\}^{2/5} \quad \psi = a^2\,\Omega\left\{\frac{(z + \gamma)\,\nu}{a^3\,\Omega}\right\}^{3/5} F\,(\zeta)$$

$$v + \Omega\,a\,V\,(\zeta) \text{ and } p = \varrho\,a^2\,\Omega^2\left\{\frac{\nu\,(z + \gamma)}{a^3\,\Omega}\right\}^{2/5} P\,(\zeta) \qquad (2.14)$$

where $\gamma/a = 0\left\{\dfrac{a^5\,\Omega^2\,a^4}{\nu^2}\right\}^{1/7}$ is a constant.

It is then found that

$$V'' + \frac{3}{5}FV' = 0, \quad P' = V^2 \text{ and } F''' + \frac{3}{5}FF'' - \frac{1}{5}F'^2 = \frac{2}{5}(P - \zeta P') \quad (2.15)$$

where primes denote differentiation with respect to ζ, and

$$V(0) = 1, \quad V(\infty) = 0, \quad F(0) = F'(0) = F'(\infty) = P(\infty) = 0. \quad (2.16)$$

The same equation as (2.15) also occurs in the theory of the shear layer at the boundary of a finite circular cylinder of mercury rotating about its axis [16]. In the present problem the equations have been integrated numerically and the results are displayed in fig. (2). It was found that

$$V'(0) = -0.318, F''(0) = 0.584, P(0) = -1.152 \text{ and } F(\infty) = 1.330.$$

The boundary layer described by (2.13) is of a different type from those discussed previously for it is driven by a self-generated pressure gradient. The mechanism is that v drives p drives u drives v. Accordingly it is not surprising that it is thicker and the axial velocity is of higher order than hitherto. The flow field descri-bed by (2.13) is not the final one as $z \to \infty$ since ultimately as mentioned earlier the bound-ary layer gets so thick that curvature terms can no longer be neglected. Although the new term in the equation of con-tinuity may be absorbed by means of MANGLER's trans-formation [9] the effect of curvature on the momentum

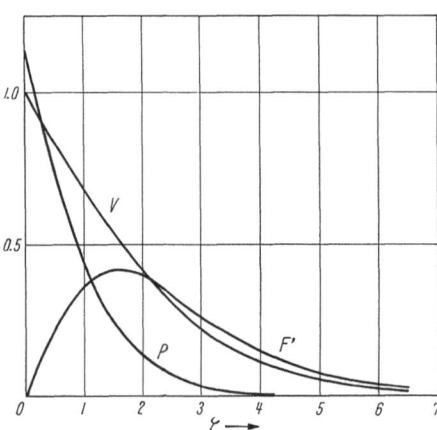

Fig. 2. Velocity and presure distributions according to (2.15)

equations is to add terms $(v/r)\partial u/\partial r$ and $(v/r)\partial v/\partial r$ respectively which eventually alter the character of the boundary layer. A similar situation occurs in the boundary layer caused by the same cylinder as here but moving in the direction of z decreasing without rotation [5],[17]. It starts at the front stagnation point, gradually thickens, almost parabolically, and weakens. Ultimately the fluid in it is practically at rest except for a relatively narrow region near the body. In the present problem the curvature terms begin to be significant when

$$z/a = 0\left(\frac{a^2\Omega}{\nu}\right)$$

and gradually dominate the flow. It is probable that ultimately the flow will be the same as that caused by a cylinder which is infinite in both directions of z and for which

$$u = w = 0, \quad v = \Omega\, a^2/r.$$

Since this solution satisfies the full equations of motion and all the boundary conditions except those in $z < 0$ there is no obvious perturbation procedure and no further progress has been possible in the asymptotic expansion.

The solution of (2.15) is also appropriate to the problem of the flow outside a semi-infinite hollow circular cylinder rotating about its axis, on setting $\gamma = 0$. Fluid is drawn towards the cylinder and swept along it from the open end in a boundary layer of thickness $0\ (\nu^{2/5})$. The solution must be modified to take account of the curvature of the cylinder when $z/a = 0\left(\dfrac{a^2 \Omega}{\nu}\right)$ and is invalid near $z = 0$ where for example it predicts a singularity in w at all values of $r > a$. The second phenomenon is similar to that occurring at the leading edge of a flat plate in a uniform stream and like it, is due to the choice of the coordinate system. It may formally be removed by using optimal coordinates in the manner of KAPLUN [7] but it must be borne in mind that the flow inside the hollow cylinder must be entirely different because the corresponding equations to (2.15) have no solution. It may be that the main body of the fluid inside the cylinder has a general circulation. If the cylinder is at rest and the fluid rotating about the axis of symmetry then a boundary layer of the type discussed above is possible inside the cylinder but not outside it.

If the cylinder is finite in length, two boundary layers are set up, one at each end, and advance towards each other. The description of the flow at the collision is similar to that in the problem of the rotating sphere. The boundary layers on rotating circular cylinders $1''$ and $\frac{1}{2}''$ in diameter and 2 ft. long enclosed in a fixed cubical box of side 2 ft. have been examined experimentally by RICHARDSON [13]. The agreement with the present theory is not good which may be due to the effect either of turbulence or of the cubical container.

3. Sphere at rest in a rotating fluid

The appropriate differential equations are now, using the notation of section 1

$$\frac{u}{a}\frac{\partial u}{\partial \theta} + w\frac{\partial u}{\partial R} - \frac{v^2}{a}\cot \theta = -\Omega^2 a \sin \theta \cos \theta + \nu\frac{\partial^2 u}{\partial R^2} \qquad (3.1)$$

$$\frac{u}{a}\frac{\partial v}{\partial \theta} + w\frac{\partial v}{\partial R} + \frac{u v}{a}\cot \theta = \nu\frac{\partial^2 v}{\partial R^2} \qquad (3.2)$$

and $\quad \dfrac{1}{a}\dfrac{\partial u}{\partial \theta} + \dfrac{\partial w}{\partial R} + \dfrac{u}{a}\cot \theta = 0 \qquad (3.3)$

5 Gortler, Grenzschichtforschung

with boundary conditions

$$u = v = w = 0 \text{ when } R = a \text{ and } u = v - \Omega a \sin \theta = 0$$

outside the boundary layer. The boundary layer now begins at the equator and spreads round the sphere towards the poles. Bödewadt[1] [2] has given a solution of the boundary layer due to a fixed disc in a rotating fluid which is the counterpart of the von Kármán solution initiating the boundary layers on the rotating sphere. If Bödewadt's solution is appropriate here however it can only describe the end of the boundary layer for according to it there is a net *outflow* and the fluid in the boundary layer is moving towards the axis of symmetry.

In order to describe the motion near the equator write

$$\psi = a^2 (\Omega v)^{1/2} \sum_{n=0}^{\infty} \left(\theta - \frac{\pi}{2} \right)^{2n+1} \Psi_n(Z), \quad v = \Omega a \sum_{n=0}^{\infty} \left(\theta - \frac{\pi}{2} \right)^{2n} V_n(Z)$$

where

$$Z = \left(\frac{\Omega}{v} \right)^{1/2} (R - a).$$

Then the leading terms of the expansions satisfy

$$\Psi_0''' + \Psi_0 \Psi_0'' - \Psi_0'^2 - V_0^2 = -1 \tag{3.4}$$

$$V_0'' + \Psi_0 V_0' = 0 \tag{3.5}$$

with boundary conditions

$$\Psi_0(0) = \Psi_0'^2(0) = \Psi_0'(\infty) = V_c(0) = 0, \quad V_0(\infty) = 1. \tag{3.6}$$

The solution is displayed in fig. (3). Its principal properties are that

$$\Psi_0''(0) = 0.953 \quad V_0'(0) = 0.463 \quad \Psi_0(\infty) = 1.33 \tag{3.7}$$

and, as required for consistency, there is an inflow of fluid to the sphere in the immediate neighbourhood of the equator to start the boundary layer. It is interesting to note that if the sphere is rotating and the fluid is otherwise at rest a boundary layer starting at the equator is impossible. To prove this note that if it existed the velocity components would satisfy (3.4), (3.5) with *both* right-hand sides equal to zero and boundary conditions

$$\Psi_0(0) = \Psi_0'(0) = \Psi_0'(\infty) = V_0(\infty) = 0, \quad V_0(0) = 1. \tag{3.8}$$

---

[1] This solution was criticized by me [*18*] because I could not recover it using step-by-step integration. Subsequently a pupil of mine, A. R. Browning, using relaxation confirmed Bödewadt's solution as did H. E. Fettis [*3*] in an independent investigation. It appears that the step-by-step method is not suitable for this problem because random errors increase exponentially and so my objection is invalid.

Integrating (3.4) twice and using the appropriate boundary conditions we have

$$\frac{1}{2} \Psi_0^2(\infty) + \int_0^\infty Z (V_0^2 + \Psi_0'^2) \, dZ = 0 \tag{3.9}$$

which is impossible. The physical interpretation is that the boundary layer if it exists must start elsewhere. In fact if the body is rotating and the fluid is otherwise at rest the boundary layer spreads outwards from the axis whereas if the fluid is rotating and the body is at rest the boundary layers converge on the poles from the equator.

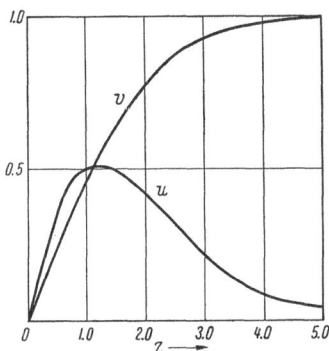

Fig. 3. Velocity functions near the equator of a fixed sphere in a rotating fluid

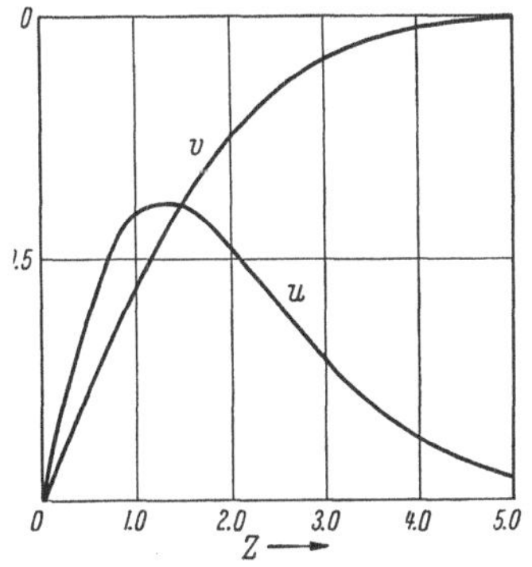

Fig. 4. Velocity functions near the edge of a fixed circular disc in a rotating fluid

The solution when the sphere is at rest may be continued in a similar way to HOWARTH's treatment of the rotating sphere. On making the same assumptions as he did it may be shown that

$$\left(\frac{\partial u}{\partial R}\right)_{R=a} = -\frac{18}{35} \frac{\delta a \Omega^2}{\nu} \sin\theta \cos\theta$$

$$\text{and} \qquad \frac{3}{4} \sin\theta \cos\theta \frac{d\delta^4}{d\theta} - \delta^4 (1 + 0.09 \cos^2\theta) = -40.7 \left(\frac{\nu}{\Omega}\right)^2 \tag{3.10}$$

where δ is the thickness of the boundary layer. Hence within the bounds of the approximation it is legitimate to take δ constant and equal to $2.52 \, (\nu/\Omega)^{1/2}$, in which case fluid is drawn into the boundary layer near the equator and expelled near the poles there being none left inside at $\theta = 0, \pi$. Although such a description seems reasonable at first sight, and agrees with the natural view that the flow near the poles should be of the kind described by BÖDEWADT [2], the method is suspect because according to it

$$\int_a^\infty (\psi u^2)_{\theta=\pi} \, dr = \int_{\pi/2}^\pi \cos\theta \, d\theta \int_a^\infty \psi (v^2 - \Omega^2 a^2 \sin^2\theta) \, dr \tag{3.11}$$

is positive so that the boundary layer cannot be empty of fluid at the poles. Further if Bödewadt's solution is correct at the poles for all bodies of revolution it means that the effect of shape and size on the boundary layer will have died out after a finite distance. Nevertheless I believe that that solution is correct at the poles and that the momentum integral method gives a reasonably accurate picture of the flow for the following reason. If the longitudinal component of the velocity is not small near the poles the viscous term in the azimuthal component of the momentum equation becomes of less importance than the inertia terms. In this case the azimuthal velocity *inside* the boundary layer $\propto (\sin \theta)^{-1}$, while remaining small outside. It acts therefore mainly like a strong *adverse* pressure gradient in (3.1) retarding the longitudinal velocity. Although breakaway is possible its mechanism is not clear because the equivalent pressure gradient is *favourable* at the wall, and the most likely outcome is a balance between the longitudinal velocity and the azimuthal velocity in which both come smoothly down to zero as required in Bödewadt's solution.

The earliest example of a solid at rest in a rotating fluid to be considered was in fact a circular disc of radius a. An approximate solution using the momentum integrals was given up by Schultz-Grunow [*14*] and he found that the boundary layer begins at the edge of the disc and its thickness is initially $\propto (a - r)^{1/4}$ where r denotes the distance from the axis of the disc and of rotation. The boundary layer spreads over the whole disc but the convergence of the series he used were not satisfactory near $r = 0$.

Defining (w, v, u) as in section 2 the appropriate boundary layer equations for this problem are

$$w \frac{\partial w}{\partial r} + u \frac{\partial w}{\partial z} - \frac{v^2}{r} = - \Omega^2 r + \nu \frac{\partial^2 w}{\partial z^2}, \quad w \frac{\partial v}{\partial r} + u \frac{\partial w}{\partial r} + \frac{w v}{r} = \nu \frac{\partial^2 v}{\partial z^2}$$

and

$$\frac{\partial}{\partial r} (r w) + \frac{\partial}{\partial z} (r u) = 0, \tag{3.12}$$

with boundary conditions

$$u = v = w = 0 \text{ on } z = 0, \ r < a$$
$$u = 0, \ v = \Omega a \text{ on } r = a, \ z > 0 \tag{3.13}$$

and $u \to 0$, $v \to \Omega r$ outside the boundary layer.

Writing $\partial \psi / \partial z = - r w$ the flow near the edge of the disc may be described by setting

$$\psi = + a^2 (\Omega \nu)^{1/2} \sum_{n=0}^{\infty} \left\{ \frac{a-r}{a} \right\}^{n + 3/4} \psi_n(X), \qquad v = \Omega a \sum_{n=0}^{\infty} \left\{ \frac{a-r}{a} \right\}^n V_n(X)$$

where

$$X = z \left(\frac{\Omega}{\nu} \right)^{1/2} \left(\frac{a-r}{a} \right)^{1/4}. \tag{3.14}$$

The leading terms of this expansion satisfy

$$\psi_0''' + \frac{3}{4}\,\psi_0\,\psi_0'' - \frac{1}{2}\,\psi_0'^2 = V_0^2 - 1\,.$$

$$V_0'' + \frac{3}{4}\,\psi_0\,V_0' = 0 \qquad (3.15)$$

with boundary conditions

$$\psi_0\,(0) = \psi_0'\,(0) = \psi_0'\,(\infty) = V_0\,(0) = 0, \quad V_0\,(\infty) = 1.$$

The solution is displayed in fig. (4). Its principal properties are that

$$\psi_0''\,(0) = 1\cdot071, \quad V_0'\,(0) = 0\cdot442, \quad \psi_0\,(\infty) = 1\cdot720 \qquad (3.16)$$

and, as required for consistency, there is an inflow of fluid to the disc near $r = a$ to start the boundary layer. It is noted that, if the disc is rotating and the fluid otherwise at rest, a similar argument to that of (3.9) may be used to show that no solution of this kind is possible. The reason is of course that the boundary layer starts elsewhere. This solution of the boundary layer equations confirms SCHULTZ-GRUNOW's view of its behaviour near $r = a$ and it is found that

$$\int_0^\infty \psi\,(v^2 - \Omega^2\,r^2)\,dz = -1\cdot05\,a^3\,\Omega^2\,\nu\,(a - r)\left[1 + 0\left(\frac{a - r}{a}\right)\right] \qquad (3.17)$$

whereas in BÖDEWADT's solution

$$\int_0^\infty \psi\,(v^2 - \Omega^2\,r^2)\,dz = +0\cdot89\,\Omega^2\,\nu\,r^4\,. \qquad (3.18)$$

Hence the effect of the edge of the disc cannot be neglected over a considerable portion of the disc. This confirms a suggestion due to MOORE [10] who also suggested that the boundary layer may erupt at the centre of the disc. I am reluctant to accept this latter view for the same reason that I gave in discussing the flow near the poles of the sphere. It is clear however from (3.17) and (3.18) that BÖDEWADT's solution is valid at most only in the immediate neighbourhood of the centre (say $0 < r/a < 0\cdot2$).

Accordingly the theory of BATCHELOR [1] and myself [18] concerning the flow between coaxial discs rotating in the same sense which make use of solutions of the Navier-Stokes equations like BÖDEWADT's are also valid only in the immediate neighbourhood of the axis at most. It is not surprising therefore that Prof. K. S. PICHA's [12] observations show considerable divergences from it. They do however agree with my theory (but not BATCHELOR's) when the discs are rotating in opposite senses, in which the boundary layers on the discs are practically independent.

References

[1] BATCHELOR, G. K.: Quart. J. of Mech. and App. Maths. **4**, 29 (1951).

[2] BÖDEWADT, U.: ZAMM, **20**, 241 (1940).

[3] FETTIS, H. E.: private communication.

[4] GLAUERT, M. B.: J. Fluid Mech. **1**, 625 (1956).

[5] GLAUERT, M. B., and M. J. LIGHTHILL: Proc. Roy. Soc. A. **230**, 188 (1955).

[6] HOWARTH, L.: Phil. Mag. Ser. 7, **42**, 1308 (1951).

[7] KAPLUN, S.: ZAMP **5**, 111 (1954).

[8] v. KÁRMÁN, T.: ZAMM, **1**, 233 (1921).

[9] MANGLER, W.: Rep. Aero. Res. Counc. 740 (1946) No. 9.

[10] MOORE, F. K.: Advances in Mechanics Vol. IV (1956) Academic Press, 166.

[11] NIGAM, S. D.: ZAMP **5**, 151 (1954).

[12] PICHA, K. S.: Thesis, University of Minnesota (1957).

[13] RICHARDSON, E. A.: Phil. Mag. Ser. 7, **11**, 1215 (1941).

[14] SCHULTZ-GRUNOW, F.: ZAMM **15**, 191 (1935).

[15] SQUIRE, H. B.: 50 Jahre Grenzschichtforschung eds. H. GÖRTLER and W. TOLLMIEN. Vieweg 1955, S. 47.

[16] STEWARTSON, K.: Quart. J. of Mech. and App. Maths. **10**, 134 (1954).

[17] STEWARTSON, K.: Quart. of App. Maths. **13**, 113 (1955).

[18] STEWARTSON, K.: Proc. Cam. Phil. Soc. **49**, 333 (1953).

Aus der Diskussion

H. GÖRTLER (Freiburg i. Br.): Meine Bemerkungen beziehen sich auf Grenzschichten an rotierenden Körpern. 1) Es gibt eine Klasse exakter Lösungen der vollen Navier-Stokesschen Gleichungen von dem folgenden Charakter (vgl. [1] oder [2]): In Achsenrichtung längs eines unendlich langen Kreiszylinders strömt die Flüssigkeit von beiden Seiten aus dem Unendlichfernen, und diese beiden Strömungen begegnen sich in einer Symmetrie-Ebene (normal zur Achse), längs der sie, radial nach außen abgelenkt, abfließen. Über das Verhalten weitab vom Zylinder bin ich nicht orientiert. Immerhin handelt es sich hier um eine exakte Klasse von Strömungen, bei denen sich zwei Reibungsschichten begegnen. Ich möchte vermuten, daß man sie unschwer auf den Fall eines rotierenden Zylinders erweitern kann. Das gäbe Gelegenheit zur Kontrolle der Größenordnungsaussagen an exakten Lösungen der Navier-Stokesschen Gleichungen. — 2) Ich möchte ferner hinweisen auf Untersuchungen von T. GEIS [3], [4], in welchen exakte Lösungen der Grenzschichtgleichungen für den Fall axialsymmetrischer, um ihre Achse rotierender Körper in sonst ruhenden Flüssigkeiten gefunden wurden. GEIS hat gezeigt, daß es für gewisse angebbare Körperformen „ähnliche" Lösungen gibt, die sich leicht berechnen lassen, deren Existenz aber auch in aller mathematischen Strenge nachgewiesen wurde. Diese Ergebnisse dürften von unmittelbarem Interesse für den behandelten Problemkreis des Vortrags von Herrn STEWARTSON sein.

Literatur:

[1] ROSENBLATT, M. A.: Solutions exactes des équations du mouvement des liquides visqueux. Mémorial des Sciences Mathématiques, Fasc. LXXII, Paris: Gauthier-Villars, 1935.

[2] BERKER, A. RATIB: Sur quelques cas d'integration des équations du mouvement d'un fluide visqueux incompressible. Paris et Lille: A. Tiffin-Lefort, 1936.

[*3*] GEIS, T.: Ähnliche Grenzschichten an Rotationskörpern. „50 Jahre Grenz-schichtforschung", Braunschweig: Vieweg, 1955, S. 294—303.
[*4*] GEIS, T.: Elementarer Existenzbeweis für die Grenzschichtströmung an einer Klasse rotierender Körper. ZAMM **36**, 222—229 (1956).

K. GERSTEN (Braunschweig): Im Zusammenhang mit den Ausführungen von Herrn Dr. STEWARTSON sei auf Ergebnisse von Messungen hingewiesen, die im Institut für Strömungsmechanik der Technischen Hochschule Braunschweig aus-geführt wurden. Ein drehsymmetrischer Halbkörper (vgl. Abb. 1), welcher sich in einem konzentrisch dazu angeordneten Mantel (Rohr) befindet, besitzt einen rotierenden Nabenkopf. Lediglich durch dessen Rotation wird Luft von vorn angesaugt und in dem Raum zwischen dem festen Halbkörperteil und dem Mantel nach hinten gefördert. Wo rotierender und fester Teil zusammenstoßen, liegt ein Gebiet mit „Rückströmung".

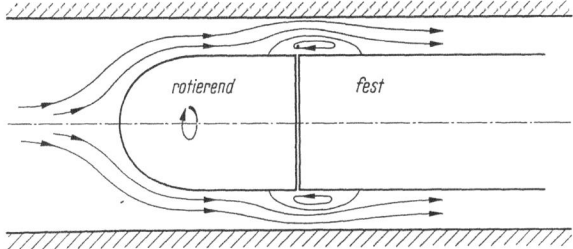

Abb. 1. Förderung von Luft an einem Halbkörper in einem dazu konzentrischen Rohr durch Rotation des Nabenkopfes

I. TANI (Tokyo): I would like to inform you that hot-wire measurements have been made by Y. KOBASHI [*1*] on the velocity field around a rotating sphere. His results were found to agree qualitatively with NIGAM's analysis, but the agreement was rather poor near the pole and the equator. It was moreover ascertained that the boundary layer fluid was expelled in the form of a thin radial jet along the equatorial plane. These results seem to confirm the view of Dr. STEWARTSON on the first problem.

Reference:

[*1*] KOBASHI, Y.: Journal of Science of the Hiroshima University, Japan, Series A, **20**, No. 3 (1957).

On laminar wall jets

By

M. B. Glauert

The University of Manchester

1. Introduction

The name "wall jet" is applied to a jet of fluid which spreads out over a surface, the fluid outside the jet being at rest. It may, for example, be produced by allowing a free jet to impinge normally on a plane surface. A discussion of the phenomenon on the basis of boundary layer theory, for both laminar and turbulent incompressible flow, has recently been given in a paper [1] by the author (1956). For laminar flow it was shown that a similarity solution exists in which the velocity profile does not vary along the jet length, and is the same for both radial and two-dimensional wall jets. In the present paper this similarity solution is obtained in a generalised form, with an additional length parameter governed by the distance which the jet profile takes to assume its final similar shape. Mr. N. RILEY has shown that the similarity solution is also applicable when the fluid is compressible or heated, if it is assumed that the coefficient of viscosity is proportional to the temperature, and his results are briefly discussed, with his permission. Finally, consideration is given to perturbations of the basic similarity solution due to a small external flow, as might arise as a consequence of entrainment of fluid into the wall jet.

2. Basic equations

For compressible laminar flow, the equations of momentum and continuity for a radial laminar wall jet with no pressure gradient are

$$\varrho\, u\, \frac{\partial u}{\partial x} + \varrho\, v\, \frac{\partial u}{\partial y} = \frac{\partial}{\partial y}\left(\mu\, \frac{\partial u}{\partial y}\right), \tag{1}$$

$$\frac{\partial}{\partial x}\left(\varrho\, x\, u\right) + \frac{\partial}{\partial y}\left(\varrho\, x\, v\right) = 0, \tag{2}$$

where x and y denote distances along and normal to the wall (x being measured from the jet axis), u and v are the corresponding velocity components, ϱ is the density and μ is the coefficient of viscosity. Eq. (2) is satisfied by the introduction of the Stokes stream function ψ such that

$$\varrho\, x\, u = \partial\psi/\partial y, \ \varrho\, x\, v = -\,\partial\psi/\partial x. \tag{3}$$

The boundary conditions are

$$u = v = 0 \ \text{at}\ y = 0, \ u \to 0 \ \text{as}\ y \to \infty\,. \tag{4}$$

In [1] a valuable relation was obtained by integrating the momentum eq. (1) across the boundary layer in a suitable manner, and making use of the continuity eq. (2) and the boundary conditions (4). If the same procedure is followed for compressible flow, Mr. RILEY has shown that

$$\frac{dF}{dx} + x^2 \int_0^\infty \mu\,\varrho\, u\, \frac{\partial u}{\partial y}\, dy = 0 \tag{5}$$

where

$$F = \int_0^\infty \varrho\, x\, u \left(\int_y^\infty \varrho\, x\, u^2\, dy \right) dy\,. \tag{6}$$

If $\mu\,\varrho = $ constant, which is satisfied for an incompressible fluid, and when the viscosity is proportional to the temperature for a compressible fluid, the integral in (5) vanishes by virtue of the boundary conditions (4), and hence

$$F = \text{constant.} \tag{7}$$

The quantity F is interpreted in [1] as the flux of exterior momentum flux, but no physical explanation of (7) has been found. Any such explanation must of necessity be somewhat complex, as it is seen to depend on the nature of the law governing the viscosity. It may be noted that F as defined here differs by a factor ϱ^2 from its definition in [1].

3. Similarity solution

We now examine the possibility that the foregoing equations shall have a solution in which the form of the velocity profile does not vary along the jet length, considering first the case of incompressible flow. Such a similarity solution was obtained in [1], the velocity and the jet width varying as powers of x. More generally, we may employ a technique used by Mr. E. J. WATSON in a study of the spread of a liquid jet over a horizontal surface, and seek a solution in which

$$u = U(x)\, f'(\eta) \ \text{where}\ \eta = y/\delta(x) \tag{8}$$

and hence, from eq. (3),

$$\psi = \varrho\, x\, U\, \delta\, f(\eta). \tag{9}$$

Eq. (9) shows that

$$x^2\, U^3\, \delta^2 = \text{constant,} \tag{10}$$

and the condition that eq. (1) should be compatible with (9) then reduces to

$$\delta^2 \, U' = \text{constant.} \tag{11}$$

From (10) and (11) it follows that

$$U = C \, (x^3 + l^3)^{-1/2}, \; \delta = D \, x^{-1} \, (x^3 + l^3)^{3/4}, \tag{12}$$

where C, D and l are arbitrary constants. If l is negligibly small compared with x, these relationships become $U \propto x^{-3/2}$, $\delta \propto x^{5/4}$, exactly as obtained in [1]. However, the fact that the velocity in (12) remains finite as $x \to 0$ means that this solution may have significance even near the axis.

The equation for $f \, (\eta)$, given by eq. (1) when (11) is satisfied, is

$$f''' + f \, f'' + 2 \, f'^2 = 0 \tag{13}$$

if the constant in (11) is chosen to have the value $- \, 2 \, \mu/\varrho$. There are arbitrary scaling factors in U and δ, so this involves no loss of generality. The boundary conditions (4) require

$$f \, (0) = f' \, (0) = 0, \; f' \, (\infty) = 0, \tag{14}$$

and the second of the scaling factors may be fixed by requiring that $f \, (\infty) = 1$. The eq. (13) was integrated explicitly in [1], the solution being given as

$$f = g^2, \; \eta = \log \frac{\sqrt{(1 + g + g^2)}}{1 - g} + \sqrt{3} \tan^{-1} \frac{\sqrt{3} \, g}{2 + g} . \tag{15}$$

Inserting the values of the scaling factors as in [1], we arrive at the final forms

$$\left. \begin{aligned} u &= \left\{ \frac{15 \, F}{2 \, \mu \, \varrho \, (x^3 + l^3)} \right\}^{1/2} f' \, (\eta), \\ \eta &= \left\{ \frac{135 \, \varrho \, F \, x^4}{32 \, \mu^3 \, (x^3 + l^3)^3} \right\}^{1/4} y \, . \end{aligned} \right\} \tag{16}$$

The solution (16) contains 2 parameters, F and l, while the solution in [1] contains only F. However, if F has been estimated (and its constancy does not depend upon the velocity profile having attained its final similar form), l can be roughly determined by equating the maximum velocity at $x = 0$ as given by (16) to the maximum initial velocity, if this is known. It is readily shown that l is of the same order as the distance in which a boundary layer on the plate, starting from the axis, would grow to the width of the whole wall jet. Thus it is likely to be of importance in any comparison with experiment. We may also note that a length l can be introduced in a similar manner into the solutions for turbulent wall jets developed in [1].

It was shown in [1] that the velocity profile is the same for plane and radial wall jets. If the calculation for a plane jet is carried out as in this paper, the only change from the solution obtained in [1] is that x is replaced everywhere by $(x + l)$. Thus l corresponds merely to a shift of virtual origin, which is obviously allowable physically. Since x appears explicitly in the continuity equation for a radial wall jet, an origin shift is impossible in this case, but (16) shows that an analogous arbitrary length can still be incorporated in the solution.

4. Compressible flow

In his study of compressible laminar wall jets, Mr. RILEY applies the VON MISES transformation in eq. (1), so that x and ψ become the independent variables. The equation becomes

$$\frac{\partial u}{\partial x} = x^2 \frac{\partial}{\partial \psi} \left(\mu \varrho \, u \, \frac{\partial u}{\partial \psi} \right). \tag{17}$$

If $\mu \varrho = $ constant, the solution of (17) in the form

$$u = u \, (x, \psi) \tag{18}$$

is independent of compressibility, and hence the solution obtained above for incompressible flow is still applicable. The similarity variable η is no longer related to the distance from the wall y by eq. (16), instead y must be found from the relation

$$y = \int \frac{\partial \psi}{\varrho \, x \, u} . \tag{19}$$

Eq. (19) can be integrated only when the temperature distribution through the boundary layer has been determined. However, the shearing-stress τ, given by

$$\tau \, (x, \psi) = \mu \frac{\partial u}{\partial y} = \mu \varrho \, x \, u \frac{\partial u}{\partial \psi} , \tag{20}$$

is unaffected by temperature if $\mu \varrho = $ constant, and hence the skin-friction is the same as for incompressible flow. Also, as shown above, the flux of exterior momentum flux F remains constant on the same hypothesis. Mr. RILEY has obtained solutions of the energy equation corresponding to a variety of wall temperature conditions, initial jet temperatures, and values of the Prandtl number σ. A particularly simple case is for $\sigma = 1$, when there is an integrated form of the energy equation

$$T + u^2/2 \, c_p = T_\infty + B \, u , \tag{21}$$

where T and T_∞ are the temperatures in and outside the jet, and B is a constant. This solution applies to a wall jet blowing over a plate maintained at the temperature of the outside air, the value of B depending on the temperature in the initial jet.

5. Departures from similarity

As a consequence of entrainment of fluid into the wall jet, there may be a small radial velocity in the fluid outside the jet. Its effect is only likely to be appreciable when x is large, so we shall consider incompressible flow and take the wall jet solution (16) with $l = 0$. The simplest case is when the fluid is bounded by an upper plate $y = d$, where d is small compared with x but large compared with the jet width. Continuity then requires that the inflow velocity outside the jet is $(\psi)_{y\,=\,\infty}/\varrho\,x\,d$, and hence is proportional to $x^{-1/4}$. The production of the wall jet usually involves the introduction of fluid near $x = 0$. This causes an additional outflow velocity proportional to x^{-1}.

For generality, suppose that

$$u \to \varepsilon\, x^c \text{ as } y \to \infty, \tag{22}$$

where c and ε are constants, ε being small. If we suppose that u has a term $\varepsilon\, x^c\, h'\,(\eta)$ in addition to that given by (16), then on equating to zero the terms linear in ε in eq. (1) we obtain

$$h''' + f\, h'' + (5 - \beta)\, f'\, h' + \beta\, f''\, h = 0, \tag{23}$$

where $\beta = \dfrac{1}{3}\,(9 + 4c)$. The pressure forces in the external flow make no contribution of order ε. The boundary conditions are

$$h\,(0) = h'\,(0) = 0,\ h'(\infty) = 1. \tag{24}$$

Eq. (23) may be solved explicitly for a few special values of β. When $\beta = 2$, the equation may be integrated to give $h'' + f\, h' + 2\, f'h = 1$, in view of (24), and hence $h''(0) = 1$. When $\beta = 0$, the equation is of second order only and can be integrated completely since $h = f'$ is a known integral (this is true for all β); by use of properties of $f\,(\eta)$ established in [1] it is readily shown that $h''(0) = -3$. When $\beta = 1, h = f + \eta f'$ is a solution of (23), and when $\beta = -3$, $h = f - 3\,\eta\, f'$ is a solution. For these two solutions $h'(\infty) = 0$ but the other two conditions of eq. (24) are satisfied, and hence we conclude that (24) requires $h''(0) = \infty$ in each case. These solutions correspond to small changes in the basic parameters F and l respectively.

Cases of practical interest are $\beta = 5/3$ and $\beta = 8/3$, as shown above. Eq. (23) was integrated numerically in each of these cases by Milne's numerical method, and it was found that when $\beta = 5/3$, $h''(0) = 1.77$ and when $\beta = 8/3, h''(0) = 0.44$. The velocity profiles are shown in fig. 1, together with the profiles for $\beta = 1$ and $\beta = -3$. For purposes of comparison, all are scaled to have gradient unity at $\eta = 0$.

It might perhaps be hoped that some approximation of the Kármán-

POHLHAUSEN type would enable the solution to be calculated with reasonable accuracy for general values of β. Eq. (23) may be integrated to give

$$h'' + f h' + \beta f' h + (4 - 2\beta) \int_0^\eta f' h' d\eta = h''(0),\qquad (25)$$

on making use of (24), and hence

$$h''(0) = (4 - 2\beta) \int_0^\infty f' h' d\eta + 1.\qquad (26)$$

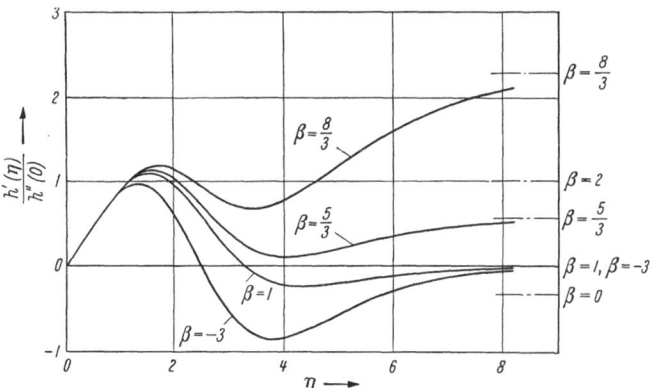

Fig. 1. Perturbation of a laminar wall jet by an external flow. Variation of velocity increment (h') with distance from the wall (η), for various values of β. Values at $\eta = \infty$ shown on the right

If one substitutes $h' = f + kf'$, which satisfies all the conditions (24) for any value of the constant k, one obtains from (26)

$$h''(0) = \frac{3 - \beta}{2\beta - 3}.\qquad (27)$$

The value for $\beta = 2$ is correct, as it clearly must be from (26), but none of the other known values is given at all accurately. No modification in the assumed form of h' or in the detailed procedure has been found to give any significant improvement. No doubt the unusual type of velocity profile, shown in fig. 1, accounts for the failure of these approximate methods.

We may now assess the effect on the wall jet of the presence of an upper plate $y = d$. In the unperturbed wall jet the skin-friction is

$$\left(\frac{125\,F^3}{216\,\mu\varrho\,x^{11}}\right)^{1/4}.\qquad (28)$$

Due to a source of strength $2\pi m$ at the axis, $\varepsilon = m/p\,d$, and the increment of skinfriction is

$$\frac{m}{d}\left(\frac{135\,F\,\mu}{32\,p^3\,x^9}\right)^{1/4} h''(0),\qquad (29)$$

where $h''(0) \doteq 1.77$ as calculated above. For entrainment into the wall jet itself, $\varepsilon = - d^{-1} (40 \, F \mu/3 \, \varrho^3)^{1/4}$, and the increment of skin-friction is

$$- \frac{1}{d} \left(\frac{15 \, F \mu}{2 \, \varrho \, x^3} \right)^{1/2} h''(0), \tag{30}$$

where $\dot{h}''(0) = 0.44$. An estimate of the position at which the wall jet separates from the surface may be obtained by equating to zero the sum of (28), (29) and (30). If (29) is ignored, either because no source is present or on the grounds that this term falls off more rapidly with x than does (30), then separation is predicted to occur when

$$x^5 = 0.27 \, (F \, \varrho \, d^4/\mu^3). \tag{31}$$

References

[1] GLAUERT, M. B. (1956): J. Fluid Mech. I, 625.

Aus der Diskussion

K. STEWARTSON (Bristol): Mr. GLAUERT's integral of the boundary layer equations may also be used to determine the decay of a steady modification to the BLASIUS profile at some station, a long way downstream. For example, suppose that we have a flat plate on $x \geqslant 0$, $y = 0$ and a uniform stream with velocity U is flowing past it. In the usual solution it is assumed that $u = U$ at $x = 0$, $y > 0$. Suppose, however, that $u = Uf'(y)$ at $x = 0$ where f is arbitrary and let us inquire how important is f when x is large. From the boundary layer equations it may be shown that

$$\psi = \sqrt{2 \nu x U} \left(F(\eta) + \frac{A}{x} \, [\eta \, F'(\eta) - F(\eta)] + \ldots \right) \tag{1}$$

when x is large where ψ is the stream function,

$$\eta = y \sqrt{\frac{U}{2 \nu x}}, \quad F''' + FF'' = 0, \quad F(0) = F'(0) = 0, \quad F'(\infty) = 1$$

and A is at present unknown. GLAUERT's integral may be used to find A for it asserts that

$$\frac{\partial}{\partial x} \int_0^\infty \psi \, (Uu - u^2) \, dy = \frac{U^2}{2} \nu$$

whence

$$\int_0^\infty \psi \, (Uu - u^2) \, dy = \frac{U^2}{2} \nu x + U^3 \int_0^\infty f \, (f' - f'^2) \, dy. \tag{2}$$

Substituting (1) into (2) it follows that

$$\nu A = - \int_0^\infty f \, (f' - f'^2) \, dy.$$

Diskussionsveranstaltung zur II. Sitzung

Contribution to the Subject

Special Solutions of the Boundary Layer Equations

By **B. Thwaites,** Winchester College, Winchester

I expect to be in a minority of one, but I confess that whenever I see, for the first time, mathematical solutions unsupported by experimental evidence, I wonder about their physical meaning and uniqueness. I would like to give three examples to illustrate why I have these doubts.

1. In the diverging flow between two plane porous walls, similar solutions of the boundary layer equations may be found if k, the strength of the suction, is greater than some minimum value k_0. But when $k = k_0$, the flow is nowhere near separation, so what will happen in experiment if $k < k_0$? Whatever happens, there is the strong suggestion that the mathematical solution is quite unrealistic. (In the case of $U \sim x^{-1/3}$ the mathematical minimum value of k does correspond to separation.)

2. In the uniform flow past a porous well-rounded body, it is theoretically to be expected that sufficient boundary-layer suction will prevent separation and that the flow will conform closely to the potential flow which may be calculated. This expectation is confounded, since in experiment such a flow is very unsteady with a relatively large wake.

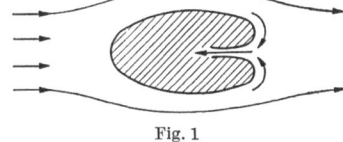
Fig. 1

3. Again it is possible, as has been shown by M. B. GLAUERT, to design a shape on which the potential-flow velocity is increasing along the surface, so that separation will not take place, as shown in the fig. 1. In fact, this flow is found to be highly unstable.

(It is interesting to note that in cases 2 and 3 above, a constraint in the form of a thin plate extending downstream of the body along the plane of symmetry is enough to stabilize the flows and create the theoretically-predicted conditions.)

I do not want to apply my remarks to the two last papers except to say that a mathematical solution does *not* necessarily correspond to a physical flow; and I do hope that it will be possible for both the rotating-flow and the wall-jet solutions to be fully justified by experiment.

M. B. GLAUERT (Manchester): I would like to tell Dr. THWAITES that the experiments on the wall jet have, in fact, been published in the latest issue of the Journal of Fluid Mechanics.

Beitrag zum Thema

Allgemeine Verfahren zur Lösung der Grenzschichtgleichungen

Von **H. Görtler,** Institut für Angewandte Mathematik der Universität Freiburg i. Br.

Neben der Frage nach speziellen exakten Lösungen der Grenzschichtgleichungen soll auch die Frage der Entwicklung von Lösungsverfahren bei allgemeinen Strömungsbedingungen Gegenstand der Diskussionsveranstaltung zur II. Sitzung sein. Da sich noch niemand zu dieser Frage zu Worte gemeldet hat und da der im Programm vorgesehene Vortrag von Dr. WALZ, der zu Diskussionen über dieses Thema angeregt hätte, wegen Erkrankung des Vortragenden auf die VIII. Sitzung verlegt werden mußte, möchte ich mit einigen Ausführungen über unsere Freiburger Arbeiten die Diskussion in Gang bringen.

In einer Anfang dieses Jahres erschienenen Arbeit [1] habe ich über ein neues Berechnungsverfahren für ebene, stationäre und inkompressible laminare Grenzschichten bei allgemeiner äußerer Geschwindigkeitsverteilung berichtet. Dieses Verfahren beruht auf einer formal exakten Lösung des allgemeinen Randwertproblems in Gestalt einer unendlichen Reihe unter Verwendung von, wie mir scheint, besonders geeigneten Variablen. Als unabhängige Veränderliche werden

$$\xi = \frac{1}{\nu} \int_0^x U(x)\, dx \quad \text{und} \quad \eta = \frac{U(x)\, y}{\nu} \Big/ \sqrt{2\,\xi}$$

benutzt, und für die Stromfunktion der Grenzschicht als der abhängigen Funktion wird

$$\psi(x, y) = \sqrt{2\,\xi}\; F(\xi, \eta)$$

geschrieben. Für $F(\xi, \eta)$ wird eine Potenzreihe

$$F(\xi, \eta) = \sum_{k=0}^{\infty} F_k(\eta)\, \xi^k$$

angesetzt.

Ausgangspunkt der Untersuchung war der Versuch, eine besser konvergierende Reihe für die Lösung zu finden als die bekannte, auf BLASIUS zurückgehende Reihe nach Potenzen der Wandbogenlänge x. Die erfahrungsgemäß oft sehr schlechte Konvergenz der BLASIUSschen Reihe schien uns wesentlich durch den Umstand bedingt zu sein, daß die äußere Randbedingung $u(x, y) \to U(x)$ nach Potenzreihenentwicklung von $U(x)$ durch gliedweises Abgleichen erfüllt wird, womit dann natürlich die Konvergenz der Lösungsreihe weitgehend von der Güte der Konvergenz oder, allgemeiner, von der Gestalt von $U(x)$ abhängt. Die neue Reihe erfüllt, wie von uns erstrebt, die äußere Randbedingung bereits exakt für alle x mit dem Gliede nullter Ordnung. Als Folge gibt das Glied nullter Ordnung bereits eine gute Approximation vom vorderen Staupunkt bis weit stromabwärts, während die an der Wand und außen verschwindenden Beiträge höherer Ordnung zum Geschwindigkeitsfeld erst weit stromabwärts ins Gewicht zu fallen beginnen. Die neue Reihe ist überdies nicht nur für Körper mit runder Nase anwendbar, sondern auch für Körper mit einem beliebigen vorderen Keilwinkel einschließlich Platten (Keilwinkel Null) bei beliebiger äußerer Druckverteilung.

Es war darüber hinaus das Ziel, ein Verfahren zu gewinnen, das, obwohl auf einer formal exakten Lösung beruhend, in der Anwendung nicht mehr, sondern eher noch weniger Arbeit erfordert als die nicht zuverlässigen Verfahren vom Typus des VON KÁRMÁN-POHLHAUSEN-Verfahrens. Auch dieses Ziel wurde erreicht, denn die Koeffizientenfunktionen $F_k(\eta)$ lassen sich als Linearkombinationen von universellen Funktionen darstellen; sind diese bis zu einer ausreichend hohen Ordnung ein für allemal vertafelt, so besteht die Anwendung des Verfahrens in jedem Einzelfall in der Durchführung der einleitend angegebenen Variablentransformation und in einigen wenigen elementaren Rechenoperationen. Solche Tafeln liegen vor [2].

Unter den Erprobungen in zahlreichen Anwendungen [1], [3], [4] sei hier nur kurz auf die Neuberechnung der erstmals von I. TANI mittels einer Reihe vom BLASIUSschen Typus berechneten Plattengrenzschichten zu äußeren Geschwindigkeiten der Form $U(x) = u_0 - u_n x^n$ $(n \geq 0)$, eine Verallgemeinerung der HOWARTHschen Reihe für linearen Druckanstieg $(n = 1)$, hingewiesen. Für die neue Reihe war dies ein besonders ungünstiges Anwendungsfeld, weil bei der vorliegenden Form von $U(x)$ die bereitstehenden universellen Funktionen überwiegend den Koeffizienten Null erhalten, mithin nur wenige Glieder der neuen Reihe zur Berechnung der Lösung ausgenutzt werden können. Es zeigte sich jedoch, daß mit diesen wenigen Gliedern bereits volle Übereinstimmung mit den TANIschen Resultaten erzielt werden konnte, soweit diese bei Verwendung einer weit größeren Zahl von Reihengliedern eine brauchbare Approximation liefern. Es war möglich, die Grenzschicht bis unmittelbar vor der Ablösungsstelle mit ausreichender Näherung zu ermitteln.

Dasselbe günstige Resultat erzielten wir bei allen bisher von uns untersuchten Plattenströmungen, wobei uns die universellen Funktionen der F_k bis zur Ordnung $k = 5$ einschließlich numerisch zur Verfügung standen. Bei der gleichen Ordnung der Approximation fielen die Resultate an Zylindern mit gerundeter Nase weniger günstig aus.

Die Methode der neuen Reihe ist inzwischen formal auf Grenzschichten mit allgemeiner Verteilung einer kontinuierlichen Absaugung an der Wand [5], auf achsensymmetrische Grenzschichten [6] und auf Grenzschichten mit äußerer Scherströmung [7] erweitert und auf die Berechnung der Temperaturgrenzschichten bei konstanten Stoffwerten übertragen worden [8]. Für die Verallgemeinerung auf kompressible Grenzschichten liegen erste Ergebnisse vor. Für alle diese Erweiterungen bereiten wir die Vertafelung der zugehörigen universellen Funktionen vor.

Literatur:

[1] GÖRTLER, H.: A new series for the calculation of steady laminar boundary layer flows. Journ. Math. Mech. **6**, 1—66 (1957).

[2] GÖRTLER, H.: Zahlentafeln universeller Funktionen zur neuen Reihe für die Berechnung laminarer Grenzschichten. Bericht Nr. 34 der DVL (1957).

[3] GÖRTLER, H., und H. WITTING: Zu den Tanischen Grenzschichten. Österr. Ing.-Archiv **11**, 111—122 (1957).

[4] GÖRTLER, H., und H. WITTING: Einige laminare Grenzschichtströmungen, berechnet mittels einer neuen Reihenmethode. ZAMP **9** (1958). Erscheint demnächst.

[5] GÖRTLER, H.: On the calculation of steady laminar boundary layer flows with continuous suction. Journ. Math. Mech. **6**, 323—340 (1957).

[6] SALJNIKOV, V.: Erscheint als DVL-Bericht.

[7] SALJNIKOV, V.: Erscheint als DVL-Bericht.

[8] WRAGE, E.: Erscheint als DVL-Bericht.

TH. VON KÁRMÁN (Paris): I would like to make a remark concerning the method that I proposed to POHLHAUSEN and that POHLHAUSEN carried out, a method now 36 years old. It is rather interesting that the wall condition we used:

$$\nu \frac{\partial^2 u}{\partial y^2}\bigg|_{y=0} = \frac{1}{\varrho} \frac{d p}{d x}$$

and according to which the second derivative of the velocity profile at the wall is fixed, was too "stark", too restrictive. It does, therefore, not work in all cases. Applications to problems of suction and injection lead to results for which δ can become imaginary. It seems that this is due to the wall condition mentioned.

H. GÖRTLER (Freiburg i. Br.): When applying methods of this type, results depend largely on the choice of the family of functions that are used to approximate the velocity profiles. Many years ago, PRETSCH applied the POHLHAUSEN method to the calculation of steady boundary layer flow along a wavy wall [1]. The results showed surprisingly strong separation even for weak waviness. This was mainly due to an inadequate choice of approximating functions. Reliable numerical evaluations of examples by A. W. QUICK and K. SCHRÖDER [2], applying a difference procedure, and a rather rigorous analytical treatment of the general problem [3, 4] lead to the result that, as predicted by PRETSCH, separation will, indeed, occur for arbitrarily weak waviness if only the flow passes a sufficient number of wave periods. However, separation occurs in a very thin sublayer of the boundary layer and, since the family of functions used by PRETSCH could not give a satisfactory approximation of this unusual type of velocity profile, his quantitative results were necessarily wrong. — It may be added that the total viscous drag of a wavy plate is smaller than that of a plane plate of equal depth [4]. This decrease is, of course, compensated by pressure drag caused by waviness.

References:

[1] PRETSCH, J.: Die Stabilität einer ebenen Laminarströmung bei Druckgefälle und Druckanstieg. Jb. dtsch. Luftfahrt-Forsch. **1941**, I 58—75.

[2] QUICK, A. W., und K. SCHRÖDER: Verhalten der laminaren Grenzschicht bei periodisch schwankendem Druckverlauf. Dtsch. Luftfahrtforsch., Zentr. wiss. Ber.-Wes., UM 1257 (1944).

[3] GÖRTLER, H.: Einfluß einer schwachen Wandwelligkeit auf den Verlauf der laminaren Grenzschichten. ZAMM **25/27**, 223—244 (1947) und ZAMM **28**, 13—22 (1948).

[4] GÖRTLER, H.: Reibungswiderstand einer schwach gewellten, längs angeströmten Platte. Arch. Math. **1**, 450—453 (1948/49).

B. THWAITES (Winchester): Can your new series method be applied with confidence up to separation?

H. GÖRTLER (Freiburg i. Br.): We have checked our results in every case of application by applying our difference procedure [1] that has been improved by Dr. WITTING for the near vicinity of a separation point [2]. We have started this procedure in a region far upstream where the series results were undoubtedly correct within very high accuracy. In all cases of flat plate flow with many different types of unfavorable outer pressure gradient, the fifth order approximation given by the truncated series was correct within the accuracy of a large drawing downstream till immediately ahead of separation. Results for separation found by a short freehand extrapolation were then always accurate within one or two per cent as compared to the results given by the difference procedure.

References:

[1] GÖRTLER, H.: Ein Differenzenverfahren zur Berechnung laminarer Grenzschichten. Ing.-Archiv **16**, 173—187 (1948).

[2] WITTING, H.: Verbesserung des Differenzenverfahrens von H. GÖRTLER zur Berechnung laminarer Grenzschichten. ZAMP **4**, 376—397 (1953).

I. TANI (Tokyo): I am happy to learn that your results agree well with those of my previous calculation [1]. I really appreciate your new analysis in view of the rapidness in convergence.

I agree with Dr. VON KÁRMÁN in that the wall condition for the second derivative of velocity is too restrictive. If this condition is abandoned and the energy equation is used instead, the KÁRMÁN-POHLHAUSEN procedure can be improved in accuracy. This has really been shown in my paper [2].

References:

[1] TANI, I.: On the solution of the laminar boundary layer equations. J. Physic. Soc. Japan **4**, 149—154 (1949).

[2] TANI, I.: On the solution of the laminar boundary layer equations. „50 Jahre Grenzschichtforschung", hrsg. von H. GÖRTLER u. W. TOLLMIEN. Braunschweig: Verlag Friedr. Vieweg u. Sohn 1955.

G. I. TAYLOR (Cambridge): The particular choice of functions for the KÁRMÁN-POHLHAUSEN method should really depend on what happens upstream.

H. L. DRYDEN (Washington): Did anybody prove that you can get any result you wanted ?

H. GÖRTLER (Freiburg i. Br.): Well, Dr. VON KÁRMÁN has told us that, at least, you can get an imaginary boundary layer thickness, even if you don't want it.

B. THWAITES (Winchester): I think the important thing to realize, in using POHLHAUSEN's general idea, is that it is quite unnecessary to work in terms of assumed velocity profiles, which only confuse the significance of the assumptions. All that is necessary for a solution of the KÁRMÁN integral equation, are three functional relationships between the skin friction, the second derivative of the velocity profile at the wall, the momentum thickness and the displacement thickness. This I have shown in my paper [1].

Reference:

[1] THWAITES, B.: Approximate calculation of the laminar boundary layer. Aero. Quart. 1, 245—280 (1949).

K. STEWARTSON (Bristol): An important reason for the failure of the KÁRMÁN-POHLHAUSEN method near separation is that while the separation point is an ordinary point of the approximation it is in general a singular point of the boundary layer equations. Any improvement on the method must take this into account.

H. GÖRTLER (Freiburg i. Br.): At separation one can not expect the boundary layer equations to give a good approximation. Moreover, what happens immediately behind separation and also influences separation according to the full Navier-Stokes equations depends largely on the behavior of the flow further backward from where the fluid moves forward towards separation. Any approximation taking only into account the history upstream as does the parabolic boundary layer approximation, can not be expected to give a correct answer since this answer is not uniquely determined by conditions upstream alone.

F. K. MOORE (Cleveland): In connection with the general discussion of limitations of the KÁRMÁN-POHLHAUSEN method, I should like to point out that the single parameter $\lambda = \delta^2 \, U'/\nu$ is an obviously incomplete description of the pressure gradient. One may place the following physical interpretation on λ: δ^2/ν (or θ^2/ν) is a measure of the time required for diffusion to adjust the entire boundary layer to a change at the boundaries of the layer. $1/U'$ is a time which is a measure of how rapidly the velocity of a fluid particle at the edge of the boundary layer changes. λ is the ratio of these times, comparing "how fast the boundary layer can adjust" and "how fast changes take place requiring adjustment".

Now, it is clear that $1/U'$ is not the only characteristic time of imposed change. A whole sequence of such times may be written down, in terms of higher derivatives: $[U^{n-1} \, d^n U/dx^n]^{-1/n}$, $n = 1, 2, 3 \ldots$ Generally, these times form a sequence of increasing magnitude, and therefore, the parameter λ enters as a first approximation to the effect of pressure gradient.

Since an analysis concerned only with λ should be considered as, in a sense, a small-perturbation analysis, it is not surprising that trouble occurs when λ approaches — 12. If λ is large, then, presumably, higher-order terms ($\delta^{2n} \, \nu^{-n} \, U^{n-1} \, d^n U/dx^n$, $n = 1, 2, 3 \ldots$) must be important. It is interesting to note that the two-parameter method proposed by I. TANI [1] may be shown to include the effect of U'' as well as of U'.

Reference:

[*1*] TANI, I.: On the approximate solution of the laminar boundary layer equations. Journ. Aero. Sci. **21**, 487—495 (1954).

B. THWAITES (Winchester): I do not think that it is true that the boundary layer development depends on *all* derivatives of the free stream velocity. The boundary layer equations (for two-dimensional incompressible flow) can be rewritten in the form

$$\frac{\partial u}{\partial x} = F\left(\frac{d U}{d x}, u(y)\right)$$

at any point, which shows:

1. that the streamwise derivative depends *only* on the velocity profile and the *first* derivative of the stream velocity, and incidentally

2. that the boundary-layer at a point depends only on what has happened upstream (which is typical of parabolic problems).

G. E. GADD (Teddington): Concerning the matter raised by Dr. STEWARTSON, that the separation point is a singularity for solutions of the boundary layer equations, it should be borne in mind that experimentally no such singularity is observed.

Furthermore, although the shape of the velocity profile at a separation point depends considerably on conditions upstream, a simple approximate formula for the position of separation does seem to apply to nearly all cases. The formula, which is due to Dr. STRATFORD, is

$$x^2 C_p \left(\frac{d C_p}{d x}\right)^2 = 0.0104 \text{ at separation,}$$

where x is the effective distance (in many cases the actual distance) from the leading edge to separation, and C_p is the pressure coefficient $(p - p_0)/\frac{1}{2}\varrho U_0{}^2$, suffix "0" denoting conditions at the pressure minimum. This relation was derived theoretically for the special case where there are abrupt adverse pressure gradients in the vicinity of separation after an extended upstream region of constant pressure. However, Dr. CURLE of the National Physical Laboratory has shown that the formula does in fact agree well with all the known exact or quasi-exact solutions of the laminar incompressible boundary-layer equations, including those of Dr. GÖRTLER and Dr. TANI.

K. STEWARTSON (Bristol): In reply to Mr. GADD's comment, I agree that according to experiment there is no singularity at separation but in the only complete numerical study of a separating boundary layer the singularity is certainly present. Further, GOLDSTEIN's study of the flow near separation leads us to expect one. It may be that the singularity can be avoided if the main stream at separation satisfies some condition which it in fact does in all practical cases.

Mechanism of transition at subsonic speeds

By

G. B. Schubauer

National Bureau of Standards, Washington, D.C.

1. Introduction

The subject of transition from laminar to turbulent flow is currently being investigated in a number of laboratories. Much of the activity has to do with the extent of laminar flow on bodies at supersonic speeds and on how the position of transition, or the Reynolds number at which transition occurs, depends on the steadiness of the stream, roughness of the surface, or heat transfer to or from the surface. Certain investigations, conducted mostly at low speeds, attempt to probe into the mechanism of transition. A recent survey of the subject has been given by DRYDEN [1].

The purpose of the present paper is to describe one such probing investigation being carried on at the National Bureau of Standards under the sponsorship of the National Advisory Committee for Aeronautics. Significant progress has been made so far only in the regime of subsonic speeds. The results obtained during the first phase have been published and are available from two sources [2]. The present paper will briefly review this work and then discuss the more recent results, comparing where possible with the results obtained by other investigators in an attempt to see what light is being thrown on the transition problem by investigations of this kind. A detailed account of the investigation must be left to a forthcoming report by the author's co-workers, P. S. KLEBA-NOFF and K. D. TIDSTROM, to be submitted to the National Advisory Committee for Aeronautics for publication. The author wishes to acknowledge the many necessary discussions with Mr. KLEBANOFF regarding the results and their interpretation and his assistance in preparing the material for this paper.

The objective throughout the investigation has been to clarify the mechanism of transition by experimental studies of the processes by which turbulence comes into being. The boundary layer on a flat plate in the low speed range (100 ft. per sec. or less) was selected as the subject when the investigation was begun. Since the conditions proved favorable for probing into the details of the mechanics of transition, it has remained the principal subject and the results to be discussed pertain to this case.

The plate, made of $\frac{1}{4}$-inch aluminum and having a sharpened leading edge, is 12 feet long and $4\frac{1}{2}$ feet wide. It is installed in the $4\frac{1}{2}$-foot wind tunnel which incorporates an adjustable side wall to eliminate pressure gradient. The arrangement is essentially the same as that used in the study of Tollmien-Schlichting waves conducted in the early 1940's [3]. A low stream turbulence, amounting to about 0.03 percent at a speed of 80 ft. per sec., permitted a laminar boundary layer five or more feet in length prior to transition, and thus provided sufficient length of layer in which to study the development of natural and artificially produced disturbances and sufficient thickness of layer in which to employ pitot and hot-wire probes.

The hot-wire anemometer proved to be the most useful tool. Variously arranged hot-wire probes were used together with conventional compensated amplifiers and measuring equipment. Extensive use was made of cathode-ray oscilloscopes for visual observation of the hot-wire signal and photographic recording on continuously-moving 35 mm film. Film speeds were usually of the order of 5 ft. per sec.

2. Identification of character of transition

The first phase of the investigation as reported in reference 2 consisted of a detailed exploration of the transition region. The initial objective was to study phenomena at the instantaneous point of change-over from laminar to turbulent flow. Various schemes of inducing transition were used, but the result was always a non-stationary character of what was believed to be the transition point, and the immediate objective was defeated. Numerous records were obtained of the alternately laminar and turbulent flow characterizing the transition region in the hope that something could be learned from them. These efforts made possible the principal contribution of this phase of the investigation, namely the identification of the turbulent portions of the record with the downstream passage of turbulent patches, more generally referred to as spots.

The identification was made possible by comparing the many records on hand with the record of the passage of an artificially created spot. The method of producing a spot was the same as that used earlier by MITCHNER [4] namely, by passing an electric spark from an electrode to the plate across the laminar layer. Fig. 1 shows the comparison. The upper trace is a sample typical of all records in the transition region obtained with a single hot-wire probe, sensitive to the longitudinal component of the fluctuations, u, when the wire is within 0.05 inch of the surface. The lower trace is typical of the passage of a spot. The kink in the trace, which results from an electrical disturbance picked up by the circuit, conveniently marks the time of firing the spark. After an interval of time the turbulent spot arrives at the probe causing the step-up in velocity and the following turbulent trace. The abrupt ending, followed by the gradual decline, marks the passage of the spot on downstream.

The characteristics of a spot, such as its shape, growth rate, and velocity of propagation, together with an associated effect, termed the "calming effect" are treated in detail in reference 2 and will not be discussed here. The characteristic features appearing in all of the records proved beyond any doubt that turbulent sections of the traces were always the result of turbulence that had begun somewhere upstream and, when observed, was a downstream-moving patch, depending for its size on growth time or a merging with neighboring patches. A continuously turbulent region merely indicated a complete merging and the wiping out of all intervening laminar flow. The theory proposed by EMMONS on the basis of his water-table observations [5] was thus confirmed.

Evidence of two other seemingly universal behavior patterns emerged from this study. The first was that the spread of turbulence was always a growth process; in other words, the breakdown of laminar flow occurred at discrete points, and the turbulence resulting therefrom was initially restricted to small spots which then grew. The second was that the genesis of a spot was intimately connected with the presence of amplified waves when the impressed disturbances were at a low level. This could readily be inferred from the work [3], but it was demonstrated in a striking manner by the calming effect by which waves could be wiped out for a short time and the formation of

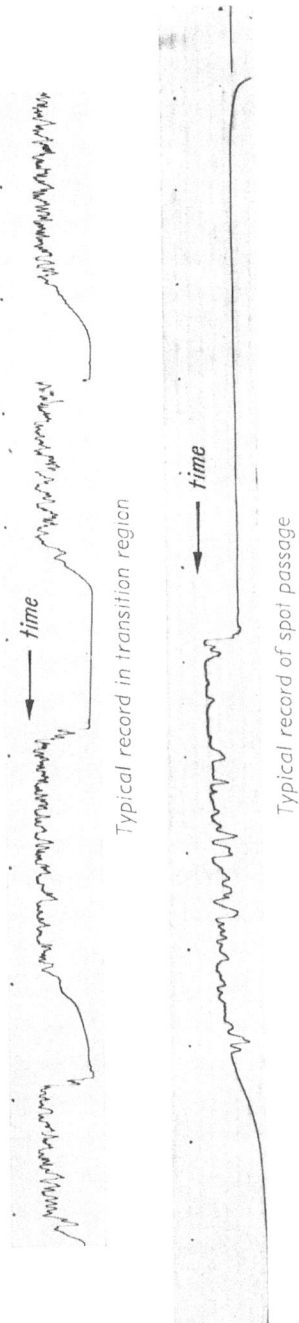

Fig. 1. Oscillograms of u-fluctuations typical of transition region on flat plate and passage of turbulent spot. Time interval between dots, 1/60 sec.

new spots thereby prevented during that interval. Further evidence was furnished in the supersonic regime by the work of Laufer and Verbalovich [6].

3. Transition from wave to turbulence

While the foregoing work brought out a number of interesting features worthy of further investigation, it appeared that questions most urgently requiring answers were: (1) By what process did amplified waves lead to an eventual breakdown of laminar flow? (2) Was the breakdown, however it occurred, an inherently point-like phenomenon?

In order to investigate the first question and to examine the second somewhat more closely, it was decided to generate waves of controlled frequency and amplitude and trace their downstream development to the onset of turbulence. It was known from earlier work [3] that reasonably

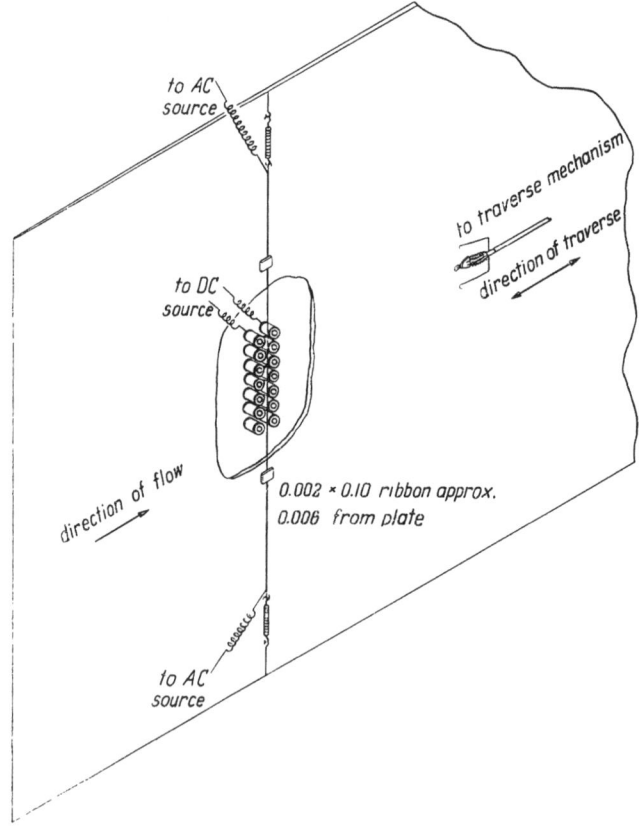

Fig. 2. Set-up for introducing waves in boundary layer of flat plate by vibrating ribbon. Dimensions where indicated, inches

two-dimensional Tollmien-Schlichting waves could be produced by the vibrating-ribbon technique. The experimental arrangement is shown schematically in fig. 2. By passing an alternating current through the ribbon in the presence of a d.c. magnetic field from electromagnets on the opposite side of the plate, the ribbon could be made to vibrate to and from the plate in a single loop. The length of the vibrating segment was varied, but was never less than 12 inches, the usual length being 13 inches. In all cases the waves obeyed small-perturbation theory when the amplitude was sufficiently small. The ribbon could be placed in any position, but for most of the work it was stationed 35 inches from the leading edge. The Reynolds number, $R_{\delta*}$ (where $R_{\delta*} = U_1 \delta*/\nu$, $U_1 = $ free-stream velocity, $\delta* = $ displacement thickness), was chosen so that according to small-perturbation theory the wave passed first through a region of amplification and then into a region of damping for properly chosen frequencies.

In many respects the procedure was the same as that followed in the work [3]. Now, however, the objective was to examine more completely the connection between the wave and transition. For this purpose provision was made for surveying in all directions over the region downstream from the ribbon. The sliding-type probe shown in fig. 2 was used for traversing in the stream direction, x, at some chosen fixed distance from the plate, y, and at a selected spanwise position, z. Other arrangements were employed for continuous traversing in the z-direction at fixed values of x and y, or in the y-direction at fixed values of x and z. The hot-wire probes were mostly of the simple normal-wire type, with a platinum wire about 0.04 inch long and 0.0001 inch in diameter, sensitive to the fluctuation velocity component, u, in the x-direction. A v-wire arrangement was also used to obtain the component w in the z-direction, or both u and w from the same probe. Because of the thinness of the layer the component v in the y-direction was not measured. In the region of interest the boundary-layer thickness, δ, was about 0.2 inch.[1]

Traverses in the z-direction revealed a feature about which most of the subsequent investigation centered. This was the almost periodic variation in intensity in the spanwise direction shown in fig. 3. The peaks and valleys occupied fixed positions and formed streets of high and low wave intensity. These developed with distance, x, irrespective of whether the initial intensity was high or low. When the initial intensity was low and the frequency was such that the wave first passed through a region of amplification and then into a region of damping, the intensity grew and then decayed at both peaks and valleys. However, if the intensity was sufficiently high initially, or became so through amplification, the

---

[1] δ is defined as the thickness in which a velocity reduction can be detected. As a working rule $\delta = 3 \delta*$.

difference between peaks and valleys was greatly accentuated, the growth
rates became very large, and growth continued to the breaking point. This
is the condition shown in fig. 3. Here the frequency was 145 c.p.s. and
$R_{\delta*}$ at the ribbon was 1625. Had the input been sufficiently small, the
wave would have first amplified and then damped with progression downstream.

The breaking occurs only at the peaks, as illustrated by the sample
oscillogram opposite one peak in fig. 3, and manifests itself as bursts of

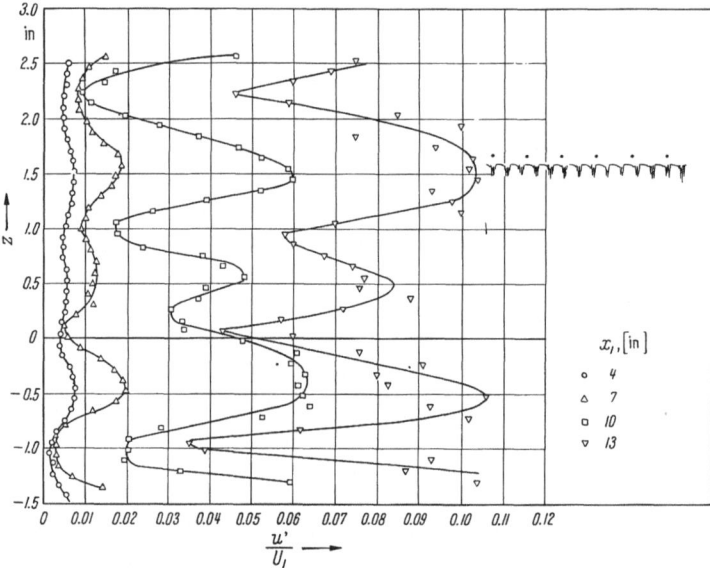

Fig. 3. Wave intensity warping into peaks and valleys. Oscillogram shows characteristic laminar
breakdown occurring at peaks. u'/U_1 = ratio root-mean-square value of u-fluctuation to free-stream
velocity. x_1 = distance from ribbon; $y/\delta = 0.23$ for curves, 0.6 for oscillogram. Ribbon 35 inches
from leading edge. $U_1 = 50$ ft. per sec. Frequency 145 c.p.s.

high-frequency fluctuations occurring once each cycle of the base wave.
The breaking begins characteristically with one or two sharp spikes appearing on the record pointing towards the lower velocities. For the case
shown in the figure this occurred somewhere between an x_1 of 10 and
13 inches. They multiply in number very rapidly to form the bunches
appearing at $x_1 = 13$ inches. The width of the bunch continues to grow,
occupying more and more of the length of the base wave until ultimately
the separate segments merge with one another. The fluctuations appear
rather regular at first, but they lose much of their regularity and take on
the irregular character of turbulence as they grow. These bursts are separate turbulent spots which here occur in regular succession because of
the steadiness of the wave fed into the boundary layer. They grow in

thickness, width, and length, producing wedge-shaped paths as they are convected downstream. Turbulence does not appear in the valleys until the spots, originating at the peaks, have spread into the valleys. The initial size of the spot and its position in the boundary layer will be discussed in Section 6.

The paths traced out by the turbulent spots can be shown by the china-clay technique. A typical surface pattern is shown by the photograph, fig. 4. Here a strip of the surface one foot wide was coated with

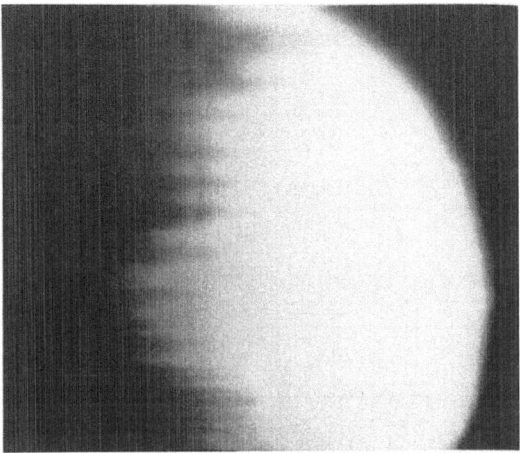

Fig. 4. Transition pattern indicated by china-clay method. Wave introduced by vibrating ribbon

china clay. It was then sprayed with oil of wintergreen and allowed to dry under chosen conditions of air flow. After about 20 minutes of running at a speed of 50 ft. per sec., during which time waves were produced, the drying pattern was obtained. The vibrating ribbon is shown by the black line (which appears slanted because of the angle at which the picture was taken). The darkened regions (wet) show laminar flow; the white regions (dry) show turbulent flow. The wedge-like white streaks coincide with paths of turbulent spots. These align themselves with streets of peak wave intensity. The wave motion itself prior to the formation of spots produced no discernible pattern.

4. Effect of flow conditions

One of the most puzzling questions was why the wave warping pattern was fixed spatially and remained the same day after day, unaffected by all changes that could readily be made about the setup. These included placing the vibrating ribbon in different regions and changing distance from surface, width, and length of vibrating span; altering the design of the electromagnet on the opposite side of the plate; changing traversing apparatus; and cleaning and waxing the surface of the plate. During the search for possible causes, a spanwise variation of mean velocity was detected in the laminar layer. This was then explored in considerable detail by traversing in the z-direction at fixed values of x and y with a small total-head tube. Streets of slightly higher and slightly lower velocity running in the x-direction were found. The differences generally amounted to a few percent, but in some spanwise locations they became as large as 10 percent. They existed throughout the thickness of the layer but disappeared in the free stream. These velocity streets were seemingly unaffected by the

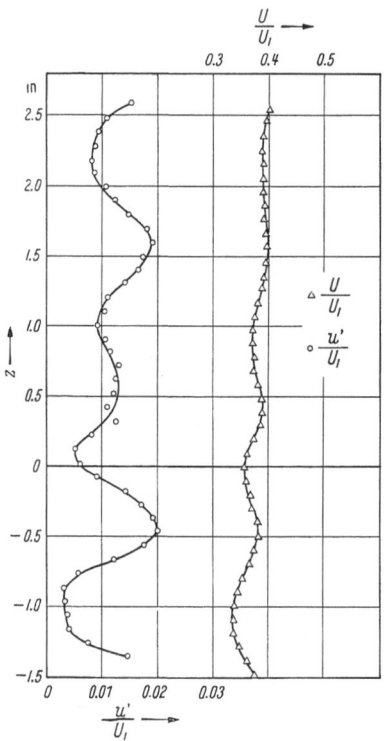

Fig. 5. Comparison of mean velocity and wave intensity variations. Distance downstream from ribbon 7 inches; $y/\delta = 0.23$. Frequency 145 c.p.s.

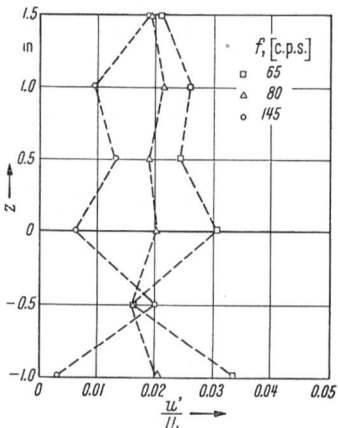

Fig. 6. Effect of frequency on wave intensity warping. Distance downstream from ribbon, 7 inches; $y/\delta = 0.23$

presence of a wave introduced by the vibrating ribbon, but when the wave was introduced, the resulting streets of high and low amplitude coincided with the streets of higher and lower velocity. This coincidence is illustrated by fig. 5.

There was thus established a definite relationship between the spatial velocity variations present in the layer and the warping of the wave. Furthermore, the effect could be accounted for in terms of changes in amplification arising from the associated variations in δ^*. As δ^* varies with spanwise position, so does the rate of amplification. Reference is here made to a typical stability diagram of the Tollmien-Schlichting type, for example fig. 27 of [3]. In the example shown in fig. 5, the frequency of 145 c.p.s. places the wave in a position to traverse the upper part of the amplification zone near branch II of the neutral curve as it advances downstream. A decrease in δ^* here produces a greater rate of amplification. Accordingly, in regions where the velocity is high and δ^* is low the wave amplitude is greater.

By using other frequencies the course of the wave may be placed in other parts of the amplification zone. The effect of this is shown in fig. 6. When the frequency is 65 c.p.s., the wave is introduced near branch I of the neutral curve, and its initial course traverses the lower part of the amplification zone. Here the effect of the variation in δ^* is the reverse of that in the upper part of the zone. When the frequency is 80 c.p.s., the wave is introduced in the lower part of the amplification zone, but nearer the center where variations in δ^* have had little effect over the course to the point of observation.

These findings raised two important questions. The first was, how did the velocity variations come to be in the boundary layer, what was their cause, and why did they maintain fixed positions? The second was, what part did they play in shaping the phenomena leading to transition? Neither of these questions have yet been fully answered, but a considerable amount of contributing information has been obtained.

When no velocity variations could be detected outside the boundary layer nor in the approaching stream, it was suspected that the variations within the layer were due to flow conditions about the leading edge or to surface conditions. In order to avoid disturbances from the leading edge, the angle of attack had to be such as to place the stagnation point slightly on the working side. No effect on the velocity variations could be observed when the angle was reduced to the limiting point where sporadic turbulence set in. The original surface was then covered over with a smooth sheet of material called "formica." This likewise had no effect. Finally, after the flow had remained unalterable for many months, the pattern suddenly changed. This followed running the tunnel at near its maximum speed. The effect was finally traced to the damping screens located in the settling chamber of the tunnel about 20 feet upstream from the plate. The nature of this effect is not yet clear, but each time something was done to the screens, such as adjusting or cleaning, the pattern was altered. In the course of the work a variety of velocity patterns was obtained. In all cases the wave warping followed the velocity pattern. Finally all six of the damping screens were given a thorough cleaning in the hope that the velocity variations could be eliminated. This hope was not realized, but marked reductions were effected over widths of 2 or 3 inches. Wave warping in these regions was delayed, but it nevertheless set in when the amplitude became large. However, the position of the peaks and valleys was now less firmly locked into fixed streets.

Certain inferences could be drawn from this work. First it appeared that the wave had a strong tendency to warp. Second it appeared that this tendency was easily influenced by any irregularity in the boundary layer, and was therefore easily locked into positions by fixed irregularities even though these were very weak. The influence could be reduced by placing the course of the wave where it was less sensitive to variations in δ^*, but still it proved to be impossible to prevent the development of peaks and valleys. In short, transition never occurred without first being preceded by a wave strongly warped into peaks and valleys. Transition then consisted of the succession of bursts, occurring at the peaks as previously described.

Since the laminar layer was never free from spatial velocity irregularities, it is not known how the wave would behave under uniform conditions. The question is probably an academic one since waves initiated by natural disturbances would be introduced in an irregular manner.

The cause and nature of the mean velocity variations still remain somewhat obscure. The damping screens contributed in some way, perhaps by introducing small velocity irregularities into the stream. These could not be found in the stream, but the effects of them could easily be measured in the boundary layer. The direction changes occurring along with the velocity distribution suggested the presence of weak streamwise vortices in the layer. Neither their strength nor their spacing was uniform. However, in spite of the many changes that were made the spatial cycle in the z-direction was always of the order of one inch. It is felt that these phenomena deserve further study, and further investigations are planned in which changes will be made in the setup and in the environmental conditions.

5. Wave growth and energy concentration

As the wave characteristics were studied in more detail, it was noted that the relative growth rates between peaks and valleys depended on distance from the wall. Distributions of root-mean-square intensity across the layer were then measured, and it was found that the distribution at a peak was like that reported in [3], essentially in agreement with small-perturbation theory, but that the distribution at a valley was distorted. The two distributions are shown in fig. 7. The distortion was found to lie in the region approximately $0.05 < y/\delta < 0.6$, and it was within this region that the marked differences in intensity between peak and valley developed. Outside of this region of y/δ the two distributions were similar and magnitudes were in approximate agreement. The effect in the inner and outer regions of differences in δ^* was apparently masked by other effects. When the wave amplitude was small, the two distributions were more nearly similar, but a remnant of the characteristic difference shown in fig. 7 still existed.

This difference becomes greatly aggravated with downstream growth

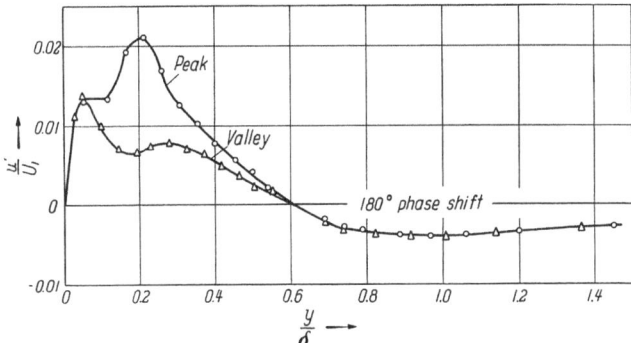

Fig. 7. Distribution of intensity across boundary layer. Ribbon frequency, 145 c.p.s.

$$\left(\frac{(x - x_0)}{\lambda} = 4, \text{ curve B, fig. 8} \right)$$

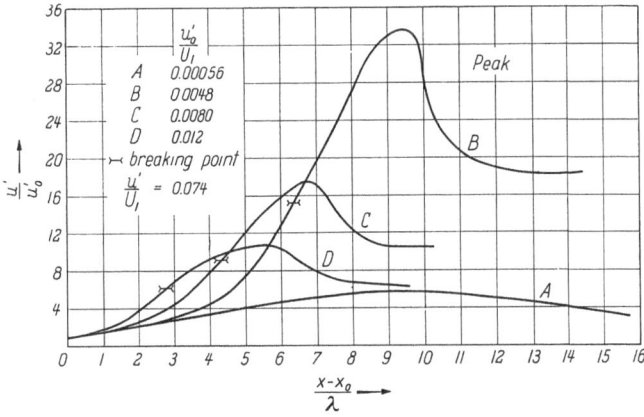

Fig. 8. Wave growth at peaks. x_0, reference position 2 inches downstream from ribbon. u'_0, root-mean-square intensity at x_0. Frequency = 145 c.p.s. λ, TOLLMIEN-SCHLICHTING wave length, = 1.46 inches, U_1 = 50 ft per sec. $R_{\delta *}$ at ribbon = 1625

when the initial amplitude is sufficiently large and accounts for a different growth rate at the peaks and valleys. If a hot-wire probe is placed at the position of the maximum of the peak distribution and traversed in the x-direction, growth curves are obtained as shown in fig. 8 for different initial amplitudes. The corresponding growth curves for the valley, with the hot-wire probe in the same y-position, are shown in fig. 9. Curves A

in both figures show the normal behaving amplification and damping for small amplitudes and is in reasonably good agreement with small-perturbation theory. When the input amplitude is no longer sufficiently small, the growth follows the normal curve for a distance and then at a given point departs with a much larger rate of growth at the peak. This is shown in the order of increasing inputs by curves B, C, and D. The departure in the valley occurs at about the same point, or a little sooner, but now is initially a decreased growth rate causing the curve to fall below curve A.

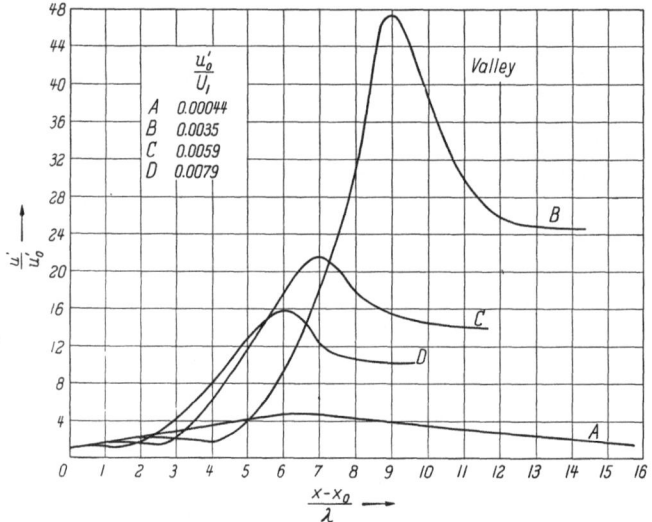

Fig. 9. Wave growth in valley. Conditions same as for fig. 8

Eventually the growth rate increases to large values and the growth curves rise sharply. Curves B, C, and D of figs. 8 and 9 show how the departure occurs at smaller x-values as the input is increased.

With reference to fig. 8, it is found that the departure occurs in each case where u'/U_1 has reached 0.01. On curves B, C, and D of fig. 8 the points of initial laminar breakdown are indicated. This occurs when u'/U_1 has reached 0.074 in each case. The only effect of initial input was to advance the points of departure and the points of breaking into turbulent spots farther upstream with increased input. It will be recalled that these values correspond to the neighborhood of the maximum in fig. 7. Since the maximum appearing in fig. 7 is 0.021, the distribution across the layer shown here corresponds to the condition soon after departure from curve A, and considerable more growth is required before breaking occurs. However, fig. 8 shows that only a few wave lengths are required for the wave to grow to the breaking point, once departure has occurred.

Following the breakdown the intensity continues to rise as shown in fig. 8 and reaches a maximum after a distance of 2 or 3 wave lengths. This is approximately the distance required for the separate spots, which occur one wave length apart, to grow to a length of one wave and connect with one another. When this has occurred the roughness associated with the passage of spots subsides, and the fluctuation level begins a decrease which continues until a steady-state turbulent flow prevails. The rapid growth ensuing in the valley following the initial depression does not result in spots. Instead the flow becomes disturbed from the neighboring streets and the spots eventually expand into this region. Here likewise the intensity level subsides as a steady-state turbulent flow ensues.

Similar curves were obtained using a 70-cycle wave. For this frequency the position of the peaks and valleys was interchanged, but the behavior was basically the same as with the 145-cycle wave. Again breaking began when u'/U_1 was very close to the original value of 0.074. The value for departure from the normal curve A was, however, 0.02 instead of the 0.01 applying to the 145-cycle wave. It may or may not be significant that the value was doubled when the wave length was approximately doubled. It is remarked that the maximum shear stress produced by the wave, and expressed by density times $\overline{uv}$, was estimated to be only about 0.7 percent of the maximum laminar shear stress at the point of departure. The estimation was made assuming $v' = 0.1\,u'$ and a correlation coefficient of 0.18. At the breaking point the $\overline{uv}$-stress was still less than the laminar shear stress, but had increased to the same order of magnitude. In none of this work were the boundary-layer conditions varied sufficiently to establish universal criteria for either departure or breaking. For example, $R_{\delta*}$ for breaking ranged only from 1730 to 2020.

One of the most difficult questions to answer was whether wave fronts remained straight across peaks and valleys or whether the fronts, like the intensities, also became warped. This led to a search for phase changes from point to point in the y- and z-directions. For TOLLMIEN-SCHLICHTING waves the only observable phase change is the 180-degree shift indicated in fig. 7. This condition prevailed at the peaks for all intensities up to near the breaking point. Very close to the breaking point the measurements were uncertain. However, in the valleys an unusual and not completely understood condition was found to have developed in the case of the 145-cycle wave after departure from the small-amplitude growth curve. In the middle region of the layer where the intensity in the valley was depressed below that at the peak an advance in phase over that at the peak was indicated. This began at a y/δ of about 0.05 and increased with increasing y out to about half the boundary-layer thickness. From there on out the motion in the valley was again in phase with that at the peak. The amount of the advances seemed to increase with x for

7 Görtler, Grenzschichtforschung

a distance of a few wave lengths and then remain constant, the ultimate shift appearing to be around 180 degrees. In the region of advancing phase, the wave length and velocity likewise increased, but apparently returned to their original value when the phase no longer increased. These phenomena occurred in the region of rapid growth up to near the value of x where breaking began at the peaks.

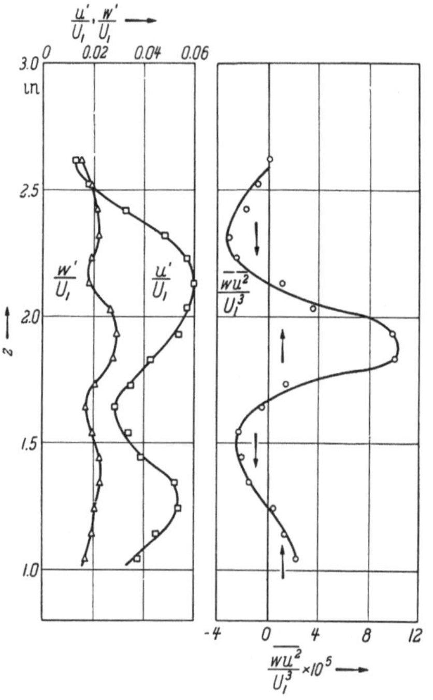

If these effects are real, the wave appears to exhibit a dual character. The in-phase condition is indicative of a straight wave front extending across peaks and valleys, while the out-of-phase condition is indicative of a wave front which in the valley is looped ahead of that at the peak, but only in a restricted interval of y/δ. It is difficult to picture a motion with these characteristics, just as it is difficult to picture a condition in which wave length increases and then decreases to its former value. A more rational interpretation is that the apparent phase shifts and changes in wave length and velocity are in reality effects arising from change of wave form. The phenomena were accompanied by distortions which gave the wave an irregular and flat-topped character. In support of this interpretation, such anomalies were not apparent for the 70-cycle wave.

Fig. 10. Energy transfer and comparison of u' and w'. Distance downstream from ribbon, 7 inches; $y/\delta = 0.23$. Frequency = 145 c.p.s. $R_{\delta*}$ at ribbon = 1625

There is much room for future clarification of these questions. For the present we must conclude that the wave fronts do not develop into a looped configuration like that reported for dye patterns in [8] and [9], and that the wave lengths and velocities remain about the same as those for small amplitudes. The wave length, λ, in figs. 8 and 9 is the small-amplitude value. These questions will be discussed further in Section 7.

Measurements of the spanwise component of the fluctuations, w, and the correlations, $\overline{uw}$ and $\overline{wu^2}$, contributed significantly to an understanding of what was taking place in the selective growth process at peaks and valleys. Whenever the wave exhibited any warping at all the w-component was present. When normal amplification and damping prevailed (curves A of figs. 8 and 9) the root-mean-square value, w', was small. However, when departure occurred, w' intensified very rapidly. A typical after-departure condition is shown by the comparison between u' and w' in fig. 10.

It is to be expected that a w-fluctuation must accompany variations in wave intensity along the wave front. The maximum amplitude of w accompanying this variation in wave intensity should also fall between peaks and valleys defined in terms of the u-fluctuations. This was borne out by observation, as shown in fig. 10. The w'-distribution across the layer had a maximum around $y/\delta \simeq 0.2$, resembling the u'-distribution at a peak in this respect.

Large values of the correlation coefficient, $\overline{uw}/u'w'$, were also found, but the value depended markedly on distance from the wall and changed from positive to negative two or more times in crossing the layer. Maximum positive and negative values of the correlation coefficient were near unity. The variation in the coefficient with y was due to a continuous phase change of w relative to u from the wall outward. In one typical case the phase shift amounted to 270 degrees; in another it was 540 degrees. While the relative magnitudes of u and w depended on wave amplitude, the correlation coefficient had the same value and behaved in the same way whether the amplitude was small or large.

The most significant property associated with w was a non-zero value of the mean energy transfer, $\overline{wu^2}$. This showed that wave energy was transferred from valleys to peaks as illustrated in fig. 10. The quantity $\overline{wu^2}/U_1^3$ is a dimensionless coefficient giving the rate of flow of energy; the arrows indicate the direction in which it is flowing. It now becomes clear that the growth curves of figs. 8 and 9 do not simply signify augmented growth at the peaks and stunted growth in the valleys, but rather that an energy concentrating mechanism is at work removing energy from the valleys and depositing it at the peaks. This at once explains why curves B, C, and D of fig. 9 fall below the normal amplification curve A. Very near the wall there was a measurable, but comparatively insignificant, transfer in the opposite direction. In the small-amplitude range, typified by curves A of figs. 8 and 9, any transfer, if present, was so small that it was lost in the uncertainty of measurement.

While the presence of the energy transfer explains why wave energy concentrates and evolves into the characteristic transition pattern, there is as yet no certain explanation for the existence of the transfer itself. The mere presence of an initial irregularity insures the presence of a w strongly correlated with u, but it does not insure a mean triple correlation of the type $\overline{wu^2}$. The mechanism giving rise to this quantity has not been determined.

It has been suggested by Görtler [7] that the concave flow curvature associated with Tollmien-Schlichting waves can produce a secondary instability and may give rise to vortices with their axis parallel to the main flow. An interesting question is therefore whether the energy transfer represented by $\overline{wu^2}$ is caused by such vortices. Direct evidence of their presence, derived from mean velocity and direction measurements, has so far been inconclusive. The situation is confused somewhat by a background of pre-existing velocity variations as described in Section 4. Intuitively one feels that when the spanwise variations of wave intensity once exist, these must themselves generate vortices lying along the stream. There is therefore a strong likelihood that vortices were present, but proof of this and of their connection with processes leading to transition must await future clarification.

Since the subject of secondary instability is being treated by H. Görtler and H. Witting ("Theorie der sekundären Instabilität der laminaren Grenzschichten", Symposium on Boundary Layer Research, University of Freiburg, 26—29. August 1957), it may be of interest to note here the minimum radius of curvature, r, and the maximum flow displacement, Δy, calculated on the basis of a sinusoidal wave with the intensities pertaining to the points of departure from curve type A of fig. 8. These quantities compared to displacement thickness, δ^*, are:

	u'/U_1	r/δ^*	$\Delta y/\delta^*$
70-cycle wave	0.02	924	45×10^{-3}
145-cycle wave	0.01	600	16×10^{-3}

6. Character of laminar-flow breakdown

The bursts shown in fig. 3 were studied at various stages of development by visual inspection of the u-fluctuations on the screen of an oscilloscope and by examining many film records. The results were totally consistent in showing the sudden onset of a violent motion, vastly different from the progressively evolving motions that preceded it. The records showed the groups of oscillations, appearing once each cycle of the parent wave, to be a slightly advanced stage resulting from an increase in number of closely spaced disturbances having the appearance of thin, sharp spikes protruding toward lower velocities. One or two such spikes marked the onset of the new regime. The progression is shown in fig. 11 which displays a series of records made by successively increasing the input wave strength. The growth into turbulent spots takes place as described in Section 3.

Similar phenomena were found without employing the vibrating ribbon, i.e., for the so-called natural transition resulting from the background of unavoidable disturbances in the stream. Now, however, the wave amplitude varied with time, as though modulated in an irregular manner, and the spikes and spots occurred sporadically, lacking the regular repetition characteristic of a steady wave.

The studies so far have not revealed the nature of the motions from which the records result. They have shown, however, that the region of initial breakdown is a highly restricted one. This is illustrated, again for

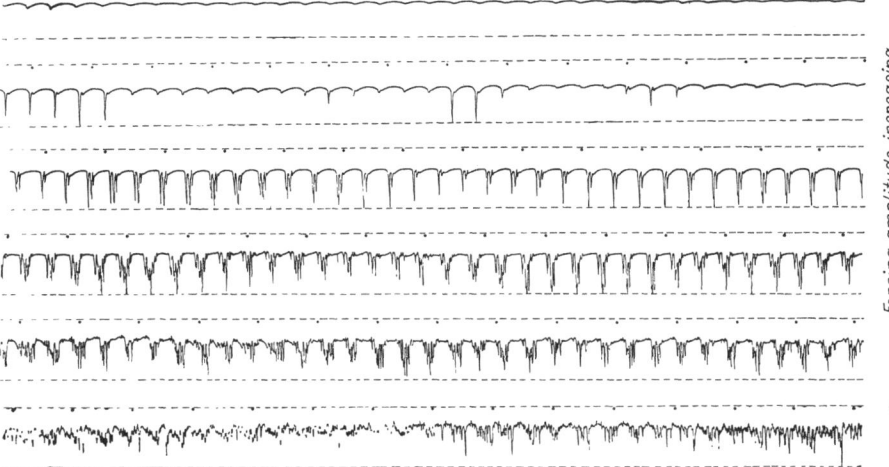

Transition from forced oscillations

Fig. 11. Progression of breaking bursts and formation of turbulent spots occurring in streets of peak wave intensity. Downward movement of trace means velocity decrease. Time interval between dots, 1/60 sec.

the wave induced by the vibrating ribbon, first for spanwise distances in fig. 12, and second for distances through the boundary layer in fig. 13. Both figures show simultaneous records at two positions obtained by using two u-sensitive probes and a dual-beam oscilloscope.

In fig. 12 the lower trace of each set shows the breaking when the probe was centered in the pattern, i.e., at the z-position of peak wave intensity. The top trace of each set shows the simultaneous response of a second probe displaced by the amount Δz from the first. When $\Delta z = 0.1$ inch, the trace shows only a slight ripple on the high-velocity side of the wave. As Δz is increased this decreases and finally disappears. It was initially thought that the ripple was characteristic of the disturbance in the valley, but it is now known that the ripple is a disturbance from a breaking pattern nearby. A displacement of 0.1 inch, or in this case half the boundary layer thickness, has placed the probe outside the breaking pattern.

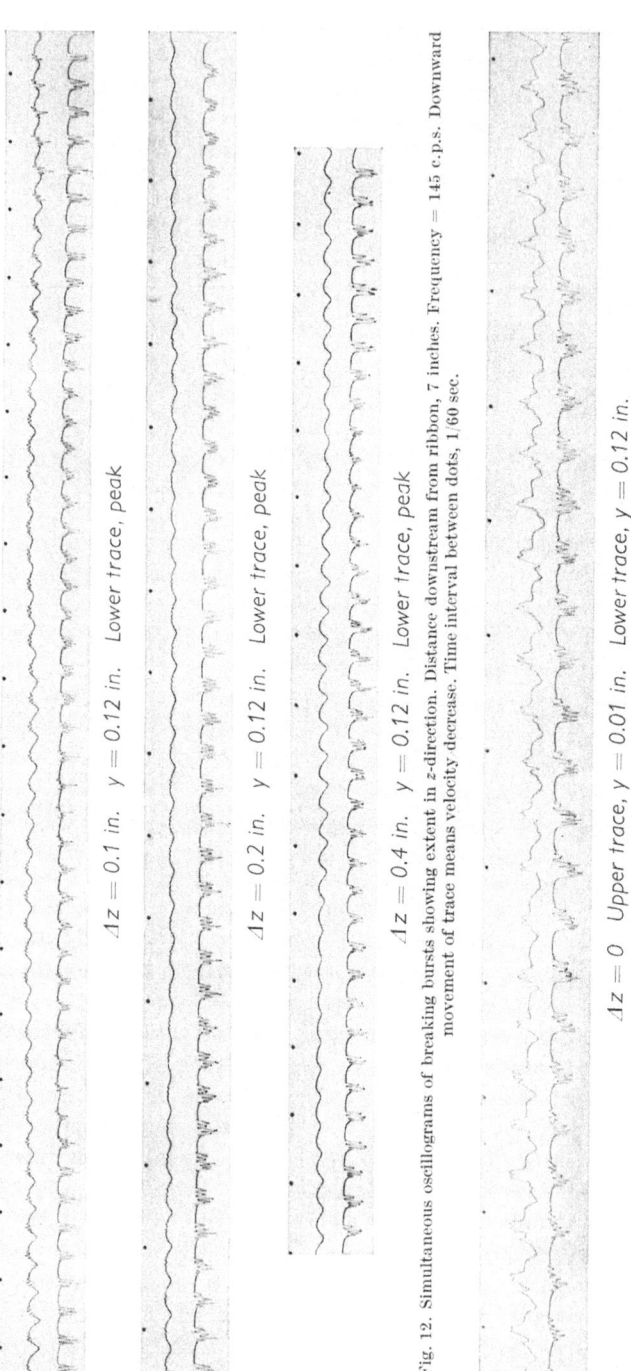

$\Delta z = 0.1$ in. $y = 0.12$ in.   Lower trace, peak

$\Delta z = 0.2$ in. $y = 0.12$ in.   Lower trace, peak

$\Delta z = 0.4$ in. $y = 0.12$ in.   Lower trace, peak

Fig. 12. Simultaneous oscillograms of breaking bursts showing extent in z-direction. Distance downstream from ribbon, 7 inches. Frequency = 145 c.p.s. Downward movement of trace means velocity decrease. Time interval between dots, 1/60 sec.

$\Delta z = 0$ Upper trace, $y = 0.01$ in.   Lower trace, $y = 0.12$ in.

$\Delta z = 0$ Upper trace, $y = 0.12$ in.   Lower trace, $y = 0.23$ in.

Fig. 13. Simultaneous oscillograms of breaking bursts showing extent in y-direction. Conditions same as for fig. 12

Fig. 13 shows simultaneous records for displacement in the y-direction. One probe has been held at $y = 0.12$ inch where the breaking pattern was centered (about 0.6 δ) while the other probe is displaced 0.11 inch towards the wall in one case and towards the outer edge in another. At the displaced positions the flow is highly perturbed, but it does not show the fluctuations typical of a breaking pattern.

Without knowing something about the physical nature of the motion it is impossible to specify the exact limits of a breaking pattern. At the stages shown in these figures it would appear that the spot occupies less than a boundary-layer thickness in both the z- and y-directions. Presumably at an earlier time it was still smaller. The breakdown appears to begin about 0.6 δ from the wall.

There was no evidence in the records that could be interpreted as indicating flow separation in connection with breaking phenomenon or the processes leading up to it. The one-sidedness of the fluctuations which amounts to brief excursions to lower velocities does not appear to involve separation. It is not impossible, however, that intermittent separation occurred without its presence being detected. If it did occur, it does not seem that it could play any decisive role in the breaking process. The characteristics of breaking are not seen at the wall, where separation would be found, but rather in the central region of the layer where the mean velocity is large, say 0.8 U_1.

Coincident with the beginning of breaking, the mean velocity profile began to change in the streets containing the breaking. This was the characteristic merging into a turbulent profile, involving velocity increases in the inner half of the boundary layer and decreases in the outer half. In the streets containing the valleys the mean velocity also changed accompanying the breaking in neighboring streets, but now the initial effect was to depress the profile in the middle region of the layer. At a later stage the velocity distribution in the valley changed in the direction of a turbulent profile.

7. Discussion and comparison with work of others

When we survey the foregoing results, we see that our instrumental methods give us only bits of information, and this only at discrete points in the field, from which we must infer the motions. Having only a partial view, we cannot fully understand what we have seen, nor can we go far toward finding the causes. Obviously more detailed probing and exploitation of other techniques will eventually reveal facts now missing.

In attempting to assess where we now stand probably the first question that comes to mind is whether the present results are truly indicative of a universal transition mechanism. Obviously there are many situations, and we would not expect the same succession of events, for example, in highly disturbed flow and in cases where transition occurs following laminar separation caused by an adverse pressure gradient. Excluding such cases, we may still question whether the flow environment plays a conditioning role such that in some cases the disturbance intensity will concentrate into streets, as it did here, and in others will remain distributed. To be basically the same the streets in question need not be stationary as they were under present conditions.

We can throw light on this question by examining the work of other investigators. The work of FALES; HAMA, LONG, and HEGARTY; and WESKE [8, 9, 10] all deal with the processes leading up to and resulting in transition on a flat plate and rely on visual methods using dye in water. Such works give us the opportunity to see the phenomena in a different perspective. The work of LEITE [11] pertains to fully-developed flow in a pipe using air and employs hot-wire techniques to measure and trace the evolution of an artificially created disturbance to the onset of turbulence.

FALES was the first to show phenomena of interest to us here. By introducing a layer of dye on the surface of a plate and towing it through still water he showed that vortex formations were made visible and that their behavior or their effects could be followed. He showed, among other things, that an initially uniform bed of dye was gathered into lines after passing over a trip wire, evidently being concentrated by the vortices shed from the trip, and that these lines subsequently warped into a sawtooth formation. Following this, diffusive motions set in. Similar effects were demonstrated in the absence of a trip by decelerating the plate.

HAMA, LONG, and HEGARTY, using a similar dye technique, again towing a flat plate through still water, found these phenomena behind a trip wire and behind a step and studied them in considerable detail in the light of stability theory. They again found that the lines of dye developed very pronounced, and regularly spaced, loop formation which stretched out in the line of flow as they advanced downstream. The loops followed one another in succession. The loops did not develop if the Reynolds number and relative height of the trip, k/δ^*, were too small; but the pattern was well developed when, for example, $R_{\delta*} = 513$ and $k/\delta^* = 0.69$. Near the forward part of each loop, but not at the very tip, bursts of turbulence occurred. A succession of turbulent spots therefore developed in streets as described in earlier sections of this paper. It is highly significant that the pattern of transition should repeat itself under the different environments represented by these experiments. HAMA, LONG, and HEGARTY believe that imperfections in the trip or its height produce variations in vorticity which propagate transversely with a wave length determined by the mean-flow field.

In WESKE's experiments [10] the water was flowing over a stationary plate. In one set of experiments two-dimensional waves were introduced by means of an oscillating cylinder at the water surface. At a distance of two or three wave lengths from the cylinder, three-dimensional patterns were observed along with a cross-stream thickening and thinning of the boundary layer. At the ridges formed by the thickening, looped vortices developed having the characteristics of the horse-shoe vortex proposed by THEODORSEN [12]. WESKE observed local separation under these loops when the excitation or the amplification was sufficiently high. Turbulence originated within the loop. Here then is another experiment in which three-dimensional motions with a characteristic cross-stream pattern precede transition.

Among the various investigations there are certain differences. For example, WESKE observes separation, whereas in the present experiment no evidence of separation was found. In the HAMA experiment separation was only associated with the tripping action and played no decisive role in transition. Whether or not separation enters into the picture may depend on the strength which the disturbance can assume before laminar breakdown occurs. In none of the dye experiments have we much knowledge of the strength of the disturbance. Estimates show that in the HAMA experiments the disturbance introduced by the trip was larger than that introduced by the vibrating ribbon in the present investigation. The occurrence of separation in WESKE's experiment would suggest an even larger input there.

A rather puzzling difference between the dye results and the hot-wire results is the warping of lines of dye into loops as opposed to the straightness of wave fronts indicated by the hot wire. If the dye marks the center of the individual vortices introduced by a trip wire, then there is indeed a fundamental difference, arising either from the method of observing or from the phenomenon itself. In the present investigation, some hot-wire observations were made behind a trip wire, but these were not sufficiently systematic to shed light on this question.

By way of speculation we might assume that this difference is not one of method of observation, but an effect associated with the kind of disturbances introduced. The vibrating ribbon may be regarded as creating a downstream propagating wave in the vorticity already present in the layer. A trip wire, which has the immediate effect of collecting dye into lines, must be dropping off discrete vortices in its wake. It would seem that a vortex would be convected by the mean flow at rates dependent on position in the gradient field and therefore would readily become warped in shape. On the other hand, we would suppose that a wave would not be so affected unless it were sufficiently strong to concentrate the vorticity into discrete eddies. Since shape distortion was not apparent, it is assumed that transition effects intervened before the line vortex stage was reached. In short, the present investigation evidently applies to a range in which the Reynolds number are higher and the perturbations are weaker than those pertaining to the dye experiments.

We turn next to the work of LEITE [11], and here we find that the character of the wave motion and the onset of transition in a pipe are unlike those on a flat plate. Small disturbances, intended to be symmetrical, were introduced by an axially-oscillating sleeve which formed a short section of the inner wall of the pipe. These decayed with distance downstream in support of the theoretical deduction that axially-symmetric perturbations are stable in fully-developed laminar pipe flow. There was, however, little resemblance between these waves and TOLLMIEN-SCHLICHTING waves. In order to introduce symmetrical disturbances large enough to produce transition at some downstream station, a ring airfoil was centrally mounted within the sleeve. By observing the resulting wave at a station 47 diameters downstream at a pipe Reynolds number of 4000 as the airfoil was oscillated at 25 c.p.s. with gradually increasing amplitude, the onset of turbulent flow was found to take place by a process of gradual wave distortion involving increasing irregularity and the progressive entrance of higher frequency components. LEITE emphasized the fact that the onset of turbulence in fully-developed pipe flow is not an abrupt phenomenon. This is an interesting case, for it suggests that there may be no opportunity for the primary wave energy to concentrate and produce the local, "explosive" type of breakdown found on a flat plate.

We see that there are a number of questions that have not been completely resolved. In addition there are other cases and circumstances under which transition occurs and about which more needs to be known. The kind of researches herein described are believed to be contributing to our fund of knowledge regarding the mechanism of transition. Perhaps these will lead the way to similar researches at supersonic speeds where there is a present and growing need for basic information.

References

[1] DRYDEN, H. L.: Transition from Laminar to Turbulent Flow at Subsonic and Supersonic Speeds. Proceedings, Conference on High-Speed Aeronautics, Polytechnic Institute of Brooklyn, Jan. 20–22, 1955, pp. 41–74.

[2] SCHUBAUER, G. B., and P. S. KLEBANOFF: Contributions on the Mechanics of Boundary-Layer Transition. Proceedings, Symposium on Boundary Layer Effects in Aerodynamics, National Physical Laboratory, England, March 31 to April 2, 1955. Also NACA Rep. 1289, 1956 (formerly NACA TN 3489,1955).

[3] SCHUBAUER, G. B., and H. K. SKRAMSTAD: Laminar Boundary-Layer Oscillations and Transition on a Flat Plate. NACA Rep. 909, 1948.

[4] MITCHNER, MORTON: Propagation of Turbulence from an Instantaneous Point Source. Readers Forum, J. Aero. Sci., 21, 350–351 (1954), No. 5.

[5] EMMONS, H. W.: The Laminar-Turbulent Transition in a Boundary Layer—Part I. J. Aero. Sci., **18**, 490–498 (1951), No. 7.

[6] LAUFER, J., and TH. VERBALOVICH: Experiments on the Instability of a Supersonic Boundary Layer. External Publ. No. 350, Jet Propulsion Laboratory, California Institute of Technology, August 27, 1956.

[7] GÖRTLER, H.: Dreidimensionale Instabilität der ebenen Staupunktströmung gegenüber wirbelartigen Störungen. 50 Jahre Grenzschichtforschung (ED. H. GÖRTLER and W. TOLLMIEN). Braunschweig: Vieweg (1955), pp. 304–314.

[8] FALES, E. N.: A New Laboratory Technique for Investigation of the Origin of Fluid Turbulence. J. Franklin Institute, Vol. 259, No. 6, June 1955, pp. 491–515.

[9] HAMA, F. R., J. D. LONG and J. C. HEGARTY: On Transition from Laminar to Turbulent Flow. J. Appl. Phys., **28**, 388–394 (1957), No. 4.

[10] WESKE, J. R.: Experimental Study of Detail Phenomena of Transition in Boundary Layers. Tech. Note BN-91, Feb. 1957, The Institute for Fluid Dynamics and Applied Mathematics, University of Maryland.

[11] LEITE, R. J.: An Experimental Investigation of the Stability of Axially Symmetric Poiseuille Flow. Report IP-188, Nov. 1956, The University of Michigan.

[12] THEODORSEN, TH.: The Structure of Turbulence. 50 Jahre Grenzschichtforschung (ED. H. GÖRTLER and W. TOLLMIEN). Braunschweig: Vieweg (1955), pp. 55–62.

Aus der Diskussion

H. GÖRTLER (Freiburg i. Br.): The data you have mentioned for $\Delta y/\delta^*$ are of much interest in connection with our theoretical results to be presented in the following lecture. If the warping effect you have observed has something to do with the "secondary instability" with which our paper is concerned, then our result giving $10^{-4} \delta^*$ as order of magnitude of the TOLLMIEN-SCHLICHTING wave amplitude leading to secondary instability in the valleys of the waves should be compared with your observed values of $\Delta y/\delta^*$.

G. B. SCHUBAUER (Washington, D.C.): Yes, I had noticed the figure $10^{-4} \delta^*$ in the abstract of your paper. That is why I mentioned our results for $\Delta y/\delta^*$. These values are so much larger by a factor of 50 or 100 or even more.

H. GÖRTLER (Freiburg i. Br.): The result $10^{-4} \delta^*$, in itself only a rough statement of order of magnitude, gives the minimum amplitude of TOLLMIEN-SCHLICHTING waves of certain wave lengths numerically investigated for which, for the first time, *neutral* longitudinal vortices (with a distinct wave length) become possible, and are possible only in the immediate vicinity of the valleys of the waves. It depends much on the type of initial disturbances in the real fluid if this pattern of vortices will appear or others that are neutral only for higher values of the wave amplitudes. Furthermore, and this is more important, it does not seem probable that the secondary vortices generating from small disturbances will be observable in their neutral stage. It is more probable that, normally, a considerable amplification will be necessary before the vortices, or any effect of these, will be observable. Therefore, it would seem quite in agreement with our prediction for neutral vortices if longitudinal vortices in the real fluid were first observed where the amplitudes of TOLLMIEN-SCHLICHTING waves are much larger than $10^{-4} \delta^*$.

H. Witting (Freiburg i. Br.): Die kritische Wellenamplitude für die Sekundärinstabilität wurde ermittelt unter Zugrundelegung der verfügbaren Daten über neutrale Tollmien-Schlichtingsche Wellen bei der Plattenströmung (H. Schlichting, Nachr. Ges. Wiss., Göttingen, 1935). Sie hängt ab von der Reynoldsschen Zahl, für die die Welle der betrachteten Wellenlänge gerade neutral ist. Der entscheidende Eigenwert der Theorie, über die ich noch vortragen werde, ist die mit Re^2 multiplizierte Wellenamplitude ε. Der Wert $\varepsilon = 10^{-4}\, \delta^*$ bezieht sich auf dasjenige der beiden von Schlichting ausgewerteten Beispiele neutraler Wellen mit der kleineren Re-Zahl. Für das andere, zu einem größeren Re-Wert gehörige Beispiel fällt die kritische Amplitude ε noch kleiner aus. Umgekehrt sind für die bei kleineren Re-Werten noch möglichen neutralen Wellen, für die mir numerische Auswertungen noch nicht zur Verfügung stehen, größere kritische Amplituden ε als $10^{-4}\delta^*$ zu erwarten, die den Daten von Herrn Dr. Schubauer entgegenkommen.

H. Görtler (Freiburg i. Br.): Aus den Bemerkungen von Dr. Witting möchte ich noch unterstreichen, daß er seinen numerischen Auswertungen *neutrale* Tollmien-Schlichtingsche Wellen zugrunde legte. In der wirklichen Strömung werden sich diese Wellen bereits im Stadium der Anfachung befinden, bis auch die in ihren Tälern sich ausbildenden Längswirbel genügend angefacht sind, daß sie beobachtbar werden.

F. X. Wortmann (Stuttgart): Dr. Schubauer bringt den von ihm beobachteten wellenartigen Verlauf der Störgeschwindigkeiten in Spannweitenrichtung der ebenen Platte in Verbindung mit den Sieben in der Beruhigungskammer. Auf Grund eigener Beobachtungen mit der Tellurmethode in den Jahren 1952/1953 möchte ich auf einen Effekt hinweisen, der möglicherweise auch hier als Ursache der wellenförmigen Verteilung der Störgeschwindigkeiten in Frage kommt. Es handelt sich dabei um den sogenannten „Moiré-Effekt". Bei Windkanalsieben kann dieser Effekt dadurch entstehen, daß das Nachlaufraster eines Siebes beim Durchströmen des nachfolgenden Siebes noch nicht vollständig abgeklungen ist. Haben beide Siebe eine annähernd gleiche Maschenweite, so kommt es durch Überlagerung der beiden Nachlaufraster zu ausgeprägten Interferenzerscheinungen, d.h., der Betrag der Strömungsgeschwindigkeit wird wellenartig moduliert. Bei großer Wellenlänge und laminarer Strömung muß die wellenförmige Störung auch noch in der Meßstrecke vorhanden sein. Meines Wissens ist diese Störquelle bislang kaum beachtet worden. Man kann diesen Effekt vermeiden, indem man entweder sehr verschiedenartige Maschenweiten einander folgen läßt oder indem man den Fadenlauf der verschiedenen Siebe um $30°$ bis $45°$ gegeneinander kreuzt. Außerdem ist es vorteilhaft, die Reihenfolge der Siebe so zu wählen, daß die Siebe mit den niedrigsten Widerstandswerten als letzte durchströmt werden.

E. A. Eichelbrenner (Châtillons-sous-Bagneux): Über den Moiré-Effekt wurde im Dezember 1956 auf der Tagung zum 10jährigen Bestehen des NLRL in St. Louis (Ht. Rhin) ausführlich vorgetragen. Der Effekt wurde u.a. zur Sichtbarmachung schnell wechselnder Strömungsvorgänge (Durchgang von Verdichtungsstößen u.ä.) benutzt. Ausführlich ist darüber in den Comptes Rendus der Tagung berichtet.

K. Nickel (Karlsruhe): In Abb. 4 des Vortrags von Herrn Dr. Schubauer wurde der laminar-turbulente Umschlag mit der china-clay-Methode sichtbar gemacht. Häufig noch zweckmäßiger als diese Methode ist die Petroleum-Ruß-

Methode (vgl. etwa [1]). Auf der untenstehenden Abb. 1 ist der Umschlag dadurch deutlich sichtbar gemacht. Er erfolgt bei a) durch eine Ablöseblase (separation bubble), bei b) durch einen aufgeklebten Turbulenzfaden. Die bei a) vor der Linie der laminaren Ablösung liegenden hellen Striche lassen sich dort (konkave Stromlinien!) als Spuren von GÖRTLER-Wirbeln deuten.

Die Daten zu Abb. 1a und 1b sind: Profil NACA: 0015; Flügeltiefe: 110 mm; Anblasgeschwindigkeit: 48 m/s (Luft); Reynoldszahl: $3,7 \cdot 10^5$; Auftriebsbeiwert:

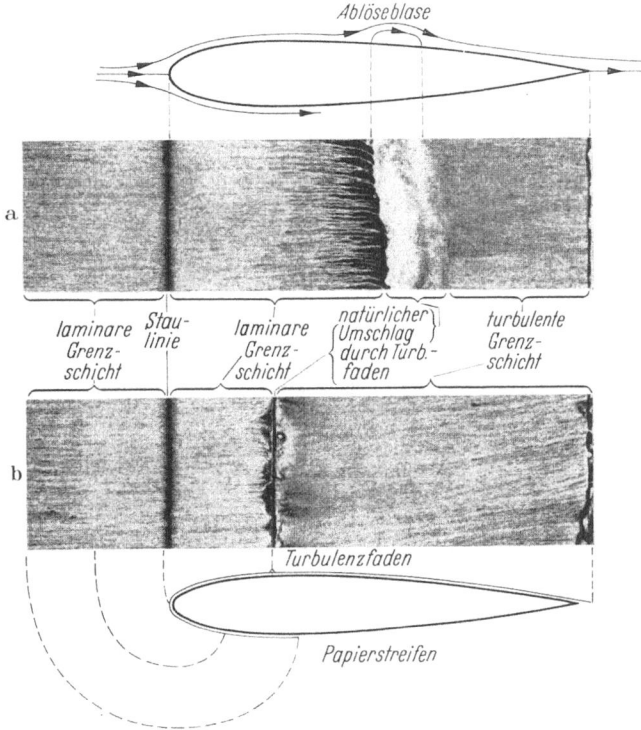

Abb. 1. Sichtbarmachung des Umschlags mit der Petroleum-Ruß-Methode

$c_a = 0$. Die Anstrichbilder der Grenzschicht auf der abgewickelten Flügeloberfläche wurden auf einem Papierstreifen gewonnen, der um den Flügel gelegt war. (Vgl. die untere Skizze von Abb. 1 und [2].)

Literatur:

[1] PRANDTL, L.: Führer durch die Strömungslehre, Braunschweig (1942), S. 131—132.
[2] GRONAU, K. H., und K. NICKEL: Experimentelle Untersuchungen an schiebenden und gepfeilten Flügeln mit Normal- und Laminar-Profilen. Deutsche Forschungsanstalt für Luftfahrt e.V., Institut für Aerodynamik, Bericht Nr. 56/14 (1956).

Theorie der sekundären Instabilität der laminaren Grenzschichten [1]

Von

H. Görtler und H. Witting

Universität Freiburg i. Br.

1. Allgemeine Grundgedanken

(Vorgetragen von H. Görtler)

Der gedankliche Ausgang der Untersuchungen, über die wir hier zu berichten beabsichtigen, ist die bekannte, zunächst theoretisch vorausgesagte [1] und dann experimentell bestätigte [2, 3, 4] Instabilität der ebenen laminaren Grenzschichtströmungen an konkaven Wänden gegenüber Störungen in Gestalt von Wirbeln, deren Längsachsen parallel zur Hauptströmungsrichtung verlaufen. Ist der Parameter $Re_\vartheta \sqrt{\vartheta/R}$ (Re_ϑ die mit der Impulsverlustdicke ϑ gebildete Reynoldssche Zahl, R der Krümmungsradius der Wand, positiv genommen an zur Strömung konkaven Wänden) größer als ein gewisser, von der Form des Geschwindigkeitsprofils praktisch unabhängiger kritischer Wert, so werden Wirbel dieser Art und geeigneter Achsenabstände in der Grenzschicht angefacht und können schließlich zum laminar-turbulenten Umschlag führen.

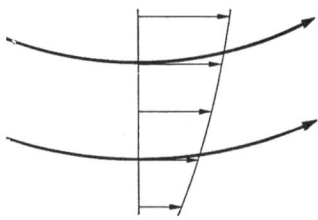

Abb. 1. Stromlinien konkav zur Normalenseite wachsender Geschwindigkeit

Schon seit 1954 habe ich in zahlreichen Vorträgen (erstmals im September 1954 an der Johns Hopkins University) die Ansicht vertreten, daß eine gleichartige Instabilität allgemein überall dort in Grenzschichtströmungen zu erwarten ist, wo die Stromlinien konkav nach der Normalenseite zunehmender Geschwindigkeiten verlaufen (Abb. 1), falls nur eine zusätzliche, der obigen Instabilitätsbedingung entsprechende und von Fall zu Fall zu ermittelnde Bedingung erfüllt ist. Es ist nicht wesentlich, ob die umströmte Wand selbst konkav ist, wesentlich ist die Konkavität (im genannten Sinne) im Stromlinienverlauf, und diese kann auch in der Nachbarschaft nichtkonkaver Wände auftreten. Mit anderen

[1] Diese Untersuchung wurde gefördert durch das Wirtschaftsministerium des Landes Baden-Württemberg.

Worten: Wesentlich ist, daß in einem Strömungsgebiet die zur Anfachung longitudinaler Wirbel führende dynamische Situation des instabilen Gleichgewichts zwischen Zentrifugalbeschleunigung und Druckgefälle normal zu den Stromlinien vorliegt, und zwar in einem Maße, das zur Überwindung der inneren Reibung ausreicht. (Vgl. auch [5, 6, 7].)

Aus der Fülle von Vorkommen der genannten dynamischen Situation bietet die ebene Staupunktströmung ein einfaches Beispiel: Diese Situation liegt in der nächsten Umgebung des Staupunkts vor. Weiter stromabwärts und bereits bevor die Stromlinien einen Wendepunkt erreichen (Abb. 2) werden die Störungen in der zähen Grenzschichtströmung im allgemeinen wieder gedämpft werden. Eine theoretische Untersuchung des idealisierten Falles der ebenen Staupunkt-

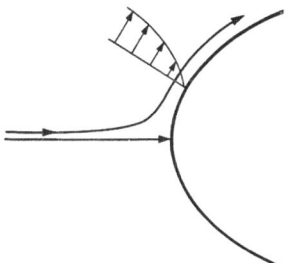

Abb. 2. Konkavität der Stromlinien in der Umgebung eines Staupunkts

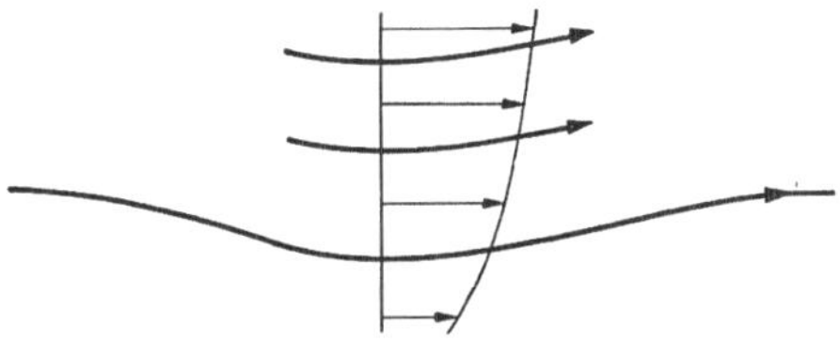

Abb. 3. Konkavität in den Tälern wellenartig verlaufender Stromlinien

strömung gegen eine unbegrenzte ebene Wand konnte in mathematischer Strenge die Möglichkeit der geschilderten Instabilität nachweisen [7, 8], aber wegen des Fehlens einer Vergleichslänge in diesem Problem kann erst die Erweiterung der Untersuchung auf die Umströmung eines Körpers mit einer endlichen Längenabmessung zu einer Aussage über einen kritischen Parameter führen.

Unter allen Möglichkeiten des Auftretens der geschilderten dynamisch instabilen Situation interessiert uns im weiteren nur die folgende Problemstellung von erheblicher praktischer Tragweite: Eine gegenüber unseren longitudinalen Wirbeln stabile Grenzschichtströmung werde wellenartig gestört (Abb. 3). Dann liegt die uns interessierende Instabilität im Bereich der Wellentäler vor, falls die Wellenamplituden eine ausreichende Konkavität der Stromlinien verursachen. Es ist zu untersuchen, wie die zusätzliche Bedingung des Überschreitens eines gewissen kritischen Parameterwerts (hier im wesentlichen durch die Wellenamplitude gekennzeichnet) lautet. Diese umfassende Fragestellung hat mein Mitarbeiter H. WITTING gründlich untersucht, und er wird anschließend über seine bemerkenswerten Ergebnisse berichten.

Die angenommene wellenartige Störung kann etwa verursacht sein durch eine Welligkeit der umströmten Wand, oder etwa durch periodische Absaugung der Grenzschicht (z. B. durch eine Folge von Schlitzen), oder aber dadurch, daß in der Grenzschicht Tollmien-Schlichtingsche Wellen vorhanden sind. Gerade der letzte Fall interessiert besonders. Die von ebenen, stromabwärts wandernden Tollmien-Schlichtingschen Wellen gestörte ebene Strömung würde dann sekundär instabil gegenüber Längswirbeln. Oberhalb der kritischen Reynoldsschen Zahl für neutrale Tollmien-Schlichtingsche Wellen würde somit die Strömung bald endgültig in eine dreidimensionale Strömung übergehen und — vermutlich auf dem Wege über Instabilitäten dritter und höherer Ordnung — schließlich zum Umschlag in die — wesentlich dreidimensionale — turbulente Strömung geführt werden. Während die kritische Reynoldssche Zahl der neutralen Wellen noch weit unterhalb der beobachteten Reynoldsschen Zahl des laminar-turbulenten Umschlags liegt, könnte vom theoretischen Standpunkte aus erwartet werden, daß die Ermittlung der sekundären Instabilität zu einer wesentlich präziseren Angabe über die Reynoldssche Zahl des Umschlags führen dürfte. Ich möchte aber hervorheben, daß selbst im unterkritischen Fall gedämpfter Tollmien-Schlichtingscher Wellen bei ausreichender Amplitude angefachte Längswirbel in den Wellentälern auftreten könnten; damit würde auch diese Strömung in eine dreidimensionale Strömung übergeführt werden, und deren weiteres Stabilitätsverhalten wäre jedenfalls grundsätzlich anders als das der gegen Tollmien-Schlichtingsche Wellen stabilen Grundströmung.

Für die mathematische Behandlung des aufgeworfenen verwickelten Problems war die Frage von Belang, ob die vermutete „sekundäre Instabilität" noch im Rahmen einer linearisierten Störungstheorie erfaßt werden kann oder erst im nichtlinearen Bereich endlicher oder gar großer Amplituden der Tollmien-Schlichtingschen oder sonstiger Wellen. Die Wittingsche Theorie ist nicht auf kleine Wellenamplituden beschränkt. Witting findet jedoch, daß bei Wellenamplituden von der Größenordnung von nur 10^{-4} Verdrängungsdicken der Grenzschicht die sekundäre Instabilität in den Wellentälern bereits auftritt. Es folgt daraus, daß die aus kleinen Störungen entstandenen Tollmien-Schlichtingschen Wellen sehr bald zur Anfachung dreidimensionaler Störungen in Form von longitudinalen Wirbeln führen müssen, das Bild der ebenen, wellenartig gestörten Strömung also nicht lange existieren kann. Das gilt analog für Strömungen längs schwach gewellten Oberflächen. Es wird auch ganz entsprechend gelten, wenn man zur Stabilisierung der Strömung gegenüber Tollmien-Schlichtingschen Wellen periodisch durch eine Folge von Schlitzen schwach absaugt. (Die experimentelle Erfahrung lehrt auch in der Tat [8], daß periodische Absaugung durch Schlitze im Vergleich zu homogener Absaugung durch poröse Wände weniger günstige Ergebnisse für die Stabilisierung der Grenzschicht liefert.)

Die oben angegebene Größenordnung der kritischen Wellenamplituden von 10^{-4} Verdrängungsdicken ist überraschend klein. Es muß bemerkt werden, daß diese das Auftreten *neutraler* Längswirbel und dieses nur in der nächsten Umgebung der Wellentäler ermöglicht, und zwar auch nur für einen bestimmten kritischen Achsenabstand der Wirbel, falls diese Störungskomponente in der in die Strömung hereingetragenen künstlich erzeugten oder zufallsartig bedingten Störung enthalten ist. Experimentell feststellbare angefachte Wirbel sind demnach erst bei einer wesentlich größeren Wellenamplitude zu erwarten.

Es scheint mir, daß hier auf der Seite der Experimentatoren einige Mißverständnisse aufgetreten sind. Die Störungstheorien bieten auf den ersten Blick ein arg simplifiziertes Bild. Wie bereits in der Theorie der TOLLMIEN-SCHLICHTINGschen Wellen und in der Theorie der Längswirbel an konkaven Wänden, so sind auch hier die untersuchten (primären und sekundären) Störungen nur „Partialstörungen" im FOURIERschen Sinne. Erst ihre lineare Überlagerung im Sinne einer FOURIER-Synthese beschreibt die Allgemeinheit der Störungen, die in dieser Theorie eigentlich betrachtet wird. So wird auch das, was man in der realen Strömung beobachtet, von Fall zu Fall und je nach der Art der Verursachung kleiner Störungen zunächst verschieden aussehen. Jedoch führt diese Analyse der Störungen und die Ermittlung der zuerst instabil und sekundär instabil werdenden Partialstörungen (Anteilen der allgemeinen realen Störung) zu den interessierenden kritischen Parametern der Instabilitäten.

Es kann im übrigen die Situation vorliegen, daß eine Strömung zunächst primär instabil wird gegenüber Längswirbeln und daß diese Wirbel dann zu einer sekundären wellenartigen Instabilität führen, die periodisch in Richtung der Längsachsen dieser Wirbel verläuft. Experimentell kennt man das bei der bekannten Instabilität der TAYLORschen Wirbel in der Strömung zwischen koaxialen rotierenden Zylindern: Ich verweise auf die schönen Aufnahmen von SCHULTZ-GRUNOW [10], aus denen man ersieht, wie diese wellenartige Verformung der TAYLOR-Wirbel als sekundäre Instabilität auftritt, und wie dann im einzelnen der Umschlag in turbulente Strömung graduell einsetzt. Bemerkenswert ist auch, daß nach vollem Turbulentwerden die TAYLOR-Wirbel nach wie vor — nur jetzt als turbulente Wirbel — deutlich sichtbar erhalten bleiben. Das wird entsprechend auch für die longitudinalen Wirbel an konkaven oder welligen Wänden gelten.

Wenn an Stelle der durch Stromlinienkrümmung bedingten Zentrifugalkräfte andere Massenkräfte mit gleichartiger Wirkung auftreten, wird man auch in diesen Strömungen eine Instabilität gegenüber Längswirbeln finden. Für den Fall des thermischen Auftriebs durch Erhitzung der Wand (oder Kühlung der Oberfläche einer Wasserschicht) ist dies längst bekannt, worauf mich Herr M. Lunc in diesen Tagen erst aufmerksam gemacht hat. P. Idrac (1921), T. Terada (1928) und später D. Avsec und M. Lunc haben schöne Aufnahmen der Längswirbel veröffentlicht. Da man damals für diese Wirbel keine Theorie entwickelte, sondern sich auf qualitative Beobachtung beschränkte, komme ich in Kürze an anderer Stelle hierauf zurück. Es besteht weitgehendes dynamisches Entsprechen zu den Wirbeln an konkaven Wänden, und es ist zu hoffen, daß man auch für letztere bald ebenso instruktive Aufnahmen haben wird, sind sie doch bisher erst indirekt durch Strichspuren in einem Wandanstrich sichtbar gemacht worden [4]. Unter den von D. Avsec 1937 publizierten Aufnahmen findet sich eine ausgezeichnete Demonstration dessen, was ich oben nach unserer heutigen Kenntnis als sekundäre (wellenartige) Instabilität der Längswirbel bezeichnet habe.

Ich möchte mit einem Hinweis schließen. Es erscheint mir im gegenwärtigen fortgeschrittenen Entwicklungsstadium der Stabilitätstheorie wichtig, daß wir unser Augenmerk auf die Rolle von Instationaritäten der Grundströmung im Mechanismus der Turbulenzentstehung richten. Wenn z.B. eine Strömung einen instationären Übergang von einem stabilen Zustand in einen anderen stabilen Zustand vollzieht, kann die Strömung im Übergangsstadium instabil werden, und es ist eine offene Frage, ob sie den gedachten stationären Endzustand laminar erreicht. Einfache Beispiele ruckartiger Übergänge von einem Bewegungszustand zu einem anderen werden in Freiburg zur Zeit untersucht. Auch der Stabilität einer zeitlich periodisch schwankenden Grundströmung wenden wir unser Interesse zu. Die Verwandtschaft mit den oben geschilderten Betrachtungen über sekundäre Instabilitäten einer primär gestörten stationären Grundströmung liegt auf der Hand. Es handelt sich nur um einen Angriff von einer anderen Seite auf das allgemeine Problem der Auswirkung der in der realen Strömung zufallsartig oder systematisch in eine Grenzschicht hineingetragenen Störungen.

2. Einzelausführungen

(Vorgetragen von H. Witting)

Im zweiten Teil dieses Berichts möchte ich Ihnen einen kurzen Einblick in die bisherigen theoretischen Untersuchungen über die sekundäre Instabilität geben. Deren Hauptergebnis hat H. Görtler bereits mitgeteilt: Eine durch Tollmien-Schlichting-Wellen gestörte laminare Grenzschichtströmung ist gegenüber sekundären Längswirbelstörungen instabil, wenn die Wellenamplitude die Größenordnung von 10^{-4} Verdrängungsdicken der Grenzschicht hat. Diese Aussage wurde für den Fall der Blasiusschen Plattengrenzschicht unter Zugrundelegung zeitlich neutraler Wellenstörungen ermittelt; sie dürfte jedoch in ähnlicher Weise auch für andere Grenzschichtströmungen sowie für zeitlich angefachte

)der gedämpfte Störungen gültig sein. Die numerischen Rechnungen ;eigen überdies, daß die kritische Wellenamplitude — wenn auch in ;chwächerem Maße — noch von der Reynoldsschen Zahl der Grund-;trömung sowie der Wellenlänge der jeweils betrachteten (partikulären) Wellenstörung abhängt.

In Anbetracht des umfangreichen Formelapparates — man hat die Aufeinanderfolge von Wellen- und Wirbelstörungen zu betrachten — muß .ch mich auf die Herausarbeitung der methodisch wesentlichen Grund-;edanken beschränken. So möchte ich insbesondere zunächst auf den Störungsansatz und die Formulierung des Eigenwertproblems etwas näher eingehen, die die Fragestellung und Vorgehensweise am besten erkennen lassen. Im Anschluß daran soll in großen Zügen der Lösungsweg ;kizziert werden. Im übrigen sei auf eine in Kürze im Archive for Rational Mechanics and Analysis erscheinende Arbeit verwiesen, in der die Theorie ler sekundären Wirbelstörungen ausführlich abgehandelt wird und in der ler Leser weitere Einzelheiten wie auch eine vollständige Wiedergabe aller benötigten Formeln findet.

2.1. Lösungsmethode. Zum Nachweis der Instabilität einer primär wellenförmig gestörten ebenen laminaren Grenzschichtströmung setzen wir in den Tälern dieser Wellen wirbelartige Sekundärstörungen an. Unser methodischer Grundgedanke ist der, die primären (wellenförmigen) Störungen als gegeben anzusehen — dieses ist bis auf einen Faktor ε der Wellenamplitude möglich — und durch Einführung neuer, auf die wellen-förmig gestörte Grundströmung bezogener Koordinaten zu berücksich-tigen. Zu behandeln ist dann nur noch der Ansatz der wirbelförmigen Sekundärstörungen; für deren Amplitudenfunktionen liefern die Navier-Stokesschen Gleichungen und die Kontinuitätsgleichung ein Eigenwert-problem, als dessen Eigenwert sich der Faktor ε der Wellenamplitude erweist.

Wir benutzen deshalb statt y (wie üblich normal zur Wand) eine neue Koordinate η, deren Koordinatenlinien die wellenförmigen Stromlinien sind. Speziell soll dabei noch η im Grenzfall verschwindender Wellen-amplitude in die alte Koordinate y übergehen. Zerlegt man überdies den Geschwindigkeitsvektor in Richtung parallel und senkrecht zu diesen wellenförmigen Stromlinien, so wird es besonders deutlich, daß es die Stromlinienkrümmung ist, die die Wirbelentstehung bewirkt.

Andererseits ist es bekanntlich so, daß in der Grenzschicht die Wellen-störungen in ihrem Phasenzustand gegeneinander verschoben sind, d.h. daß z.B. die Täler nicht auf einer Geraden $x = $ const normal zur um-strömten Wand liegen, sondern auf einer besonders in Wandnähe von dieser Geraden abweichenden Kurve $\xi(x, y) = $ const. Da wir nun die sekundären Wirbelstörungen in den Wellentälern vermuten und deshalb hier den Störungsansatz untersuchen, ist es zweckmäßig, statt x diese

Größe ξ als Koordinate zu benutzen. Speziell soll dabei noch $\xi = 0$ ein Wellental charakterisieren.

Übrigens zeigt sich im späteren Verlauf der Rechnungen, daß der Einfluß dieser Phasenverschiebung auf die Wirbelentstehung außerordentlich klein ist, so daß sich dieser als Störungskorrektur des nicht-phasenverschobenen Problems in numerisch sehr einfacher Weise berechnen läßt.

2.2. Vereinfachende Annahmen über die Grundströmung. Um für die Amplitudenfunktionen des Störungsansatzes ein Eigenwertproblem gewöhnlicher Differentialgleichungen zu erhalten, hat man über die Grundströmung geeignete Annahmen zu machen.

In der Tollmien-Schlichtingschen Theorie wellenförmiger Störungen ist es bekanntlich so, daß man die Grundströmung als lokal unabhängig von der Koordinate x in Hauptströmungsrichtung ansehen muß. Nur dann werden die Koeffizienten der linearisierten Störungsgleichungen von x unabhängig, d.h., nur dann kann die Stromfunktion der Wellenstörung angesetzt werden in der Form

$$\psi_{\bar{x}} = \Re\left\{\varphi(y)\, e^{i(\alpha x - \beta t)}\right\}.$$

In gleicher Weise muß auch H. Görtler bei seiner Theorie der Längswirbel an einer konkaven Wand lokal-konstante Wandkrümmung sowie eine Grundströmung annehmen, deren Geschwindigkeitsprofil in der Umgebung der betrachteten Stelle von der Wandbogenlänge x unabhängig ist. Nur dann werden die Koeffizienten der linearisierten Störungsgleichungen von x unabhängig, so daß die Geschwindigkeitskomponenten der wirbelartigen Störungen in der Form

$$\hat{a} = a_1(\eta)\cos\sigma\zeta\, e^{\gamma t}$$
$$\hat{b} = b_1(\eta)\cos\sigma\zeta\, e^{\gamma t}$$
$$\hat{c} = c_1(\eta)\sin\sigma\zeta\, e^{\gamma t} \tag{1}$$
$$\hat{p} = p_1(\eta)\cos\sigma\zeta\, e^{\gamma t}$$

(ζ in Richtung der Erzeugenden) angesetzt werden können, wobei $a_1(\eta)$, ... aus einem Eigenwertproblem gewöhnlicher Differentialgleichungen zu bestimmen sind.

Das sind natürlich Einschränkungen, auch wenn wir z.B. im Fall der Görtler-Wirbel die konkave Wand an einer Stelle lokal-konstanter Krümmung betrachten; denn in der Umgebung dieser Stelle wird sich die Krümmung der Stromlinien sicher in zweiter Ordnung mit x ändern. Diese Krümmungsänderung aber muß vernachlässigt werden, um auf ein (einfaches) System gewöhnlicher Differentialgleichungen zu kommen.

Die gleiche Situation liegt auch bei unserem Problem der Sekundärwirbel vor. Wir werden deshalb eine erste Lösung unter Vernachlässigung der Änderung der Stromlinienkrümmung bei Zugrundelegung des Störungsansatzes (1) angeben. Damit erreichen wir im Falle der durch eine Wandwelligkeit hervorgerufenen Primärwellen mindestens dieselbe Strenge wie in der GÖRTLERschen Theorie der Wirbelinstabilität an einer konkaven Wand. Tatsächlich ist hier die Genauigkeit sogar noch größer, da jetzt von der Vernachlässigung der Krümmungsänderung erst an sehr viel späterer Stelle Gebrauch gemacht werden muß. Insbesondere liefert diese Theorie z. B. auch eine Abhängigkeit des kritischen Achsenabstands der Wirbelstörungen von der Wellenlänge der zugrunde liegenden Störung.

Es sei hier noch die Möglichkeit einer strengeren Behandlung der wirbelartigen Sekundärstörungen der Strömung an einer welligen Wand erwähnt, bei der die bisher vernachlässigte Krümmungsänderung berücksichtigt wird. Hierzu sind die Amplituden der Wirbelstörungen nicht nur von η, sondern auch von ξ abhängig anzusehen und somit aus einem Eigenwertproblem partieller Differentialgleichungen zu bestimmen.

Aus Gründen des Arbeitsaufwandes wird man sich dabei auf die Berücksichtigung der Krümmungsänderung in zweiter Ordnung mit ξ be schränken; in entsprechender Weise kann man dann auch die Störamplituden als Polynome zweiter Ordnung in ξ mit von η abhängigen Koeffizienten ansetzen. So geht das System der partiellen Differentialgleichungen in ein gegenüber dem früheren allerdings wesentlich umfangreicheres System gewöhnlicher Differentialgleichungen über, das sich jedoch in methodisch ähnlicher Weise lösen läßt.

Indem man die Störamplituden als von ξ abhängig ansetzt, läßt sich das Modell noch dadurch wesentlich verbessern, daß man auch die Anfachung der Störungen γ von der Koordinate ξ abhängig annimmt. So läßt sich berücksichtigen, daß die Wirbel um so mehr angefacht werden, je näher sie dem Wellental sind. Da keine elektronische Rechenanlage zur Verfügung stand, konnten die Rechnungen zu dieser zweiten Lösung noch nicht systematisch durchgeführt werden. Proberechnungen bestätigten jedoch die aus der ersten Behandlungsmethode folgende Größenordnung der kritischen Wellenamplitude von 10^{-4} Grenzschichtdicken.

Im Falle einer primär durch TOLLMIEN-SCHLICHTING-Wellen gestörten Strömung wird die Untersuchung der Sekundärwirbel dadurch wesentlich schwieriger, daß hier in der Nähe der kritischen Schicht geschlossene sowie sich schneidende Stromlinien und dadurch bedingt Singularitäten im Eigenwertproblem auftreten. Im Grenzfall verschwindender Wellenamplituden ($\varepsilon = 0$) ergibt sich jedoch ein reguläres Eigenwertproblem und zwar genau dasjenige, das sich auch an der welligen Wand für $\varepsilon \to 0$ ergibt.

2.3. Stromfunktion der Wellenstörung. Da die wellenförmig gestörte Strömung die Grundströmung für die sekundären Wirbelstörungen ist und somit insbesondere die neu einzuführenden Koordinaten festlegt, wollen wir hier den analytischen Ausdruck ihrer Stromfunktion im einzelnen angeben. Zunächst muß, wie bereits erwähnt, in der TOLLMIENschen Theorie wellenförmiger Störungen die Geschwindigkeitsverteilung der ungestörten Grundströmung in der speziellen Form $u = u_0 (y)$, $v = 0$, $w = 0$ angenommen werden; ihre Stromfunktion lautet also

$$\psi_0(y) = \int\limits_0^y u_0(y)\, dy. \tag{2}$$

Bei der außerordentlich kleinen kritischen Wellenamplitude kann die primäre Wellenstörung einer linearisierten Theorie entnommen werden. Wir brauchen daher nur Partialstörungen zu betrachten und können die Stromfunktion der Tollmien-Schlichting-Wellen ansetzen in der Form

$$\begin{aligned}
\varepsilon\psi_{\mathfrak{X}}(x,y,t) &= \varepsilon\Re\left\{\varphi(y)\, e^{i(\alpha x - \beta t)}\right\} \\
&= \varepsilon\left[\varphi_r(y)\cos(\alpha x - \beta t) - \varphi_i(y)\sin(\alpha x - \beta t)\right]
\end{aligned} \tag{3}$$

α, β reell, wobei $\varphi(y) = \varphi_r(y) + i\,\varphi_i(y)$ Lösung der Sommerfeld-Orrschen Differentialgleichung ist. Als Lösung einer linearen homogenen Differentialgleichung unter homogenen Randbedingungen ist $\varphi(y)$ nur bis auf einen Faktor bestimmt; wir denken uns $\varphi(y)$ in einer beliebigen, aber festen Normierung, setzen dafür aber in (3) die Stromfunktion der Wellenstörung in der Form $\varepsilon\psi_{\mathfrak{X}}$ mit einer zunächst nicht näher festgelegten Größe ε an. Gerade diese Größe wird sich später als Eigenwert erweisen.

In formal gleicher Weise (3) läßt sich auch die Stromfunktion einer Wellenstörung ansetzen, die durch eine Wandwelle der Amplitude ε bedingt ist. Es ist dann $\beta = 0$ und $\varphi(y)$ die Lösung eines inhomogenen, linearen Randwertproblems.

Da ε für das Auftreten neutraler Sekundärwirbel nur sehr kleiner Werte bedarf, ist der linearisierte Ansatz (3) nachträglich gerechtfertigt. Ebenso können alle während der weiteren Rechnungen auftretenden Glieder höherer Ordnung in ε vernachlässigt werden; ihr Einfluß kann aber auch leicht mit Hilfe der Störungsrechnung abgeschätzt werden oder (dieses jedoch bei wachsendem Aufwand) bis zu einer beliebig hohen Ordnung berücksichtigt werden.

Im Falle Tollmien-Schlichtingscher Primärwellen ($\beta > 0$) haben wir dagegen, abgesehen vom Grenzfall $\varepsilon = 0$, in der Nähe der kritischen Schicht eine Singularität, so daß die Vernachlässigung der Glieder höherer Ordnung in ε nicht ohne weiteres gerechtfertigt ist. Da jedoch in der Nähe der kritischen Schicht die Glieder höherer Ordnung in ε beim Ansatz der Primärstörungen wie auch die Abhängigkeit der Sekundärstörungen von der Koordinate in Strömungsrichtung wesentlich sein können, haben wir im Rahmen unserer lokalen Theorie erster Ordnung bisher auf eine genauere Untersuchung der Singularität an der kritischen Schicht verzichtet und uns auf die Betrachtung des Grenzfalles $\varepsilon = 0$ beschränkt.

Wir betrachten nun zunächst den etwas einfacheren Fall der Sekundärwirbel an einer welligen Wand, der sich als formal weitgehend analog zum alten Problem der GÖRTLER-Wirbel an einer konkaven Wand konstanter Krümmung erweist, und der uns für die Lösung des Eigenwertproblems der Sekundärinstabilität wichtige Hinweise gibt.

2.4. Einführung neuer Koordinaten. Die Stromfunktion der Grundströmung für die sekundären Wirbelstörungen lautet nach (2) und (3) (bis auf Glieder höherer Ordnung in ε) bei Verwendung der Abkürzungen

$$q(y) = \sqrt{\varphi_r^2(y) + \varphi_i^2(y)} \,, \qquad j(y) = \operatorname{arctg} \frac{\varphi_i(y)}{\varphi_r(y)}$$

$$\psi(x, y, \varepsilon) = \psi_0(y) + \varepsilon\, q(y) \cos(\alpha x + j(y)) \qquad\qquad (4)$$

$$= \psi_0\!\left(y + \varepsilon \frac{q(y)}{u_0(y)} \cos(\alpha x + j(y))\right).$$

Damit erfüllen gerade die Größen

$$\xi = x + \frac{1}{\alpha}\, j(y)$$

$$\eta = y + \varepsilon\, Q(y) \cos(\alpha x + j(y)), \qquad\qquad Q(y) = \frac{q(y)}{u_0(y)} \qquad (5)$$

$$\zeta = z$$

$$\tau = t$$

die früher an die neuen Koordinaten gestellten Forderungen. Insbesondere stellen (vgl. Abb. 4) $\xi = \mathrm{const}$ die Linien gleichen Phasenzu-

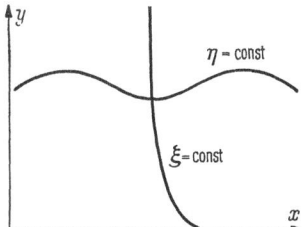

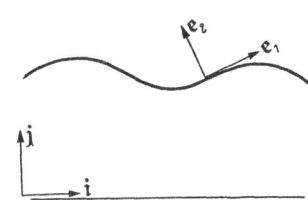

Abb. 4. Linien gleichen Phasenzustands und Stromlinien als Koordinatenlinien $\xi(x, y) = \mathrm{const}$, $\eta(x, y) = \mathrm{const}$.

Abb. 5. Vektorzerlegung bezogen auf die Wellenstörung

stands und $\eta = \mathrm{const}$ die Stromlinien dar. Überdies ist $\varepsilon\, Q(y)$ gerade die Amplitude der Wellenstörung.

2.5. Vektorzerlegung in einem auf die Wellenstörung bezogenen Dreibein. Ebenso zweckmäßig ist es, den Geschwindigkeitsvektor bzw. die Navier-Stokesschen Gleichungen bezüglich eines auf die wellenförmig gestörte Grundströmung bezogenen Dreibeins zu zerlegen. Demgemäß führen wir in jedem Raumpunkt ein neues orthogonales Dreibein $\mathfrak{e}_1$, $\mathfrak{e}_2$, $\mathfrak{e}_3$ ein, und zwar so, daß $\mathfrak{e}_1$ immer tangential zu den Stromlinien, d. h. in Richtung des Geschwindigkeitsvektors $\mathfrak{v}_0$ zeigt; $\mathfrak{e}_2$ wählen wir immer normal zu den Stromlinien, $\mathfrak{e}_3$ wird senkrecht zur Ebene der Grundströmung angenommen (vgl. Abb. 5). Außerdem wählen wir die Länge von $\mathfrak{e}_1$ so, daß der Geschwindigkeitsvektor $\mathfrak{v}_0$ der wellenförmig gestörten Grundströmung gerade die Größe $u_0\,(\eta)\,\mathfrak{e}_1$ hat. Da $u_{\mathfrak{J}}\,(y)\,\mathfrak{i}$ nach (2) der Geschwindigkeitsvektor der ungestörten Grundströmung ist, gilt mit der Bezeichnung $\varepsilon\mathfrak{v}_{\mathfrak{X}}$ für die Störung des Geschwindigkeitsvektors auf Grund der Wellenstörung (3)

$$\mathfrak{v}_0 = u_0\,(y)\,\mathfrak{i} + \varepsilon\,\mathfrak{v}_{\mathfrak{X}} = u_0(\eta)\,\mathfrak{e}_1\,. \tag{6}$$

Den Druck in der wellenförmig gestörten Grundströmung bezeichnen wir entsprechend mit $p_0 = p_0\,(\xi, \eta)$.

Bezüglich des so eingeführten Dreibeins läßt sich ein beliebiger Vektor zerlegen; wir bezeichnen die Komponenten mit a, b und c gemäß

$$\mathfrak{v} = a\,\mathfrak{e}_1 + b\,\mathfrak{e}_2 + c\,\mathfrak{e}_3$$

2.6. Allgemeiner Störungsansatz und Formulierung des Eigenwertproblems. Zur Untersuchung der Instabilität der wellenförmig gestörten Grundströmung gegenüber sekundären Wirbelstörungen machen wir in den Tälern der Wellen den Störungsansatz

$$\mathfrak{v} = \mathfrak{v}_0 + \hat{\mathfrak{v}}\,; \qquad\qquad p = p_0 + \hat{p}\,. \tag{7}$$

Die Komponenten der Geschwindigkeitsstörung und die Druckstörung haben dabei die Gestalt

$$\begin{aligned}
\hat{a} &= a_1\,(\xi, \eta)\cos\sigma\zeta\,e^{\gamma\,(\xi)\,\tau}\\
\hat{b} &= b_1\,(\xi, \eta)\cos\sigma\zeta\,e^{\gamma\,(\xi)\,\tau}\\
\hat{c} &= c_1\,(\xi. \eta)\sin\sigma\zeta\,e^{\gamma\,(\xi)\,\tau}\\
\hat{p} &= p_1\,(\xi, \eta)\cos\sigma\zeta\,e^{\gamma\,(\xi)\,\tau}\,.
\end{aligned} \tag{8}$$

Den Anfachungsgrad $\gamma\,(\xi)$ wählen wir im Rahmen unseres Approximationsgrades von der speziellen Form $\gamma\,(\xi) = \gamma_0 + \gamma_2\cos a\,\xi$, $\gamma_2 > 0$.

Durch den Störungsansatz (8) ist einerseits die Abhängigkeit von der Koordinate ζ in Strenge erfüllt, zum anderen der Charakter der Störungen als Längswirbeln mit Achsen in Hauptströmungsrichtung festgelegt. Unter der Annahme kleiner Störamplituden a_1, b_1, c_1 und p_1, die eine Linearisierung gestatten, ergibt sich für diese aus den Navier-Stokesschen Differentialgleichungen und der Kontinuitätsgleichung das System der Differentialgleichungen

$$\gamma\,a_1 + \frac{\partial a_1}{\partial \xi}\,u_0 + u_0'\,b_1 = -\frac{\partial p_1}{\partial \xi} + \frac{1}{Re}\left(\frac{\partial^2 a_1}{\partial \xi^2} + \frac{\partial^2 a_1}{\partial \eta^2} - \sigma^2\,a_1\right) + \varepsilon\,M_1$$

$$\gamma\,b_1 + \frac{\partial b_1}{\partial \xi}\,u_0 + \varepsilon\,2\,u_0\,\alpha^2 Q\cos\alpha\,\xi\,a_1$$

$$= -\frac{\partial p_1}{\partial \eta} + \frac{1}{Re}\left(\frac{\partial^2 b_1}{\partial \xi^2} + \frac{\partial^2 b_1}{\partial \eta^2} - \sigma^2\,b_1\right) + \varepsilon\,M_2$$

$$\gamma\,c_1 + \frac{\partial c_1}{\partial \xi}\,u_0 \qquad = \sigma\,p_1 + \frac{1}{Re}\left(\frac{\partial^2 c_1}{\partial \xi^2} + \frac{\partial^2 c_1}{\partial \eta^2} - \sigma^2\,c_1\right) + \varepsilon\,M_3$$

$$\frac{\partial a_1}{\partial \xi} + \frac{\partial b_1}{\partial \eta} + \sigma\,c_1 \qquad = \varepsilon\,M_0\,. \tag{9}$$

Dabei sind die Glieder mit dem Faktor ε in den Symbolen $\varepsilon\,M_0, \ldots, \varepsilon\,M_3$ am Schluß der Gleichungen zusammengefaßt worden; nur der dritte Term links in der zweiten Gleichung ist gesondert hingeschrieben worden, da er die anderen mit ε behafteten Glieder größenordnungsmäßig überwiegt. Es ist dies gerade der Term, durch den der Krümmungseinfluß berücksichtigt wird.

Zu den Differentialgleichungen (9) treten noch die Randbedingungen, die das Verschwinden der Störungen am Rande des Strömungsbereichs besagen

$$a_1 = b_1 = c_1 = 0 \quad \text{für} \quad \eta = 0 \quad \text{und} \quad \eta \to \infty. \tag{10}$$

Der Störungsansatz führt also auf Differentialgleichungen und Randbedingungen, die linear und homogen sind und somit ein Eigenwertproblem darstellen. Hierbei tritt offenbar die Größe ε als Eigenwert auf, denn die Reynoldssche Zahl ist durch Vorgabe einer speziellen Wellenstörung und damit der Grundströmung bereits festgelegt.

2.7. Vernachlässigung der ξ-Abhängigkeit; Störungsrechnung. Bei einer ersten Lösung des Eigenwertproblems (9), (10) können wir von der Tatsache Gebrauch machen, daß sich die Koeffizienten der Störungsgleichungen im Wellental nur schwach mit ξ ändern; entweder sind nämlich die ξ-abhängigen Koeffizienten mit ε, das sich als klein erweist, multipliziert und haben somit nur einen geringen Einfluß, oder die ξ-Abhängigkeit wird durch den Faktor $\cos \alpha\,\xi$ vermittelt, der sich im Wellental nur in zweiter Ordnung mit ξ ändert. Somit können wir bei der ersten Lösung die ξ-Abhängigkeit vernachlässigen.

Die Gln. (9), (10) lauten nach Elimination von p_1 und c_1 und der Ersetzung von $Re\,b_1$ durch b_1 für neutrale Störungen ($\gamma = 0$)

$$a_1'' - \sigma^2 a_1 = u_0' b_1 \qquad\qquad + \varepsilon\,L_1\,[a_1, b_1]$$

$$b_1^{IV} - 2\,\sigma^2\,b_1'' + \sigma^4\,b_1 = -\,\varepsilon\,\sigma^2\,2\,Re^2\,\alpha^2\,Q u_0\,a_1 + \varepsilon\,\sigma^2\,L_2\,[a_1, b_1] \tag{11}$$

$$a_1 = b_1 = b_1' = 0 \quad \text{für} \quad \eta = 0 \quad \text{und} \quad \eta \to \infty. \tag{12}$$

Dabei haben die in $\varepsilon\,L_1$ und $\varepsilon\,L_2$ zusammengefaßten Glieder nur einen sehr geringen Einfluß, so daß wir sie zunächst vernachlässigen können. Bezeichnen wir den Eigenwert des so vereinfachten Systems mit ε_0, die Eigenfunktionen mit a_{10} bzw. b_{10}, so können wir zur strengeren Lösung von (11), (12) allgemein ansetzen

$$\varepsilon = \varepsilon_0 \ + \varepsilon_1\,\varepsilon_0 \ \dotplus \ \varepsilon_2\,\varepsilon_0{}^2 + \ldots$$

$$a_1 = a_{10} + a_{11}\,\varepsilon_0 + a_{12}\,\varepsilon_0{}^2 + \ldots \tag{13}$$

$$b_1 = b_{10} + b_{11}\,\varepsilon_0 + b_{12}\,\varepsilon_0{}^2 + \ldots$$

Der Koeffizientenvergleich (einen von (13) etwas verschiedenen Störungsansatz findet man in der erwähnten ausführlicheren Darstellung) ergibt in nullter Ordnung das System

$$a_{10}'' - \sigma^2 a_{10} = u_0' b_{10}$$

$$b_{10}^{IV} - 2 \sigma^2 b_{10}'' + \sigma^4 b_{10} = -\varepsilon \sigma^2 2 Re^2 \alpha^2 Q u_0 a_{10} \tag{14}$$

$$a_{10} = b_{10} = b_{10}' = 0 \quad \text{für} \quad \eta = 0 \quad \text{und} \quad \eta \to \infty. \tag{15}$$

Der Vergleich der Potenzen erster Ordnung liefert

$$a_{11}'' - \sigma^2 a_{11} = u_0' b_{11} + L_1 [a_{10}, b_{10}]$$

$$b_{11}^{IV} - 2 \sigma^2 b_{11}'' + \sigma^4 b_{11} = -\varepsilon_0 \sigma^2 2 Re^2 \alpha^2 Q u_0 a_{11} \tag{16}$$

$$- \varepsilon_1 \sigma^2 2 Re^2 \alpha^2 Q u_0 a_{10} + \sigma^2 L_2 [a_{10}, b_{10}]$$

$$a_{11} = b_{11} = b_{11}' = 0 \quad \text{für} \quad \eta = 0 \quad \text{und} \quad \eta \to \infty. \tag{17}$$

Diese sog. Störungsrechnung gestattet nun eine vereinfachte Lösung des Eigenwertproblems (11), (12). Mit Hilfe der zu den linken Seiten von (14) unter den Randbedingungen (12) gehörenden Greenschen Funktionen $G(\eta, \hat{\eta})$ und $H(\eta, \hat{\eta})$ läßt sich nämlich das Eigenwertproblem nullter Ordnung in ein System von Integralgleichungen überführen

$$a_{10}(\eta) = -\int_0^\infty G(\eta, \hat{\eta}) u_0'(\hat{\eta}) b_{10}(\hat{\eta}) d\hat{\eta}$$

$$b_{10}(\eta) = \varepsilon_0 \int_0^\infty H(\eta, \hat{\eta}) \sigma^2 2 Re^2 \alpha^2 Q(\hat{\eta}) u_0(\hat{\eta}) a_{10}(\hat{\eta}) d\hat{\eta}. \tag{18}$$

Der kleinste Eigenwert läßt sich dann etwa nach dem Iterationsverfahren berechnen. Da sowohl $Q(\eta)$ als auch $u_0(\eta)$ und $u_0'(\eta)$ in $0 < \eta < \infty$ positiv sind, haben die Eigenfunktionen $a_{10}(\eta)$ und $b_{10}(\eta)$ nach einem Satz von R. Jentzsch aus der Theorie der Integralgleichungen dort keinen Vorzeichenwechsel, was dem Auftreten einkerniger Wirbel entspricht.

Das System (16), (17) dagegen stellt im eigentlichen Sinn kein Eigenwertproblem mehr dar; es handelt sich vielmehr um ein inhomogenes Differentialgleichungssystem (16) für die Funktionen $a_{11}(\eta)$ und $b_{11}(\eta)$, das unter den homogenen Randbedingungen (17) zu lösen ist; dieses ist aber nur für genau einen Wert des Parameters ε_1 möglich. Überführt man nämlich mit Hilfe der Greenschen Funktionen $G(\eta, \hat{\eta})$ und $H(\eta, \hat{\eta})$ auch das System (16), (17) in Integralgleichungen, so lassen sich aus diesen die Eigenfunktionen $a_{11}(\eta)$ und $b_{11}(\eta)$ eliminieren und somit eine Bestimmungsgleichung für die Größe ε_1 gewinnen. Bezeichnen $r_0(\eta)$ und $s_0(\eta)$ die Eigenfunktionen des zu (18) transponierten Systems von Integralgleichungen, so ergibt sich die relative Korrektur des Eigenwerts zu

$$\varepsilon_1 = \frac{\varepsilon_0^2}{I} \int\limits_0^\infty \int\limits_0^\infty \int\limits_0^\infty r_0\,(\eta)\,H\,(\eta,\hat\eta)\,\sigma^2\,2\,Re^2\,\alpha^2\,Q\,(\hat\eta)\,u_0\,(\hat\eta)\,G\,(\hat\eta,\tilde\eta)$$

$$L_1\,[a_{10}, b_{10}]\,d\eta\,d\hat\eta\,d\tilde\eta + \frac{\varepsilon_0}{I} \int\limits_0^\infty \int\limits_0^\infty r_0\,(\eta)\,H\,(\eta,\hat\eta)\,\sigma^2\,L_2\,[a_{10}, b_{10}]\,d\eta\,d\hat\eta \qquad (19)$$

$$\text{mit} \quad I = \int\limits_0^\infty r_0\,(\eta)\,b_{10}\,(\eta)\,d\eta\,.$$

Diese war in keinem der durchgerechneten Beispiele größer als 4⁰/₀₀, so daß auf die Berücksichtigung der quadratischen Glieder in ε verzichtet werden konnte.

Wie bereits früher erwähnt, wird der Einfluß der Phasenverschiebung erstmalig in ε_1 berücksichtigt. Da er sehr klein ist, ließe er sich ohne Anwendung der Störungsrechnung nur sehr schwer numerisch ermitteln. In gleicher Weise kann man — wie auch bei allen verwandten Problemstellungen, die auf Eigenwertaufgaben führen — den Einfluß von Störgrößen leicht berechnen.

2.8. Tollmien-Schlichting-Wellen als Primärstörungen. Wir betrachten nun den Fall einer primär gegenüber TOLLMIEN-SCHLICHTING-Wellen instabilen Strömung. Die Formel (3) zeigt, daß sich die Zustände gleicher Phase mit der Geschwindigkeit β/α in x-Richtung bewegen. Es liegt deshalb nahe, zu einem derartig bewegten System überzugehen. Bezeichnen x, y, z, t wieder die zugehörigen Koordinaten, so lautet die nun von der Zeit t unabhängige und somit die Bahnkurven beschreibende Stromfunktion

$$\psi\,(x, y, \varepsilon) = \psi_0\,(y) - \beta/\alpha\,y + \varepsilon \cdot q\,(y)\,\cos\,(\alpha x + j\,(y)). \qquad (20)$$

Man folgert aus (20), daß es in der Nähe der kritischen Schicht $y = y_c$ (definiert durch $u_0\,(y_c) = \beta/\alpha$) geschlossene Stromlinien gibt, die zu dem bekannten Katzenaugenbild führen. Außerhalb dieser Katzenaugen ergibt sich das erwartete Bild wellenförmiger Stromlinien, die oberhalb der kritischen Schicht in positiver, unterhalb in negativer x-Richtung durchströmt werden, und deren Wellentäler (beim Durchgang durch die kritische Schicht längs einer Linie konstanter Phase) in Wellenberge übergehen und umgekehrt. Da mit diesem Übergang von Wellentälern in Wellenberge ein Nulldurchgang der Geschwindigkeit (bezüglich des bewegten Systems) verbunden ist, haben wir längs der Linie $\xi = x + j\,(y)/\alpha = 0$ sowohl oberhalb als unterhalb der kritischen Schicht das für die Wirbelentstehung notwendige instabile Gleichgewicht. Wir wollen deshalb auch die Umgebung dieser Linie durchweg als „Wellental" ansprechen.

Ähnlich wie in (4) und (5) können wir jede Stromlinie durch den y-Wert kennzeichnen, den sie im Wellental annimmt; für die geschlossenen

Stromlinien im Innern der Katzenaugen ist dieser nicht reell. Bezeichnen wir ihn mit η, so gilt

$$\psi\,(x,\,y,\,\varepsilon)\,=\,\psi_0\,(\eta)\,-\,\frac{\beta}{\alpha}\,\eta\,+\,\varepsilon\,q\,(\eta) \tag{21}$$

Zweckmäßigerweise führen wir η neben den durch (5) definierten Größen ξ, ζ und τ als neue Koordinate ein.

Um die durch die Stromlinienkrümmung bedingte Zentrifugalbeschleunigung in unseren Gleichungen explizit auftreten zu lassen, führen wir wieder eine auf die Wellenstörung bezogene Zerlegung der Vektoren durch. Bezeichnen $\hat{u}_0$ und $\hat{v}_0$ die Geschwindigkeitskomponenten dieser wellenförmig gestörten Strömung im bewegten System, so definieren wir das Dreibein jetzt gemäß

$$e_1 = \frac{\hat{u}_0\,\mathfrak{i} + \hat{v}_0\,\mathfrak{j}}{\sqrt{\hat{u}_0{}^2 + \hat{v}_0{}^2}}\;;\,e_2 = \frac{-\,\hat{v}_0\,\mathfrak{i} + \hat{u}_0\,\mathfrak{j}}{\sqrt{\hat{u}_0{}^2 + \hat{v}_0{}^2}}\;;\,e_3 = \mathfrak{k}, \tag{22}$$

wobei jeweils der positive Wert der Wurzel zu nehmen ist. Diese Vektoren sind, abgesehen von den Schnittpunkten der die Katzenaugen berandenden singulären Stromlinien (in denen $\hat{u}_0 = \hat{v}_0 = 0$ ist), eindeutig erklärt. Beim Durchgang durch die Katzenaugen drehen sie sich stetig mit der Neigung der Stromlinien, wechseln also hierbei, dem Vorzeichenwechsel von $\hat{u}_0$ entsprechend, im wesentlichen ihre Orientierung.

Führt man nun den Störungsansatz (1) in die in dieser Weise umgeformten Navier-Stokesschen Gleichungen ein, so ergibt sich, wie bereits erwähnt, im Grenzfall $\varepsilon = 0$ das Eigenwertproblem (14), (15) bzw. (18), falls man in diesem den Ausdruck $Q\,(\eta)\,u_0\,(\eta)$ gemäß (5) durch $q\,(\eta)$ ersetzt. Da $q\,(\eta)$ auch hier durchweg positiv ist, haben die Funktionen $a_{10}\,(\eta)$ und $b_{10}\,(\eta)$ wieder keinen Vorzeichenwechsel. In Anbetracht des außerordentlich kleinen Zahlenwerts von ε kann deshalb angenommen werden, daß auch die Störamplituden $a_1\,(\eta)$ und $b_1\,(\eta)$ keinen Vorzeichenwechsel haben. Beim Durchgang durch die relativ dünnen Katzenaugen dreht sich also der Störamplitudenvektor $\mathfrak{v}_1 = a_1\,e_1 + b_1\,e_2$ im wesentlichen starr mit dem Dreibein e_1, e_2, e_3.

Geht man nun von dem Dreibein e_1, e_2, e_3 zurück zu den alten Basisvektoren $\mathfrak{i}$, $\mathfrak{j}$, $\mathfrak{k}$, so entspricht dem Wirbelansatz (1) bzw. (8) gemäß (22) ein formal gleicher Ansatz für die Geschwindigkeitsstörungen $\hat{u}$, $\hat{v}$, $\hat{w}$. Der Drehung des Vektors $\mathfrak{v}_1$ entspricht dann gerade ein Vorzeichenwechsel der Störamplituden $u_1\,(\eta)$ und $v_1\,(\eta)$. Unsere Untersuchungen lassen also vermuten, daß oberhalb und unterhalb der kritischen Schicht zwei getrennte, gegensinnig orientierte Wirbel auftreten, die möglicherweise noch durch kleine Wirbel im Innern der Katzenaugen ergänzt werden.

2.9. Ergebnisse. Die Kurve des kleinsten Eigenwerts $\varepsilon_0\,(\sigma)$ neutraler Störungen hat die in Abb. 6 wiedergegebene Gestalt. Sie trennt den Bereich zeitlich anwachsender Störungen von dem zeitlich gedämpfter

Störungen. Bei der Wellenamplitude ε^* sind erstmalig zeitlich neutrale Wirbelstörungen möglich; der Abstand ihrer Achsen ist $\lambda = \dfrac{2\,\pi}{\sigma^*}$, wenn σ^* die Stelle des Minimums der Kurve $\varepsilon_0\,(\sigma)$ und $\varepsilon^* = \varepsilon_0\,(\sigma^*)$ ist. Vergleicht man diesen Wert mit dem Abstand der Straßen, wie er im vorigen Vortrag von G. B. SCHUBAUER in Abb. 10 als Resultat seiner Messungen mitgeteilt wurde, so stellt man eine sehr gute Übereinstimmung fest (Abweichungen bleiben unter 10%). Die kritische Wellenamplitude ist, wie bei allen theoretischen Stabilitätsrechnungen, kleiner als die experimentell gemessene Größe.

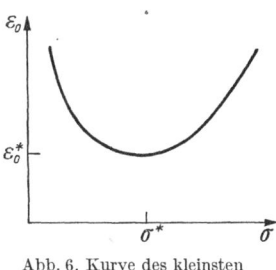

Abb. 6. Kurve des kleinsten Eigenwertes $\varepsilon_0\,(\sigma)$

Die Ergebnisse lassen vermuten, daß der kritische Wert σ^* mit wachsendem α anwächst, d. h., daß mit wachsender Wellenlänge der Primärstörung der kritische Achsenabstand der TAYLOR-GÖRTLER-Wirbel größer wird. Außerdem liegt noch eine schwache Abhängigkeit der in (18) formal als Eigenwert auftretenden Größe $\varepsilon\,Re^2$ von der Reynoldsschen Zahl vor, entsprechend der Tatsache, daß die Amplitudenverteilung $Q\,(\eta)$ von dieser abhängt.

Literatur

[1] GÖRTLER, H.: Über eine dreidimensionale Instabilität laminarer Grenzschichten an konkaven Wänden. Nachr. Ges. Wiss. Göttingen, Math.-Phys. Klasse, Neue Folge **I**, 2, 1—26 (1940).
Englische Übersetzung: On the three-dimensional instability of laminar boundary layers on concave walls. NACA Techn. Memorandum **1375** (June 1954), 32 pp.

[2] LIEPMANN, H. W.: Investigations on laminar boundary-layer stability and transition on curved boundaries. NACA Wartime Report (1943) W-107; Investigation of boundary layer transition on concave walls. NACA Wartime Report (1945) W-87.

[3] DRYDEN, H. L.: Recent advances in the mechanics of boundary layer flow. "Advances in Applied Mechanics" **1**, 1—40 (1948).

[4] GREGORY, N., and W. S. WALKER: Rep. Memor. aero. Res. Coun., London, **2779** (1950).

[5] GÖRTLER, H.: Boundary layer effects in aerodynamics. Bericht über ein Symposium. Z. f. Flugwiss. **3**, 159—164 (1955).

[6] GÖRTLER, H.: Dreidimensionales zur Stabilitätstheorie laminarer Grenzschichten. ZAMM **35**, 362—364 (1955).

[7] GÖRTLER, H.: Dreidimensionale Instabilität der ebenen Staupunktströmung gegenüber wirbelartigen Störungen. „50 Jahre Grenzschichtforschung" hrsg. v. H. GÖRTLER u. W. TOLLMIEN. Braunschweig: Vieweg (1955). S. 304—314.

[8] HÄMMERLIN, G.: Zur Instabilitätstheorie der ebenen Staupunktströmung; ebenda. 315—327.

[9] SCHLICHTING, H.: Absaugung in der Aerodynamik. WGL-Jahrbuch 1956, S. 19—28.

[*10*] Schultz-Grunow, F., und H. Hein: Beitrag zur Couette-Strömung. Z. f. Flugwiss. **4**, 28—30 (1956).

[*11*] Schlichting, H.: Amplitudenverteilung und Energiebilanz der kleinen Störungen bei der Plattenströmung. Nachr. Ges. Wiss. Göttingen, Math.-Phys. Klasse, Neuge Folge **I**, 1, 47—78 (1935).

Aus der Diskussion

R. Timman (Delft): Sie brauchen doch das Geschwindigkeitsprofil der Grundströmung zur Berechnung Ihrer Greenschen Funktion?

H. Witting (Freiburg i. Br.): Die Greenschen Funktionen lassen sich unabhängig von $u_0(y)$ berechnen. Es handelt sich um lineare Differentialoperatoren mit konstanten Koeffizienten.

G. Hämmerlin (Freiburg i. Br.): Bei der strengen Lösung des Görtlerschen Problems der Instabilität der Strömung längs einer konkaven Wand in der ursprünglichen Fassung ergibt sich eine Kurve für den kritischen Parameter $Re_\vartheta \sqrt{\vartheta/R}$, die für $\sigma \to 0$ einen endlichen Grenzwert besitzt, und die für wachsende σ monoton ansteigt, vgl. [*1*]. Für die Beschreibung der Verhältnisse bei sehr kleinen Werten von σ, also bei sehr großen Achsenabständen der Wirbel, wobei solche Wirbel sich auch entsprechend weit hinaus in die äußere Strömung außerhalb der Grenzschicht erstrecken würden, sind gewisse Vernachlässigungen, die dem ursprünglichen Görtlerschen Modell zugrunde liegen, nicht mehr möglich. Eine verfeinerte Behandlung dieses Grenzfalls sehr großer Wirbel [*2*] hat gezeigt, daß die Kurve des kritischen Parameters $Re_\vartheta \sqrt{\vartheta/R}$ der neutralen Wirbel in Abhängigkeit von σ ein Minimum hat, wie physikalisch zu erwarten ist und wie Görtler ursprünglich, aber ohne diese Rechtfertigung, behauptet hatte. Die Lage dieses Minimums (und damit der Achsenabstand der ersten mit wachsendem Re möglichen neutralen Wirbel) ist schwach von ϑ/R abhängig, jedoch ist diese schwache Abhängigkeit von keinem praktischen Interesse.

Literatur:

[*1*] Hämmerlin, G.: Über das Eigenwertproblem der dreidimensionalen Instabilität laminarer Grenzschichten an konkaven Wänden. Journ. Rat. Mech. Analysis **4**, 279—321 (1955).

[*2*] Hämmerlin, G.: Zur Theorie der dreidimensionalen Instabilität laminarer Grenzschichten. ZAMP **7**, 156—164 (1956).

Numerische Beiträge zur Stabilitätstheorie der Grenzschichten

Von

J. A. Zaat

Nationaal Luchtvaartlaboratorium, Amsterdam

1. Einleitung

Der Umschlag laminar-turbulent einer Grenzschichtströmung wird in vielen Fällen durch die Instabilität der laminaren Grenzschicht gegen kleine Störungen verursacht. Die mathematischen Grundlagen der Theorie sind schon lange bekannt und sind in 1940 durch die sehr bekannten Untersuchungen von SCHUBAUER und SKRAMSTAD experimentell bestätigt [1].

Die Theorie über diesen Gegenstand ist an mehreren Stellen [2, 3, 4] umfassend bearbeitet worden und wird deshalb hier nicht mehr in Einzelheiten wiederholt. Herr Prof. TIMMAN gab eine numerische Lösungsmethode, welche verschieden ist von den Methoden, die in anderen Arbeiten angewandt sind. Eine Darlegung dieser numerischen Methode [5], die von Herrn ZAAT bearbeitet worden ist, wird hier gegeben. Dabei wird von der sehr bekannten Differentialgleichung ausgegangen, die das Phänomen beschreibt. Als Grenzschichtprofile werden derartige Profile einer einparametrigen Familie verwendet, wie sie in der Arbeit [6] zur Berechnung der laminaren Grenzschicht gegeben sind.

Die Berechnungen wurden an dem Rechenbüro des N.L.L. unter der Leitung von Herrn T. BURGERHOUT, der sich weitgehend an der Ausarbeitung beteiligt hat, ausgeführt.

Die Arbeit wurde erteilt und finanziert vom N.I.V. (Niederländisches Institut für Flugzeugentwicklung).

2. Ergebnisse der Stabilitätstheorie der laminaren Grenzschichten

Wir betrachten eine zweidimensionale Grenzschichtströmung an einer Wand. Nach Einführung der Koordinaten x und y in der Richtung an der Wand und senkrecht dazu, werden die Geschwindigkeitsprofile eines Querschnittes $x = $ konstant durch die Gleichung

$$u = \frac{\bar{u}}{U} = f(y/\delta, \lambda) \qquad (1)$$

gegeben. U ist die örtliche freie Stromgeschwindigkeit, δ ist eine Länge proportional mit der Grenzschichtdicke, und λ ist der Profil-Formparameter.

Wir setzen voraus, daß die Grundströmung von einer zweidimensionalen, von der Zeit abhängigen Störung überlagert wird. Diese Störung denken wir uns aus einzelnen Partialschwingungen aufgebaut. Jede Partialschwingung ist eine in der x-Richtung fortschreitende Welle, bestimmt durch eine Stromfunktion

$$\psi(x, y, t) = \varphi(y)\, e^{i\,\bar{\alpha}\,(x - \bar{c}\,t)}, \tag{2}$$

wo $\bar{\alpha}$ und $\bar{c}$ reell sind. $\bar{c}$ bedeutet die Störungsfortpflanzungsgeschwindigkeit in der x-Richtung. $\beta_n = \bar{\alpha}\,\bar{c}$ ist die Kreisfrequenz der Schwingung. Für einen bestimmten Zeitpunkt t ist diese Störung eine Welle mit konstanter Amplitude $\varphi(y)$ und mit einer Wellenlänge $2\pi/\bar{\alpha}$. Die Amplitudenfunktion $\varphi(y)$ befriedigt, nach Einführung der dimensionslosen Größen

$$c = \bar{c}/U, \quad \eta = y/\delta, \quad \alpha = \bar{\alpha}\,\delta,$$

die Sommerfeldgleichung (Lit. [2], S. 282)

$$(u - c)(\varphi'' - \alpha^2\varphi) - u''\varphi = -\frac{i}{\alpha R}(\varphi^{\mathrm{IV}} - 2\,\alpha^2\,\varphi'' + \alpha^4\,\varphi). \tag{3}$$

Die Akzente bedeuten Differentiation nach η.

$R = \dfrac{U\delta}{\nu}$ ist die Reynoldssche Zahl gebildet mit der freien Strömungsgeschwindigkeit und der Grenzschichtdicke.

Eine allgemeine Lösung φ setzt sich aus den Partikular-Lösungen φ_1, φ_2, φ_3 und φ_4 der Gl. (3) zusammen,

$$\varphi(\eta) = A_1\,\varphi_1 + A_2\,\varphi_2 + A_3\,\varphi_3 + A_4\,\varphi_4. \tag{4}$$

Sie befriedigt die Randbedingungen

$$\varphi = 0 \qquad \varphi' = 0 \qquad \text{für} \qquad \eta = 0 \tag{5}$$

$$\varphi \to 0 \qquad \varphi' \to 0 \qquad \text{für} \qquad \eta \to \infty. \tag{6}$$

Aus der Theorie geht hervor, daß für große Werte der Reynoldsschen Zahl R eine gute Annäherung für zwei Partikular-Lösungen φ_1 und φ_2 erhalten wird, falls für φ_1 und φ_2 zwei unabhängige Lösungen der reibungslosen Störungsdifferentialgleichung

$$(u - c)(\varphi'' - \alpha^2\varphi) - u''\varphi = 0 \tag{7}$$

gewählt werden, sofern nötig, unter Heranziehung der Reibungseinflüsse. Die Funktionen φ_3 und φ_4 sind für große Werte von R asymptotisch darzustellen als Lösungen einer einfachen Gleichung, die aus dem Verhalten der Gl. (3) in der Umgebung des Punktes $\eta = \eta_0$, wo $u = c$ ist, erhalten wird. Hier ist $u - c \approx (\eta - \eta_0)\, u_0'$, und die Näherungsgleichung ist, nach Einführung einer neuen Veränderlichen,

$$z = (\alpha\, R\, u_0')^{1/3}\, (\eta - \eta_0) \tag{8}$$

$$i\, \frac{d^4\, \varphi}{d\, z^4} + z\, \frac{d^2\, \varphi}{d\, z^2} = 0\,. \tag{9}$$

Mit Rücksicht auf die Randbedingung (6) kommt für die Gl. (9) nur die Lösung

$$\varphi_3\,(z) = \int\limits_\infty^z \int\limits_\infty^z \eta^{1/2}\, H_{1/3}^{(1)} \left[\frac{2}{3}\,(i\,\eta)^{3/2}\right] d\,\eta \tag{10}$$

in Frage. Jede andere Lösung φ_4 ist unendlich für $z \to \infty$ und gibt keinen Beitrag zu der allgemeinen Lösung, folglich ist in Gl. (4) $A_4 = 0$. Wählen wir φ_1 so, daß $\varphi_1\,(\infty) = 0$, dann ist in der Gl. (4), wegen $\varphi\,(\infty) = A_1\,\varphi_1\,(\infty) + A_2\,\varphi_2\,(\infty) = 0$, auch $A_2 \equiv 0$. Die übrigen Koeffizienten $A_{1,3}$ lassen sich bestimmen aus dem System

$$\begin{aligned} A_1\,\varphi_1\,(0) + A_3\,\varphi_3\,(0) &= 0 \\ A_1\,\varphi_1'\,(0) + A_3\,\varphi_3'\,(0) &= 0. \end{aligned} \tag{11}$$

Eine Lösung ist für nicht-verschwindende Koeffizienten $A_{1,3}$ nur dann möglich, falls c und α Werte haben, wofür

$$\begin{vmatrix} \varphi_1\,(0) & \varphi_3\,(0) \\ \varphi_1'\,(0) & \varphi_3'\,(0) \end{vmatrix} = 0. \tag{12}$$

Die Werte von c und α, die die Gl. (12) befriedigen, geben dargestellt als Funktionen von R die gewünschten Stabilitätskurven.

3. Die reibungslose Lösung

Nach Einführung der Geschwindigkeitsprofile der Grenzschicht $f\,(\eta)$ geht die reibungslose Differentialgleichung (7) über in

$$\varphi'' = \left[\alpha^2 + \frac{f''\,(\eta)}{f\,(\eta) - f\,(\eta_0)}\right]\varphi\,. \tag{13}$$

Dabei ist η_0 der Wert von η für den $f\,(\eta_0) = c$. Für die Funktion $f\,(\eta)$ wählen wir, wie bei der Pohlhausenschen Impulsgleichungsmethode, eine einparametrige Funktionenfamilie, die aber mit dem asymptotischen Verhalten der Lösungen der Grenzschichtgleichungen arbeitet

$$f\,(\eta) = 1 - \left[\lambda + \frac{2}{3\,\sqrt{\pi}}\,(1 - \lambda)\,\eta + \frac{1}{14}\,(2\,\lambda + 5\,\lambda^2)\,\eta^2\right]e^{-\eta^2} -$$
$$- \frac{2}{\sqrt{\pi}}\,(1 - \lambda)\int\limits_\eta^\infty e^{-t^2}\,dt\,. \tag{14}$$

Die Funktion $f(\eta)$ ist eine Modifikation des Geschwindigkeitsprofiles der Grenzschicht, gewählt nach Lit. [6]. Das Glied mit η^2 ist hinzugefügt, um die Diskontinuität in dem Übergang der beschleunigten zu der verzögerten Strömung zu vermeiden. Der Profil-Formparameter λ ergibt sich aus der Impulsgleichung. Im Staupunkt ist $\lambda = -0{,}412$. In dem Ablösungspunkt, wo die Schubspannung an der Wand verschwindet, d.h. wo $f'(0) = 0$ ist, ist $\lambda = 1$. Für genügend große Entfernung von der Wand verschwinden die Ableitungen der Grundströmung, und die Gl. (13) vereinfacht sich zu

$$\varphi'' = \alpha^2 \, \varphi.$$

Die Lösung φ, welche zusammen mit ihren Ableitungen für $\eta \to \infty$ verschwinden muß, kann angenähert werden durch die Funktion

$$\varphi_1(\eta) = e^{-\alpha \eta}. \tag{15}$$

Diese Funktion kann in dem Bereich, wo sie gültig ist, als Ausgangsfunktion für die numerische Integration der reibungslosen Gl. (13) genommen werden. Die numerische Integration wurde mit der Adams-Methode ausgeführt, die mit zentralen Differenzen und Extrapolation arbeitet. Diese Methode ist brauchbar bis in die Nähe des Poles $\eta = \eta_0$ des Koeffizienten von φ in der Differentialgleichung (13).

4. Die Integration durch den Pol hindurch

Die Gl. (13) für die reibungslose Lösung hat eine singuläre Stelle für $\eta = \eta_0$. Aus der allgemeinen Theorie ist bekannt, daß die Differentialgleichung zwei Lösungen hat. Die erste Lösung ist regulär an der Stelle η_0, die zweite hat eine logarithmische Singularität. Weil die hier zu berechnende Funktion $\varphi_1(\eta)$ eine lineare Zusammensetzung dieser zwei Lösungen ist, muß sie auch einen logarithmischen Teil enthalten. Unter der Voraussetzung, daß $f(\eta) - f(\eta_0)$ an der Stelle $\eta = \eta_0$ eine einfache Nullstelle hat, läßt sich der Koeffizient von φ in der Gl. (13) in eine Reihe entwickeln

$$F(\eta) = \alpha^2 + \frac{f''(\eta)}{f(\eta) - f(\eta_0)} = (\eta - \eta_0)^{-1} \sum_{K=0}^{\infty} a_K (\eta - \eta_0)^K. \tag{16}$$

Die Koeffizienten a_K, die durch aufeinanderfolgende Differentiationen von $(\eta - \eta_0) F(\eta)$ erhalten werden, können mit Hilfe eines Differenzen-Schemas der tabulierten Funktion bestimmt werden. Darauf wird bezüglich der allgemeinen Theorie für φ_1 eine in der Umgebung von $\eta = \eta_0$ gültige Reihenentwicklung angenommen

$$\varphi_1(\eta) = p \sum_{n=0}^{\infty} \alpha_n (\eta - \eta_0)^n + \{ p \, l n \, (\eta - \eta_0) + q \} \sum_{n=0}^{\infty} \beta_n (\eta - \eta_0)^n. \tag{17}$$

Substitution in die Differentialgleichung (13) gibt die folgenden Beziehungen für die Berechnung der unbekannten Koeffizienten α_n und β_n

$$n(n-1)\beta_n = \sum_{K=1}^{n-1} a_{K-1}\beta_{n-K} \qquad\qquad \beta_0 = 0 \qquad \beta_1 = 1$$

$$n(n-1)\alpha_n + (2n-1)\beta_n = \sum_{K=1}^{n} a_{K-1}\alpha_{n-K} \qquad \alpha_0 = \frac{1}{a_0} \qquad \alpha_1 = 0.$$

Die Werte p und q werden bestimmt durch die Werte der Funktion (17) und ihrer Ableitungen für einen kleinen positiven Wert von $\eta - \eta_0$ und Vergleich mit den durch numerische Integration erhaltenen Werten φ_1 und φ_1'. Eine Prüfung der Genauigkeit der numerischen Integration kann hier gegeben werden durch Vergleich der Ergebnisse in zwei Nachbarpunkten η und $\eta + \Delta\eta$. Sind die Koeffizienten der Reihenentwicklung bekannt, dann können die Funktionen $\varphi_1(\eta)$ und $\varphi_1'(\eta)$ an der anderen Seite des Poles bestimmt werden. Dort ist $\eta - \eta_0$ negativ, folglich wird der Logarithmus komplex. Die Zweideutigkeit im Vorzeichen des imaginären Teiles wird in der allgemeinen Theorie auf Grund einer Betrachtung der Reibungskorrektur (Lit. [2], S. 292) behoben. Für $\eta - \eta_0 < 0$ ist zu nehmen

$$\ln(\eta - \eta_0) = \ln|\eta - \eta_0| - \pi i.$$

Hiernach wird die numerische Integration fortgesetzt bis zu $\eta = 0$, wo der Quotient $\dfrac{\varphi_1'(0)}{\varphi_1(0)}$ berechnet ist.

Ein zweites Integrationsverfahren durch die komplexe Ebene um den Pol herum mit Hilfe analytischer Fortsetzung führt zu genau denselben Ergebnissen. Diese Methode ist im Anhang gegeben.

5. Die Bestimmung der neutralen Schwingungen

Die Werte $\dfrac{\varphi_1'(0)}{\varphi_1(0)}$ sind berechnet für jedes Geschwindigkeitsprofil, charakterisiert durch einen Wert der Profilparameter λ, und für mehrere Werte von c und α. Die Bestimmung des Wertepaares (c, α) der neutralen Schwingungen ergibt sich aus

$$\frac{\varphi_1'(0)}{\varphi_1(0)} = \frac{\varphi_3'(0)}{\varphi_3(0)}.$$

Die Funktion φ_3 ist nach (8) und (10) als eine Funktion von $z = (\alpha R u_0')^{1/3}(\eta - \eta_0)$ gegeben [7].

Gewöhnlich geben die Tabellen und Diagramme die Funktion

$$\frac{D(z)}{z} = -\frac{\varphi_3(z)}{z\, d\varphi_3/dz}.$$

In den Diagrammen sind mit $z_0 = -\eta_0 (\alpha R\, u_0')^{1/3}$ als Parameter die reellen und imaginären Teile der Funktion $D(z_0)/z_c$ als Koordinatenachsen aufgetragen. Um nun die Gleichung

$$\frac{\varphi_1(0)}{\eta_0\, \varphi_1'(0)} = \frac{D(z_0)}{z_0}$$

zu berechnen, werden für verschiedene Werte von α die Kurven $\dfrac{\varphi_1(0)}{\eta_0\, \varphi_1'(0)}$
in dem gleichen Diagramm als Funktionen von η_0 (oder c) aufgetragen. Die Schnittpunkte der zwei Kurven geben die Werte α, c und z_0 für die Punkte der Indifferenzkurven. Aus $z_0 = -\eta_0 (\alpha R\, u_0')^{1/3}$ kann die Reynoldssche Zahl ermittelt werden.

6. Ergebnisse und Schlußfolgerungen

Die Diagramme sind berechnet für verschiedene Parameterwerte λ des Geschwindigkeitsprofils der laminaren Grenzschicht

$$\lambda = -0{,}4;\ 0;\ 0{,}4;\ 0{,}8\ .$$

Die korrespondierenden Geschwindigkeitsprofile sind in Abb. 1 gegeben.

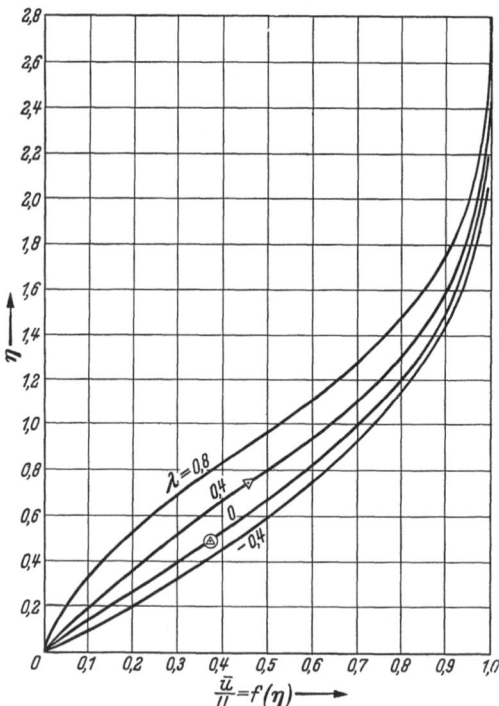

Abb. 1. Die laminaren Grenzschichtprofile, für welche die Indifferenzkurven berechnet sind. λ ist der Profilparameter

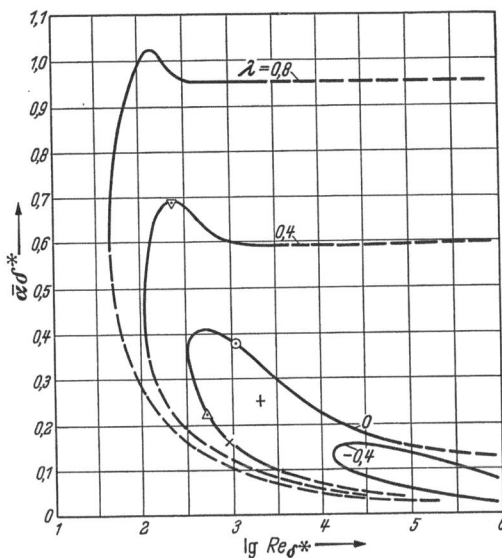

Abb. 2. Die Indifferenzkurven für die Störungswellenlängen $\overline{\alpha}\,\delta^*$ der laminaren Grenzschichtprofile mit Druckabfall ($\lambda < 0$) und Druckanstieg ($\lambda > 0$) in Abhängigkeit von der Reynoldsschen Zahl. δ^* ist die Verdrängungsdicke

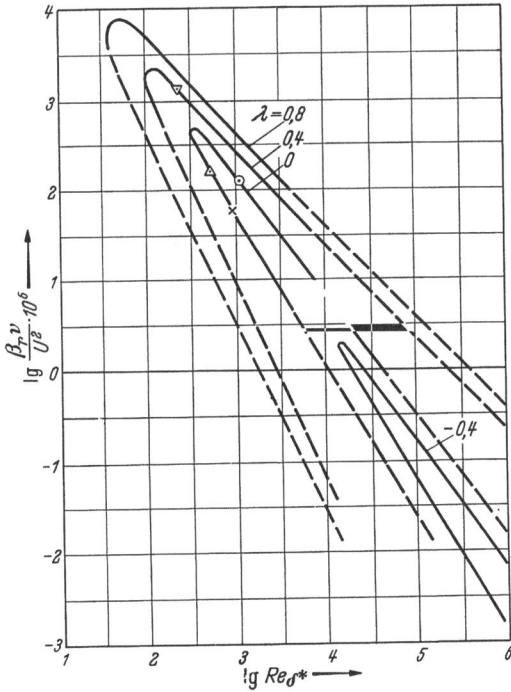

Abb. 3. Die Indifferenzkurven für die Störungsfrequenzen β_r der laminaren Grenzschichtprofile mit Druckgradient in Abhängigkeit von der Reynoldsschen Zahl

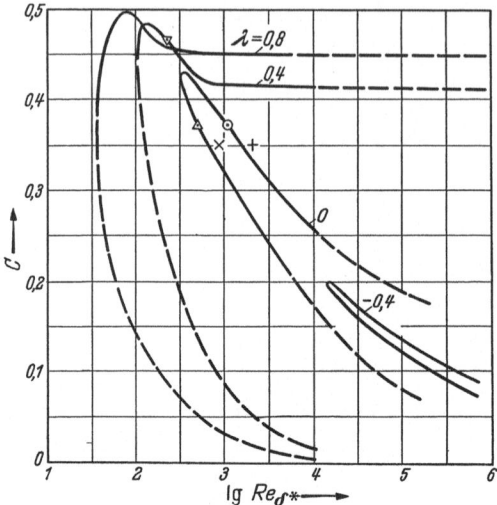

Abb. 4. Die Indifferenzkurven für die Wellenfortpflanzungsgeschwindigkeiten der laminaren Grenzschichtprofile mit Druckgradient als Funktion der Reynoldsschen Zahl

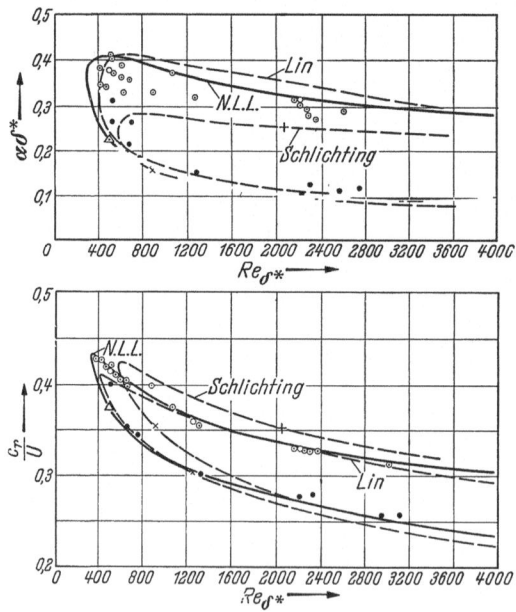

Abb. 5. Ein Vergleich zwischen den in verschiedener Weise berechneten Indifferenzkurven für die
Wellenlängen und Wellenfortpflanzungsgeschwindigkeiten der neutralen Störungen im Falle des
Blasius-Profils. Die ausgezogenen Linien sind die theoretischen Werte. Die gemessenen Werte sind mit
Kreisen und Punkten gegeben. Die kritischen Werte sind:

$$Re_{\delta*} \text{ Krit} = 321 \text{ NLL}, 420 \text{ Lin}, 575 \text{ Schlichting}$$

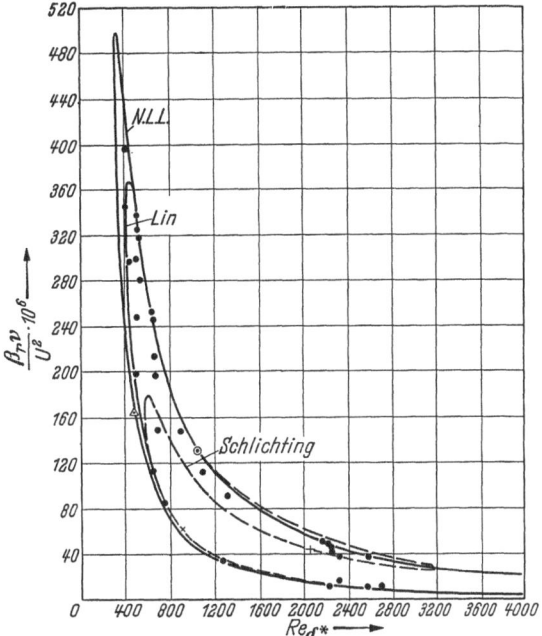

Abb. 6. Ein Vergleich zwischen den in verschiedener Weise berechneten Frequenzen der neutralen Störungen im Falle des Blasius-Profils und die Messungen von SCHUBAUER und SKRAMSTAD. Die letzteren Werte sind durch Punkte gegeben

	λ	η_0	$Re_{\delta}*$	c_r/U	$\tilde{\alpha}\delta*$	$\frac{\beta_r v}{U^2} \cdot 10^6$
△	0	0,50	507	0,372	0,224	164
⊙	0	0,50	1070	0,372	0,371	131
▽	0,4	0,75	226,5	0,466	0,685	1409

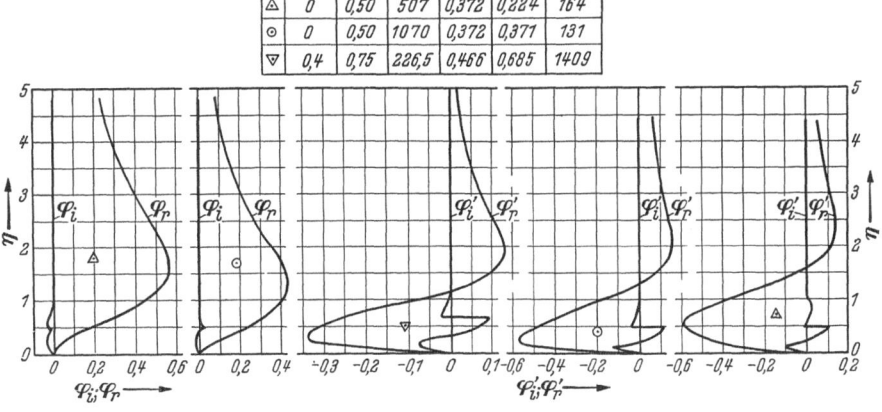

Abb. 7. Die Funktionen $\varphi_i(\eta)$, $\varphi_r(\eta)$, $\varphi_i'(\eta)$ und $\varphi_r'(\eta)$ für die Punkte △, ⊙ und ▽ aus den Abbildungen 1 bis 4

Für diese Profile sind in der Abb. 2 die Indifferenzkurven für die Störungswellenlängen $\bar{\alpha}\,\delta^*$ in Abhängigkeit von der Reynoldsschen Zahl $Re_\delta^* = \dfrac{U\,\delta^*}{\nu}$ aufgetragen. Die korrespondierenden Frequenzen

$$\frac{\beta_n\,\nu}{U^2} = \frac{\bar{\alpha}\,\bar{c}\,\nu}{U^2} = \bar{\alpha}\,\delta^*\,\frac{c}{Re_\delta^*},$$

und die Wellenfortpflanzungsgeschwindigkeit $c = \bar{c}/u$ sind in den Abb. 3 und 4 gegeben (die gestrichelten Linien sind durch Extrapolation der berechneten Werte erhalten).

Die Instabilitätsbereiche, eingeschlossen von den Indifferenzkurven, wachsen in dem Bereich des positiven Druckgradienten, besonders in der Nähe des Ablösungspunktes. Für das Blasiusprofil $\lambda = 0$ sind die Ergebnisse mit den Ergebnissen von SCHLICHTING und LIN verglichen (Abb. 5 und 6). Die hier berechneten Werte stimmen am besten überein mit den Werten von LIN. Die Übereinstimmung mit den Messungen von SCHUBAUER und SKRAMSTAD ist sehr gut. Für die unterste Grenze der kritischen Reynoldsschen Zahl erhielt SCHLICHTING $Re_\delta^* = 575$, LIN 420, aus der Kurve hier folgt der Wert 320. Die Funktionen $\varphi\,(\eta)$ und $\varphi'\,(\eta)$ der Geschwindigkeitsprofile der neutralen Störungen sind berechnet für zwei Punkte der Indifferenzkurve des Blasius-Profils ($\lambda = 0$) und für einen Punkt der Indifferenzkurve des mit $\lambda = 0,4$ korrespondierenden Profils. Die Punkte sind in den Abb. 1 bis 4 mit $\triangle$, $\odot$ und $\triangledown$ bezeichnet. Die Werte

$$\varphi\,(\eta) = \varphi_1\,(\eta) + A_3\,\varphi_3\,(\eta)$$

und

$$\frac{d\,\varphi}{d\,\eta} = \frac{d\,\varphi_r}{d\,\eta} + i\,\frac{d\,\varphi_i}{d\,\eta},$$

mit $A_3 = -\dfrac{\varphi_1(0)}{\varphi_3(0)}$, sind als Funktionen von η in Abb. 7 aufgetragen. Damit sind die Amplitudenverteilungen $k\,|\,\varphi\,|$ und $k\,\left|\,\dfrac{d\,\varphi}{d\,\eta}\,\right|$ für diese neutralen Störungen innerhalb der Grenzschicht bestimmt. Die theoretischen Amplitudenverteilungen von $\tilde{u} = \dfrac{\partial\,\psi}{\partial\,\eta}$ sind als Funktionen von y/δ^* in Abb. 8 aufgetragen und mit den experimentellen Werten, Lit. [1], S. 21, verglichen. Obwohl die Bedingungen der gemessenen Werte ($+$ und $\times$ in Abb. 2) nicht völlig mit den Bedingungen der berechneten Punkte [$\triangle$ und $\odot$ in Abb. 2] korrespondieren, zeigen die Ergebnisse eine gute Übereinstimmung. In dem Bereich der verzögerten Strömung bleibt die Gestalt der Amplitudenverteilung dieselbe. Schließlich ist in Abb. 9 die kritische Reynoldssche Zahl für die Geschwindigkeitsprofile der laminaren Grenzschichtströmungen mit Druckgradienten als eine Funktion des Formparameters

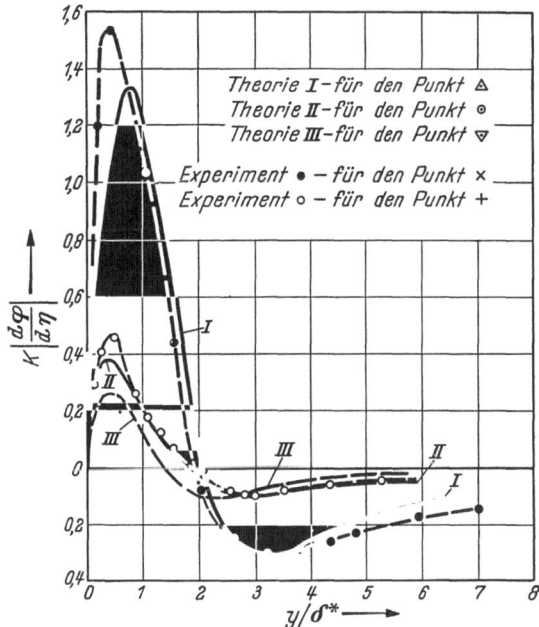

Abb. 8. Die Amplitudenverteilung der Störungen in der Grenzschicht für die Punkte △, ⊙, ▽, +
und × in der Abbildung 2

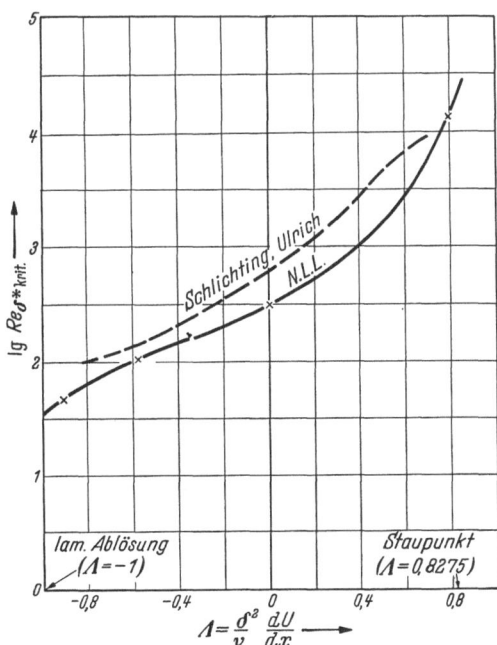

Abb. 9. Die kritische Reynoldssche Zahl der Grenzschicht mit Druckgradient in Abhängigkeit von dem
Formparameter $\varLambda = \dfrac{\delta^2}{\nu}\dfrac{d\,U}{d\,y}$

$$\Lambda = \frac{\delta^2}{\nu}\frac{dU}{dx} = -\left(\frac{d^2 f}{d\eta^2}\right)_{\eta=0} = \frac{5}{7}\lambda^2 - \frac{12}{7}\lambda$$

dargestellt. Dieses gibt ein Mittel, um unmittelbar die kritische Reynolds-sche Zahl bei einfachen Grenzschichtberechnungen zu erhalten. Zum Vergleich sind in der Abb. 9 auch die Ergebnisse von SCHLICHTING und ULRICH [8] aufgetragen. Die Abhängigkeit der Parameter λ_{P_6} und Λ erhält man aus der Beziehung

$$\left(\frac{2}{7} - \frac{\lambda_{P_6}}{105}\right)^2 \lambda_{P_6} = \frac{\delta^{*2}}{\nu}\frac{dU}{dx} = \frac{\delta^{*2}}{\delta^2}\frac{\delta^2}{\nu}\frac{dU}{dx} = \Delta_1^2\,\Lambda\,,$$

wo $\delta^* = \delta \cdot \Delta_1 = \int\limits_0^\infty (1 - f)\, d\, y$ die Verdrängungsdicke und δ eine Länge proportional zu der Grenzschichtdicke sind.

Literatur

[1] SCHUBAUER, G. B., und H. K. SKRAMSTAD: Laminar Boundary-Layer Oscillations and Transition on a Flat Plate. NACA Report No. 909 (1948).
[2] SCHLICHTING, H.: Grenzschichttheorie. Karlsruhe: G. Braun (1950), S. 273 bis 303 mit Literatur.
[3] LIN, C. C.: On the Stability of Two-Dimensional Parallel Flows. Quarterly Appl. Math. **3**, 117—142 (1945), **3**, 218—234 (1945), **3**, 277—301 (1946).
[4] DRYDEN, H. L.: Recent Advances in the Mechanics of Boundary-Layer Flow. Adv. in Appl. Mech., **1**, 8—23 mit Literatur (1948).
[5] TIMMAN, R., J. A. ZAAT and TH. J. BURGERHOUT: Stability diagrams for laminar boundary layer flow. N. L. L. Report F. 193 (1956).
[6] TIMMAN, R.: A One-Parameter Method for the Calculation of Laminar Boundary Layers. N. L. L. Report F. 35 (1949).
[7] HOLSTEIN, H.: Über die äußere und innere Reibungsschicht bei Störungen laminarer Strömungen. ZAMM **30**, 25—49 (1950).
[8] SCHLICHTING, H., und A. ULRICH: Zur Berechnung des Umschlages laminar-turbulent. Jahrbuch d. dt. Luftf. Forschung, Bd. 1 (1942), S. 54—71.

Anhang

Eine Integrationsmethode mit Hilfe analytischer Fortsetzung

Die Funktion $f(\eta)$ zur Beschreibung des Geschwindigkeitsprofils ist auf der ganzen reellen Achse regulär. Die Funktion, die in der komplexen Ebene fortsetzbar ist, wird durch $f(\zeta)$ gegeben mit $\zeta = \eta + i\,\nu$. Setzt man $\varphi = \chi + i\,\psi$, dann ändert sich die Differentialgleichung der reibungslosen Lösungen in

$$\frac{\partial^2 \varphi}{\partial \eta^2} - [\alpha^2 + G(\zeta)]\,\varphi(\zeta) = 0 \qquad (\nu = \text{konstant}) \qquad \text{(A. 1)}$$

$$\frac{\partial^2 \varphi}{\partial \nu^2} + [\alpha^2 + G(\zeta)]\,\varphi(\zeta) = 0 \qquad (\eta = \text{konstant}). \qquad \text{(A. 2)}$$

Der Integrationsweg in der komplexen Ebene ist in Abb. 10 gezeichnet. Der Weg fängt im Unendlichen an, durchläuft die reelle Achse, geht um den Pol herum an den Seiten eines Rechtecks entlang und endet schließlich in dem Ursprung. In dem Falle einer Richtungsänderung folgen die

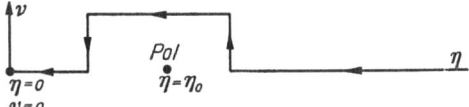

Abb. 10. Integrationsweg in der komplexen Ebene

Anfangsbedingungen, die für numerische Integration nötig sind, aus den CAUCHY-RIEMANNschen Beziehungen

$$\chi_\eta = \psi_\nu \qquad \psi_\eta = -\chi_\nu$$

$$\chi_{\eta\eta} = -\chi_{\nu\nu} \qquad \psi_{\eta\eta} = -\psi_{\nu\nu}$$

$$\chi_{\eta\eta\eta} = -\psi_{\nu\nu\nu} \qquad \psi_{\eta\eta\eta} = \chi_{\nu\nu\nu}, \text{ usw.}$$

Der Vorteil dieser Methode ist, daß die Funktion $G(\zeta)$ den ganzen Integrationsweg entlang regulär ist. Schwierige Reihenentwicklungen wie im Abschnitt 4 werden in dieser Weise vermieden. Ist der Integrationsweg hinreichend weit von dem Pol entfernt, dann brauchen die Integrationsschritte im allgemeinen nicht verkleinert zu werden, verglichen mit den Schritten auf dem rechten Teil des Integrationsweges.

Diskussionsveranstaltung zur III. Sitzung

Contribution to the Subject

Stability of the Laminar Boundary Layer

By J. Laufer, Jet Propulsion Laboratory,
California Institute of Technology, Pasadena, Cal.

In any wind tunnel investigation of laminar boundary layer instability, the free stream turbulence level of the wind tunnel is of primary concern. Dr. SCHUBAUER, in his very excellent lecture, pointed out that even in his wind tunnel, where the turbulence level is indeed low, some local disturbance in the stream might influence the turbulent spot formation.

We are especially concerned now with this problem in our supersonic boundary layer stability investigation. Last year at the IXth International Congress of Applied Mechanics, we reported the successful detection of the stability zone of a boundary layer on a flat plate at a free stream Mach number 2.2. In trying to extend these measurements to higher Mach numbers, considerable difficulty was encountered. This difficulty was traced to the high disturbance level in the free stream, the level growing very rapidly with increasing Mach number. We spent considerable time and effort in studying the nature of the fluctuations and their origin. It soon became clear that they did not originate from the supply section. Here the measurements indicated only rotational velocity (vorticity) fluctuations to be present; their magnitude was of the order of 1% of the local mean velocity. Fluctuations due to temperature variation or noise were negligibly small. Simple theoretical considerations and some experiments have indicated that the effect on the free stream turbulence level of the vorticity fluctuations present in the supply section decreases rapidly as the tunnel Mach number is increased. Indeed, analyzing the test section fluctuations using the KOVASZNAY-MORKOVIN method, it was found that the disturbances are mainly of the "sound mode" type. This would indicate that the fluctuations must originate downstream of the supply section in the nozzle throat and along the tunnel walls. Since the wall boundary layer is turbulent everywhere in this region, we conjecture the disturbances measured in the free stream originate from the turbulent wall boundary layers. If this conjecture proves to be correct, the realization of a high Mach number wind tunnel having a low turbulence level would be extremely difficult.

Contribution to the Subject

Transition

By **W. Pfenninger,** Northrop Aircraft, Inc., Hawthorne, Cal.

When Dr. SCHUBAUER published his experiments on the growth of turbulent spots we became interested as to what causes the growth of turbulent spots. In order to obtain a better understanding of the flow phenomena involved we conducted laminar flow experiments in the inlet length of an 8" i.d. tube of 40-feet length at low speeds. The flow was made visible by means of smoke. Turbulent spots were produced by blowing out air through a hole. At the edge of these spots the formation of new vortices close to the surface was observed, at least at the upstream and downstream sides of the turbulent spots. In many cases these vortices had the appearance of horseshoe vortices. G. RAETZ in my group has suggested that these new vortices are continuously formed by induction from the already existing vortices within the turbulent spot.

Later we became concerned with the question whether or not boundary layer suction would be effective in reducing the growth of turbulent spots. Dr. KUETHE of the University of Michigan carried out experiments on the growth of turbulent spots with suction, using one of our suction models. With increasing suction quantities the growth of turbulent spots was reduced.

Contribution to the Subject

Transition

By **H. L. Dryden,** National Advisory Committee
for Aeronautics, Washington, D. C.

I believe that the growth of a turbulent "spot" proceeds by a progressive breakdown of the laminar flow adjoining the spot as a result of flow disturbances induced by the spot in nearby regions. As is well known, the breakdown of laminar flow is very sensitive to stream turbulence, the transition Reynolds number decreasing rapidly as the turbulence level is increased.

If this explanation is correct, spots should not grow in a boundary layer below the critical Reynolds number for flow instability. SCHUBAUER has demonstrated that this is true for spots artificially produced by a spark discharge through the boundary layer. The same phenomenon was shown many years ago at the National Physical Laboratory for the spreading of the wake behind a rivet head at low Reynolds number. At first the vortices from the rivet head remained in two parallel lines downstream and only when the Reynolds number of the boundary layer increased did the turbulent wake spread in the typical wedge of turbulence.

Beitrag zum Thema

<h2 align="center">Umschlag</h2>

Von R. Wille, Hermann-Föttinger-Institut für Strömungstechnik,
Technische Universität Berlin

Die von H. GÖRTLER theoretisch begründeten Wirbel mit Achsen in Strömungs-richtung konnten auch in der turbulenten Randzone eines Freistrahls qualitativ nachgewiesen werden. Zum Nach-weis wurde ein einfaches „Drallmeß-rädchen" benutzt, wie es von S. EIKE und W. v. TREUENFELS [1] für Drall-messungen hinter Krümmern ver-wendet wurde. Die Achse des Drall-meßrädchens und seine ebenen, unverwundenen Flügel wurden nach der Strahlachse ausgerichtet; die beobachteten Drehungen entspre-chen dann Rotationen, deren Vektor in der Strahlachse liegt. Die Lage der Drallzonen im Freistrahl ist in Abb. 1 angegeben: Längswirbel

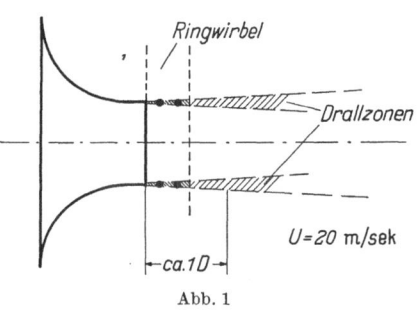

Abb. 1

traten nur stromab hinter dem von Ringwirbeln beherrschten laminar-turbulenten Übergangsgebiet auf. Weitere Ausführungen hierzu werden in dem Beitrag von O. WEHRMANN und R. WILLE in der VIII. Sitzung dieses Symposiums folgen.

Literatur:

[1] EIKE, S., und W. v. TREUENFELS: Forschg. Ing.Wesen **8** (1937).

H. GÖRTLER (Freiburg i. Br.): Betrachtet man Geschwindigkeitsfeld und Strom-linien eines einzelnen geraden Wirbels oder auch eines Wirbelrings, so ergeben die Überlegungen, über die ich ausführlich vorhin in meinem Vortrag berichtet habe, daß eine Instabilität gegenüber Wirbeln mit Achsen in Richtung der Stromlinien des betrachteten Einzelwirbels dort zu erwarten ist, wo außerhalb des Wirbelkerns die Geschwindigkeit wieder abzunehmen beginnt (und falls in diesem Außenbereich der radiale Gradient der Geschwindigkeit einen hinreichend großen Betrag erreicht). Es ist eine interessante Frage, ob und wann diese Instabilität im Spiel der gegen-seitigen Beeinflussung von stabilem Kern und zur Instabilität neigendem Außen-bereich wirklich eintritt. Da wir uns, wie ich am Schlusse meines Vortrags andeutete, auch für den Einfluß von Instationaritäten auf die Stabilität interessieren, unter-suchen wir zur Zeit die Stabilität des linearen zähen Einzelwirbels (dessen Ge-schwindigkeitsfeld nach OSEEN exakt bekannt ist).

G. I. Taylor (Cambridge): Since Dr. Görtler has referred to the stability of vortex rings, I might report that in the outside region where velocity decreases with radius the instability referred to by Dr. Görtler may perhaps be stabilized by a density gradient, for the centrifugal acceleration produces instability when $r \cdot u$ decreases with r whereas the density gradient in a ring produced from hot material will have a stabilizing effect against this instability. The result is that rings produced, say, from an explosion may carry a sharply defined ring of smoke to very great distances. This question has been broached by Mr. Turner in a recent communication [1].

Reference:

[1] Turner, J. S.: Buoyant vortex rings. Proc. Roy. Soc. A **239**, 61—75 (1957).

Beitrag zum Thema

Umschlag

Von **N. Scholz**, BMW-Triebwerksbau-GmbH, München

In einer kürzlich erschienenen Dissertation [1] von E. G. Feindt wurde ein bemerkenswerter experimenteller Beitrag zum Problem des laminar-turbulenten Umschlags geliefert. Für eine von Beginn an unter einem linearen Druckverlauf stehende laminare Grenzschicht ($dp/dx=$const) wurde der Umschlagpunkt bei verschiedener Rauhigkeit der Wand ermittelt.

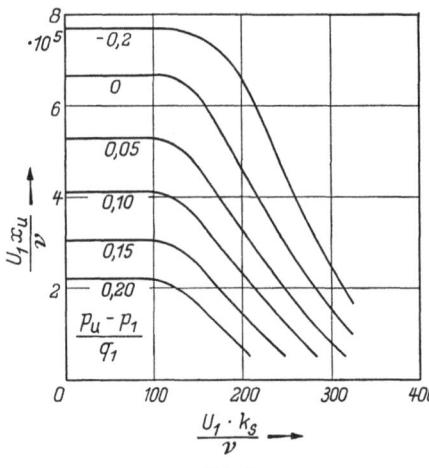

Abb. 1

Für die homogene Sandrauhigkeit der Größe k_s ergibt sich aus Ähnlichkeitsbetrachtungen die Lage des Umschlagpunktes aus den drei Kenngrößen $Re_x = U_1 x_u/\nu$, $Re_k = U_1 k_s/\nu$ und $(p_1-p_u)/q_1$. (Es bezeichnen: U_1 Geschwindigkeit, p_1 Druck und q_1 Staudruck am Beginn der Grenzschicht; x_u Lauflänge der Grenzschicht bis zum Umschlagpunkt; p_u Druck am Umschlagpunkt; ν kinematische Zähigkeit.) Das Ergebnis der Messungen zeigt Abb. 1. Wie zu erwarten, nimmt mit steigendem Druck die kritische Reynoldssche Zahl schnell ab, während sie umgekehrt mit fallendem Druck vergrößert wird. Die Rauhigkeit wirkt in jedem Falle erniedrigend auf die kritische Reynoldssche Zahl, sobald sie eine kritische Grenze überschritten hat. Im Bereich der hydraulisch glatten Wand ($U_1 k_s/\nu < \sim 120$) und auch noch bei schwacher Rauhigkeit ist außerdem ein Einfluß des Turbulenzgrades der Zuströmung vorhanden, der für größere Rauhigkeitswerte verschwindet.

Ähnliche Untersuchungen wurden auch für zweidimensionale Rauhigkeiten (Kreiszylinder) durchgeführt, für die auf die Originalarbeit hingewiesen sei.

Literatur:

[1] FEINDT, E. G.: Untersuchungen über die Abhängigkeit des Umschlages laminarturbulent von der Oberflächenrauhigkeit und der Druckverteilung. Diss. T. H. Braunschweig, 1956. (Referent: Prof. Dr. H. SCHLICHTING.) Jahrb. d. Schiffbautechn. Ges. **50**, 179—203 (1956).

Contribution to the Subject

Stability of Laminar Flows

By **T. Tatsumi,** University of Kyoto

Hydrodynamic stability of a two-dimensional jet flow has been investigated and the result is discussed here in brief.

According to the usual procedure of the linear stability theory, the two-dimensional flow pattern is approximated by a parallel flow: $U(y) = \mathrm{sech}^2 y$. The solution of the ORR-SOMMERFELD equation is assumed in an ascending power series of $i\,a\,R$ (regardless of $i\,a\,R$ which is involved in the parameter $\beta^2 = a^2 - i\,a\,Rc$), which is confirmed to be uniformly convergent. The neutral curve is calculated for the antisymmetric disturbance (even function of y). The following results are obtained: 1. Asymptotic behaviour of the lower branch of the neutral curve is given by $R = 1.12\,a^{-1/2}$, $c = 1.20\,a^2$. 2. Critical Reynolds number is $R_{cr} = 4.0$.

For such a small value of Reynolds number the fundamental assumption of the stability theory that the basic flow is well approximated by the parallel flow is no longer valid. Therefore the above result seems to cast a doubt to the applicability of the usual process of the linear stability theory to unlimited laminar flows.

On the instability of laminar flow and its transition to turbulence[1]

By

C. C. Lin

Massachusetts Institute of Technology, Cambridge, Mass.

1. Introduction

In recent years, there has been much progress in our understanding of the mechanism of transition from laminar flow to turbulent flow, but the problem remains not completely solved. Theoretically, the most easily tractable mechanism of transition is that following the development of the disturbances in the laminar motion caused by its inherent instability. When the disturbances are small, the equations describing the motion can be linearized, and the mathematical theory can be developed in great detail. Even in this case, it is well-known that the theory of the instability of parallel flows has involved a great deal of controversy, largely due to the difficulty of the mathematical problems involved.

In trying to develop the theory of the instability of the laminar motion into a theory of transition, it is clear that disturbances of finite amplitudes must be considered. In all physical problems, whatever its nature may be, there are two principal processes associated with oscillations of finite amplitudes. The first is the generation of harmonics, the second is the change of the mean condition. In the problem of hydrodynamic stability, the consequences of the second effect has been considered by MEKSYN and STUART (1951, and later by others), but the first effect has not yet been theoretically treated in any detail. Clearly, if one wants to develop the instability theory into a theory of turbulence, the harmonic components must be considered. One of the purposes of this paper (§ 3) is to bring out the remarkable fact, that *for disturbances in a parallel flow, all the harmonic components of the oscillation simultaneously become important around the critical layer, before the amplitude of the fundamental component is large enough to cause any significant distortion of the mean flow*. On the other hand, for convective instability or flow between rotating cylinders, the distortion of the mean flow appears to be more important. Indeed, there is experimental support of

[1] This work is sponsored by the Office of Naval Research of the United States Navy and may be reproduced for any purpose of the United States Government.

the results recently obtained by STUART (1957) based on a consideration of the *dual distortion process* between the mean flow and the fundamental disturbance (§ 2).

In § 4, we shall give a discussion of the multiplicity of the non-linear effects. The complicated nature of the situation leads one to wonder whether one should now attempt to carry out a general analysis of non-linear disturbances without paying special attention to any particular effect at the outset. The solution of the general equations should automatically include all of them. Such a point of view has been advocated by Professor THEODORE VON KÁRMÁN during a private discussion at the conference.

The analysis of § 3 also underlines the need for the recent improvement of the mathematical aspects of the theory of small oscillations. In order to give an adequate treatment of the non-linear effects in the critical layer, it is desirable to have an adequate representation of the gradient of the vorticity of the disturbance. Mathematically speaking, we must have uniformly valid asymptotic solutions of the Orr-Sommerfeld equation up to and including the *third-order* derivatives of the solution functions. In the earlier theories of TOLLMIEN (1947) and WASOW (1953) the solutions are proved to be valid only up to and including the derivatives of the *first* order. There are also other limitations in their theories. More recently, LANGER (1957) on the one hand and RABENSTEIN and the present writer (LIN, 1957) on the other, independently developed theories which are free from these restrictions. In the theory of LANGER, the hydrodynamical problem belongs to a class of singular cases. Special treatment has therefore to be adopted; the theory becomes more complicated and the results are more difficult to apply. In our treatment, the hydrodynamical problem is just an ordinary special case of the general theory. The explicit formulae obtained for a fundamental system of four solutions are given in § 5. A preliminary version of our work, including some of the mathematical details, was reported at the Ninth International Congress for Applied Mechanics[1].

In the Appendix, there is included an account of some recent work by L. N. HOWARD dealing with a problem in the theory of the stability of the jet. This is another example of the variety of problems occurring in the theory of hydrodynamic stability. There has also been much recent work on the instability of fluid motion involving the interface of two fluids or a free surface. Clearly, the large number of investigations made on the general subject of hydrodynamic stability makes it necessary to select and restrict the subject matter for this report; and only some of the problems relevant to the basic mechanism of transition are discussed here.

---

[1] I am indebted to Professor THEODORE VON KÁRMÁN for presenting that paper on our behalf.

2. Oscillations of finite amplitudes

The theory of small oscillations can only take account of the beginning of instability. In such a framework, both two-dimensional and three-dimensional disturbances have been treated. The next step in the development of the theory is clearly the study of disturbances of finite amplitudes. Such studies have been made by MEKSYN and STUART (1951), STUART (1956a, b, 1957), MALKUS and VERONIS (1957) and others. To fix our ideas let us first consider the case of a parallel flow.

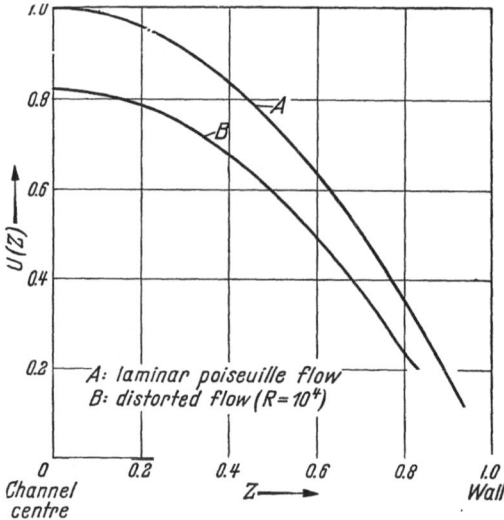

Fig. 1. Distortion of the mean velocity distribution in a channel by the Reynolds stresses due to a wavy disturbance (after STUART, 1957)

The examination of general three-dimensional oscillations is a difficult process, and consequently MEKSYN and STUART limited themselves to the discussion of the two-dimensional case. They assumed that in the case of flow between parallel plates, the non-linear effect is prominent in the distortion of the velocity profile of the mean flow (fig. 1). This in turn produces a distortion of the amplitude distribution of the oscillation. From such an investigation they concluded that the non-linear influence is essentially to make the flow less stable.

More recently, STUART (1957) developed a simpler analysis of the non-linear effect by considering the energy balance for oscillations of finite amplitude. In this analysis, the distortion of the mean flow is considered, but the consequent change in the amplitude distribution of the oscillation is neglected. (Justification of this process is given by examining some numerical examples.) Thus, the theory is applicable only to "supercritical" disturbances, i.e., to conditions where the flow is unstable on the basis of the linear theory. The non-linear effect is then mainly a limiting influence when the amplitude of the oscillation becomes of the order of ten per cent of the maximum velocity in the channel, as is observed in fully developed turbulent flow.

STUART also applied his theory to the case of flow between rotating cylinders, and calculated the resultant increase in torque required to maintain a steady motion with a finite secondary flow. The results are in agreement with TAYLOR's experimental observations (fig. 2).

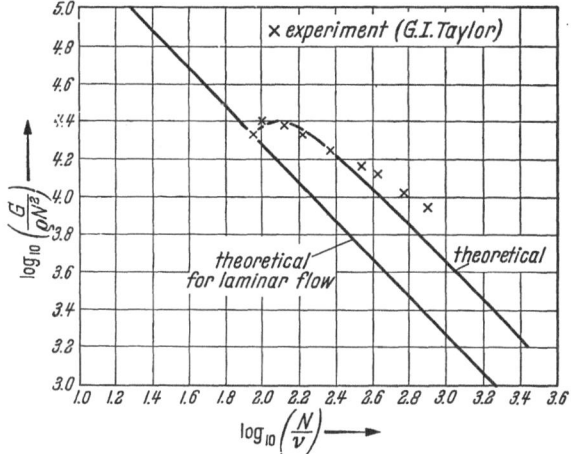

Fig. 2. Torque required to maintain a steady relative rotation of two concentric cylinders (after STUART, 1957)

MEKSYN and STUART made some analysis to show that the production of harmonics plays a less important role, and this process is therefore neglected in their theory. However, in their examination of the non-linear terms, the rapid variation of the amplitude in the critical layer is not considered. In the next section, we shall carry out a more careful examination of the non-linear terms in the critical layer, and show that the production of harmonics in that layer occurs at amplitudes *lower* than that required to give appreciable distortion of the mean motion. Thus, while the work of MEKSYN and STUART shows very clearly some important features of non-linear oscillations, the non-linear effect is only partly taken into account (cf. LIN, 1952)[1]. On the other hand, in the case of flow between rotating cylinders, there is no critical layer, and the agreement

---

[1] This was also noticed at that time by Professor LESTER LEES (private communication). The new feature in the present discussion is that the harmonics are generated in the critical layer at amplitudes of the fundamental *lower* than that giving appreciable distortion of the mean flow.

between theory and experiment lends strong support to STUART's simplified analysis.

Since the production of higher harmonics is one of the main problems under consideration we shall now substantiate the above statement with some more detailed analysis. In particular, we wish to emphasize, from such an analysis, the desirability of obtaining adequate representation of the third-order derivative of the solution function. This gives further motivation to the mathematical theory discussed in § 5.

3. Non-linear effects in the critical layer

To examine the effect of the non-linear terms, let us consider the vorticity equation

$$\frac{\partial \zeta}{\partial t} + u \frac{\partial \zeta}{\partial x} + v \frac{\partial \zeta}{\partial y} = \nu \triangle \zeta, \qquad \zeta = \triangle \psi \tag{3.1}$$

where x, y are distances parallel and perpendicular to the plates respectively, t is the time, $\psi (x, y, t)$ is the stream function,

$$u = \frac{\partial \psi}{\partial y}, \qquad v = -\frac{\partial \psi}{\partial x} \tag{3.2}$$

are the components of velocity,

$$\zeta = \frac{\partial u}{\partial y} - \frac{\partial v}{\partial x} = \triangle \psi \tag{3.3}$$

is the vorticity, and ν is the kinematic coefficient of viscosity. If we decompose the motion into a mean flow (mean taken with respect to x) and a disturbance

$$\left. \begin{aligned} \psi &= \bar\psi (y, t) + \psi' (x, y, t), \\ \zeta &= \bar\zeta (y, t) + \zeta' (x, y, t), \end{aligned} \right\} \tag{3.4}$$

and substitute these expressions into (3.1), we obtain

$$\frac{\partial \bar\zeta}{\partial t} = \nu \frac{\partial^2 \bar\zeta}{\partial y^2} - \frac{\partial}{\partial y} \overline{v'\zeta'}$$

$$\frac{\partial \zeta'}{\partial t} + \bar u \frac{\partial \zeta'}{\partial x} + v' \cdot \frac{\partial \bar\zeta}{\partial y} = \nu \triangle \zeta' + \frac{\partial}{\partial y} \overline{v'\zeta'} - \frac{\partial}{\partial x} (u'\zeta') - \frac{\partial}{\partial y} (v'\zeta'). \tag{3.5}$$

In the linear case, one considers periodic wavy solutions of the form

$$\psi' (x, y, t) = Re \left\{ \varPhi (y) e^{i\alpha (x - ct)} \right\}, \tag{3.6}$$

and obtains from (3.5) the ORR-SOMMERFELD equation

$$\varphi^{IV} - 2\alpha^2 \varphi'' + \alpha^4 \varphi = i (\alpha/\nu) [(u - c) (\varphi'' - \alpha^2 \varphi) - u'' \varphi]. \tag{3.7}$$

The critical layer is the neighborhood around the point where $u - c = 0$, in which the amplitude distribution varies rapidly, and viscous forces are important.

Let us now examine the importance of the non-linear terms. Some care has to be exercised, especially in the critical layer[1]. First, one should notice that

$$\bar{\zeta} = \frac{\partial^2 \bar{\psi}}{\partial y^2}, \quad \overline{v' \zeta'} = \overline{v' \left(\frac{\partial u'}{\partial y} - \frac{\partial v'}{\partial x} \right)} = \frac{\partial}{\partial y} \left(\overline{u' v'} \right) \tag{3.8}$$

so that (3.5) may be integrated to give

$$\frac{\partial \bar{\psi}}{\partial t} + A(t) y + B(t) = \nu \frac{\partial^2 \bar{\psi}}{\partial y^2} - \overline{u' v'} = \nu \frac{\partial \bar{u}}{\partial y} - \overline{u' v'}, \tag{3.9}$$

where $A(t)$ and $B(t)$ are "constants of integration." Thus, if one is interested in the effect of the non-linear terms on $\dfrac{\partial \bar{u}}{\partial y}$, one should use the form (3.9). On the other hand, if one is interested in the rate of diffusion of vorticity, the form (3.5) should be used. The reason for this distinction is the rapid variation of the amplitude function $\varphi(y)$ in the critical layer. Thus, if a denotes a typical magnitude of $\varphi(y)$ in the critical layer, then

$$u' = O(a/\varepsilon), \quad v' = O(a), \tag{3.10}$$

where $\varepsilon = (\nu/a)^{1/3}$, since each differentiation with respect to y increases a quantity by a factor of the order ε^{-1}. Thus, if we base our estimate of the non-linear effect on (3.9), we have to consider the ratio

$$\nu \frac{\partial \bar{u}}{\partial y} : \overline{u' v'} \approx \varepsilon^3 : \varepsilon^{-1} a^2 = \varepsilon^4 : a^2. \tag{3.11}$$

Thus, the influence in the critical layer of the Reynolds stress becomes important when

$$a = O(\varepsilon^2). \tag{3.12}$$

On the other hand, if (3.5) is used as a basis, we would have

$$\nu \frac{\partial^2 \bar{\zeta}}{\partial y^2} : \frac{\partial^2}{\partial y^2} \overline{u' v'} \approx \varepsilon^3 : \varepsilon^{-2} (\varepsilon^{-1} a^2) = \varepsilon^6 : a^2. \tag{3.13}$$

Thus, the vorticity transfer is already important compared with the viscous diffusion when

$$a = O(\varepsilon^3). \tag{3.14}$$

Again, both (3.12) and (3.14) fail to apply as an estimate outside of the critical layer, where a y-differentiation does not change the order of magnitude of a quantity. In that case

$$\nu \frac{\partial^2 \bar{\zeta}}{\partial y^2} : \frac{\partial^2}{\partial y^2} \overline{u' v'} \approx \varepsilon^3 : a^2, \tag{3.15}$$

---

[1] It is true that the form (3.6) does not hold in the general non-linear case; but if the disturbance originates from a small oscillation of that type, there is still a region where the same order-of-magnitude relation holds.

so that we arrive at the conclusion

$$a = O\,(\varepsilon^{3/2}). \tag{3.16}$$

Thus, if we consider only conditions *outside of the critical layer*, when the non-linear terms become important in the first equation of (3.5) governing the mean flow, its effect in the second equation is only of the order of

$$a^2 = O\,(\varepsilon^3). \tag{3.17}$$

This gives the order of magnitude of the second harmonics. Since the reaction of the second harmonic with the fundamental depends on their product, the influence on the fundamental is of the order of ε^3, and may be neglected. This is why MEKSYN and STUART considered the modification of the fundamental only through the change of the profile of the mean flow.

On the other hand, *inside of the critical layer*, the viscous terms in (3.5) are of paramount importance, and the non-linear terms, such as $\dfrac{\partial}{\partial y}\,(v'\,\zeta')$ become important when it is comparable with $v\,\triangle\,\zeta'$. Now

$$v\,\triangle\,\zeta' : \frac{\partial}{\partial y}\,(v'\,\zeta') \approx \varepsilon^3 \cdot \varepsilon^{-2}\,\zeta' : \frac{a}{\varepsilon}\,\zeta' \tag{3.18}$$

Thus, the transport of vorticity becomes important when

$$a = O\,(\varepsilon^2); \tag{3.19}$$

that is, when

$$\zeta' = O\,(a\,\varepsilon^{-2}) = O\,(1). \tag{3.20}$$

This criterion agrees with (3.12), which is based on a consideration of the change of velocity gradient of the mean flow inside of the critical layer. However, to get a realistic comparison, one must remember that the Reynolds stress-$\varrho\,\overline{u'\,v'}$ occurring in (3.9) and (3.11) vanishes like y^4 near the wall[1], and is actually much less than that given by (3.11), whereas no comparable reduction occurs in the vorticity transfer term in (3.18). Thus, we have reached the conclusion that as the amplitude of the disturbance grows, *the non-linear effect first shows up in the generation of harmonic modes in the critical layer* even before the distortion of the mean flow is noticeable.

A closer examination shows that the harmonics are modifying the fundamental to an appreciable extent. Indeed, if we use a subscript to denote harmonics, we have

$$v\,\triangle\,\zeta'_2 \approx \frac{\partial}{\partial y}\,(v'_1\,\zeta'_1). \tag{3.21}$$

---

[1] cf. Lin (1955), eq. (4. 5. 12), p. 63.

Since
$$v_1' = O(a) = O(\varepsilon^2),$$
eq. (3.21) shows that
$$\zeta_2' = O(\zeta_1') = O(1). \tag{3.22}$$

Since the change in $v \triangle \zeta_1'$ due to the second harmonic terms is of the order of $\dfrac{\partial}{\partial y}(v_1' \zeta_2')$, clearly the change in ζ_1' is of the order of $\zeta_2' = O(1)$.

The important conclusion is that once the amplitude of the disturbance is so large that the *vorticity of the fundamental component becomes of the order of unity* in the critical layer, the vorticity for *all* the harmonic components becomes of the *same* order, and consequently the same holds true also for the velocity components in the critical layer. Thus, one would expect to observe *many higher harmonic components in the critical layer even before the distortion of the mean flow is noticeable.* Under these circumstances, the non-linear influence is also negligible outside of the critical layer.

The simultaneous occurrence of all higher harmonics of the same order of magnitude makes it difficult to carry out the analysis in a simple manner[1]. At the same time the importance of the non-linear term $\dfrac{\partial}{\partial y}(v' \zeta') = v' \dfrac{\partial \zeta'}{\partial y} + \dfrac{\partial v'}{\partial y} \zeta'$ clearly indicates the desirability of getting an accurate representation of $\varphi'''(y)$ in the critical layer. This gives a strong motivation for the development of the theory to be described below (§ 5). The solution oft he boundary value problem in the linear case requires only an adequate approximation for φ and φ', which is much less stringent than that imposed by the non-linear theory.

4. Some general remarks about disturbances in laminar flows

The above discussions show clearly that the theory of oscillations of finite amplitude is very complicated, even for two-dimensional disturbances. In reality, the disturbances are known to be three-dimensional. The three-dimensional nature of the disturbance may be expected, if for no other reason, from the mere fact that they become unstable at sufficiently high Reynolds numbers. To make this position tenable, one has to explain why the *free* oscillations observed by SCHUBAUER and SKRAMSTAD (1947) conform so well to the two-dimensional theory. Indeed, on the basis of the stability theory, one would expect two-dimensional oscillations to develop to a maximum amplitude at a condition close to the second branch of the curve of neutral stability. If observations are now made at a given location in the boundary layer, with no special effort made to control the oscillation, the randomly ex-

---

[1] To get an idea of the nature of the problems involved, see STUART (1956 b), p. 34.

152   C. C. LIN

citated oscillations will be amplified in a selective manner, and the most
dominant frequency would be that corresponding to maximum amplification. The frequency of the free oscillations observed indeed conforms to
this idea, if two-dimensional disturbances are assumed. Why are the
three-dimensional disturbances not observed?

The answer is not difficult to find if one examines the amplification
characteristics of the three-dimensional disturbances. Indeed they have

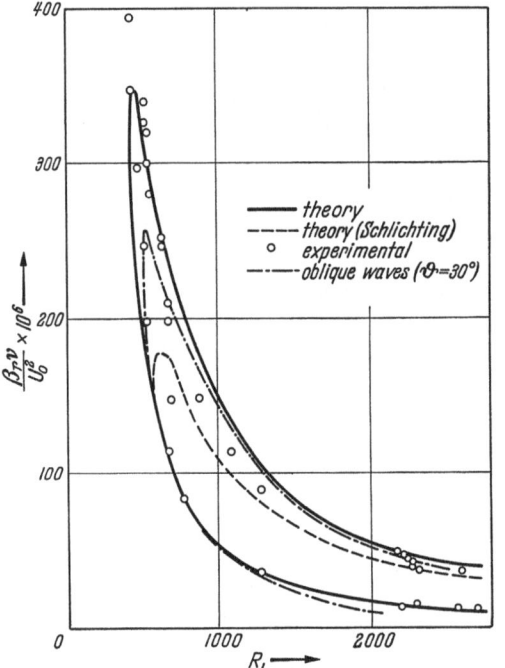

Fig. 3. Stability characteristics for boundary layer over a flat plate. Free oscillations are observed
for $R_1 \approx 2000$

nearly the *same* second branch for the neutral curve in the range of the
experimental data (fig. 3). Since only the frequency of the oscillation
was observed by SCHUBAUER and SKRAMSTAD, the two-dimensional and
three-dimensional oscillations cannot be distinguished from each other.
Thus, one would indeed expect the oscillations in a boundary layer to
become progressively three-dimensional as the Reynolds number of the
motion increases in a downstream direction.

Another reason for expecting three-dimensional disturbances has been
discussed by GÖRTLER and WITTING at the present conference. A remark
may perhaps be added as to the nature of the disturbance studied by
them. Since the instability discussed by GÖRTLER and WITTING is based

on that associated with a steady three-dimensional motion, one can get a clear physical understanding by viewing the "cat's-eye" picture (fig. 4) described by Lord KELVIN for a two-dimensional wavy disturbance in a shear flow, i.e., by considering an observer moving with the wave so that a stationary pattern is observed. Clearly then there would be *two cells* of secondary flow in each period, one above the corner of the eye and one below the corner of the eye. In each case, the centrifugal force associated with the curvature of the stream lines would tend to throw the high speed particles toward the critical layer where the fluid is at rest relative to the moving observer. This double-cell structure of the secondary flow,

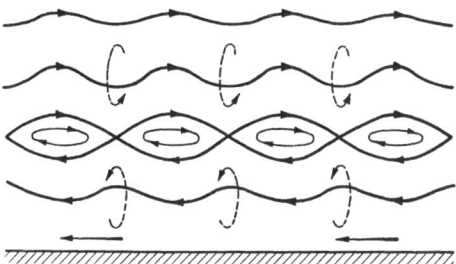

Fig. 4. Stream line as observed by an observer moving with the wave speed

if it does occur, may account for the multiple change of sign of the correlation coefficient in the velocity fluctuations reported by SCHUBAUER in the previous session.

The complication of the picture by the three-dimensional nature of the disturbances does not alter the qualitative conclusion reached in § 3 that many of the harmonics of the oscillation should become noticeable in the critical layer before they are noticeable elsewhere in the boundary layer, and before the distortion of the profile of mean flow can be detected. In addition to this non-linear effect deduced from a two-dimensional theory, there is a stretching of vortex lines associated with three-dimensional disturbances of finite amplitudes. One is therefore led to wonder whether one should simply carry out a general analysis of non-linear disturbances without focusing particular attention to any one of the individual effects considered above. The solution of the non-linear equations should automatically yield these effects. Such a point of view has been advocated by Professor VON KÁRMÁN during a private discussion at the conference.

There is still one more point of general interest that should be considered here. Some years ago, LANDAU (1944) proposed a theory of successive instabilities as a possible mechanism for transition. The theory of GÖRTLER and WITTING is certainly in line with these ideas. The disturbances developed in the cases of convective instability and the instability of the flow between rotating cylinders also appear to conform to such a concept. On the other hand, the development of harmonics on the basis of the non-linear effects does not appear likely to be marked by critical values of the amplitude of the oscillation. But the final breakdown of a disturbed flow with observable harmonics into a random turbulent motion may yet agree with LANDAU's concepts.

5. Mathematical theory of two-dimensional oscillations of small amplitudes

The theory of stability of a two-dimensional parallel flow with respect to two-dimensional disturbances have been well established from a physical point of view. There are, however, certain inadequacies in the methods of solution used which limit the applicability of the theory, and also tend to lead to controversy[1]. In general, the difficulty lies in the lack of uniformity of the asymptotic methods used. This purely mathematical issue will now be discussed.

The basic equation for small disturbances in a parallel flow is the equation of ORR and SOMMERFELD:

$$\varphi^{IV} - 2\,a^2\,\varphi'' + a^4\,\varphi = i\,a\,R\,[(w-c)\,(\varphi'' - a^2\,\varphi) - w''\,\varphi], \quad (5.1)$$

where $w\,(y)$ is the basic velocity distribution in dimensionless form, R is the Reynolds number, and the disturbance is given by the stream function

$$\varPsi'\,(x,\,y) = Re\,\{\varphi\,(y)\,\exp\,[ia\,(x-ct)]\}. \quad (5.2)$$

Most existing work deals with the case of large Reynolds number R, and asymptotic solutions for large aR are usually used. Such methods are, however, found to be inadequate for the study of profiles of the jet type. In fact, HOWARD (see Appendix) has shown that along the lower branch of the neutral curve, $aR \to 0$ as $R \to \infty$; thus, the theory of asymptotic solutions for large aR is even inadequate for large Reynolds numbers of the flow, and a method must be developed for dealing with cases of small aR. Since aR as a parameter enters (5.1) in a regular manner, one idea would be to try solutions of the form[2]

$$\varphi = \varPhi^{(0)} + aR\varPhi^{(1)} + \cdots \quad (5.3)$$

HOWARD however developed another approach based on the conversion of the boundary value problem into an integral equation, and satisfactory numerical results are obtained. An outline of the main steps used is described by HOWARD in an appendix to this paper.

---

[1] For a discussion of the various issues, see LIN (1955), Chapter 8.

[2] At the conference, TATSUMI presented results based on such a method. They are in substantial agreement with HOWARD's.

For large values of aR, the non-uniformity of simple asymptotic solutions has long been realized and there is a large body of literature on the problem of getting uniformly valid solutions for large aR. It has long been recognized that the application of simple asymptotic methods to (5.1) leads to formal solutions with branch behavior at the *critical* point y_c where $w - c = 0$. Indeed the solutions take on the forms

$$\varphi = \varphi^{(0)}(y) + (aR)^{-1}\varphi^{(1)}(y) + \cdots \qquad (5.4)$$

and

$$\varphi = \exp\left[\pm (aR)^{1/2} Q(y)\right] \cdot \left\{f^{(0)}(y) \pm (aR)^{-1/2} f^{(1)}(y) + \cdots\right\} \qquad (5.5)$$

where

$$Q(y) = \int_{y_c}^{y} [i(w-c)]^{1/2}\, dy, \quad f^{(0)}(y) = (w-c)^{-\frac{5}{4}}, \ldots \qquad (5.6)$$

The initial approximation $\varphi^{(0)}(y)$ in (5.4) satisfies the inviscid equation

$$(w - c)(\varphi'' - a^2 \varphi) - w'' \varphi = 0. \qquad (5.7)$$

One of the solutions of (5.7) (commonly denoted by $\varphi_1^{(0)}$) is regular, while another (commonly denoted by $\varphi_2^{(0)}$) has a logarithmic behavior. Such solutions are obviously inadequate near the critical point y_c, and other asymptotic solutions were developed in the earlier theoretical work for the neighborhood of y_c. The clarification of the nature of the asymptotic solutions (5.4) and (5.5) and their proper use led to a great deal of controversy, which are clearing up only in recent years[1]. The most effective method of settling the difficulty and of getting solutions adequate for all cases with aR large is to obtain asymptotic solutions which are indeed uniformly valid in a neighborhood containing the critical point. Solutions of this general nature have been obtained by TOLLMIEN (1947), WASOW (1953), and LANGER (1957). However, each one of these solutions has its own limitations. Both TOLLMIEN's and WASOW's solutions are limited to a first approximation and there is no obvious algorism for obtaining the higher approximations. Also, they only proved the adequacy of their solutions up to and including their first derivatives, and there is no indication whether the higher derivatives are adequately given by their formulae. Besides, TOLLMIEN's solution is limited to the case of neutral oscillations; while one of WASOW's solutions still has a singularity[2]. LANGER's theory is free from such difficulties, and indeed deals with a general class of differential equations which includes (5.1) as a special case. However, the hydrodynamical problem belongs to a class of sin-

---

[1] See LIN (1955), Chapter 8.

[2] This singularity can be removed by a slight modification of his method (DI-PRIMA, 1954). Another minor difficulty can also be removed by a transformation of variable.

gular cases in his theory, and the solution of (5.1) would depend on the coefficients of the expansion of $w - c$ as a power series of $y - y_c$ in a rather lengthy manner. An explicit representation of the solution is not obtainable in terms of easily definable quantities. A theory developed by RABENSTEIN and the present writer is free from the above-mentioned difficulties. It yields a fundamental system of solutions of the ORR-SOMMERFELD equation in the form

$$\varphi = K_0 u + K_1 \dot{u} + \lambda^{-2} [K_2 \ddot{u} + K_3 \dddot{u}]. \tag{5.8}$$

In this formula, u is a solution of the equation

$$\ddot{u} + \lambda^2 (z\ddot{u} + \beta u) = 0, \tag{5.9}$$

where

$$z = \left[\frac{3}{2} \int_{y_c}^{y} (w - c)^{1/2} \, dy\right]^{2/3} \tag{5.10}$$

and

$$\lambda^2 = i a R, \qquad \beta = - w_c''/(w_c')^{4/3}. \tag{5.11}$$

(A dot over a quantity denotes differentiation with respect to z whereas a dash denotes differentiation with respect to y.) The coefficients K_i are given by the following formulae:

$$K_0 = (w_c')^{-7/6} [k H_0^{(1)} (2\beta^{1/2} z^{1/2} e^{-\pi i}) \varphi_1^{(0)} + J_0 (2\beta^{1/2} z^{1/2}) \varphi_2^{(0)}], \tag{5.12}$$

$$K_1 = (w_c')^{-7/6} 1/\beta (\beta z)^{1/2} [k H_1^{(1)} (2\beta^{1/2} z^{1/2} e^{-\pi i}) \psi_1^{(0)} - J_1 (2\beta^{1/2} z^{1/2}) \psi_2^{(0)}], \tag{5.13}$$

$$K_2 = K_0/z + K_3 - z^{5/4} (w - c)^{-5/4}, \tag{5.14}$$

$$K_3 = K_1/z; \tag{5.15}$$

where $\varphi_1^{(0)}$ and $\varphi_2^{(0)}$ are the regular and singular solutions of the inviscid eq. (5.7) with constant factors specified as follows

$$\varphi_1^{(0)} = (y - y_c) \{1 + \cdots\} \tag{5.16}$$

$$\varphi_2^{(0)} = 1 + \cdots + \frac{w_c''}{w_c'} \varphi_1^{(0)} \ln (y - y_c). \tag{5.17}$$

The constant k is given by

$$k = - \pi i w_c''/w_c', \tag{5.18}$$

and is such that the functions K_0 and K_1 are both regular. Regularity of K_2 and K_3 is also thus assured despite of the factors $1/z$ appearing in (5.14) and (5.15). The solution (5.8), together with derivatives up to the third order, are accurate up to a factor of the form $1 + O(\lambda^{-4/3} \ln \lambda)$.

A preliminary report of this theory was presented at the 9th International Congress of Applied Mechanics. At that time, the proof of the existence theorem was not yet carried out. It was also realized later that the coefficients of the reference eq. (5.9) has to be modified by higher order terms if the higher approximations are to be carried out[1]. However, the main ideas of the theory are not changed. This modified form is given in another report (LIN, 1958). For practical purposes, it is sufficient to use the form (5.8) given here on the basis of the simpler eq. (5.9).

The uniformly valid solutions have various applications in the stability theory. Their generalization to the supersonic case would allow better results to be obtained. They also give hope of meeting the challenge formulated[2] for applied mathematicians by Professor J. L. SYNGE (1938) some twenty years ago; namely to prove the stability or instability of the plane Poiseuille motion without heavy numerical calculations. These are interesting problems for further investigations. But by far the most important application would be their use for the study of the non-linear theory. Thus they represent the end of the long trail of the mathematical investigations of the linear theory and at the same time the beginning of another long trail of the theory of non-linear oscillations.

Appendix[3]

Stability of the two dimensional jet

By

L. N. Howard

Massachusetts Institute of Technology, Cambridge, Mass.

The ORR-SOMMERFELD equation can be written in the form

$$(D^2 - \alpha^2) (D^2 - \omega^2) \Phi = i \, \alpha \, R \, [w \, (D^2 - \alpha^2) \, \Phi - w'' \, \Phi] \qquad (1)$$

where $\omega^2 = \alpha^2 - i \, \alpha \, R \, c$, $D = d/dy$, $w = w \, (y) = \operatorname{sech}^2 y$ and otherwise the notation is standard (cf. LIN's book). Vanishing velocity components at $y = \pm \infty$ give for boundary conditions

$$\alpha \, \Phi \, (\pm \infty) = \Phi' \, (\pm \infty) = 0 \qquad (2)$$

and we are interested here in *even* eigenfunctions $\Phi \, (y)$. By regarding the right-hand side of (1) as an inhomogeneous term, one can solve the constant-coefficients differential equation that remains, using (2), to get a representation of Φ in terms of the right-hand side of (1). The term in Φ''

---

[1] There is also a slight change of notation. The independent variable z used in the earlier paper (denoted by z^* below) is related to the present variable z by the formula $z^* = z \exp (7 \, \pi \, i/6)$.

[2] Also pointed out by Professor S. GOLDSTEIN in his early papers on this subject.

[3] Dieser Anhang wurde auszugsweise von C. C. LIN in der Diskussionsveranstaltung zur V. Sitzung vorgetragen, und zwar im Anschluß an den Diskussionsbeitrag von T. TATSUMI. — D. Hrsg.

can be reduced by some integration by parts, again using (2), to one in Φ alone, and in this way one converts the original problem into the following integral equation ($Re\ \omega \geq 0$)

$$\Phi(y) = i\,\alpha\,R \int_{-\infty}^{\infty} \left\{ -\frac{1}{2\,\omega} e^{-\omega|y-y'|} w(y') + \right.$$
$$\left. + \frac{1}{\omega^2 - \alpha^2} (e^{-\alpha|y-y'|} - e^{-\omega|y-y'|}) \frac{|y-y'|}{(y-y')} w'(y') \right\} \Phi(y')\,dy'. \tag{3}$$

One application of this formulation of the problem is to give a simple proof of the following

Theorem: There is no point on the neutral curve with

$$R < [1 + \pi\sqrt{2/3}]^{-1} \cong 0.28.$$

The idea of the proof is as follows: By using the fact that c is real on the neutral curve one shows that $|\omega| \geq Re\ \omega \geq \alpha$ and then some simple estimates show that the kernel of (3) is in absolute value

$$\leq \frac{R}{2}[1 + 2\,|\,y - y'\,|]\,w(y').$$

We thus have

$$|\Phi(y)| \leq \frac{R}{2} \int_{-\infty}^{\infty} [1 + 2\,|\,y - y'\,|]\,w(y')\,|\,\Phi(y')\,|\,dy'. \tag{4}$$

Setting $N^2 = \int_{-\infty}^{\infty} w(y)\,|\,\Phi(y)\,|^2\,dy$, multiplying (4) by $w\,|\,\Phi(y)\,|$ and integrating, and using the SCHWARTZ inequality repeatedly, one obtains $N^2 < RN^2\,[1 + \pi\sqrt{2/3}]$, from which the result follows.

The behavior of the lower branch of the neutral curve can be studied as follows. Consider R fixed and try to solve the problem by a perturbation about $\alpha = 0$. At first glance this appears to be a regular perturbation problem, for no derivatives are lost in (1) on setting $\alpha = 0$, but it is really a singular perturbation problem for it turns out that the series for Φ in powers of α, $\Phi = \Phi_0 + \alpha\,\Phi_1 + \alpha^2\,\Phi_2 \ldots$ is not a uniformly (in y) valid asymptotic expansion, the non-uniformity occurring at $y = \pm\,\infty$. It is thus not possible to get such a perturbation solution of the differential equation because the boundary conditions at $\pm\,\infty$ cannot be used. This suggests that one has a "boundary layer at infinity" and that one should use two expansions with suitable matching conditions as in ordinary boundary layer problems. This double expansion can be avoided here, however, by using the integral equation formulation (3). This is because the boundary conditions have already been used in obtaining (3), and the error in Φ, *at large y only*, committed by using the truncated expansion $\Phi = \Phi_0 + \alpha\,\Phi_1 + \cdots$ under the integral sign, does not cause trouble because of the factors w and w' (which fall rapidly to zero as $y \to \pm \infty$) in the kernel. In this way it is possible to find the terms in the expansion in powers of α and with them the eigenvalue relation between $\alpha, R,$ and c. More specifically the procedure is as follows. It turns out that the proper expansion to assume for ω is $\omega = \alpha \, \omega_0 + \alpha^2 \, \omega_1 + \cdots$ and putting this into the kernel of (3) one gets an expansion of the kernel in powers of α (for fixed R) of the form:

$$K\,(y, y') = K_0 + \alpha \, K_1 + \alpha^2 \, K_2 + \cdots \tag{5}$$

here K_n is a known function of y, y', R and the coefficients $\omega_c, \omega_1, \ldots \omega_n$, but of no higher ω_k's. Putting the expansions of K and φ into (3) and separating the various powers of α gives a sequence of integral equations:

$$\Phi_0\,(y) = \int_{-\infty}^{\infty} K_0\,(y, y') \, \Phi_0\,(y') \, d \, y' \tag{6}$$

$$\Phi_1\,(y) = \int_{-\infty}^{\infty} K_0 \, \Phi_1 \, d \, y' + \int_{-\infty}^{\infty} K_1 \, \Phi_0 \, d \, y' \tag{7}$$

$$\Phi_2\,(y) = \int_{-\infty}^{\infty} K_0 \, \Phi_2 \, d \, y' + \int_{-\infty}^{\infty} K_1 \, \Phi_1 \, d \, y' + \int_{-\infty}^{\infty} K_2 \, \Phi_0 \, d \, y' \tag{8}$$

etc.

Besides these we have for convenience the normalizing condition $\Phi\,(0) = \Phi_0\,(0) + \Phi_1\,(0)\,\alpha + \cdots = 1$. The integral eq. (6) then determines Φ_0 and the eigenvalue parameter ω_0. With the normalization it turns out that in fact $\Phi_0 = 1$. Using these in (7) we then get a relation which determines ω_1, and, with the normalizing condition, Φ_1. This process can be continued as far as desired, although the algebraic complications soon become very great. Once the coefficients ω_k (functions of R only) are determined one easily gets from them the corresponding coefficients in the expansion of the wave speed $c = c_0 \, \alpha + c_1 \, \alpha^2 + \cdots$, and then by setting Im $c = 0$ one gets the equation of the neutral curve. For large values of R one can write simple formulas, and the results are as follows: Using only c_0 and c_1 one gets for the neutral curve:

$$\alpha = \frac{8}{9 \, R^2} + \frac{16}{R^4} + \cdots$$

Using c_0, c_1 and c_2 this is somewhat modified:

$$\alpha = \frac{.954}{R^2} + \frac{5.4}{R^4} + \cdots$$

Using more c_k's will no longer affect the coefficient of R^{-2}, but that of R^{-4} will be changed further. From this it is clear that as $R \to \infty$ on the lower branch, $\alpha \, R \to 0$.

For smaller values of R it is necessary to compute the neutral curve numerically. The computations are simple in principle but difficult in practice because they involve complicated algebraic formulas with complex numbers. The calculations have not been completed yet but indications are that using c_0, c_1 and c_2 leads to a curve with a minimum critical Reynolds number of about 4, and that this will probably not be modified too much by the inclusion of higher terms.

The results of the computations and the mathematical details of the method will be given elsewhere. The main results of this investigation are:

1. As $R \to \infty$ along the lower branch of the neutral curve, Φ does *not* tend to a solution of the non-viscous equation of hydrodynamic stability;

2. In fact, $\alpha R \to 0$ like $1/R$.

3. The neutral curve is definitely bounded away from $R = 0$.

4. A method is given for computing the lower branch of the neutral curve, and this seems to be applicable down to the minimum critical Reynolds number, which is roughly 4.

References

DIPRIMA, R. C. (1954): J. Math. Phys. **33**, 249.

LANDAU, L. (1944): C. R. Acad. Sci. U.R.S.S. **44**, 311.

LANGER, R. E. (1957): Trans. Am. Math. Soc. **84**, 144–191.

LIN, C. C. (1952): Math. Rev. **13**, 792.

LIN, C. C. (1955): *Hydrodynamic Stability* (Camb. Univ. Press).

LIN, C. C. (1957): Proc. 9th Int. Congr. Appl. Mech. (Brussels 1956), pp. 136–148.

LIN, C. C. (1958): To appear in Proceedings of Symposium on Naval Hydrodynamics, U.S. Office of Naval Research (Washington, D. C., 1956).

MALKUS, W. V. R., and G. VERONIS (1957): Paper submitted to the Journal of Fluid Mechanics.

MEKSYN, D., and J. T. STUART (1951): Proc. Roy. Soc. **A 208**, 517.

SCHUBAUER, G. B., and H. K. SKRAMSTAD (1947): J. Aero. Sci. **14**, 69–78.

SHEN, S. F. (1954): J. Aero. Sci. **21**, 62–64.

STUART, J. T. (1956a): J. Aero. Sci. **23**, 86, 894.

STUART, J. T. (1956b): Z. angew. Math. Mech. Sonderheft (1956), p. 532.

STUART, J. T. (1957): To appear in J. Fluid Mech.

SYNGE, J. L. (1938): Semi-centennial publications of Am. Math. Soc. **2**, 227–269

TOLLMIEN, W. (1947): Z. angew. Math. Mech. **25/27**, 33–50, 70–83.

WASOW, W. (1953): Ann. Math. **58**, 222–252.

Aus der Diskussion

H. GÖRTLER (Freiburg i. Br.): 1. Referring to fig. 4, the "cat's-eye" picture, it was, of course, exactly this streamline picture as seen by an observer moving with the wave speed that Dr. WITTING had in mind when introducing his coordinates ξ, η. 2. I also appreciate Dr. LIN's reference to the theory of successive instabilities proposed by LANDAU in 1944. The general idea of transition as result of a succession of instabilities is probably very old. This idea was brought to my mind in a course on boundary layer flow given by PRANDTL at Göttingen in the 1930's. However, I always had the impression that the secondary and higher-order instabilities were expected to result only as non-linear effects. Dr. WITTING's paper shows that an assumption of this kind is not correct, at least not in the case of a boundary layer with a wavy primary disturbance.

Vergleich von theoretischen Ansätzen zur Bestimmung des Umschlags laminar-turbulent in drei Dimensionen mit Versuchen im Windkanal der O.N.E.R.A. zu Cannes

Von

E. A. Eichelbrenner und R. Michel[1]

O.N.E.R.A., Châtillon-sous-Bagneux

1.

1.1. Bei dreidimensionalen Hindernissen, deren Oberfläche keine Gebiete mit zu kleinen Krümmungsradien aufweist, führt die Berechnung der laminaren Grenzschicht mit Hilfe des „Prävalenzprinzips" zu zufriedenstellenden und mit den Versuchen gut übereinstimmenden Ergebnissen [*1, 2*].

1.2. Bekanntlich besteht das Prävalenzprinzip in der Zerlegung der in speziellen Koordinaten geschriebenen Gleichungen der dreidimensionalen Grenzschicht in ein zweidimensionales System für die Primärströmung und eine Zusatzgleichung für die zunächst vernachlässigte Querströmung in der Grenzschicht. Das spezielle Koordinatensystem ist quasiorthogonal, gebildet aus der örtlichen Richtung der Potentialstromlinien (s), der Äquipotentiallinien (z) und der Normalen zur Oberfläche des Hindernisses (n).

1.3. Selbst in der Nähe der Ablösung, die von der Theorie mit großer Genauigkeit vorausgesagt wird, werden dadurch die primäre und die sekundäre Grenzschicht rechnerisch in gutem Einklang mit den Beobachtungen, z. B. im Wassertunnel zu Châtillon, erhalten.

1.4. Man hat damit eine Methode zur Hand, die es gestattet, auch in bereits technisch interessanten Fällen (z. B. an angestellten oder abgeplatteten Rotationskörpern) die laminare Grenzschicht mit ausreichender Sicherheit zu berechnen. Leider aber bleibt bei den für die Technik wichtigen Konfigurationen die Grenzschicht in den seltensten Fällen bis zur Ablösung laminar. Es ist daher ein wünschenswertes, wenn auch noch fernes Ziel, turbulente Grenzschichten in drei Dimensionen einigermaßen verläßlich berechnen zu können.

---

[1] Vorgetragen von E. A. Eichelbrenner.

1.5. Als erster Schritt dazu, der den Vorteil hat, von den recht verläßlichen Ergebnissen der laminaren Theorie ausgehen zu können, soll hier der Versuch gemacht werden, den Umschlag laminar-turbulent an einem angestellten Rotationskörper vorauszuberechnen und die Ergebnisse dieser Rechnungen durch Versuche zu kontrollieren.

2.

2.1. Leider aber liefert die Theorie noch kein wirkliches Umschlagskriterium, und zwar auch nicht im zweidimensionalen Fall. Man muß also empirische und halbempirische Kriterien heranziehen und die Möglichkeit ihrer Verallgemeinerung auf drei Dimensionen untersuchen.

Bis etwa 1951 beschränkten sich die meisten (zweidimensionalen) Umschlagskriterien auf die Angabe einer kritischen Reynoldszahl R_s oder R_{δ_2}, deren Wert bekanntlich auch noch mit dem betrachteten Profil und den Gegebenheiten des Problems wechselte, vgl. z. B. [3], Kapitel II und XVI.

In den kritischen Reynoldszahlen bedeutet dabei δ_2 die Impulsverlustdicke, s die Lauflänge in der Grenzschicht.

2.2. Es ist einleuchtend, daß ein solches Kriterium keine Verallgemeinerung auf drei Dimensionen zuläßt, jedenfalls solange man nicht weiß, wie sich die erwähnten Formeinflüsse gegenüber den Methoden dreidimensionaler Grenzschichtrechnungen (z. B. dem Prävalenzprinzip oder der verallgemeinerten Mangler-Transformation) verhalten.

2.3. Im Jahre 1951 hat R. Michel ein allgemeines Umschlagskriterium für zwei Dimensionen angegeben [4]: Er stellte die ihm zugänglichen Versuchsergebnisse durch Punkte in der (R_s, R_{δ_2})-Ebene dar, die sich recht gut um eine einzige Kurve gruppieren. Repräsentiert man dann die Entwicklung der laminaren Grenzschicht eines Hindernisses durch eine Kurve $R_{\delta_2} (R_s)$ in dieser Ebene, so ergibt ihr Schnittpunkt mit der Kurve von Michel den Umschlagspunkt an dem betrachteten Hindernis.

2.4. Dies Kriterium wurde von A. O. Smith untersucht [5], der zeigen konnte, daß es im Rahmen der Genauigkeit der zugänglichen Beobachtungen mit der Kurve $\int \beta_i \, dt = $ const. zusammenfällt, die einer festen Anfachung der nach Tollmien-Schlichting jeweils kritischsten Frequenz in der Grenzschicht entspricht: Unter der Annahme nämlich, daß sich die Strömung um ein Hindernis lokal durch eine Strömung nach Hartree mit konstantem m ersetzen läßt, liefern die Pretschschen Tafeln eine Beziehung zwischen m, R_{δ_2} und $\int \beta_i \, dt$, durch die für gegebenes $\int \beta_i \, dt = $ const. jedem R_{δ_2} ein Wert R_s zugeordnet werden kann, der für const. $= 9$ mit dem durch die Kurve von Michel vermittelten R_s

im wesentlichen zusammenfällt. Unabhängig davon hat fast gleichzeitig I. v. INGEN [6] ein ähnliches Kriterium angegeben, bei dem er const. $= 7,8$ wählt.

Damit ist das Kriterium von MICHEL durch eine präzise Formel mit der TOLLMIEN-SCHLICHTINGschen Stabilitätstheorie verknüpft worden, berechtigt also zu der Hoffnung, mathematischen Transformationen gegenüber gewisse Invarianzeigenschaften aufzuweisen.

2.5. In der Tat konnte A. O. SMITH in der angegebenen Arbeit dies Kriterium auf nicht angestellte Rotationskörper dadurch ausdehnen, daß er die (R_s, R_{δ_2})-Ebene einer speziellen MANGLER-Transformation unterwarf: Bekanntlich bleibt bei dieser Transformation eine Konstante willkürlich, da die Koordinaten der ebenen Ersatzströmung, die einem Rotationskörper vom Radius r (s) entspricht, durch

$$\bar{s} = \frac{1}{L^2} \int\limits_0^s r^2 (s) \, ds \quad \text{uud} \quad \bar{n} = \frac{r(s)}{L} \cdot n \qquad \text{mit beliebigem } L$$

gegeben werden.

A. O. SMITH wählt nun, mit einer Art „Variation der Konstanten", L so, daß für jedes s der Wert von $R_{\bar{s}}$ mit dem von R_s übereinstimmt. Er behält also die mit der Lauflänge gebildete Reynoldszahl bei und bildet $R_{\bar{\delta}_2}$ für jedes s nach MANGLER mit dem so bestimmten lokalen L.

2.6. Der Vergleich der so erhaltenen Umschlagspunkte mit Versuchsergebnissen führt zu annehmbaren Resultaten, die allerdings weniger befriedigend sind als im ebenen Fall; immerhin halten sich die Abweichungen in der Größenordnung von etwa 15% des Wertes von s_{trans}.

3.

3.1. Man kann infolgedessen hoffen, daß das Kriterium von MICHEL bei Hindernissen, die nicht zu sehr von Rotationskörpern abweichen (also z.B. bei schwach angestellten oder nur wenig abgeplatteten Rotationskörpern), auch dann noch gültig bleibt, wenn man es auf einen nach der verallgemeinerten MANGLERschen Umformung transformierten Stromstreifen an der Oberfläche des Hindernisses, vgl. [1], anwendet. Man hat

$$\bar{s} = \frac{1}{L^2} \int\limits_0^s e^2 (s) \, ds \quad \text{und} \quad \bar{n} = \frac{e(s)}{L} \cdot n \,,$$

wo e (s) die örtliche Breite eines hinreichend schmal gewählten Stromstreifens ist.

3.2. Wieder wählt man das lokale L so, daß $R_{\bar{s}} = R_s$ wird; die Kurve $R_{\delta_2} (R_s)$ ergibt dann in der (R_s, R_{δ_2})-Ebene den Umschlagspunkt durch ihren Schnitt mit der Kurve $\int \beta_i \, dt = 9$.

3.3. Selbstverständlich muß dabei die Impulsverlustdicke unter Berücksichtigung nicht nur der primären, sondern auch der sekundären Grenzschichtströmung in dem betrachteten Stromstreifen berechnet werden. Man könnte dabei etwa folgendermaßen vorgehen: Man bestimmt den Impuls der Flüssigkeitsmenge, die in der Zeiteinheit durch ein Flächenelement der Höhe dn senkrecht zur lokalen Richtung der Strömung in der Grenzschicht fließt. Dieselbe Flüssigkeitsmenge hätte in der Außenströmung einen entsprechend davon verschiedenen Impuls gehabt. Der Unterschied dieser beiden Impulse entspricht dem einer Schicht von der Dicke $d\,\delta_2$ in der Außenströmung.

3.4. Praktisch fällt das so bestimmte $\delta_2 = \int\limits_{n=0}^{\infty} d\,\delta_2$, außer in unmittelbarer Nähe der laminaren Ablösung, mit der Impulsverlustdicke der Primärströmung zusammen; es wird daher im allgemeinen genügen, diese letztere zugrunde zu legen, womit man den weiteren Vorteil gewinnt, eines der auf der Berechnung der Impulsverlustdicke beruhenden globalen Verfahren im Zweidimensionalen, z. B. das POHLHAUSENsche, unmittelbar anwenden zu können.

4.

4.1. Die im vorstehenden auseinandergesetzte Verallgemeinerung des Kriteriums von MICHEL wurde auf ein Rotationsellipsoid vom Achsenverhältnis 1:6 für 0° und 10° Anstellung bei den Reynoldszahlen

$$R_a = \frac{2a\,U_\infty}{v} = 2, 4, 6 \cdot 10^6 \qquad (a = \text{halbe große Achse des Ellipsoids})$$

angewendet. Die entsprechenden Versuche wurden im Windkanal der O.N.E.R.A. in Cannes bei Windgeschwindigkeiten zwischen 15 und 45 m/sec durchgeführt. Diese relativ niedrigen Windgeschwindigkeiten wurden gewählt, um Vergleiche mit den Rechnungen, die im Inkompressiblen durchgeführt wurden, zu erleichtern. Als Nachteil mußten relativ große Abmessungen der Modelle ($a = 1$ m) in Kauf genommen werden.

4.2. Es wurden die folgenden Untersuchungen vorgenommen:

1. Kontrolle der nach K. MARUHN [7] berechneten Druckverteilung.

2. Nachprüfung der nach dem Prävalenzprinzip berechneten Geschwindigkeitsprofile und Grenzschichtdicken im laminaren Gebiet.

3. Sichtbarmachung der Umschlagslinie an der Körperoberfläche mit Hilfe der Kaolin-Methode.

Die ersten beiden Punkte (Druckverteilung und Geschwindigkeitsprofile) ergaben, vgl. Abb. 1 und 2, eine sehr gute Übereinstimmung zwischen Theorie und Experiment. Von einer Ausnahme (Abb. 2c, $x/l = 0,5$) abgesehen, auf die wir noch zurückkommen, können also etwaige Abweichungen im Verlauf der berechneten und der beobachteten Umschlagslinie nicht den der Rechnung zugrunde liegenden Ausgangsdaten zugeschrieben werden.

4.3. Abb. 3 gibt einen ersten Vergleich zwischen den Ergebnissen der vorgeschlagenen Theorie und den Versuchen in Cannes: Es wird die Lage des Umschlagspunktes auf Oberseite, Unterseite und an der Flanke des

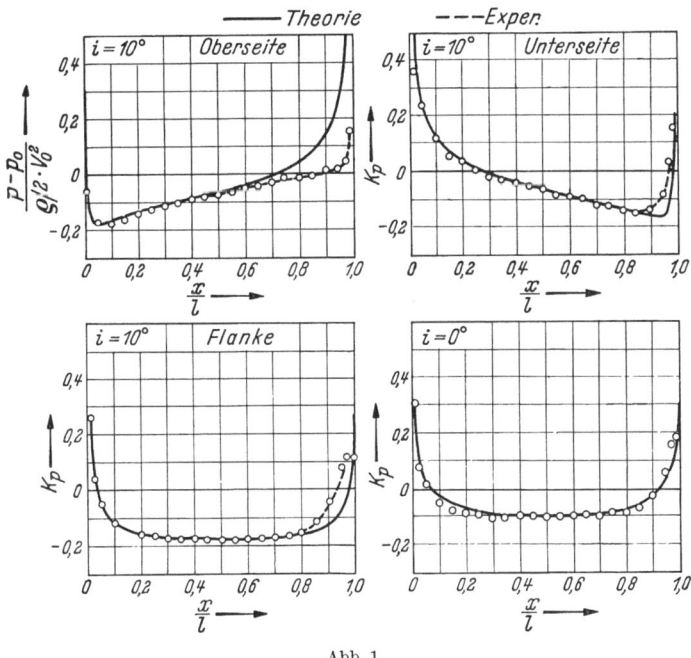

Abb. 1

Ellipsoids als Funktion der Reynoldszahl R_a wiedergegeben, und zwar für Anstellwinkel von 0° und von 10°.

4.4. Die Übereinstimmung zwischen Rechnung und Versuch ist durchaus annehmbar für das nicht angestellte Ellipsoid: Die Abweichungen halten sich in dem von A. O. SMITH angegebenen Rahmen und können noch ohne weiteres durch den verhältnismäßig hohen Turbulenzgrad (nahe 1%) des Windkanals sowie durch die geringe Präzision der Kaolin-Methode erklärt werden.

Das Kriterium von MICHEL wurde in der Tat mit Hilfe von Versuchen in Kanälen mit möglichst niedrigem Turbulenzgrad aufgestellt, um so Verhältnissen zu entsprechen, die möglichst nahe bei denen der freien Atmosphäre liegen; es ist also anzunehmen, daß in Windkanälen mit vergleichsweise hohem Turbulenzgrad der Umschlag vor dem berechneten auftreten wird. Immerhin geben die Versuche am nicht angestellten Ellipsoid ein Maß für die so noch zu erklärenden Abweichungen, das das von A. O. Smith angegebene nicht wesentlich überschreitet.

4.5. Dagegen werden die Abweichungen erheblich größer — und damit nicht mehr auf die angegebene Weise erklärbar — für das angestellte Ellipsoid, und zwar vor allem an dessen Flanke.

Abb. 4 zeigt dies recht deutlich in einer Seitenansicht des Hindernisses. Man kann sofort die beiden folgenden Feststellungen machen:

1. Das verallgemeinerte Kriterium von Michel ergibt den Umschlag an der Flanke des Ellipsoids ganz erheblich zu spät.

2. Das Kriterium gibt brauchbare Werte auf der Oberseite und — in geringerem Maße — auf der Unterseite des Ellipsoids; zusammen mit dem für das nicht angestellte Ellipsoid Gesagten läßt sich das Kriterium also in Gebieten mit einem Druckgradienten in Querrichtung $\frac{\partial p}{\partial z} \sim 0$ als näherungsweise gültig ansehen.

4.6. Ein Sonderfall von gegenüber den Rechnungen verspätetem (also entgegengesetzt zur

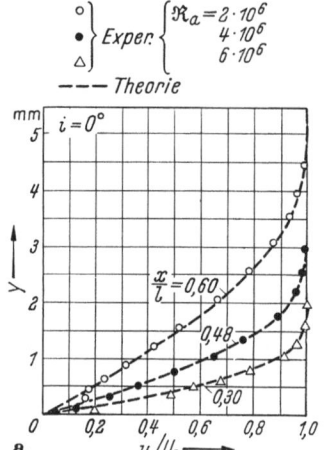

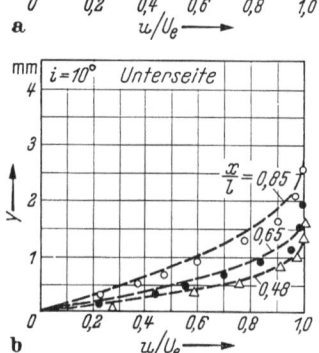

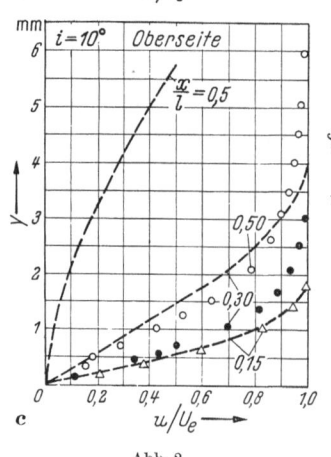

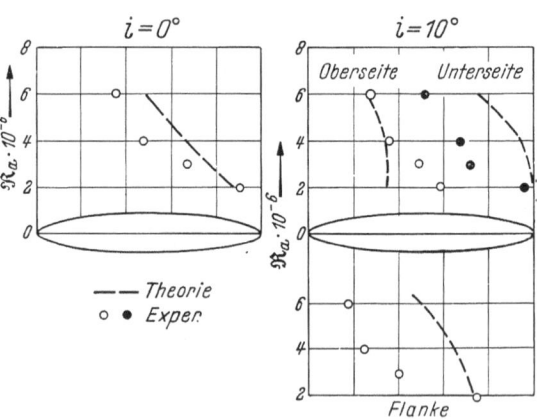

Abb. 2　　　　　　　　　　Abb. 3

üblichen Tendenz von diesem abweichenden) Umschlag tritt dabei für niedrige Reynoldszahlen auf der Oberseite des Ellipsoids auf; er wird durch das gemessene Grenzschichtprofil (vgl. Abb. 2c) bestätigt, das als einziges erheblich von dem berechneten abweicht. Bei Verwendung des aus dem gemessenen Profil sich ergebenden δ_2 führt übrigens das Kriterium von MICHEL zu einer mit den Ver-

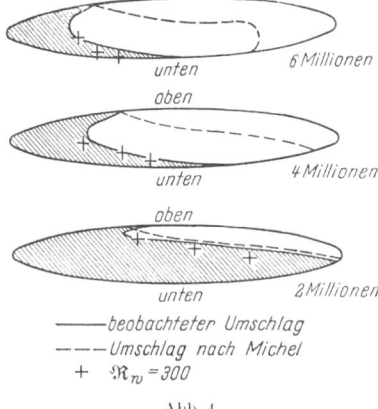

suchen besser übereinstimmenden Lage des Umschlags. Die Erklärung für diesen Sonderfall ist wahrscheinlich in dem zu suchen, was am Schluß über eine zusätzliche Form der Instabilität in drei Dimensionen gesagt werden soll. In diesem Zusammenhang sei schon jetzt darauf hingewiesen, daß dieser Sonderfall für *niedrige* Reynoldszahlen auftritt und für hohe verschwindet.

————beobachteter Umschlag
— — —Umschlag nach Michel
+ $\mathfrak{R}_{v} = 300$

Abb. 4

5.

5.1. Das bis jetzt verwendete Kriterium berücksichtigt zwar die Querströmung, aber nur bei der Berechnung der Impulsverlustdicke, wo ihr Einfluß ohnehin nur gering ist. Außerdem berücksichtigt es noch die veränderliche Breite der Stromstreifen an der Oberfläche des Hinder-

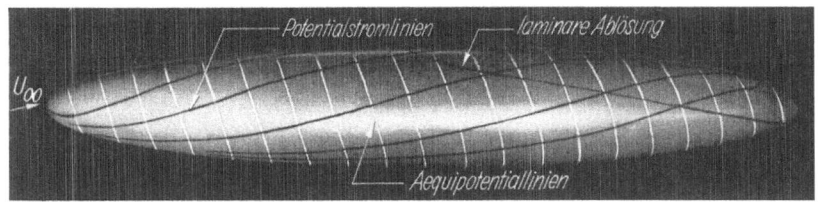

Abb. 5

nisses, vernachlässigt dagegen völlig die räumliche Windung dieser Stromstreifen unter der Wirkung eines Druckgradienten $\dfrac{\partial p}{\partial z} \neq 0$.

Abb. 5 zeigt aber, wie groß diese räumliche Windung an der Flanke des Ellipsoids ist.

5.2. OWEN und RANDALL [8, 9] haben 1952 in Verfolgung RAYLEIGHscher Gedankengänge darauf hingewiesen, daß diese Windung der Stromstreifen eine destabilisierende Wirkung auf die laminare Grenzschicht ausübt, die durch die Anwesenheit von mindestens einem Wende-

punkt im Profil der sekundären Grenzschicht (die an der Wand und
— mit ihrer Ableitung — beim Übergang zur Außenströmung verschwindet) erklärt wird und proportional zur Amplitude dieses
Profils ist.

5.3. Auf Grund experimenteller Ergebnisse an Pfeilflügeln schlagen sie
daher ein Kriterium für diesen Einfluß vor, in dem die Maximal-
amplitude der Querströmung auftritt: Umschlag tritt (und zwar unabhängig von der longitudinalen Instabilität nach TOLLMIEN-SCHLICH-
TING) dann ein, wenn eine kritische Reynoldszahl

$$R_w = \frac{|w_{\max}| \cdot \delta}{\nu}$$

den Wert 125 erreicht. Dabei ist δ die Grenzschichtdicke und w die Geschwindigkeitskomponente der Querströmung. Das Maximum ist bei
jeweils festem s und z bezüglich n im Intervall $0 \leqslant n \leqslant \delta$ zu nehmen.

Bei Polynomansätzen nach Art von POHLHAUSEN für die Geschwind-
heitsprofile in der Grenzschicht findet man

$$R_w = \sqrt{R_a} \cdot \frac{\Lambda}{27} \cdot \left[\sqrt{R_a}\, \frac{\delta}{a} \right] \cdot \frac{u_e}{U_\infty} \cdot \frac{\dfrac{\partial p}{\partial z}}{\dfrac{\partial p}{\partial s}},$$

eine Form, die die Bedeutung des Verhältnisses von transversalem zu
longitudinalem Druckgradienten hervorhebt. Λ ist dabei der POHL-
HAUSENsche Formparameter für die primäre Grenzschicht und u_e/U_∞ die
dimensionslose Außengeschwindigkeit im betrachteten Stromstreifen.
Zu bemerken ist noch, daß die eckige Klammer $\left[\sqrt{R_a}\, \dfrac{\delta}{a} \right]$ von der Reynoldszahl nicht abhängt; $R_w/\sqrt{R_a}$ hängt also außer von Formeinflüssen
und der Außenströmung nur noch von dem erwähnten Verhältnis der
Druckgradienten ab.

5.4. Dies Kriterium läßt den Umschlag für $\dfrac{\partial p}{\partial z} = 0$, wo $R_w = 125$ nie
erreicht werden kann, unverändert. An der Flanke des Ellipsoids dagegen
wandert der Umschlagspunkt gegenüber dem Kriterium von MICHEL
erheblich nach vorn.

5.5. Dies entspricht qualitativ gut den Beobachtungen, wie sie an
einem ähnlichen Rotationskörper mit schwacher Anstellung übrigens
auch schon von GREGORY, STUART und WALKER gemacht worden sind [10].
Auch quantitativ ist die Übereinstimmung für alle untersuchten Reynoldszahlen und mehrere Stromstreifen, die in verschiedenen Höhen an
der Flanke des Ellipsoids gewählt wurden, *gleichmäßig* gut, wenn man
$R_w \sim 300$ als kritischen Wert wählt, wie die Kreuze in Abb. 4 zeigen.

5.6. Der Unterschied im kritischen Wert von R_w für Pfeilflügel und für das Ellipsoid erklärt sich leicht daraus, daß das neue Kriterium noch die unzweckmäßige Gestalt der zweidimensionalen Kriterien vor MICHEL hat und also Formeinflüssen unterliegt. Es wird zweckmäßig sein, R_w mit einem anderen Grenzschichtparameter so zu verknüpfen, wie es das Kriterium von MICHEL mit R_{δ_2} und R_s tut. Diese Aufgabe soll bei der O.N.E.R.A. demnächst in Angriff genommen werden.

6.

6.1. Zusammenfassend sei daran erinnert, daß der Umschlag ein komplexes Phänomen ist, das durch verschiedene miteinander konkurrierende Einflüsse bestimmt wird, von denen die beiden hier behandelten zwar besonders wichtig erscheinen, aber sicherlich keineswegs die einzigen sind.

6.2. Allerdings schließt der vorliegende Fall eines konvexen Rotationskörpers TAYLOR-GÖRTLER-Instabilitäten — die konkave Wände voraussetzen — als *primäre* Ursache des Umschlags aus. Die wahrscheinlichsten Ursachen für einen Umschlag sind dann offenbar:

1. Longitudinale Instabilität nach TOLLMIEN-SCHLICHTING, die durch das Kriterium von MICHEL erfaßt wird.

2. Transversale Instabilität nach RAYLEIGH, die durch das Kriterium von OWEN-RANDALL wiedergegeben werden kann.

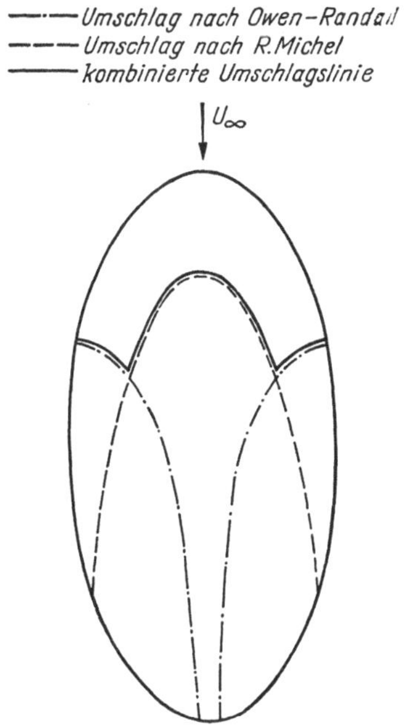

6.3. In erster Näherung scheint also eine vielversprechende Lösung des dreidimensionalen Umschlagsproblems eine Kombinierung des verallgemeinerten

Abb. 6

Kriteriums von MICHEL mit dem (eventuell nach 5.6 verbesserten) Kriterium von OWEN und RANDALL zu sein.

6.4. Abb. 6 zeigt in einer Ansicht der Oberseite schematisch, wie die Linie des dreidimensionalen Umschlags aus der Verknüpfung der jeweils weiter stromauf liegenden Äste der Umschlagslinie nach OWEN-RANDALL (die asymptotisch zur Erzeugenden der Oberseite verläuft) und der nach MICHEL (die gerade auf dieser Erzeugenden den Umschlag rechtzeitig angibt) erhalten werden kann. Auf diese Weise ist zum Beispiel Abb. 4 mit ihren recht guten Werten berechnet worden.

6.5. Die so erhaltene Linie führt zu qualitativ und im Falle des Rotationsellipsoids auch quantitativ recht ansprechenden Resultaten. Eine Tendenz der Umschlagslinie zur Bildung ein- und ausspringender „Zungen" scheint damit jedenfalls verständlich gemacht zu werden. Trotzdem ist diese Erklärung noch nicht ausreichend:

In der Tat scheint schon beim nicht angestellten, mehr noch beim angestellten Rotationsellipsoid eine dritte Form der Instabilität wirksam

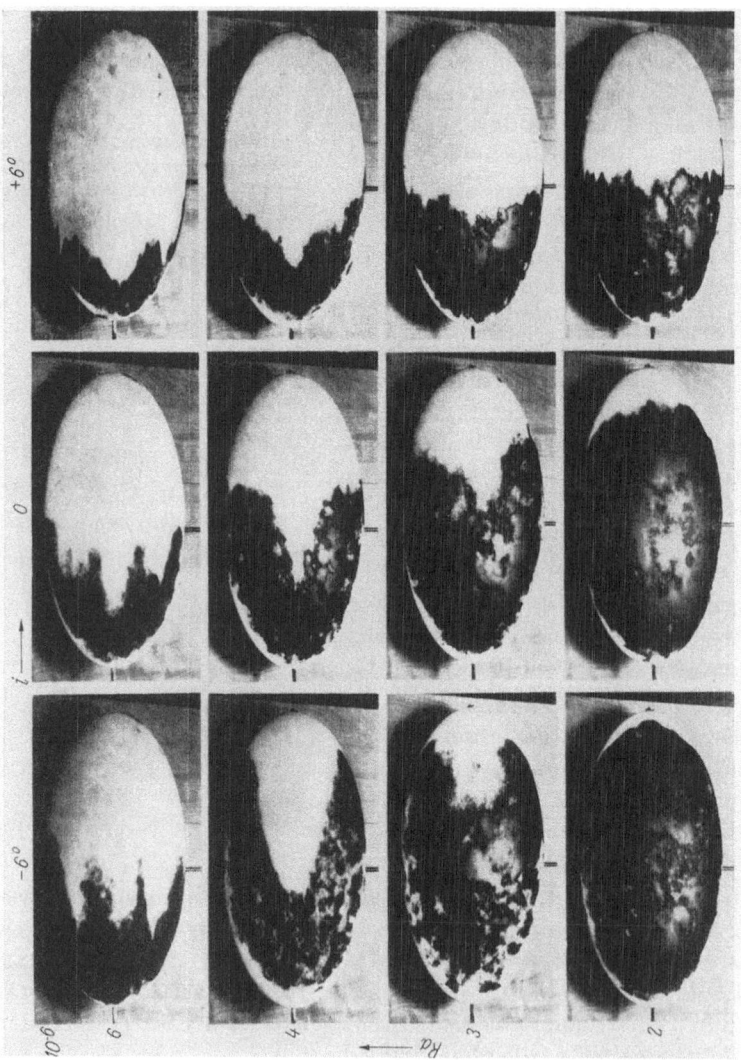

Abb. 7

zu sein, die zu einem wellenförmigen Verlauf der Umschlagslinie führt. Die Amplitude wie die Zahl dieser Wellen scheinen dabei nicht monoton von R_a abzuhängen.

6.6. In der Tat wurde beim angestellten Rotationsellipsoid bereits auf eine Unregelmäßigkeit für niedrige Reynoldszahlen hingewiesen, die sehr wahrscheinlich auf eine solche Wellenstruktur zurückgeht. Beim abgeplatteten Ellipsoid, von dem Abb. 7 einige Versuchsergebnisse zeigt, sind es dagegen besonders die höheren Reynoldszahlen, bei denen das erwähnte Phänomen auftritt. Endlich ist noch ein englischer Bericht von Burrows über Messungen an Pfeilflügeln [11] zu erwähnen, wo diese Welligkeit der Umschlagslinie bei mittleren Reynoldszahlen auftritt, während sie bei hohen und bei niedrigen Reynoldszahlen verschwindet bzw. nur noch abgeschwächt erscheint (vgl. Abb. 35 des erwähnten Berichts).

7.

7.1. War bei den in Cannes beobachteten „Zungen" zunächst an die Auswirkung wellenförmiger Störungen sehr kleiner Amplitude im Einlauf des Kanals gedacht, die zu ähnlichen Phänomena führten, wie die in dem Vortrag von G. Schubauer erwähnten wellenförmigen Intensitätsschwankungen mit entsprechendem Verlauf der Umschlagslinie, so handelte es sich bei den englischen Versuchen jedoch um Beobachtungen im freien Flug; die auch hier bei einigen Reynoldszahlen in nicht monotoner Weise auftretenden Wellen der Umschlagslinie können also jedenfalls nicht durch Störungen eines Windkanals erklärt werden.

7.2. Das Auftreten der erwähnten Zacken nur bei bestimmten Reynoldszahlen läßt an die Möglichkeit einer Interpretation als Eigenwertproblem denken; in diesem Sinne sollen weitere Experimente in Cannes bei der O.N.E.R.A. untersucht werden.

Literatur

[1] Eichelbrenner, E. A., und A. Oudart: Méthode de calcul de la couchelimite tridimensionelle. O.N.E.R.A. — publication No. 76 — (1955).

[2] Eichelbrenner, E. A.: Décollement laminaire en trois dimensions sur un obstacle fini. O.N.E.R.A. — publication No. 89 — (1957).

[3] Schlichting, H.: Grenzschichttheorie. Karlsruhe: G. Braun (1951).

[4] Michel, R.: Détermination du point de transition et calcul de la traînée des profils d'ailes en incompressible. O.N.E.R.A. — publication No. 58 — (1952).

[5] Smith, A. O.: Transition, pressure gradient and stability theory. Douglas Aircraft Co. — Rep. No. ES 26388 — (1956).

[6] van Ingen, I. L.: A suggested semi-empirical method for the calculation of the boundary layer transition region. Report V.T.H. 74 — Delft (Nederland) — (1956).

[7] Maruhn, K.: Druckverteilung auf den gleichförmig-gradlinig bewegten dreiachsigen Ellipsoidkörper. ZWB — Forschungsbericht No. 1174 — (1940).

[8] Owen, P. R., und D. G. Randall: Boundary layer transition on a sweptback wing. R.A.E. Farnborough — Techn. Memor. No. Aero 277 — (1952).

[9] Owen, P. R., und D. G. Randall: Boundary layer transition on a sweptback wing: a further investigation. R.A.E. Farnborough — Techn. Memor. No. Aero 330 — (1953).

[10] Gregory, N., I. T. Stuart und W. S. Walker: On the stability of three-dimensional boundary layers. N.P.L. Teddington — Symposium on Boundary Layer Effects — (1955).

[11] Burrows, F. M.: A theoretical and experimental study of the boundary layer flow on a 45° sweptback wing. The College of Aeronautics Cranfield — Rep. No. 109 — (1956).

A method for visualizing periodic boundary layer phenomena

By

H. Bergh

National Aeronautical Research Institute (N.L.L.), Amsterdam

In the course of an experimental investigation of the periodic motion of the transition point on an oscillating wing, it became clear that the method used could be applied also to visualize the periodic boundary layer phenomena on steady aerofoils. The method uses the fact that periodic motions seem to come to rest if they are illuminated by a stroboscopic light source, flashing with the right frequency. By introducing smoke into the boundary layer of a flat plate or an aerofoil and assuming that periodic motions occur, they might be studied easily with stroboscopic illumination.

In order to investigate the usefulness of the method, it has been applied to visualize the transition process on a two-dimensional wing with aerofoil section NACA 0012.

The wing with a chord of 0.30 m. was mounted horizontally between the walls of a normal low speed wind-tunnel, having a turbulence degree of about 0.1 percent. The test section was 1.10 m. high and 0.80 m. wide. The smoke was introduced into

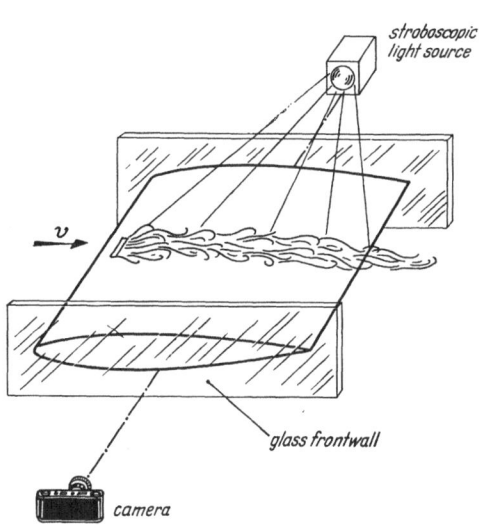

Fig. 1. The test set-up

the boundary layer through a slit in the surface, located at 10 percent of the chord and 0.25 m. from the glass frontwall. The dimensions of the slit were about 3 by 10 mm. The smoke was produced by an electrically heated wire, embedded in asbestos fibres, to which an adjustable amount of kerosene was supplied. The best way to illuminate the oil vapour is to place the stroboscopic light source opposite the observer or camera and a little above the plane of the wing (fig. 1). Then it looks like a fine white smoke, showing many details of the flow structure, especially if the flash duration is very short (in this investigation it was about 1 μ sec.).

The first observations were performed at an airspeed of 4 m./sec. and the model at zero incidence. With steady enlightening, the smoke seemed

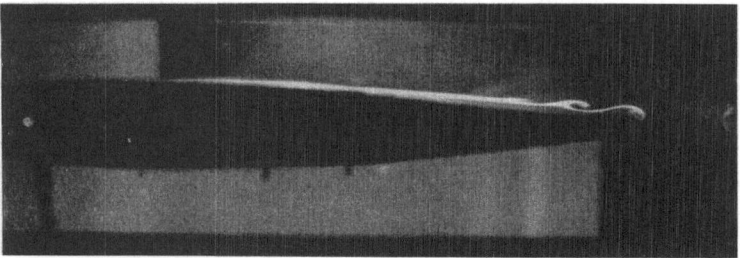

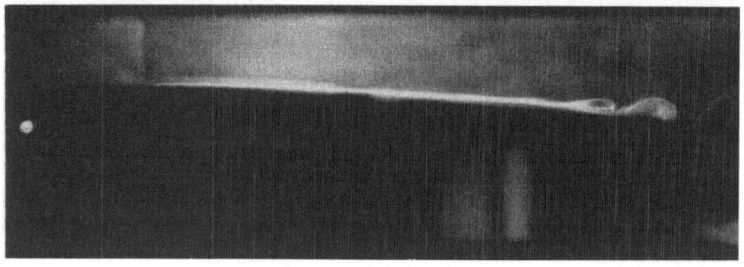

Fig. 2, a Without artificial excitation U_∞ = 4 m/sec. 63 c.p.s. 1 flash
b As in the upper fig., but 7 flashes together

to indicate an undisturbed laminar boundary layer. However, when the stroboscopic light was switched on and the frequency increased up to 60 c.p.s., the picture shown in fig. 2 became visible. The phenomenon was definitely periodical and could be reproduced as can be seen in the lower picture of fig. 2, where 7 pictures are taken on one film frame with a time-interval of $1/_{60}$ sec. Increasing the airspeed also increased the frequency of the travelling waves and the whole region was shifted upstream.

In order to vary frequency and airspeed independently, artificial disturbances were introduced by a loudspeaker placed at the top of the test-section. These sound disturbances were able to produce regular waves in the boundary layer at chordwise locations where without sound an undisturbed laminar boundary layer existed. Even it was easy to suppress the originally present waves by waves, having a frequency equal to that of the sound excitation. Especially at the lower frequencies, it became possible to get complete transition from laminar to turbulent flow by the sound excitation. Moreover, the intensity of the sound

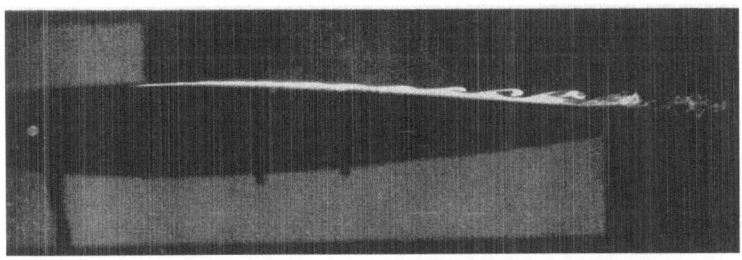

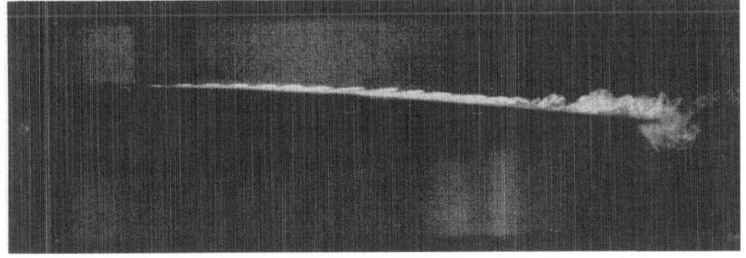

Fig. 3. a Excitation by sound $\beta_r =$ 75 c. p. s. $\Big\}$ $U_\infty = 4$ m./sec.
b Excitation by sound $\beta_r =$ 160 c. p. s.

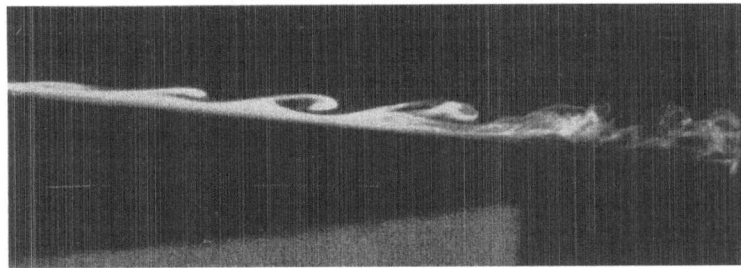

Fig. 4. Enlargement of a part of fig. 3a

influenced the position of the transition point. In fig. 3, examples are given of the flow picture at two different frequencies. An enlargement of a part of the upper picture is shown in fig. 4.

With the artificial excitation it became possible to produce waves in a large range of frequencies. Due to distortion of the loudspeaker, frequencies lower than about 30 c.p.s. could not be generated. At frequencies higher than 180 c.p.s., there was only a small amplification of the disturbances over the first part of the chord, without giving transition to turbulence, while during the second part of the chord, these small waves died out. However, in that case the original waves of fig. 2 were also present.

The influence of the airspeed is shown in fig. 5. It can be seen, that already a small increase of the airspeed gives an appreciable shift of the transition region.

As could be expected, the angle of incidence plays an important part.

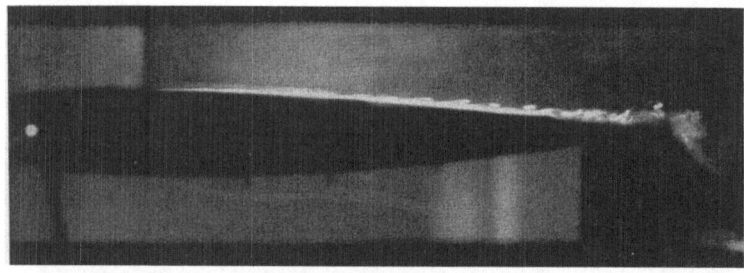

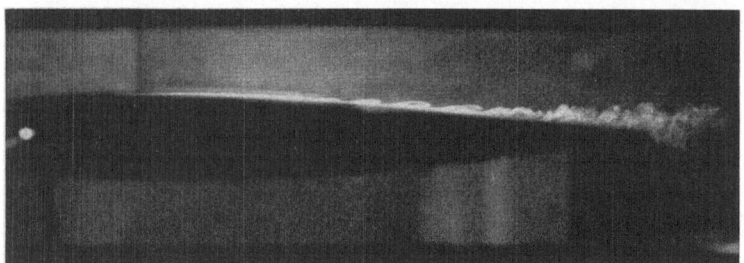

Fig. 5. a Excitation by sound $U_\infty = 4$ m./sec. $\Big\}$ $\beta_r = 125$ c. p. s.
b Excitation by sound $U_\infty = 5.7$ m./sec.

This is illustrated by the next figure (fig. 6), which shows the flow pictures for various angles of incidence at a frequency of 145 c.p.s.

Observations by eye have given us the impression that in this case the waves in the laminar part are developing themselves to horseshoe vortices, inclined with the flow direction. A three-quarter view of the smoke at an angle of incidence of 2 degrees is shown in fig. 7, where an open loop can be seen, due to the horseshoe character. An enlargement of a part of the middle picture of fig. 6, giving the same smoke picture in side-view, is shown in fig. 8.

Though the purpose of this preliminary investigation was to verify the practical value of the described method, an attempt has been made to correlate the results with the stability theory. Therefore the experiments are roughly compared with the theoretical results of TIMMAN and

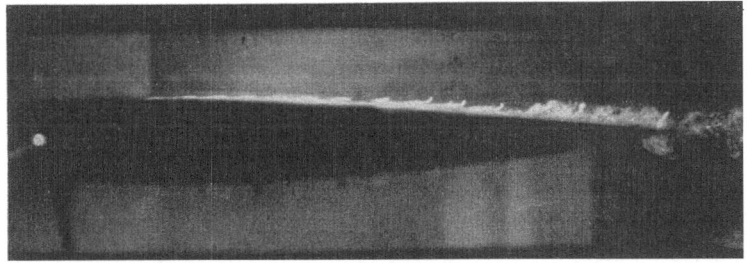

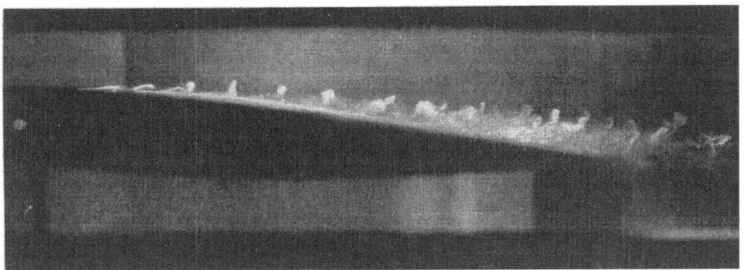

Fig. 6. a $\alpha = 0°$ } Excitation by sound
       b $\alpha = 2°$ } $U\infty$ = 4 m.'sec.
       c $\alpha = 4°$ } β_r = 145 c.p.s.

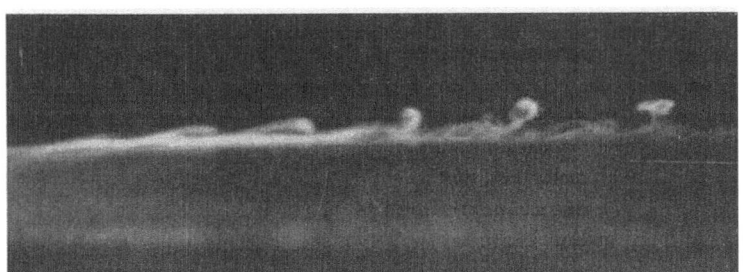

Fig. 7. Three-quarter view of fig. 6b

ZAAT. Fig. 9 shows that the experimental points all fell within the stability curve for the velocity profile for $b = 0.8$.

Summarizing, I think it is shown that the described method can be used in addition to the conventional methods to obtain more information about periodic boundary layer phenomena and especially about the transition process.

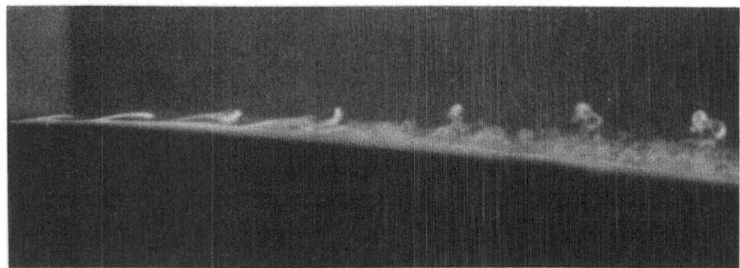

Fig. 8. Enlarged side-view of fig. 6 b

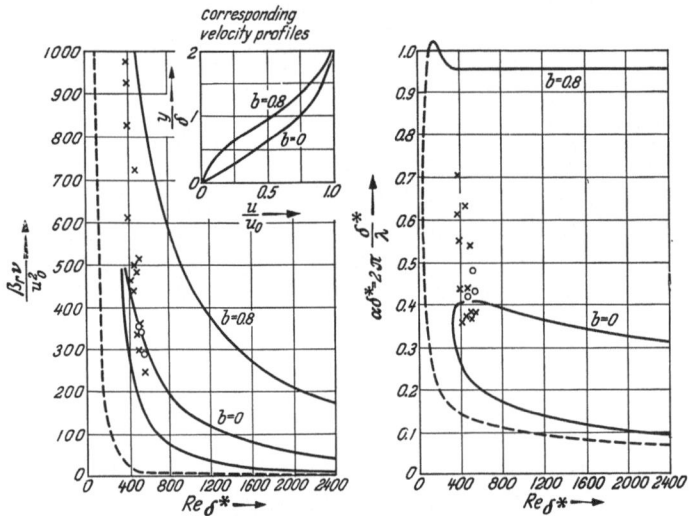

Fig. 9. Comparison of measured frequency β_r and wavelength λ with stability theory of TIMMAN and ZAAT for the case of zero and non-zero pressure gradients ($b = 0$ and $b = 0.8$ resp.)

Aus der Diskussion

F. X. WORTMANN (Stuttgart): Konnten Sie bei fester Amplitude allein durch Variation der Frequenz die Entwicklung der Wirbel und die Lage des Umschlagspunkts verändern? Dies wäre ein Kriterium dafür, daß es sich um Eigenschwingungen handelt.

H. Bergh (Amsterdam): Dies haben wir getan. Es gelang sehr gut für Frequenzen unter 100 Hz. Bei höheren Frequenzen hatten wir Schwierigkeiten mit dem Lautsprecher. Bei derselben Intensität des Lautsprechers kam nicht die gleiche Intensität in die Strömung.

O. Wehrmann (Berlin): In meinem Vortrag auf der DVL/WGL-Tagung im vergangenen Frühjahr habe ich über die akustische Beeinflussung der Ringwirbelerzeugung in einem Freistrahl berichtet. Bei den nach den Methoden der Hitzdrahtmeßtechnik vorgenommenen Messungen konnten folgende Ergebnisse erzielt werden: 1. Ähnlich wie bei den Messungen von Herrn Bergh, konnte eine Stabilitätskurve ermittelt werden, die mit derjenigen von Tollmien, Schlichting und Lin verglichen werden kann. 2. Die Abhängigkeit der Steigerung der Wirbelintensität vom aufgeprägten Schalldruck konnte bestimmt werden. Es ergab sich hierbei zunächst ein linearer Bereich, sodann Zustreben zu einem Sättigungswert. Das Verhältnis von Schallenergie zu Strahlenergie an der Düsenmündung betrug dabei maximal $1 : 10^4$. 3. Durch Phasenmessungen mittels zweier Hitzdrähte konnte festgestellt werden, daß im Falle der sich ohne akustische Beeinflussung einstellenden Frequenz die Wirbelfortpflanzungsgeschwindigkeit ein Extremum hat. 4. Durch eine Rückkopplung, bei der die vom Hitzdraht gelieferten Signale dem Lautsprecher wieder zugeführt wurden, konnte innerhalb der Stabilitätskurve eine Kurve der maximalen Erregungsfrequenz bestimmt werden.

B. Thwaites (Winchester): Recently, a theory was put forward by R. A. Shaw which he calls "Acoustic Theory." I wonder if, in your experiments, you have observed any frequencies which are of the order in question.

H. Bergh (Amsterdam): I have discussed this theory with Mr. Shaw but I have not yet received a copy of his report. However, I intend to look into this matter in the near future.

Diskussionsveranstaltung zur IV. Sitzung

Contribution to the Subject
Stability of Laminar Flow

By **W. Pfenninger,** Northrop Aircraft, Inc., Hawthorne, Cal.

The development and the stability of the laminar boundary layer on swept laminar suction wings was calculated by the BLC group at Northrop. The boundary layer development on such wings was determined by Raetz by means of finite difference methods in the form of the Crocco transformation and with IBM equipment. The stability limit Reynolds number on swept laminar suction wings was calculated by Brown and Sayre by a direct integration of the Orr-Sommerfeld disturbance equation under cross-flow conditions from the wall to the outer edge of the boundary layer. The values α, R, c were chosen in such a manner as to match this solution at the outer edge with the solution of the disturbance equation obtained beyond the outer edge of the boundary layer.

(The results of calculations of the boundary layer development on a swept wing with various amounts of suction were presented. The calculated stability limit

Reynolds number was shown for the cross-flow profiles at the leading and trailing edge of swept laminar suction wings as well as for the rotating disc critical profile and for the BLASIUS case.)

The agreement with the experiment has been satisfactory for the cross-flow profile at the leading edge of a swept wing. The stability limit Reynolds number of boundary layer cross-flow profiles is strongly dependent on the shape of these profiles.

Contribution to the Subject
Boundary Layer Transition[1]

By **E. R. van Driest,** North American Aviation, Inc., Downey, Cal.[2]

1. Criterion for Minimum Transition

a) Pipe Flow. Upon interpretation of data on velocity distribution for turbulent flow near a wall at subsonic speeds [1] (see fig. 1, in which $u_* = u/\sqrt{\tau/\varrho}$ and $y_* = \varrho \sqrt{\tau/\varrho} \cdot y/\mu$, where u is velocity, ϱ is density, τ is wall shear stress, μ is viscosity, and y is distance normal to the wall), the writer proposes the following criterion for minimum transition, viz., $y_{*T} = \varrho \sqrt{\tau/\varrho} \cdot y_T/\mu = 60$. Here y_T represents the

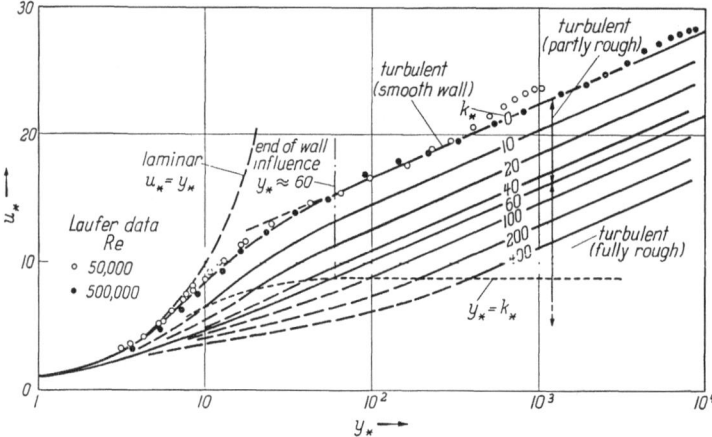

Fig. 1. Semilogarithmic plot of velocity profiles for incompressible turbulent flow near smooth and rough walls

inner viscous subregion in which originates the eddying motion (and therefore Reynolds stresses) that characterizes the onset of fully turbulent flow. When this criterion is applied to the Poiseuille (parabolic) velocity profile for laminar flow in a pipe of diameter D, it is readily found that $Re_D = \varrho \, u_{ave} \, D/\mu = 1800$. This mini-

---

[1] This research was supported by the United States Air Force, through the Office of Scientific Research of the Air Research and Development Command.

[2] Dieser Diskussionsbeitrag lag beim Symposium vorbereitet vor, der Verfasser war jedoch dann unerwartet an der Teilnahme verhindert. — D. Hrsg.

mum value of transition agrees well with pipe friction data (see fig. 2 at $\log_{10} 1800$ $= 3.26$). It should be noted that a value of $y*_T = 63$ would yield $Re_D = 2000$, or $\log_{10} 2000 = 3.30$. This value of transition then represents the beginning of non-linear turbulence bursts in a pipe, i. e., the appearance of Reynolds stresses and the consequent deviation from the laminar-flow law. When the frequency of

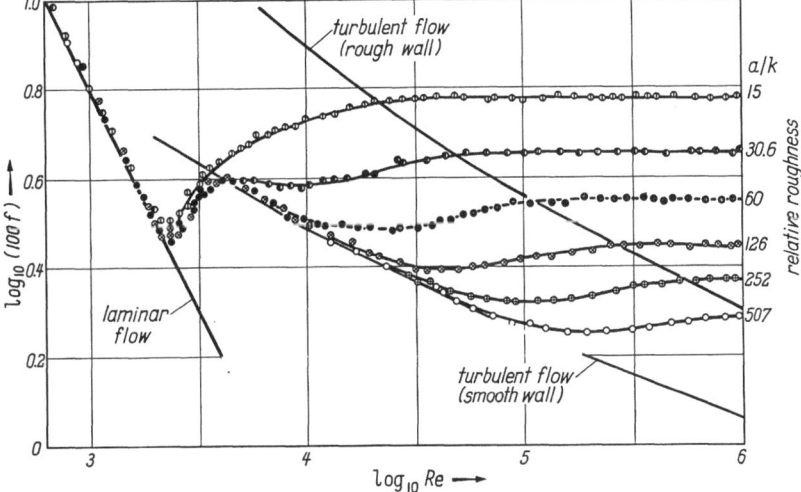

Fig. 2. Friction factor for incompressible flow in a pipe

the bursts becomes great enough, the bursts intermingle as fully developed turbulence and then cause the flow to follow the smooth-wall fully-turbulent law at $Re_D \approx 5000$, or $\log_{10} 5000 = 3.70$.

b) Flat-Plate (Blasius) Boundary-Layer Flow. When the above criterion $y*_T =$ $= 60$ is applied to the laminar boundary layer of a flat plate (Blasius profile), where y_T is set equal to the thickness δ of the layer, then it is found that for a smooth plate $Re_x = \varrho\, u_\delta\, x/\mu = 160,000$, where x is the distance from the leading edge of the plate. This minimum transition Reynolds number again seems to agree well with experiment when artifical turbulence is produced upstream of the plate [2].

c) Stagnation-Point Flow. Application of the criterion to the exact laminar-flow solution for incompressible flow on a smooth sphere yields $Re_\Theta = \varrho\, u_\delta\, \Theta/\mu = 150$, where Θ is the momentum thickness of the boundary layer. This value should apply roughly near the stagnation point behind the shock wave of a high-speed vehicle. It will correspond approximately to a value of $\varrho\delta\, u_\delta\, x/\mu\delta = 400,000$ along the surface for $Re_{D\infty}$ (where D = sphere diameter here) $= \varrho_\infty\, u_\infty\, D/\mu_\infty = 20,000,000$ and undisturbed stream Mach number $M_\infty = 10$. Incipient transition would then occur at $x/D = 0.08$, or at an angle of about 9 degrees off the stagnation point.

d) Roughness Criterion. According to fig. 1, the trip Reynolds number for effective tripping (i.e., effective generation of turbulence) is $\varrho\, u_\delta\, k/\mu = (u_\delta/\sqrt{\tau/\varrho}) \cdot$ $\cdot (\varrho\, \sqrt{\tau/\varrho}\, k/\mu) = 9 \cdot 60 = 540$, where k is roughness height.

2. Control of transition by Cooling

One effective means of delaying transition is by cooling the laminar layer [3]. Fig. 3 is a set of schlieren photographs that clearly shows the delay of transition on a smooth 10-degree (apex angle) cone at local free-stream Mach number 3.65 as a result of steady-state internal cooling. The boundary layer has been magnified 20

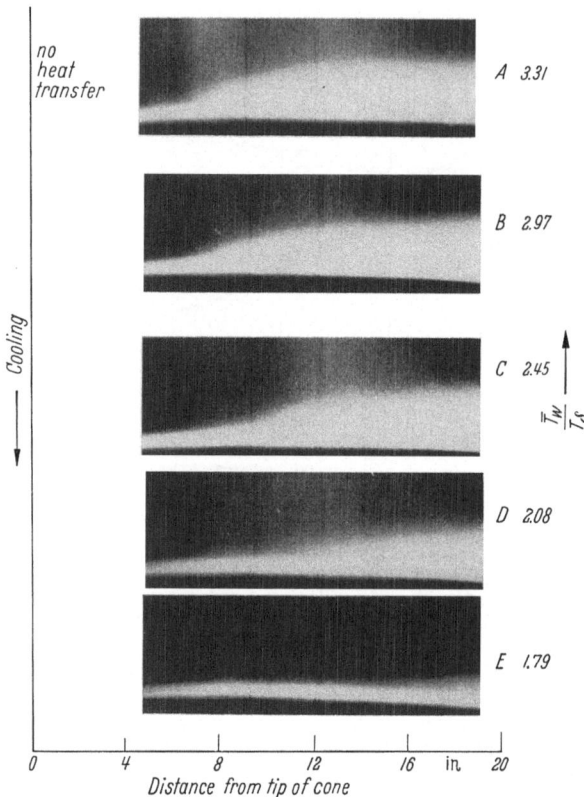

Fig. 3. Typical magnified schlieren photos showing the delay of transition by cooling for a smooth 10-degree (apex angle) cone. $M_\delta = 3.65$

times normal to the flow. The column on the right is the ratio of the average wall temperature $\bar{T}_w$ to the local free-stream temperature T_δ for successive stages of cooling. While fig. 3 represents the case for a cone, it is expected that the favorable phenomenon would occur on the curved leading surfaces of re-entering aircraft where negative pressure gradients exist.

Specific data on the delay of transition Reynolds number $Re_{\delta T}$ by cooling for the smooth cone are shown in fig. 4 for three local Mach numbers M_δ equal to 1.90, 2.70 and 3.65. It is seen in the figure that cooling is increasingly more critical at the lower temperature ratios (higher cooling rates).

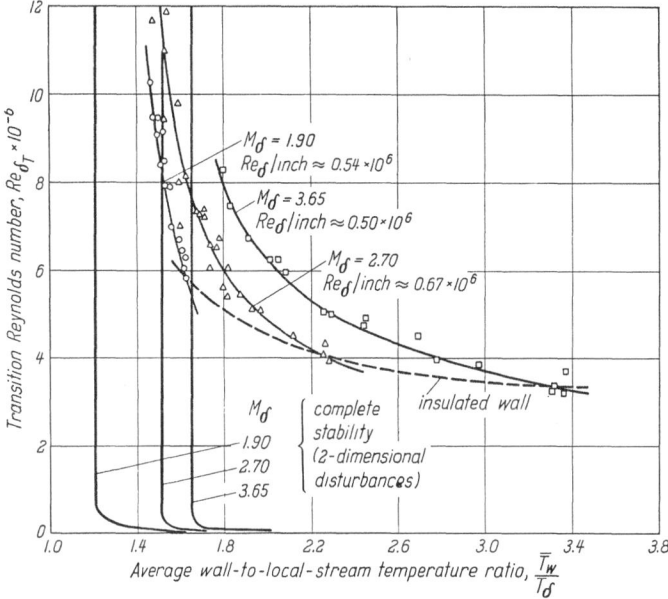

Fig. 4. Effect of surface cooling on transition Reynolds number for several Mach numbers for a smooth 10-degree (apex angle) cone. (0.4 percent supply turbulence)

3. Effect of Roughness on transition

Fig. 5 gives recent data showing the effect of Mach number on transition [4]. The data are for wire trips 3 inches from the tip of the 10-degree (apex angle) cone with zero heat transfer. The ordinate is the ratio of the transition Reynolds number Re_{δ_T} with trips to the transition Reynolds number Re_{δ_0} for the smooth body, and the abscissa is the ratio of roughness height k to boundary-layer displacement

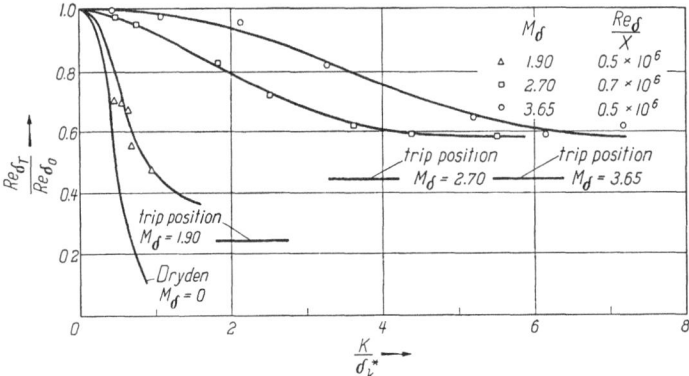

Fig. 5. Effect of Mach number on transition. Two-dimensional trips at 3-inch station. Zero heat transfer. 10-degree (apex angle) cone

thickness δ_k^* at the trip position. It is concluded that an increase in Mach number decreases the effect of roughness in promoting transition.

Fig. 6 indicates the effect of trip-position Reynolds number on transition for local Mach number 1.90 [4]. It is seen that an increase in trip-position Reynolds number apparently decreases the effect of a wire in promoting transition, and therefore the trip-position Reynolds number also appears to be an important parameter.

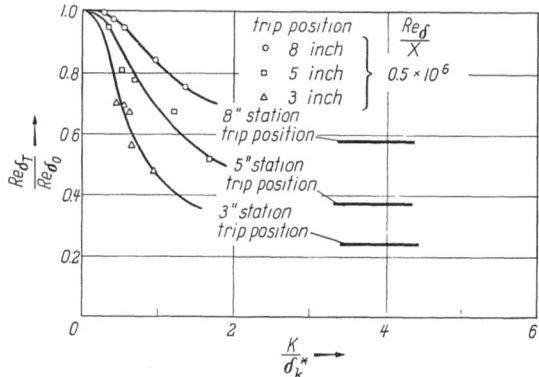

Fig. 6. Effect of trip-position Reynolds number on transition. Zero heat transfer. Local Mach number 1.90. 10-degree (apex angle) cone

References:

[1] VAN DRIEST, E. R.: On Turbulent Flow Near a Wall. J. Aero. Sci., **23**, 1007—1011 (1956), No. 11.

[2] VON KÁRMÁN, TH.: Turbulence and Skin Friction. J. Aero. Sci., **1**, 1—20 (1934), No. 1.

[3] VAN DRIEST, E. R., and J. CHRISTOPHER BOISON: Research on Stability and Transition of the Laminar Boundary Layer. OSR-TN-55-465, September 1, 1955.

[4] Unpublished data obtained by the author and W. D. MCCAULEY.

Internal viscous flows with body forces

By

Simon Ostrach

Lewis Flight Propulsion Laboratory, National Advisory Committee for Aeronautics,
Cleveland, Ohio

Summary

A series of problems dealing with the flow of confined viscous fluids subject to body forces are discussed and it is shown that numerous factors can appreciably affect the flow and heat transfer. In particular, the effects of the confining walls, heating the fluid from below, frictional heating, and rotating the fluid container are studied. The configurations considered simulate heat sinks in turbine blades, nuclear reactors, and high-speed reentry vehicles.

Introduction

The development of lightweight, efficient, and simple heat sinks is one of the formidable problems posed by the rapid technological advances in the fields of jet propulsion, nuclear power, and hypersonic flight.

It has been suggested that the requirements for such heat sinks might be fulfilled by applications of natural-convection phenomena, wherein fluid motion is generated by gravitational, centrifugal, and electro-magnetic body forces. E. SCHMIDT in [1] presents one of the first practical applications of this idea where he describes a turbine which was successfully operated with gas temperatures of 2560° R. To absorb sufficient heat to keep the blades from failing water was circulated by natural convection generated by the large centrifugal force through cylindrical cavities in the blades. However, relatively little information on natural convection exists, especially for internal flows. When it is also noted that in new applications the three primary factors governing natural convection, namely, the body force, the fluid volumetric expansion coefficient, and the temperature differences all can greatly exceed previously considered bounds it is clear that even known information must be reevaluated. In addition, a study of contemplated practical configurations indicates that a multiplicity of interacting effects can influence a conventional internal natural-convection flow. By a conventional, or basic, flow is here meant one in which a density or tem-

perature gradient is transverse to the body force. What types of interactions can be expected in internal flows? A fluid in a container could,
for example, be heated from below as well as from the sides so that, in a
sense, another density variation exists which is parallel to the body force.
It is known that meteorologists and astrophysicists have found various
instabilities associated with heating from below. Therefore, the influence of this instability on a conventional flow has to be evaluated.
Also, the container could be rotated about one of its axes so that the
basic flow is subject to additional (centrifugal and Coriolis) body forces.

The purpose of this paper is, therefore, to discuss a series of analyses
made not only to determine the effects of confining walls on natural-
convection flow and heat transfer but also to find the influence of some
of the new conditions and interacting factors on the basic flow.

1. General analysis

The first step toward the understanding of a phenomenon usually is to
find the dimensionless parameters pertaining thereto. Accordingly, the
differential equations expressing the conservation of momentum and
energy for a viscous, compressible, heat-conducting fluid with constant
viscosity and thermal conductivity coefficients subject to a body force
are nondimensionalized by letting

$$u_i = U_i/\overline{U}, \quad \tau = \theta/\theta_w, \quad x_i = X_i/d \quad \text{and} \quad p = P/\varrho \, \overline{U}^2 \tag{1}$$

where the symbols are all defined in the appendix. The reference velocity $\overline{U}$
should denote a unique velocity which characterizes the flow. For
combined forced- and natural-convection flows $\overline{U}$ is usually taken to be
the free stream velocity U_∞ so that the equations are

$$u_j \frac{\partial u_i}{\partial x_j} = \frac{Gr}{Re^2}\,\tau + \frac{1}{Re}\left[\frac{\partial}{\partial x_j}\left(\frac{\partial u_i}{\partial x_j} + \frac{\partial u_j}{\partial x_i}\right) - \frac{2}{3}\frac{\partial}{\partial x_i}\left(\frac{\partial u_j}{\partial x_j}\right)\right] - \frac{\partial p}{\partial x_i} \tag{2}$$

and

$$\frac{1}{\gamma}\, u_j \frac{\partial \tau}{\partial x_j} = \frac{1}{Pr\,Re}\,\frac{\partial^2 \tau}{\partial x_j \partial x_j}$$
$$- E\,p\,\frac{\partial u_j}{\partial x_j} + \frac{E}{Re}\left[\frac{\partial u_i}{\partial x_j}\left(\frac{\partial u_i}{\partial x_j} + \frac{\partial u_j}{\partial x_i}\right) - \frac{2}{3}\left(\frac{\partial u_j}{\partial x_j}\right)^2\right] \tag{3}$$

Thus, solutions of the dynamic and thermodynamic problems for forced-
convection flows with body forces depend on four parameters; namely,
the Reynolds number Re, Prandtl number Pr, Eckert number E, and
Froude number, which is Re^2/Gr, where Gr is the Grashof number which
denotes the ratio of body to viscous forces. The Froude number compares
inertia and body forces. The Eckert number is defined as

$$E = \frac{U_\infty{}^2}{c_p\,\theta_w} = 2\,\frac{\theta_a}{\theta_w} \tag{4}$$

and is the parameter that determines the influence of the compression work and frictional heating.

For the case of natural-convection flows there exists no unique prescribed characteristic velocity like U_∞ so that the Eckert and Reynolds numbers lose their significance. The dimensional equations still hold, but now a new reference velocity must be used which involves the three physical quantities (β, f_X, and θ_w) that govern the natural-convection phenomenon. If now

$$\overline{U} = \sqrt{\beta\, f_X\, \theta_w\, d} \tag{5}$$

the nondimensional momentum and energy equations for natural convection are

$$u_j\, \frac{\partial u_i}{\partial x_j} = \tau + \frac{1}{\sqrt{Gr}} \left[\frac{\partial}{\partial x_j} \left(\frac{\partial u_i}{\partial x_j} + \frac{\partial u_j}{\partial x_i} \right) - \frac{2}{3} \frac{\partial}{\partial x_i} \left(\frac{\partial u_j}{\partial x_j} \right) \right] - \frac{\partial p}{\partial x_i} \tag{6}$$

and

$$\frac{1}{\gamma} u_j\, \frac{\partial \tau}{\partial x_j} = \frac{1}{Pr\sqrt{Gr}} \frac{\partial^2 \tau}{\partial x_j\, \partial x_j}$$

$$- \overline{K}\, p\, \frac{\partial u_j}{\partial x_j} + \frac{\overline{K}}{\sqrt{Gr}} \left[\frac{\partial u_i}{\partial x_j} \left(\frac{\partial u_i}{\partial x_j} + \frac{\partial u_j}{\partial x_i} \right) - \frac{2}{3} \left(\frac{\partial u_j}{\partial x_j} \right)^2 \right]. \tag{7}$$

Comparison of eqs. (2) and (3) with (6) and (7) shows that the Grashof number is analogous to the Reynolds number and the new dimensionless group $\overline{K} = \beta\, f_X\, d/c_p$ is analogous to the Eckert number. In fact, Gr is simply the square of the Reynolds number based on $\overline{U}$ as given in eq. (5) and $\overline{K}$ is the Eckert number based on $\overline{U}$ as given in eq. (5). The appearance of the parameter $\overline{K}$ marks the first important deviation from existing information on natural convection, because the effects of the compression work and frictional heating are determined by it and not the Eckert number. Previously, these effects were neglected on the basis of arguments showing that E was always small; in order to do so a guess had to be made of the magnitude of the reference velocity. When it is realized that in new applications the body force and volumetric expansion coefficient which are in the numerator of $\overline{K}$ can easily be many times the values usually associated with this type of heat transfer (for example, f_X could be as much as 10^5 g in a centrifugal field associated with rotating machinery and β approaches infinity near the critical state of a liquid) it is clear that compression work and frictional heating can influence natural-convection flow and heat transfer.

If the frictional heating is taken into consideration an interesting characteristic of the natural-convection phenomenon becomes evident. The frictional heating is added to the physically imposed heat and should act as a heat source in the fluid and, hence, tend to increase flow velocities.

2. Fully-developed flows

The nonlinearity of the basic equations governing natural convection and the interaction of the dynamics and thermodynamics in this phenomenon make theoretical studies very difficult. Consideration of compression work and frictional heating further complicate the problem. Simplifying assumptions are, therefore, necessary to obtain tractable problems. In this section, a series of idealized problems will be discussed which lead to less complicated equations, but which, nevertheless, retain the essential physical behavior of the natural-convection process. In this way detailed velocity and temperature distributions can be determined for confined flows and the effects of frictional heating can be studied.

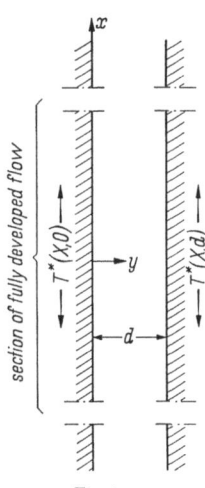

Fig. 1.
Schematic sketch of vertical channel configuration for fully-developed flows

Accordingly, consideration is given herein to the fully-developed laminar flow of a viscous fluid between two vertical planes and subject to a vertical body force (see fig. 1). By fully developed is meant that the velocity components are independent of the axial (vertical) distance or, in other words, that the solutions obtained are valid away from the ends and, hence, apply to channels with large length-gap ratios. The temperatures of the two surfaces are, in general, taken to be functions of the vertical coordinate and can differ by a constant. It is further assumed that the physical properties of the fluids are constant except that the essential influence of density variations on the flow is accounted for by the introduction of the fluid volumetric expansion coefficient in the body force term. The other influences of variable density and the variation of the expansion coefficient with temperature are assumed negligible.

2.1. Basic equations. Under the conditions stated it follows from the continuity equation that the only nonzero velocity component is the axial one and that it is a function only of the transverse (horizontal) coordinate y. Further the temperature can be expressed as

$$T^* = A X + T (Y) \tag{8}$$

where $A = \partial T^*/\partial X$ is the (constant) axial temperature gradient.

The nondimensional momentum and energy equations thus become

$$u'' + \tau = C K \tag{9}$$

$$\tau'' - Ra_A u + (u')^2 = 0 \tag{10}$$

where $u = U \sqrt{K \, Pr/c_p \, \theta_R}$, $\tau = K \theta/\theta_R$, $y = Y/d$, and the modified Rayleigh number $Ra_A = Pr \, Gr_A$. The primes denote differentiations with respect to y. The continuity equation is identically satisfied. Justification of the assumptions and the explicit development of the above equations including also the effects of heat sources in the fluid are presented in [2] to [4]. The terms (from left to right) in eq. (9) denote the viscous, buoyancy, and axial pressure forces and in eq. (10) represent the conduction, convection, and frictional heating. The parameter C merely specifies the temperature level or axial pressure gradient.

Elimination of τ between eqs. (9) and (10) yields a fourth-order nonlinear ordinary differential equation

$$u^{\mathrm{IV}} + Ra_A\, u - (u')^2 = 0 \tag{11}$$

which applies to the three problems treated in [2] to [4]. The specific problems are defined by the parameter Ra_A through the parameter A (see eq. (8)).

2.2. Boundary conditions.

The no-slip condition for viscous fluids is imposed in all cases, i.e.,

$$u\,(0) = u\,(1) = 0 \tag{12}$$

In the first problem, [2], we specify that the surface temperatures are constant but that they can differ. Hence $A = 0$. If the definition of τ for this case is noted above and eq. (9) is used, the thermal boundary conditions can be written as

$$u''\,(0) = (C - 1)\,K \tag{13a}$$

$$u''\,(1) = m\,(C - 1)\,K \tag{14a}$$

where m which is essentially a measure of the difference between the two wall temperatures is

$$m = (C - \theta_{w_1}/\theta_{w_0})/(C - 1) \tag{15a}$$

The thermal boundary conditions for linearly varying wall temperature problems are

$$u''\,(0) = CK \tag{13b}$$

$$u''\,(1) = m\,CK \tag{14b}$$

where $\qquad\qquad m = 1 - \theta_{w_1}/CAd \tag{15b}$

The cases $Ra_A > 0$ and $Ra_A < 0$ (the latter corresponding to heating from below) are treated in [3, 4] and [5] respectively. Note that Ad replaces θ_{w_0} in K and elsewhere in the varying wall temperature cases.

2.3. Results.

The problems specified by the nonlinear equation and appropriate boundary conditions are first solved approximately by an analytical iteration technique [2, 3, 4]. The zeroth-order solution (neglecting frictional heating) is given by

$$u_0 = \frac{(C-1)\,K}{6}\left[-(m+2)\,y + 3\,y^2 + (m-1)\,y^3\right] \tag{16}$$

for the constant wall temperature case. For the variable wall temperature cases the solutions depend on the sign of the modified Rayleigh number, Ra_A, and are given by

$$u_0 = \frac{C\,K}{2\sqrt{Ra_A}}\left[(m-1)\,v_{01} + (m+1)\,v_{02}\right] \tag{17}$$

where, for positive Rayleigh numbers,

$$\left.\begin{aligned}
v_{01} &= \frac{1}{\cos^2 \bar{R} - \cosh^2 \bar{R}}\,(\cosh \bar{R} \sin \bar{R} \sinh \bar{R}\eta \cos \bar{R}\eta \\
&\qquad\qquad\qquad\qquad - \sinh \bar{R} \cos \bar{R} \cosh \bar{R}\eta \sin \bar{R}\eta) \\
v_{02} &= \frac{1}{\cosh^2 \bar{R} + \cos^2 \bar{R} - 1}\,(\cosh \bar{R} \cos \bar{R} \sinh \bar{R}\eta \sin \bar{R}\eta \\
&\qquad\qquad\qquad\qquad - \sinh \bar{R} \sin \bar{R} \cosh \bar{R}\eta \cos \bar{R}\eta)
\end{aligned}\right\} \tag{18}$$

For negative Rayleigh numbers

$$\left.\begin{aligned}
v_{01} &= -\frac{1}{2}\left(\frac{\sinh \sqrt{2}\,\bar{R}\eta}{\sinh \sqrt{2}\,\bar{R}} - \frac{\sin \sqrt{2}\,\bar{R}\eta}{\sin \sqrt{2}\,\bar{R}}\right) \\
v_{02} &= -\frac{1}{2}\left(\frac{\cosh \sqrt{2}\,\bar{R}\eta}{\cosh \sqrt{2}\,\bar{R}} - \frac{\cos \sqrt{2}\,\bar{R}\eta}{\cos \sqrt{2}\,\bar{R}}\right)
\end{aligned}\right\} \tag{19}$$

In eqs. (18) and (19), one has

$$\bar{R} = (Ra_A)^{1/4}\big/ 2\sqrt{2}, \qquad \eta = 2\,y - 1$$

where Ra_A in these equations is always positive. The corresponding zeroth-order temperature distributions can be obtained from eq. (9) by using eqs. (16) and (17). Higher-order approximations are presented in [2] and [3].

The first objective of this series of problems, namely, to find detailed information on the flow and heat transfer of confined fluids subject to a body force, is fulfilled by the solutions given by eqs. (16) and (17).

2.3.1. Heating from below. An interesting aspect of natural convection becomes evident if it is noted from eqs. (18) and (19) that the solutions for negative Rayleigh numbers are quite different from those for positive Rayleigh numbers. In particular, the negative Rayleigh number solutions become meaningless for $Ra_A = (k\pi)^4$ where k denotes integers; for $k = 1$ and 2 the critical values of Ra_A are 97.41 and 1558.55, respectively. It can further be seen from fig. 2 that the flow for negative Ra_A is of a different character for Rayleigh numbers on opposite sides of the critical values. To explain these seemingly unusual results we note that the modified Rayleigh number can change sign if either the axial temperature gradient A or the volumetric expansion coefficient changes sign or if the body force direction changes. If, for example, we assume that the sign change of Ra_A is due to a change in A, the physical interpretation of the unusual result becomes clearer. A negative axial temperature gradient implies that the fluid is heated from below and this situation leads to an instability due to the piling of heavy fluid on lighter. Analogous situations, of course, follow directly for changes in body force direction or in sign of β. Natural-convection flows heated from below have been studied in some detail theoretically in [6] and experimentally in [7] and [8]. A more complete bibliography and discussion can be found in [9]. The studies show that for horizontal layers the fluid motion does

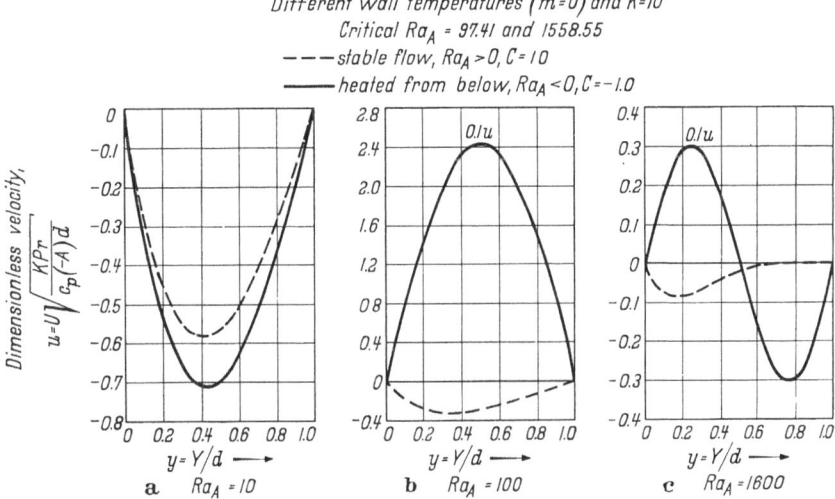

Fig. 2a. Representative velocity distributions for stable flows ($Ra_A > 0$) and flows heated from below ($Ra_A < 0$). Walls at different temperatures ($m = 0$) and $K = 10$. Critical $Ra_A = 97.41$ and 1558.55

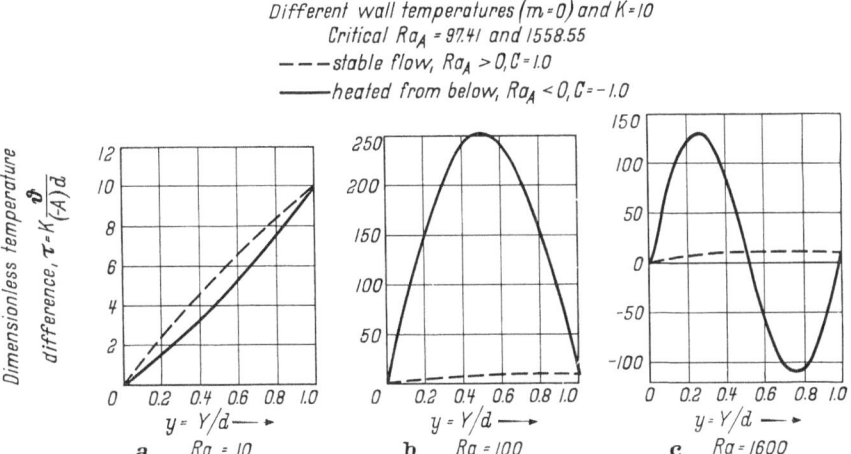

Fig. 2b. Representative temperature distributions for stable flows ($Ra_A > 0$) and flows heated from below ($Ra_A < 0$). Walls at different temperatures and $K = 10$. Critical $Ra_A = 97.41$ and 1558.55

indeed depend on critical values of the Rayleigh number. The fluid remains at rest until a critical value is attained. It then has a columnar mode of motion which changes to a cellular mode (the well-known Bénard cells) at another critical Rayleigh number. Hence, stability characteristics similar to those found for horizontal fluid layers are displayed in the present problem. The whirling instability of a rotating shaft is shown in [5] to be a mechanical analogy to the flow problem treated here.

A representative set of profiles computed for a positive Rayleigh number and all other conditions identical to those for the negative

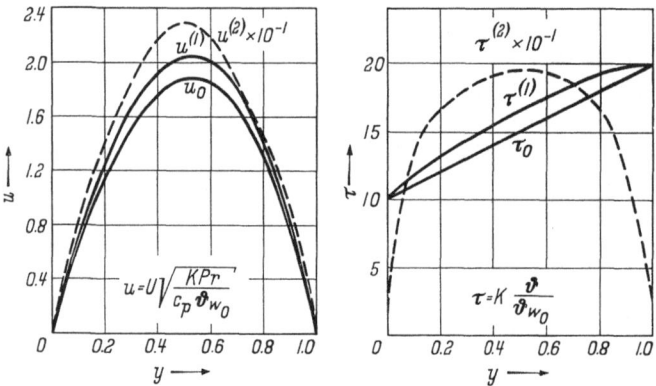

Fig. 3. Representative velocity and temperature distributions with and without frictional heating. Ratio of the constant wall temperatures $m = 2$ and $K = 10$

Rayleigh number case presented in fig. 2 a, 2 b are superposed on that figure. Comparison of the profiles for the two cases shows that the flow and heat transfer associated with the heating from below is greater and the heat transfer at the walls can be in opposite directions in the two problems.

2.3.2. Frictional heating. Let us now proceed to discuss the second objective of this series of problems, namely, how does the frictional heating affect the flow. As a check of the numerical accuracy of the analytical iteration method, referred to above and discussed in [2] and [3], the same nonlinear problems (specified by eq. (11) and the appropriate boundary conditions) were solved by numerical integration on a high-speed computing machine. It was then discovered that each problem had either a pair of solutions or no solution and that one member of each pair was approximated by the analytical solution (see fig. 3). A somewhat analogous situation was noted by HARTREE [10] in his study of boundary layer flows with adverse pressure gradients. The second solutions to HARTREE's problem are presented and further discussed in [11] and [12].

Examination of the representative profiles in fig. 3 shows that the frictional heating does increase velocities as was expected and alters the temperature gradients at the wall and, hence, the heat transfer. The analytical solutions denoted by u_0 are computed by neglecting frictional heating. Other examples of profiles computed including frictional heating are given and discussed in [2] to [4].

What is the meaning of the other interesting results obtained by considering frictional heating, namely that (1) beyond certain parametric limits no fully-developed solutions exist and (2) distinct pairs of solutions are found for a given set of conditions? A study of the machine solutions indicates that the probability of obtaining any solutions decreases as parameters like the frictional heating parameter increases, which implies more heat input to the channel. Therefore, it is suggested that the critical condition beyond which no solutions can be found might be associated with thermal choking in the channel. From fig. 3 we see that the second solution denoted by $u^{(2)}$ and $\tau^{(2)}$ represents a flow with velocities and temperature differences more than 10 times the first. These second solutions are intimately connected with the regenerative action of the frictional heating in natural convection wherein the frictional heating acts like a heat source in the fluid and thus tends to increase flow velocities. The less intense solution is the one ordinarily expected in practice because there is less frictional heating associated with that solution. The more intense solution most likely would have to be obtained by some artificial means analogous to the "swallowing" of the shock in a Laval nozzle. The existence of pairs of solutions to this problem might be analogous to the situation in a Laval nozzle although the two flows here are not necessarily sub- and supersonic. A stability analysis must still be made to determine whether the more intense solution will persist or attenuate to the less intense solution.

The combined effects of heating from below and frictional heating can be seen from fig. 4. The approximate solution neglecting frictional heating closely approximates one of the two solutions including frictional heating (because $\overline{K} < 1$) except near the critical point. Further the approximate solution corresponds to a different member of each pair of exact solutions on either side of the critical Ra_A, namely, the one with the smaller velocity extremum. The question then arises: Since the exact solutions do not display a discontinuity (with Ra_A) is the flow actually of a different type on either side of the critical value indicated by the linearized solutions? Although a more general analysis of the present configuration would be necessary (perhaps relaxing the condition of fully-developed flow) to answer this question conclusively, it seems reasonable that the actual flow is in fact of a different type on either side of the critical Ra_A. The reason for this statement is that the linearized equations are approximate only in that the frictional heating is neglected. In starting a flow, the frictional heating is not important until velocities

of appreciable magnitude are encountered. Therefore, since the larger
velocity flows result from the regenerative action of frictional heating
it is felt that for a given set of conditions the flow first established at any
Rayleigh number would be that with the smaller velocities.

The appreciable quantitative and qualitative differences between the
approximate and exact solutions near the critical Ra_A shown in fig. 4

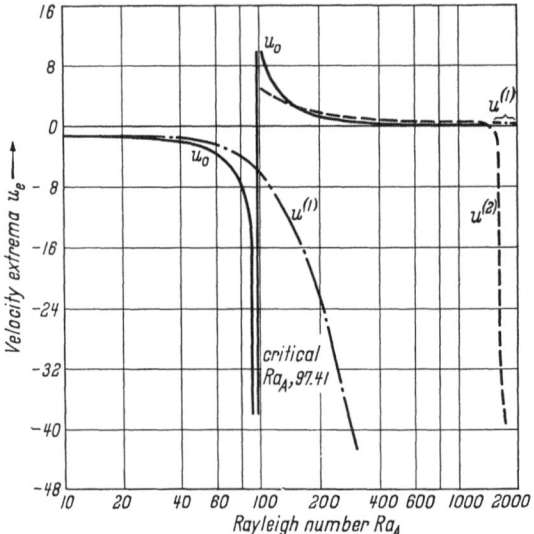

Fig. 4. Velocity extrema with and without frictional heating as functions of the Rayleigh number for
flows heated from below. Identical wall temperatures ($m = 1$), $K = 10$, and $C = -1$

indicate that near that value the effects of frictional heating are very
important. These effects may even play an important role in the transition from one type of flow to the other.

Although pairs of solutions were obtained for three somewhat different
problems if frictional heating effects are considered, all the analyses
discussed above are restricted to fully-developed flows of fluids which
are only "quasi-incompressible." Recently, MASLEN (in a paper to be
published soon) considered the same problems as are discussed above
where, however, the fluid is taken to be compressible and to have
property variations. The basic equations for such fluids are then reduced
to equivalent incompressible equations by a modified Howarth transformation [13] such as

$$\eta = -1 + 2 \frac{\int_0^y dy/\mu}{\int_0^d dy/\mu} \tag{20}$$

where the normalization is introduced so that internal flows can be treated. Viscosity-temperature laws appropriate for gases and liquids are used in eq. (20). The resulting nonlinear problems (i.e., including frictional heating) are treated by an integral method and pairs of solutions are again obtained. Thus, two flows and heat transfer states are theoretically predicted for fully-developed combined forced- and natural-convection flows of certain real fluids (with property variations). A stability analysis and definitive experiments are necessary to establish definitely the existence of the second state of flow and heat transfer indicated here.

3. Flow in a closed-end tube

The flow in partially- or fully-enclosed regions can be treated as special cases of the solutions of the preceding section by specifying that there is no net mass flow in the channel [3] and [4]. However, the effects of rigid surfaces at the channel ends are not clearly evident in this way. Therefore, to pursue further the effects of confining walls, LIGHTHILL [14] analyzed the flow in a closed-end cylindrical tube (see fig. 5) with the walls at constant temperature and the body force acting toward the closed end. An orifice which supplies cool fluid is assumed at the open end. His motivation for this problem was the previously-mentioned turbine blade cooling problem of SCHMIDT.

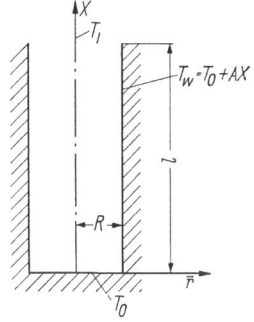

Fig. 5. Schematic sketch of closed-end tube configuration

The type of flow in the tube depends primarily on the length-radius ratio for given Prandtl and Rayleigh numbers. For very small values of this l/R ratio the flow is just like the free convection about a vertical plate, i.e., the effect of the confining walls is negligible if the boundary-layer thickness is much smaller than the tube radius. For somewhat larger l/R values, however, these effects are no longer negligible. The three flow regimes to be expected if the walls influence each other are shown in fig. 6. The first for small values of l/R is just the free-convection boundary-layer flow modified to account for a downward flow outside the boundary layer equal to the upward flow. This regime breaks down at a value of l/R for which there is no longer a maximum volume flow of cool fluid at the orifice cross section. When l/R exceeds this critical value the boundary layer mixes with the central flow and, when steady state is reached, the profiles fill the whole tube. This type of flow may be difficult to perceive intuitively so let us consider the extreme case $l/R \gg 1$. In this case the tendency of the upward layer to thicken with axial distance disappears. Then the velocity and temperature distributions are similar at each cross section, only their scale increasing as the orifice is approached. Therefore, in the intermediate l/R range, the velocity and temperature profiles fill the tube completely but vary along the tube. Since the closed-end tube is a more complicated configuration than was considered in the previous section, LIGHTHILL used an integral method to obtain solutions that behave as was just described. For the similarity regime (large l/R) he found that the flow fills the entire tube only for one value of l/R. If l/R is greater than this critical value the

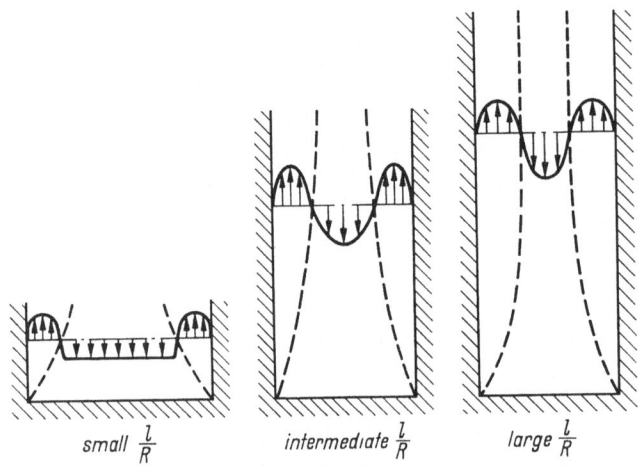

small $\frac{l}{R}$ intermediate $\frac{l}{R}$ large $\frac{l}{R}$

Fig. 6. The three flow regimes in a closed-end tube

motion stops near the closed end such that the effective l/R becomes equal to the critical. This result leads to one of the interesting and, at the same time, distressing aspects of natural convection in an enclosed region because the stagnation of part of the fluid renders it ineffective as a coolant.

To study further the stagnation of fluids in closed-end tubes consideration is given in [5] to a configuration identical with LIGHTHILL's except that now the temperature is taken to vary along the tube wall. This generalization of the wall-temperature condition may be more realistic since, for example, the temperature in a turbine blade varies in the spanwise direction. Examination of the mathematics of the problem will indicate the details of LIGHTHILL's analysis as well as the new results. The basic equations expressing conservation of mass, momentum, and energy are just the same as for free-convection boundary-layer flows except that the wall rather than the hydrostatic condition is used as a reference. The integral method employed is to satisfy integrated forms of the differential equations in the tube interior and the equations themselves at the tube axis and walls. The integrated equations are, respectively,

$$\int_0^1 r u \, dr = 0 \tag{21}$$

$$\frac{1}{Pr} \frac{\partial}{\partial x} \int_0^1 r u^2 \, dr = - \int_0^1 r (\tau - \tau_w) \, dr + \frac{1}{2} \left(\frac{\partial u}{\partial r} - \frac{\partial^2 u}{\partial r^2} \right)_{r=1} \tag{22}$$

$$\frac{\partial}{\partial x} \int_0^1 r u \tau \, dr = \left(\frac{\partial \tau}{\partial r} \right)_{r=1} \tag{23}$$

and at the wall and axis

$$\left(\frac{\partial^2 \tau}{\partial r^2} + \frac{1}{r} \frac{\partial \tau}{\partial r} \right)_{r=1} = 0 \tag{24}$$

$$\left(u \frac{\partial \tau}{\partial x} \right)_{r=0} = \left(\frac{\partial^2 \tau}{\partial r^2} + \frac{1}{r} \frac{\partial \tau}{\partial r} \right)_{r=0} \tag{25}$$

$$\frac{1}{Pr} \left(u \frac{\partial u}{\partial x} \right)_{r=0} = - (\tau - \tau_w)_{r=0} + \left(\frac{\partial^2 u}{\partial r^2} + \frac{1}{r} \frac{\partial u}{\partial r} \right)_0^1 \tag{26}$$

where $u = \dfrac{R^2}{\varkappa l} U$, $\tau = \dfrac{\beta f_X R^4}{\nu \varkappa l} (T_0 - T)$, $x = X/l$, $r = \bar{r}/R$ and $\tau_w \equiv \tau(x,1)$.
The zero subscript denotes conditions at the closed end. The boundary conditions are

$$u (x,1) = v (x,1) = u (0,r) = v (0,r) = 0 \tag{27}$$

$$\tau (0,r) = 0, \ \tau (x,1) = - Ra_A x \tag{28}$$

$$\tau (1,0) = Ra (R/l) \tag{29}$$

The effects of frictional heating are disregarded in this discussion.

Because we are particularly interested in a configuration wherein the flow can stagnate, we will restrict ourselves to only the large but finite l/R case. Examination of the equations shows that u and τ must both vary linearly with x and on the basis of symmetry and the boundary conditions at $x = 0$, u and τ must be of the form $xf(r^2)$ where $f(r^2)$ denotes third degree polynomials in r^2. For $Ra_A = 0$ these equations reduce to an eigenvalue problem (homogeneous equations and boundary conditions) which is LIGHTHILL's problem. The eigensolution obtained is valid only for a single value of $Ra (R/l)$. Therefore, the condition that the axial temperature should rise from its value T_1 at the orifice to T_0 at the closed end determines the one l/R, previously discussed, for this flow. In the variable wall-temperature case there is an additional parameter in the problem so that an eigenrelation between the two parameters is obtained. This relation and its derivation is presented in [15]. Accordingly, similar flows are possible for a range of parametric values rather than for discrete values as in LIGHTHILL's case.

Representative velocity and temperature distributions are given in fig. 7. For each $Ra\,(R/l)$ it can be seen that there is an Ra_A for which a "similar" flow exists in the entire tube. LIGHTHILL's result appears as the special case with $Ra_A = 0$. Fig. 7 shows that velocities greater than those obtained for constant wall temperatures are indicated with positive values of Ra_A and the associated smaller values of $Ra\,(R/l)$. The total

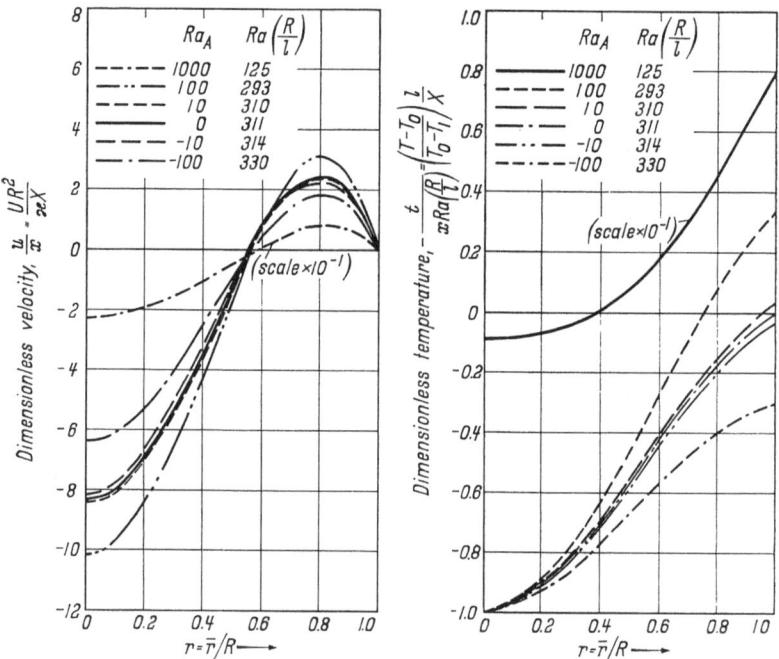

Fig. 7. Representative velocity and temperature distributions in a closed-end tube with linearly varying wall temperature

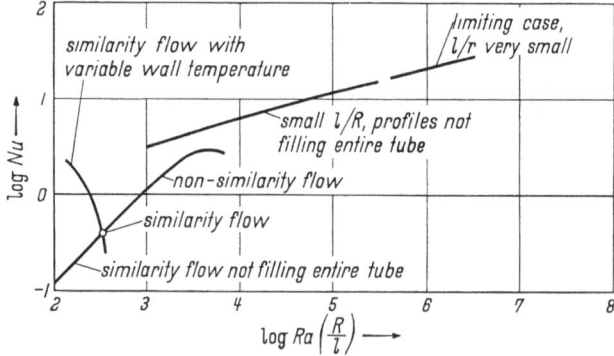

Fig. 8. Nusselt number as function of the product of the Rayleigh number and the radius-length ratio for closed-end tube flows

Nusselt number over the tube length is given as a function of Ra (R/l) in fig. 8 for the three flow regimes treated by LIGHTHILL. The result for variable wall temperatures and similar flows is superposed on this figure. Thus, higher heat transfer rates will be obtained for large l/R if the wall temperature is not constant.

The results of the generalization of LIGHTHILL's problem thus indicate that it is possible to have a "similar" flow in the whole tube for large l/R and the relation between the parameters to make this so is determined. The profiles for variable wall-temperature cases are found and the effects of the parameters on these profiles are given. Further, it is shown that higher heat transfer rates are obtained for large l/R if the wall temperature is not constant.

Experimental verification of the three flow regimes is given in [16] and LIGHTHILL's heat transfer results presented in fig. 8 are also essentially confirmed therein.

The results of this section are consistent with those of the previous section because for zero net mass flow in the channel and the walls at the same constant temperature (conditions which simulate the closed-end tube) the fully-developed solutions become trivial. The fully-developed solutions for the linear wall-temperature cases become eigensolutions under conditions simulating the closed-end tube. This behavior is somewhat like that described above for $Ra_A \neq 0$.

4. Flows in rotating containers

The body force in the analyses of the previous sections was always taken to be a constant. However, the problems associated with the cooling of high-speed vehicles reentering the earth's atmosphere by internal flows generated by body forces provide numerous configurations in which these forces can be variable. To gain some insight into problems of this kind let us study the motion and heating of such bodies. Accordingly, calculations of the heat flux and retardation of such vehicles computed from equations in [17] are presented in fig. 9. It is evident that if the deceleration force is to drive the fluid, this body force is time-dependent. Further, a lag between the heating of the body and the appearance of a large retardation force can be seen in fig. 9. One might then wonder whether the heat transfer to the internal fluid might be increased during the period when the decelerating force is small by rotating the vehicle about its axis. Centrifugal and Coriolis forces would then be the motion generating body forces as in the turbine cooling problem and these, of course, depend on the radius of rotation. The rotational effects are of interest for another reason. The decelerating force acts along the axis of the body. Therefore, the internal flow due to deceleration near the external stagnation point of a blunt body would be thermally unstable because of heating from below, i.e., the heating at the stagnation point imposes a negative temperature gradient parallel to the deceleration force. Thus, the internal fluid tends to stay at rest and is ineffective as a coolant until the critical Rayleigh number is reached. Therefore, it would appear that a large body force (such as a centrifugal

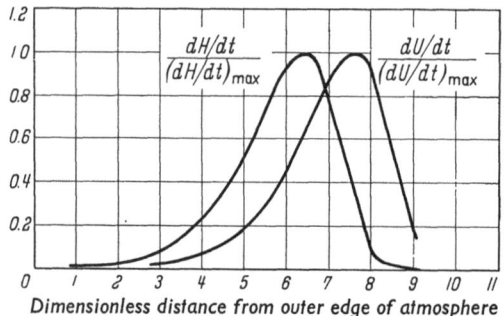

Fig. 9. Heat flux and deceleration as function of altitude for a representative reentry-type vehicle

force) transverse to the temperature gradient would increase the effectiveness of the heat sink by starting the motion sooner.

In order to avoid the complications introduced by unsteady effects we shall assume that the deceleration force is constant and concentrate here on the effects of rotation of the vehicle on the internal flow. We, therefore, will consider two cases: the first where the rotation predominates, i.e., where the internal fluid motion is essentially due to the rotation of the container, and the second where the deceleration force predominates.

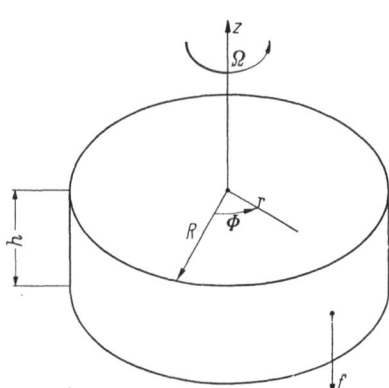

Fig. 10. Schematic sketch of configuration for flows in rotating containers

4.1. Predominant rotation. The first case simulates conditions encountered by an internal fluid, both during the time shortly after reentry when the deceleration force is negligible and in the stagnation region before a flow has been established by the instability. To analyze these situations let us consider the flow and heat transfer of a liquid contained in a circular cylinder with a relatively small height which is rotating about its axis (see fig. 10). The governing equations are just those for unsteady, compressible, viscous, heatconducting fluids subject to a body force.

If we assume axially symmetric flow and let

$$U = \varepsilon\,\Omega\,Ru, \quad V = \Omega\,R\,(r + \varepsilon\,v), \quad W = \varepsilon\,\Omega\,hw, \quad \bar{r} = Rr, \quad Z = hz$$
$$P = \varrho_0\,\Omega^2\,R^2\,(p_0 + \varepsilon\,p), \quad \varrho = \varrho_0\,(1 - \varepsilon\tau), \quad \tau = (T - T_0)/(T_w - T_0)$$
$$\Phi = \varphi + \Omega\,t \tag{30}$$

where $\varepsilon = \beta\,(T_w - T_0)$, those equations become to zeroth order in ε for steady flows

$$\frac{\partial p_0}{\partial r} = r; \quad \frac{\partial p_0}{\partial z} = -\frac{f\,h}{(\Omega R)^2}$$

which imply a solid body rotation of the fluid when there is no heating ($\varepsilon = 0$) and to first order in ε, i.e., the perturbations from the solid body rotation due to heating

$$\frac{\partial}{\partial r}\,(ru) + \frac{\partial}{\partial z}\,(r\,w) = 0 \tag{31}$$

$$\frac{Gr_\Omega}{Re_\Omega^2}\left(u\,\frac{\partial u}{\partial r} + w\,\frac{\partial u}{\partial z} - \frac{v^2}{r} + 2\,\tau\,v\right) + r\tau - 2\,v$$
$$= -\frac{\partial p}{\partial r} + \frac{1}{Re_\Omega}\left[\left(\frac{h}{R}\right)^2\left(\frac{\partial^2 u}{\partial r^2} + \frac{1}{r}\,\frac{\partial u}{\partial r} - \frac{u}{r^2}\right) + \frac{\partial^2 u}{\partial z^2}\right] \tag{32}$$

$$\frac{Gr_\Omega}{Re_\Omega^2}\left(u\,\frac{\partial v}{\partial r} + w\,\frac{\partial v}{\partial z} + \frac{uv}{r} - 2\,\tau\,u\right) + 2\,u$$
$$= \frac{1}{Re_\Omega}\left[\left(\frac{h}{R}\right)^2\left(\frac{\partial^2 v}{\partial r^2} + \frac{1}{r}\,\frac{\partial v}{\partial r} - \frac{v}{r^2}\right) + \frac{\partial^2 v}{\partial z^2}\right] \tag{33}$$

$$\frac{Gr_\Omega}{Re_\Omega^2}\left(u\,\frac{\partial w}{\partial r} + w\,\frac{\partial w}{\partial z}\right)$$
$$= -\left(\frac{R}{h}\right)^2\frac{\partial p}{\partial z} + \frac{f}{\Omega^2 h}\,\tau + \frac{1}{Re_\Omega}\left[\left(\frac{h}{R}\right)^2\left(\frac{\partial^2 w}{\partial r^2} + \frac{1}{r}\,\frac{\partial w}{\partial r}\right) + \frac{\partial^2 w}{\partial z^2}\right] \tag{34}$$

$$\frac{1}{\gamma}\,\frac{Gr_\Omega}{Re_\Omega^2}\left[u\,\frac{\partial \tau}{\partial r} + w\,\frac{\partial \tau}{\partial z} - \frac{K_\Omega}{Re_\Omega}\,(\Phi)\right]$$
$$= -\frac{1}{Pr\,Re_\Omega}\left[\left(\frac{h}{R}\right)^2\left(\frac{\partial^2 \tau}{\partial r^2} + \frac{1}{r}\,\frac{\partial \tau}{\partial r}\right) + \frac{\partial^2 \tau}{\partial z^2}\right] \tag{35}$$

where $Gr_\Omega = \beta\,\Omega^2\,h^4\,(T_w - T_0)/v^2$, $Re_\Omega = \Omega\,h^2/v$ and $\bar{K}_\Omega = \beta\,\Omega^2\,R^2/c_p$. The ε dependence in eqs. (30) was determined from a more general dependence to be the only consistent representation.

The Froude number, Re^2_Ω/Gr_Ω, is large in this problem. In fact from the definitions of the Reynolds and Grashof numbers based on Ω the reciprocal of the Froude number is just ε. Therefore, the dominant inertia terms are the centrifugal and Coriolis terms. The convection and frictional heating terms are also negligible so that the heat transfer is not affected by the motion and is merely that due to conduction. This implies that large velocities cannot be obtained by applying a temperature gradient transverse to the body force in such a configuration. To understand this seemingly unusual result let us solve the equations that result by omitting the negligible terms and assuming that $h/R << 1$.

Eq. (31) is unchanged and eqs. (32) to (35) become, respectively,

$$\frac{1}{Re_\Omega} \frac{\partial^2 u}{\partial z^2} = \frac{\partial p}{\partial r} + r\tau - 2v \tag{36}$$

$$\frac{1}{Re_\Omega} \frac{\partial^2 v}{\partial z^2} = 2u \tag{37}$$

$$\frac{\partial p}{\partial z} = 0 \tag{38}$$

$$\frac{\partial^2 \tau}{\partial z^2} = 0 \tag{39}$$

The boundary conditions on the upper and lower surfaces ($z = \pm 1/2$) are, conventionally

$$u = v = w = 0$$

At the upper surface also

$$\tau = 0$$

and the thermal boundary condition at the lower surface will be discussed a posteriori. We can essentially close the ends of the cylinder by specifying that at $r = 1$, $u = 0$. The other velocity components will satisfy inviscid conditions. Further, at the center $r = 0$, $u = 0$, and $v = 0$; the latter ensures finite angular momentum. Separation of variables yields the solutions

$$u = \frac{2}{\alpha} J_1(\alpha r) \left(C_1 \sinh \sqrt{Re_\Omega}\, z \, \cos \sqrt{Re_\Omega}\, z \right.$$
$$\left. - C_2 \cosh \sqrt{Re_\Omega}\, z \, \sin \sqrt{Re_\Omega}\, z \right) \tag{40}$$

$$v = \frac{2}{\alpha} J_1(\alpha r) \left(C_1 \cosh \sqrt{Re_\Omega}\, z \, \sin \sqrt{Re_\Omega}\, z \right.$$
$$\left. + C_2 \sinh \sqrt{Re_\Omega}\, z \, \cos \sqrt{Re_\Omega}\, z - \frac{1}{2} z \right) \tag{41}$$

$$w = \frac{1}{\sqrt{Re_\Omega}} J_0(\alpha r) \left[C_1 \left(\sinh \frac{\sqrt{Re_\Omega}}{2} \sin \frac{\sqrt{Re_\Omega}}{2} + \cosh \frac{\sqrt{Re_\Omega}}{2} \cos \frac{\sqrt{Re_\Omega}}{2} \right) \right.$$
$$+ C_2 \left(\cosh \frac{\sqrt{Re_\Omega}}{2} \cos \frac{\sqrt{Re_\Omega}}{2} - \sinh \frac{\sqrt{Re_\Omega}}{2} \sin \frac{\sqrt{Re_\Omega}}{2} \right)$$
$$- C_1 \left(\sinh \sqrt{Re_\Omega}\, z \, \sin \sqrt{Re_\Omega}\, z + \cosh \sqrt{Re_\Omega}\, z \, \cos \sqrt{Re_\Omega}\, z \right)$$
$$\left. - C_2 \left(\cosh \sqrt{Re_\Omega}\, z \, \cos \sqrt{Re_\Omega}\, z - \sinh \sqrt{Re_\Omega}\, z \, \sin \sqrt{Re_\Omega}\, z \right) \right] \tag{42}$$

$$\tau = \frac{2}{\alpha r} J_1(\alpha r) \left(\frac{1}{2} - z \right) \tag{43}$$

where J_0 and J_1 are Bessel functions and $J_1(\alpha) = 0$. Further

$$C_1 = \frac{\cosh \dfrac{\sqrt{Re_\Omega}}{2}\,\sin \dfrac{\sqrt{Re_\Omega}}{2}}{4\left(\sinh^2 \dfrac{\sqrt{Re_\Omega}}{2} + \sin^2 \dfrac{\sqrt{Re_\Omega}}{2}\right)}; \; C_2 = \frac{\sinh \dfrac{\sqrt{Re_\Omega}}{2}\,\cos \dfrac{\sqrt{Re_\Omega}}{2}}{4\left(\sinh^2 \dfrac{\sqrt{Re_\Omega}}{2} + \sin^2 \dfrac{\sqrt{Re_\Omega}}{2}\right)}.$$

The temperature of the lower plate is from eq. (43) equal to $(2/\alpha r)J_1(\alpha r)$. Any other temperature distribution could be specified at the lower surface by expanding $(\alpha r/2)\,\tau_w$ in a series in J_1.

If this solution is to apply to the internal flow in the vehicle's stagnation region no end conditions need be specified. In this case, eqs. (40) to (43) still apply where, however, J_1 is replaced by r and J_0 by a constant. The lower surface temperature thus is also a constant.

Examination of the last two terms in eq. (36) (the buoyancy due to centrifugal force and the Coriolis force, respectively) which are the driving force for the flow reveals why the velocities remain small. From eqs. (41) and (43) it can be seen that the part of the buoyancy term that is z-dependent is negated by part of the Coriolis term. The part that varies with r alone is balanced by the pressure gradient so that the net driving force is merely the remaining part of the Coriolis term. This net force increases with small Re but decreases for large Re. The u and w components must then behave in a similar manner and, thus, the inertia and convection effects are small even for large rotations.

It may be possible that other flows of a cellular type might be obtained in this configuration if h/R is not assumed to be negligibly small and the condition of axial symmetry is relaxed. However, it is doubtful that even then the heat transfer would be affected by the fluid motion. Thus, rotating a fluid container like that just considered does not significantly affect the heat transfer when the deceleration force is inoperative. In fact, the rotation may even be detrimental. CHANDRASEKHAR [18] shows that in an unstable configuration the fluid motion is delayed by the action of Coriolis forces.

Since the Coriolis force due to the circumferential velocity component opposes the radial buoyancy force and thus negates convection effects and also may impede the unstable motion, it appears that significant convection could be obtained only if the Coriolis term becomes negligible. The insertion of radial vanes would make the circumferential velocity component (and hence the corresponding Coriolis term) small relative to the radial velocity. The buoyancy effect due to the centrifugal force would, in this way, be uninhibited and could lead to large heat convection as in the Schmidt turbine blades.

4.2. Predominant deceleration force. The effects of rotation on a flow generated by the deceleration force must still be evaluated because the reentry vehicle may be rotating about its axis for aerodynamic stability reasons. The deceleration force will establish flows in a configuration like that in fig. 10 after the critical Rayleigh number is surpassed. Also if the retardation force is transverse to a temperature gradient as it would be, for example, in a curved container like a spherical shell it would immediately generate a conventional natural-convection flow. Let us, therefore, analyze the case where the flow is essentially due to the retardation force. Accordingly substitute

$$U = \sqrt{\varepsilon f h}\, u, \quad V = \Omega R r + \sqrt{\varepsilon f h}\, v, \quad W = \frac{h}{R}\sqrt{\varepsilon f h}\, w, \quad \bar{r} = R r, \quad Z = h z$$

$$P = \varrho_0 f h\,(p_0 + \varepsilon p), \quad \varrho = \varrho_0\,(1 - \varepsilon \tau), \quad \tau = (T - T_0)/T_w - T_0, \quad \varPhi = \varphi + \Omega t$$

into the basic equations in cylindrical coordinates. There then results to zero order in ε (no heating)

$$\frac{\partial p_0}{\partial r} = \omega^2 r, \quad \frac{\partial p_0}{\partial z} = 1$$

and to first order in ε

$$\frac{\partial}{\partial r}(r u) + \frac{\partial v}{\partial \varphi} + \frac{\partial}{\partial z}(r w) = 0 \tag{44}$$

$$\frac{h}{R}\left(u\,\frac{\partial u}{\partial r} + \frac{v}{r}\,\frac{\partial u}{\partial \varphi} + w\,\frac{\partial u}{\partial z} - \frac{v^2}{r} - \frac{2\omega v}{\sqrt{\varepsilon}} + \omega^2 r\tau \right)$$
$$= -\frac{h}{R}\,\frac{\partial p}{\partial r} + \frac{1}{\sqrt{Gr}}\left[\left(\frac{h}{R}\right)^2\!\left(\frac{\partial^2 u}{\partial r^2} + \frac{1}{r}\,\frac{\partial u}{\partial r} + \frac{1}{r^2}\,\frac{\partial^2 u}{\partial \varphi^2} - \frac{u}{r^2} - \frac{2}{r^2}\,\frac{\partial v}{\partial \varphi}\right) + \frac{\partial^2 u}{\partial z^2} \right] \tag{45}$$

$$\frac{h}{R}\left(u\,\frac{\partial v}{\partial r} + \frac{v}{r}\,\frac{\partial v}{\partial \varphi} + \frac{u v}{r} + w\,\frac{\partial v}{\partial z} + \frac{2\omega u}{\sqrt{\varepsilon}} \right)$$
$$= -\frac{h}{R r}\,\frac{\partial p}{\partial \varphi} + \frac{1}{\sqrt{Gr}}\left[\left(\frac{h}{R}\right)^2\!\left(\frac{\partial^2 v}{\partial r^2} + \frac{1}{r}\,\frac{\partial v}{\partial r} + \frac{1}{r^2}\,\frac{\partial^2 v}{\partial \varphi^2} + \frac{2}{r^2}\,\frac{\partial u}{\partial \varphi} - \frac{v}{r^2}\right) + \frac{\partial^2 v}{\partial z^2} \right] \tag{46}$$

$$\left(\frac{h}{R}\right)^2\!\left(u\,\frac{\partial w}{\partial r} + \frac{v}{r}\,\frac{\partial w}{\partial \varphi} + w\,\frac{\partial w}{\partial z} \right)$$
$$= -\frac{\partial p}{\partial z} + \tau + \frac{h}{R}\,\frac{1}{\sqrt{Gr}}\left[\left(\frac{h}{R}\right)^2\!\left(\frac{\partial^2 w}{\partial r^2} + \frac{1}{r}\,\frac{\partial w}{\partial r} + \frac{1}{r^2}\,\frac{\partial^2 w}{\partial \varphi^2}\right) + \frac{\partial^2 w}{\partial z^2} \right] \tag{47}$$

$$\frac{1}{\gamma}\,\frac{h}{R}\left(u\,\frac{\partial \tau}{\partial r} + \frac{v}{r}\,\frac{\partial \tau}{\partial \varphi} + w\,\frac{\partial \tau}{\partial z} \right)$$
$$= \frac{1}{Pr\,\sqrt{Gr}}\left[\left(\frac{h}{R}\right)^2\!\left(\frac{\partial^2 \tau}{\partial r^2} + \frac{1}{r}\,\frac{\partial \tau}{\partial r} + \frac{1}{r}\,\frac{\partial^2 \tau}{\partial \varphi^2}\right) + \frac{\partial^2 \tau}{\partial z^2} \right] + \frac{\overline{K}}{\sqrt{Gr}}\,\varPhi \tag{48}$$

where

$$\omega = \frac{\Omega R}{\sqrt{f h}}.$$

The parameter ω or $\omega/\sqrt{\varepsilon}$ is proportional to the Froude number. However, the Froude number (or ω) in this case can be made to take on any desired value because the numerator depends on the rotation and the denomenator on the retardation force. In the previous problem, recall that the inertia and buoyancy effects both depended only on the rotation. Therefore, if $\omega \ll 1$ the inertia and convection terms in eqs. (44) to (48) having ω as a coefficient are at most of the same order as the other like terms. The equation then looks like the conventional boundary-layer equations and the inertia and convection effects are important. The heat transfer will then be greater than that due to conduction alone. In that case also the buoyancy effect due to rotation (as determined from the term $\omega^2 r \tau$ in eq. (45) is negligible. When ω becomes of unit order or larger, the Coriolis and centrifugal terms become dominant as in the previous case and the other inertia and convection terms are negligible. Eqs. (44) to (48) then reduce to a system like eqs. (36) to (39); this reduction is identical if it is also assumed that $h/R \ll 1$ and the flow is axially symmetric. Thus, rotating the fluid container decreases the heat transfer to an internal heat sink like that in fig. 10.

The results of the last two sections, therefore, indicate that for internal heat sinks in reentry-type vehicles the rotations of the body should be small so that the parameter ω is less than unity or that radial vanes be inserted in the container for large rotations.

5. Concluding remarks

The discussion in this paper of a series of analyses made to determine the effects of confining surfaces on viscous flows with body forces under conditions appropriate to present-day applications has shown that numerous interacting factors can appreciably affect the resultant flow and heat transfer. To determine the feasibility of using an internal heat sink in a given problem careful consideration should be given to such factors as the effects of heating the fluid from below, frictional heating, thermal boundary conditions, and rotating the fluid container.

Appendix

The following notation is used in this paper. See each section for the specific definitions of the nondimensional quantities.

A	axial temperature gradient, $\partial T^*/\partial X$
C	constant, $\dfrac{1}{\mu}\left(\dfrac{\partial P}{\partial X} + \varrho_R \, f_X\right) \sqrt{Pr d^4/c_p\,\theta_R\,K}$
d	characteristic length
E	Eckert number, defined by eq. (4)
f	negative of Z-component of body force per unit mass
f_X	negative of X-component of body force per unit mass
Gr	Grashof number, $\beta f \theta_w d^3/\nu^2$ or $\beta f_X \theta_w d^3/\nu^2$
Gr_Ω	Grashof number due to rotation, $\beta \Omega^2 h^4 (T_w - T_0)/\nu^2$
g	acceleration of gravity
H	average heat transferred per unit area
h	height

$\overline{K}$	frictional heating parameter, $\beta f d/c_p$ or $\beta f_X d/c_p$
K	$Pr Gr \overline{K}$
k	thermal conductivity coefficient
l	length
P	pressure
p	nondimensional pressure
Pr	Prandtl number, $c_p \mu/k$
R	radius
Ra	Rayleigh number, $Pr Gr$
Re	Reynolds number, $U_\infty d/\nu$
Re_Ω	Reynolds number due to rotation, $\Omega h^2/\nu$
$\overline{r}$	radial coordinate
r	nondimensional radial coordinate
T, T^*	temperature
T_R	reference temperature, T_s for constant wall temperature case and T_{w_0} for linear wall temperature cases
t	time
U_i	velocity components, $i = 1, 2, 3$
$\overline{U}$	reference velocity
u_i	nondimensional velocity components, $i = 1, 2, 3$
U, V, W	velocity components
u, v, w	nondimensional velocity components
X_i	coordinates
x_i	nondimensional coordinates
X, Y, Z	coordinates
x, y, z	nondimensional coordinates
β	fluid volumetric expansion coefficient, $\varrho \left[\dfrac{\partial (1/\varrho)}{\partial T} \right]_P$
γ	ratio of specific heats
ε	constant, $\beta (T_w - T_0)$
θ	temperature difference, $T - T_R$
θ_R	reference temperature difference, $T_{w_0} - T_s$ for constant temperature case and Ad for linear wall temperature cases
$\varkappa$	thermal diffusivity, $k/\varrho c_p$
μ	absolute viscosity coefficient
ν	kinematic viscosity coefficient
ϱ	density
τ	dimensionless temperature difference
Φ	dissipation function
φ	angular coordinate
Φ	angular coordinate in inertial coordinate system
Ω	angular velocity

Subscripts:

A	denotes conditions based on the axial temperature gradient
a	denotes adiabatic condition
R	denotes a reference condition, the hydrostatic value for constant wall temperature case and the value at the wall at $y = 0$ for the linear wall temperature case
s	denotes the hydrostatic condition
w_0	denotes conditions at the wall at $y = 0$
w_1	denotes conditions at the wall at $y = 1$

References

[1] SCHMIDT, E. H. W.: Heat Transmission by Natural Convection at High Centrifugal Acceleration in Water-Cooled Gas Turbine Blades, Proceedings of the General Discussion on Heat Transfer, Sept. 11-13, 1951.

[2] OSTRACH, S.: Laminar Natural-Convection Flow and Heat Transfer of Fluids with and without Heat Sources in Channels with Constant Wall Temperatures, NACA TN 2863, 1952.

[3] OSTRACH, S.: Combined Natural- and Forced-Convection Laminar Flow and Heat Transfer of Fluids with and without Heat Sources in Channels with Linearly Varying Wall Temperatures, NACA TN 3141, 1954.

[4] OSTRACH, S.: Unstable Convection in Vertical Channels with Heating from Below, Including Effects of Heat Sources and Frictional Heating, NACA TN 3458, 1955.

[5] OSTRACH, S.: On the Flow, Heat Transfer, and Stability of Viscous Fluids Subject to Body Forces and Heated from Below in Vertical Channels, 50 Jahre Grenzschichtforschung, Braunschweig: Vieweg 1955.

[6] RAYLEIGH, Lord: On Convection Currents in a Horizontal Layer of Fluid, when the Higher Temperature is on the Under Side, Philosophical Magazine, vol. 32, Dec. 1916.

[7] BÉNARD, H.: Revue générale des Sciences, XII, 1261, 1309 (1900).

[8] CHANDRA, K.: Instability of Fluids Heated from Below, Proceedings of the Royal Society of London, Series A, 164, 231-242, (1938).

[9] OSTRACH, S.: Convection Phenomena in Fluids Heated from Below, Transactions of the ASME, 79, 299-307, (1957), No. 2.

[10] HARTREE, D. R.: On an Equation Occurring in Falkner and Skan's Approximate Treatment of the Equations of the Boundary Layer, Proceedings Cambridge Philosophical Society, 33, 223-237 (1937), No. 2.

[11] STEWARTSON, K.: Further Solutions of the Falkner-Skan Equation, Proceedings Cambridge Philosophical Society, 50, 454-465 (1954), No. 3.

[12] COHEN, C. B., and E. RESHOTKO: Similar Solutions for the Compressible Laminar Boundary Layer with Heat Transfer and Pressure Gradient, NACA TN 3325, 1955.

[13] HOWARTH, L.: Concerning the Effect of Compressibility on Laminar Boundary Layers and their Separation, Proceedings of the Royal Society of London, Series A, 194, 16-42 (1948), No. A 1036.

[14] LIGHTHILL, M. J.: Theoretical Considerations on Free Convection in Tubes, Quarterly Journal of Mechanics and Applied Mathematics, 6, 398-439 (1953).

[15] OSTRACH, S., and P. R. THORNTON: On the Stagnation of Natural-Convection Flows in Closed-End Tubes, Paper No. 57-SA-2, Presented at the Semi-Annual Meeting of the American Society of Mechanical Engineers, San Francisco, Calif., June 9-13, 1957.

[16] MARTIN, B. W.: Free Convection in an Open Thermosyphon, with Special Reference to Turbulent Flow, Proceedings of the Royal Society of London, Series A, 230, 502-531 (1955).

[17] ALLEN, H. J., and A. J. EGGERS JR.: A Study of the Motion and Aerodynamic Heating of Missiles Entering the Earth's Atmosphere at High Supersonic Speeds, NACA RM A 53 D 28, 1953.

[18] CHANDRASEKHAR, S.: The Instability of a Layer of Fluid Heated Below and Subject to Coriolis Forces, Proceedings of the Royal Society of London, Series A, 217, 306-327 (1953).

Aus der Diskussion

C. J. Peirce (Cleveland): In your slides you showed a large number of velocity distributions. Is it your intention to check some of these by experiment?

S. Ostrach (Cleveland): I have no measurements to check these distributions. However, I do not believe that they are so unusual that you would not expect them. For the second solution, if this is what you mean, I would like to obtain it experimentally, but the others are just pipe flow type distributions that you would expect to get.

G. Kuerti (Cleveland): It seems to me that the term u'^2 in your basic equation gives this equation a shape that is somewhat similar to Burgers' first turbulence model. Here, in a similar mathematical situation, if a parameter exeeds a certain value, you get two solutions instead of one. Because of this similarity, you might be able to use Burgers' results.

Zur Bestimmung der Temperaturverteilung in dünnen Platten mit anliegender Überschallgrenzschicht

Von

H. Schuh

SAAB, Linköping

1. Einleitung

Bei kompressiblen Grenzschichten sind bekanntlich Geschwindigkeits- und Temperaturfelder wegen der Temperaturabhängigkeit der Stoffwerte voneinander abhängig. Manchmal lassen sich für das Temperaturprofil einfache Randbedingungen an der Wand angeben: z. B. wenn hier die Temperaturen vorgegeben sind oder der Wärmefluß verschwindet. In vielen Fällen gehört aber die Wand, die den einen Rand der Grenzschicht bildet, zu einem festen Körper, dessen Temperaturverteilung einschließlich Wandtemperatur sich aus den Wärmeströmen in seinem Inneren bestimmt. Im allgemeinen sind also nicht nur die Grenzschichtgleichungen, sondern auch die Wärmeleitungsgleichung im angrenzenden Körper zu lösen, vor allem, wenn die Wandtemperatur selbst interessiert. Im Beharrungszustand hängt die durch Reibungswärme hervorgerufene Eigentemperatur des Körpers (gleich seiner Temperatur bei verschwindendem konvektivem Wärmeübergang) nur sehr wenig von dem Druckfall längs des Körpers ab, aber auch nur wenig von der Strömungsform: laminar oder turbulent. Deshalb treten im stationären Zustand nur dann größere Unterschiede in der Wandtemperatur auf, wenn die Wärmestrahlung bedeutend wird. Anders bei instationären Zuständen, mit denen wir uns im folgenden in erster Linie befassen wollen. Der Beharrungszustand in der Geschwindigkeitsverteilung tritt in der Grenzschicht schon nach Zeiten ein, die von der gleichen Größenordnung sind wie die, die ein Gasteilchen braucht, um die Strecke vom Anfangspunkt der Grenzschicht bis zur betrachteten Stelle zurückzulegen. Deshalb genügt es, mit Rücksicht auf die meisten Anwendungen in der Grenzschicht mit stationären Zuständen zu rechnen und nur die Wärmeströmung im anliegenden Körper instationär zu behandeln. Da es uns im folgenden nur auf wesentlich Neues ankommt, so ersetzen wir den Einfluß der Wandtemperatur auf die entsprechenden Grenzschichtgrößen durch im voraus gewählte Mittelwerte, so daß wir dann nur die Wärmeleitungsgleichung

im festen Körper mit bekanntem äußerem Wärmeübergang zu lösen
haben. Dies kann auch als erster Schritt eines Iterationsverfahrens zur
simultanen Lösung aller Grundgleichungen aufgefaßt werden.

Zur weiteren Vereinfachung nehmen wir an, daß der angrenzende Körper konstante Dicke und konstante Temperatur über den Querschnitt
besitze. Die Wärmeleitungsgleichung ist dann eindimensional. Physikalisch bedeutet dies, daß wir dünne Platten und Überschall- bzw. Hyper-
schallströmungen mit nicht allzu hohen Machzahlen betrachten. Weiter
nehmen wir in den behandelten Beispielen die Anfangstemperaturver-
teilung in der Platte konstant an und denken uns die Strömung (oder die
Platte) augenblicklich auf eine konstante Geschwindigkeit gebracht. Für
viele Anwendungen gibt diese Annahme bereits die wesentlichen Züge
wieder. Wir untersuchen nun, wie die Änderungen der Wärmeübergangszahl längs der Oberfläche die Temperaturverteilung in der Platte be-
einflussen. Probleme dieser Art wurden von PARKER [1], EMMONS [2],
HEAPS [3] und TIDEMAN [4] behandelt. In diesen Arbeiten wird laminare
Strömung an einer längs angeströmten ebenen Platte angenommen, bei
HEAPS auch turbulente Strömung längs der ganzen Platte. Der Einfluß
einer örtlich veränderlichen Wandtemperatur auf die konvektive Wärmeübergangszahl wird nur bei EMMONS berücksichtigt, aber aus dem einzigen dort angegebenen Beispiel läßt sich über dessen Größe wenig
entnehmen, außerdem scheint die angewandte Rechenmethode sehr
zeitraubend zu sein. PARKERS analytische Methode ergibt brauchbare
numerische Lösungen nur in einem begrenzten Bereich der Veränderlichen.
HEAPS verwendet grobe Approximationen für die Änderung der Wärmeübergangszahl mit dem Abstand von der Vorderkante, die in deren Nähe
grundsätzlich falsche Lösungen ergeben. Nach TIDEMAN erhält man im
ganzen Bereich Lösungen mit ausreichender Genauigkeit.

An umströmten Körpern sind die Grenzschichten häufig teils laminar,
teils turbulent. Bei hohen Geschwindigkeiten, insbesondere beim Flug in
großen Höhen und bei laminaren Grenzschichten, kann die durch Strahlung von der Körperoberfläche an die Umgebung abgegebene Wärme
nicht mehr vernachlässigt werden, wie dies bisher geschah [1–4]. Fälle
dieser Art lassen sich kaum mit analytischen Methoden, wohl aber mit
dem Differenzenverfahren behandeln, das in seinen Anwendungen sehr
vielseitig ist. Man hat gegen dieses Verfahren oft eingewendet, daß die
Lösungen nicht stabil und die numerischen Rechnungen zeitraubend
seien. Für die Wärmeleitungsgleichung sind die Stabilitätsbedingungen
seit längerer Zeit bekannt [5, 6], und in letzter Zeit konnte gezeigt
werden, daß schon bei einer sehr groben Teilung der Abstandskoordinate
sich genügend genaue Lösungen ergeben, und damit läßt sich die Rechenarbeit in erträglichen Grenzen halten [7]. Wir zeigen im folgenden die
Anwendung des Differenzenverfahrens auf die obenerwähnten Probleme

und untersuchen einige für Wandtemperaturberechnungen wichtige Fragen an typischen Einzelfällen. Dabei beschränken wir uns auf längs angeströmte ebene Platten und auf dünne Flügelprofile. Wegen des großen Unterschiedes in der Wärmeübergangszahl bei laminarer und turbulenter Grenzschichtströmung ändert sich in der Nähe des Umschlagspunktes die Temperatur sehr stark. Hier muß die Platte in sehr kleine Abschnitte geteilt werden, damit die Lösungen der Differenzenrechnung genügend genau werden. Wegen der großen Temperaturgradienten eignet sich dieser Fall gut dazu, den Einfluß von Wandtemperaturänderungen auf die Wärmeübergangszahl zu untersuchen. Schließlich wird auch ein Beispiel mit und ohne Wärmeabgabe durch Strahlung behandelt und dabei auch eine „Linearisierung" der Grundgleichung mit Wärmestrahlung erprobt.

Die Wandtemperatur ist nicht nur als Randwert für das Temperaturprofil, sondern auch für die Beurteilung der Stabilität der laminaren Grenzschicht wichtig, die theoretisch sehr stark von Richtung und Intensität des Wärmeflusses zwischen Wand und Umgebung abhängt. Beim Flug mit sehr hohen Geschwindigkeiten interessiert die Temperaturverteilung in Flugzeugwänden für Fragen der mechanischen Festigkeit wegen der Temperaturabhängigkeit der Festigkeit vieler Baustoffe (z. B. Aluminiumlegierungen) und der durch Temperaturunterschiede verursachten Wärmespannungen. Auf diese Anwendungen soll aber hier nicht näher eingegangen werden.

2. Annahmen für die Grenzschichtströmung und Aufstellung der Differenzengleichung

Bei der Erwärmung von Platten ändert sich die Wandtemperatur sowohl mit dem Abstand von der Vorderkante als auch mit der Zeit. Da die Wandtemperatur ein Randwert für das Temperaturprofil ist, so müßte man genaugenommen die Entwicklung der Grenzschichten zu jeder Wandtemperaturverteilung kennen. Wir verwenden hier hauptsächlich Wärmeübergangszahlen für konstante Wandtemperaturen und untersuchen in einem Fall den durch diese Vereinfachung verursachten Fehler. In den hier behandelten Fällen ist die Abhängigkeit der Wärmeübergangszahl vom Verhältnis der Wandtemperatur zur Eigentemperatur nur gering, so daß wir hier davon abgesehen haben; übrigens könnte man diese bei der Differenzenrechnung leicht berücksichtigen. Bei der Plattenströmung ist die turbulente Wärmeübergangszahl aus Experimenten nur für Wandtemperaturen nahe der Eigentemperatur sicher bekannt [8], weshalb wir diese Werte hier verwendeten. Neuerdings scheint eine semiempirische Methode [9, 10, 11] brauchbare Werte auch für andere Wandtemperaturen zu ergeben. Berücksichtigt man den Wandtemperatureinfluß nach diesen Arbeiten, so sind die hier verwendeten turbulenten

Wärmeübergangszahlen ungefähr um 10% zu niedrig. Diese wurden berechnet nach der Formel:

$$\alpha = \varrho_1 \, c_{p_1} \, u_1 \, c_H, \tag{1}$$

wobei ϱ_1 die Dichte, c_{p_1} die spezifische Wärme bei konstantem Druck, u_1 die Geschwindigkeit und alle Werte am Außenrand der Grenzschicht genommen sind. Für turbulente Grenzschichten und Luft ($Pr = 0,7$) wird die Stantonzahl $c_H = 1,22 \, c_f/2$ gesetzt und der Koeffizient des Reibungswiderstandes c_f nach Wilson berechnet [12]. Für die laminare Grenzschicht wurde c_H der Arbeit von van Driest [13] bei einer Wandtemperatur gleich der Eigentemperatur entnommen, um die gleichen Verhältnisse wie bei der turbulenten Grenzschicht zu haben. Durch diese Annahme für die Wandtemperatur ergibt sich hier bei $M = 2,5$ ein unbedeutender Fehler von 2%. Bei dem Beispiel mit Wärmeübergang durch Strahlung wurde die laminare Wärmeübergangszahl für eine mittlere Wandtemperatur von $1/2 \, (T_f + T_i)$ berechnet (T_i Anfangs- und T_f Endtemperatur). Bei dem hier behandelten dünnen Profil mit 5% Dickenverhältnis wurde die Wärmeübergangszahl einfach wie bei der ebenen Platte angenommen, aber mit den lokalen Werten von ϱ_1 und u_1 in Gl. (1). Bei laminarer Strömung erhält man damit gute Näherungen, wie durch hier nicht wiedergegebene Untersuchungen nachgewiesen wurde.

Die Differentialgleichung für die Wärmeleitung in einer Platte mit äußerem Wärmeübergang durch Konvektion und Strahlung lautet:

$$\varrho_w \, c_w \, d \, \frac{\partial \, T_w}{\partial \, t} = \lambda_w \, d \, \frac{\partial^2 \, T_w}{\partial \, x^2} + \alpha \, (T_f - T_w) - \varepsilon \, C_s \, (T_w{}^4 - T_1{}^4), \tag{2}$$

wobei ϱ_w die Dichte, c_w die spezifische Wärme, λ_w die Wärmeleitfähigkeit, d die Dicke, T_w die Temperatur der Platte und x die Koordinate parallel zu deren Oberfläche bedeutet. T_f ist die Eigentemperatur der Grenzschicht, T_1 die statische Temperatur der Luft, ε das Emissionsverhältnis zwischen der Gesamtstrahlungszahl der Plattenoberfläche und der eines idealen schwarzen Körpers C_s. An der einen Plattenoberfläche findet Wärmeübergang mit der Kennzahl α statt, während die andere thermisch isoliert ist. Wir nehmen konstante Temperatur über den Plattenquerschnitt an und beschränken uns deshalb in den Anwendungen auf nicht zu hohe Werte von α und d.

Wir wandeln Gl. (2) in eine Differenzengleichung um von der Form der Gl. (74) in [7], aber vermehrt um das zusätzliche Strahlungsglied:

$$T_{w,n,m+1} = \frac{p}{1+r} \, (T_{w,n+1,m} + T_{w,n-1,m}) + \frac{1-2\,p-r}{1+r} \, T_{w,n,m} +$$

$$+ \frac{r}{1+r} \, (T_{f,m} + T_{f,m+1}) - \frac{q}{1+r} \, (T_{w,n,m})^4 + \frac{q}{1+r} \, (T_1)^4. \tag{3}$$

Dabei bedeuten die Indizes n und m den n-ten Abschnitt in der Koordinate x (siehe z. B. Abb. 1) bzw. den m-ten Zeitschritt beim Differenzenverfahren; ferner ist:

$$p = \frac{\lambda_w \, (\Delta t)_N}{\varrho_w \, c_w \, [(\Delta x)_N]^2}, \qquad r = \frac{\alpha \, (\Delta t)_N}{2 \, \varrho_w \, c_w \, d}, \qquad q = \frac{\varepsilon \, C_s \, (\Delta t)_N}{\varrho_w \, c_w \, d} \qquad (4)$$

mit Δt und Δx als Größen des Zeitschrittes bzw. der x-Koordinate. Wir lassen die Möglichkeit offen, Δx und Δt in Teilen der Platte verschieden groß zu wählen, und deuten dies durch einen entsprechenden Index N an. Im übrigen verläuft die Differenzenrechnung in gewohnter Weise: Aus einer bekannten Temperaturverteilung zur Zeit $t = m \, \Delta t$ werden mit Hilfe der Gl. (3) die Temperaturen nach dem nächsten Zeitschritt Δt, also zur Zeit $t = (m + 1) \, \Delta t$, berechnet usw.

Um stabile Lösungen zu erhalten, darf der Zeitschritt nicht zu groß gewählt werden [5, 6, 7]. Stabilitätsbedingungen kann man nach SCHUH [7] auch aus physikalischen Überlegungen herleiten. Eine Formulierung von Bedingungen dieser Art, die sich leicht auf erweiterte Wärmeleitungsgleichungen, wie z. B. Gl. (3), anwenden läßt, lautet: Die Lösungen sind dann stabil, wenn die Temperaturkoeffizienten derart sind, daß der Wert für $T_{w, n, m+1}$ bei beliebigen Kombinationen von positiven oder verschwindenden Werten für $T_{w, n+1, m}$, $T_{w, n, m}$ und $T_{w, n-1, m}$ niemals negativ werden kann. Diese zunächst nur als notwendig abgeleitete Bedingung ist erfahrungsgemäß auch hinreichend. Dabei umgehen wir die Schwierigkeiten durch das Strahlungsglied in Gl. (3) folgendermaßen: Wir ersetzen das Glied $-\dfrac{q}{1 + r} \, (T_{w, n, m})^4$ durch den stets gleichen oder kleineren Term $-\dfrac{q}{1 + r} \, [(T_{w, n, m})_{\max}]^3 \, T_{w, n, m}$. Faßt man nun alle Koeffizienten von $T_{w, n, m}$ zusammen und setzt man weiter

$$1 - 2 \, p - r - q \, [(T_{w. n, m})_{\max}]^3 \geqq 0 \,, \qquad (5)$$

so sind alle Koeffizienten in Gl. (3) positiv, und damit ist obige Stabilitätsbedingung erfüllt. Da Δx meist mit Rücksicht auf die Genauigkeit der Lösungen [7] und die Geometrie des Körpers gewählt wird, so folgt mit Gl. (4) aus (5) der größtmögliche Zeitschritt für Δt. Nach Möglichkeit wird dieser oder zumindest ein ihm möglichst nahekommender kleinerer Wert hier gewählt, weil die Rechenarbeit um so geringer ist, je größer Δt.

An scharfen Vorderkanten geht die Wärmeübergangszahl wie $1/\sqrt{x}$ gegen ∞, und deshalb sind an dieser Stelle große Temperaturänderungen zu erwarten. Genaugenommen ist Gl. (2) eine Wärmeleitungsgleichung mit inneren Wärmequellen, deren Intensität an einer Stelle gegen unendlich geht. Deshalb sind hier besondere Überlegungen notwendig. Am einfachsten ist es, von einer Lösung für eine bestimmte Abschnittsbreite Δx ausgehend, die Differenzenrechnung mit kleineren Abschnittsteilungen zu wiederholen und durch Vergleich der Lösungen die für die gewünschte Genauigkeit notwendige Abschnittsbreite zu finden.

Bei der endlichen Platte mit verschwindendem Wärmestrom an beiden Enden sind die Temperaturgradienten und damit auch die Fehler des Differenzenverfahrens kleiner als bei der einseitig unendlich langen Platte. Bei der turbulenten Grenzschicht sind ebenfalls Temperaturgradienten und Fehler kleiner als bei der laminaren Grenzschicht, weil sich bei jener die Wärmeübergangszahl weniger stark mit der Entfernung von der Vorderkante ändert als bei dieser. Es genügt also, die Genauigkeit des Differenzenverfahrens an der einseitig unendlichen ebenen Platte bei Laminarströmung zu untersuchen. Betreffs weiterer Einzelheiten siehe Beginn des nächsten Abschnittes.

Schon eine grobe Abschätzung der Lösungen zeigt, daß die feine Abschnittsteilung nur in der Nähe der Vorderkante nötig ist; dadurch ist aber nach Gln. (4) und (5) ein kleiner Zeitschritt bedingt, der bei der gröberen Teilung weiter hinten gar nicht nötig ist, sondern nur überflüssige Arbeit verursacht. Wir wählen daher hier auch einen größeren Zeitschritt. Da wir nun im Teil I mit kleinerem Δx und Δt die Temperaturen auch in Zeitpunkten erhalten, in denen keine entsprechenden Werte im angrenzenden Teil II mit gröberem Zeitschritt zur Verfügung stehen, so lassen wir beide Teile einander überlappen derart, daß die Abschnittsteilung von Teil I mit einer entsprechenden Anzahl Hilfspunkte über die Grenze von Teil I hinaus fortgesetzt wird, wie dies z.B. in Abb. 1a geschehen ist. Diese Hilfspunkte des Teiles I, k an Zahl, betrachten wir als frei, d.h., von einer bekannten Temperaturverteilung zur Zeit t_1 ausgehend, lassen wir im Teil I nach dem ersten Zeitschritt den äußersten Hilfspunkt unbestimmt, nach dem zweiten Zeitschritt den nächsten usw. Nach dem k-ten Zeitschritt sind alle Hilfspunkte unbestimmt. Nun wird auch im Teil II, der die grobe Teilung besitzt, ein Zeitschritt durchgeführt, der natürlich k-mal so groß sein wird wie derjenige in Teil I. Nun, zur Zeit $t_1 + k\,(\Delta t)_\mathrm{I}$, ist wiederum die Temperatur in der ganzen Platte bekannt, und damit folgen aus der Randbedingung zwischen Teil I und II (Kontinuität der Temperatur und ihres Gradienten) die Temperaturen der k Hilfspunkte. Dann kann die Rechnung wiederum wie oben weitergehen. Im übrigen erfüllen wir Randbedingungen durch einen Hilfspunkt außerhalb der Plattenteile (z.B. 0 oder 5″ in Abb. 1a). Temperaturgradienten drücken wir stets durch „zentrale" Differenzen aus, z.B. in Abb. 1a den Gradienten an der Stelle $\xi = 0{,}5$ durch Differenzen zwischen Punkt 4 und 7′ im Teil I bzw. zwischen 5″ und 6 im Teil II [1].

---

[1] Dabei wurden die Punkte 4 und 7′ in Teil I verwendet, um die Temperaturdifferenzen über Abstände zu bilden, die denen in Teil II möglichst nahekommen, wodurch die Fehler durch Bildung endlicher Differenzen weitgehend vermindert werden.

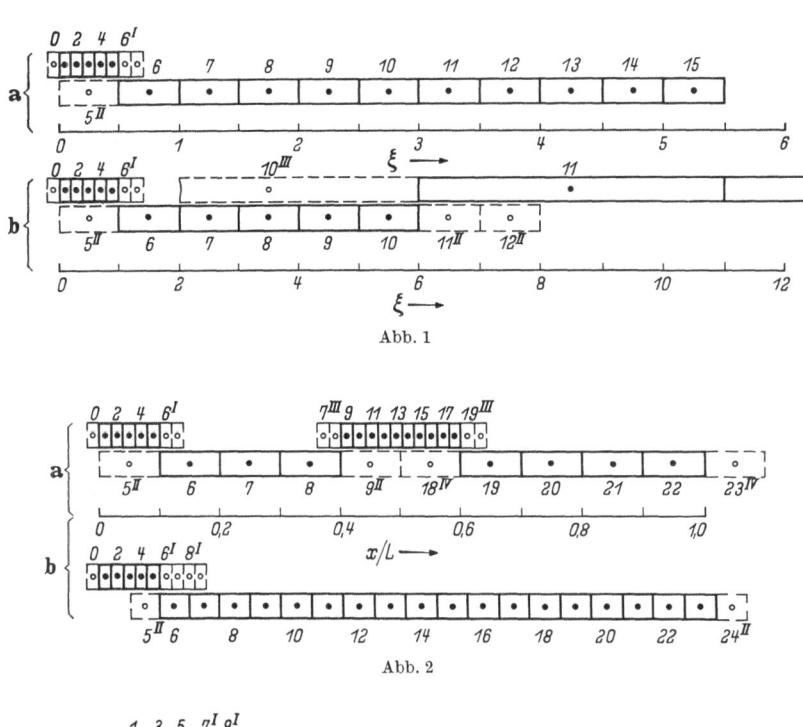

Abb. 1

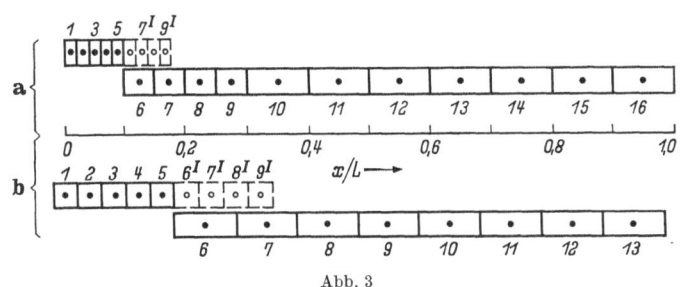

Abb. 2

Abb. 3

Abb. 1 bis 3. Abschnittsteilung der Platten für die Differenzenrechnung. Abb. 1: Bei der einseitig unbegrenzten ebenen Platte mit laminarer Grenzschicht. Abb. 2: Bei einer Platte an einem 5% dicken symmetrischen Parabelprofil mit teils laminarer, teils turbulenter Grenzschicht und einem Umschlagspunkt bei 50% Flügeltiefe. Abb. 3: Wie in Abb. 2, aber mit laminarer Grenzschicht und bei Berücksichtigung des Wärmeüberganges durch Strahlung. In allen Abbildungen: Abschnitte voll ausgezogen und durchlaufend numeriert. Hilfsabschnitte gestrichelt und durch Striche an den Nummern gekennzeichnet

3. Beispiele und Schlußfolgerungen

Zunächst sehen wir vom Wärmeübergang durch Wärmestrahlung ab und streichen die entsprechenden Glieder in Gln. (2) und (3). Als erstes Beispiel behandeln wir die einseitig unbegrenzte ebene Platte mit laminarer Strömung bei konstantem Druck. Dafür läßt sich die Wärmeübergangszahl bei inkompressibler und kompressibler Strömung wie folgt darstellen:

$$\alpha = \varrho_1 c_{p1} u_1 \frac{Z}{\sqrt{Re_{1x}}} \,. \tag{6}$$

wobei Z ein Koeffizient ist, der vom Temperaturverhältnis T_w/T_1 abhängt und Re_{1x} die Reynoldszahl ist, gebildet mit dem Abstand von der Vorderkante und den Stoffwerten bei der Temperatur am Außenrande der Grenzschicht. Z hängt nur in geringem Maße von der Temperatur ab, und es genügt wohl meist, einen konstanten Mittelwert anzunehmen. Weiter nehmen wir plötzliche Beschleunigung auf eine konstante Geschwindigkeit an. Wir führen die dimensionslosen Größen ein:

$$\xi = \frac{x}{d} M \quad \text{und} \quad \tau = \frac{a_w t}{d^2} M^2 \tag{7}$$

mit

$$M = \left(\frac{u_1 d}{\nu_1}\right)^{1/3} \left(\frac{\lambda_1}{\lambda_w} Pr_1 Z\right)^{2/3} . \tag{8}$$

Dabei ist ν_1 die kinematische Zähigkeit, Pr_1 die Prandtlzahl und u_1 die Geschwindigkeit am Außenrande der Grenzschicht. λ und a sind eine Wärme- bzw. Temperaturleitzahl mit Index w auf das Plattenmaterial und mit Index 1 auf die Luft bezogen; d ist die Plattendicke, t die Zeit und x der Abstand von der Vorderkante.

Wir führen noch ein Temperaturverhältnis

$$\Theta = \frac{T_w - T_f}{T_i - T_f} \tag{9}$$

mit T_w der Wandtemperatur, T_i der konstanten Anfangstemperatur und T_f der Eigentemperatur der Grenzschicht ein. Damit lautet die Differentialgleichung (2)

$$\frac{\partial \Theta}{\partial \tau} = \frac{\partial^2 \Theta}{\partial \xi^2} - \frac{\Theta}{\sqrt{\xi}} \tag{10}$$

mit den Rand- und Anfangsbedingungen

$$\tau = 0 ; \quad \Theta = 1$$
$$\xi = 0 \quad \text{und} \quad \xi \to \infty ; \quad \frac{\partial \Theta}{\partial \xi} = 0 \,. \tag{11}$$

Zunächst wurden vorbereitende Rechnungen mit zwei verschiedenen, aber über die ganze Platte gleichen Abschnittsbreiten von $\Delta \xi = 0{,}5$ bzw. 0,25 durchgeführt. Auf Grund der Ergebnisse dieser Rechnungen wurden die Abschnittsteilungen in Abb. 1a für die endgültigen Rechnungen mit $(\Delta \xi)_I = 0{,}1$ bzw. $(\Delta \xi)_{II} = 0{,}5$ und den Zeitkonstanten $(\Delta \tau)_I = 0{,}005$ bzw. $(\Delta \tau)_{II} = 0{,}01$ im vorderen (Index I) und im hinteren Teil (Index II) festgelegt. Für Zeiten $\tau \geq 0{,}16$ konnte man zu größeren Teilungen nach Abb. 1b und größeren Zeitschritten übergehen:

$(\varDelta\,\xi)_\mathrm{I} = 0,2$, $(\varDelta\,\xi)_\mathrm{II} = 1,0$ und $(\varDelta\,\xi)_\mathrm{III} = 5$ mit den zugehörigen Zeitschritten $(\varDelta\,\tau)_\mathrm{I} = 0,0195$, $(\varDelta\,\tau)_\mathrm{II} = 0,039$ bzw. $(\varDelta\,\tau)_\mathrm{III} = 0,078$. Dabei
wurden die Zeitschritte im vordersten Teil nahe dem größtmöglichen
Wert nach Gl. (5), im hin-
teren Teil mit Rücksicht auf
die Hilfspunkte gewählt.

Vergleicht man die Lö-
sungen der Differenzenrechnung nach Abb. 4 mit der
analytischen Lösung von
TIDEMAN [4], so ergeben
sich nur Abweichungen von
höchstens 3%, bezogen auf
die Temperaturdifferenz
zwischen Anfangs- und Endzustand.

Als nächstes Beispiel
wurde ein 5% dickes sym-
metrisches Parabelbogen-
profil mit einer Deckplatte
aus 10 mm Aluminium
($\varrho_w = 2,85\,\mathrm{kg/dm^3}$, $c_w = 0,21$
kcal/kg °C und $\lambda_w = 0,0375$
kcal/m s °C) behandelt. Es
wird konstante Anfangs-

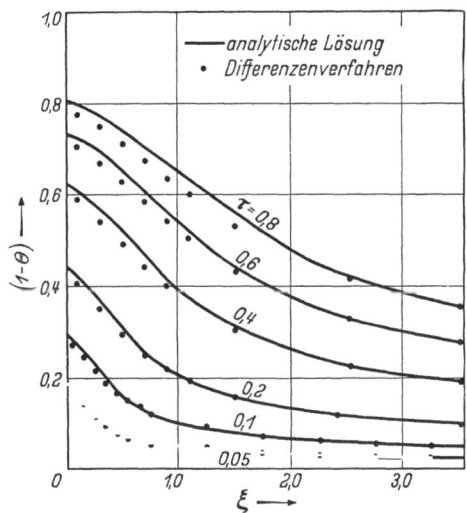

Abb. 4. Temperaturverteilung in einer einseitig unbegrenzten ebenen Platte nach der analytischen Methode von
TIDEMAN [4] und nach der Differenzenrechnung. ξ dimensionsloser Abstand von der Vorderkante nach Gl. (7) und
τ dimensionslose Zeit nach Gl. (8). Kein Wärmeübergang
durch Strahlung

temperatur bei einem Flug mit $M = 0,9$ in 20 km Höhe und plötzliche Beschleunigung im Horizontalflug auf eine $M = 2,5$ entsprechende
Geschwindigkeit angenommen. Im Vorderteil soll die Grenzschicht
laminar und im Hinterteil turbulent sein. Drei Fälle wurden berechnet
mit dem Umschlagspunkt in einem Abstand von der Vorderkante
von 20, 50 und 80% der Flügeltiefe. Die entsprechenden kritischen
Reynoldszahlen sind 2,7, 6,8 und $10,8 \cdot 10^6$, und dies kann unter gewissen Verhältnissen (glatte Oberfläche, Wärmestrom zur Wand gerichtet
usw.) auch der Wirklichkeit entsprechen [17, 18]. Am Umschlagspunkt
beginnt die turbulente Grenzschicht mit einer endlichen Dicke, die sich
aus der Gleichheit der Impulsdicken der laminaren und turbulenten
Grenzschicht ergibt. Diese Annahmen sind natürlich Idealisierungen der
wirklichen Verhältnisse. Vor allem liegt der Umschlagspunkt häufig nicht
fest, sondern seine Lage ist in den meisten Fällen Schwankungen unter-
worfen. Wegen der starken Temperaturänderungen in der Nähe des Umschlagspunktes wurde hier die gleiche feine Abschnittsteilung wie bei der
Vorderkante gewählt, siehe z. B. Abb. 2a und b, die die Abschnittsteilung
bei einem Umschlagspunkt in 50% Flügeltiefe wiedergeben. Dabei waren

die Zeitschritte 20 und 40 s für die feine bzw. grobe Teilung in Abb. 2a, und dem entsprechen bei Abb. 2b die Werte 20 bzw. 80 s.

Wie die Lösungen der Differenzenrechnung, Abb. 5, 6 und 7, zeigen, sind in der Nähe der Vorderkante die Temperaturen sehr ähnlich denen

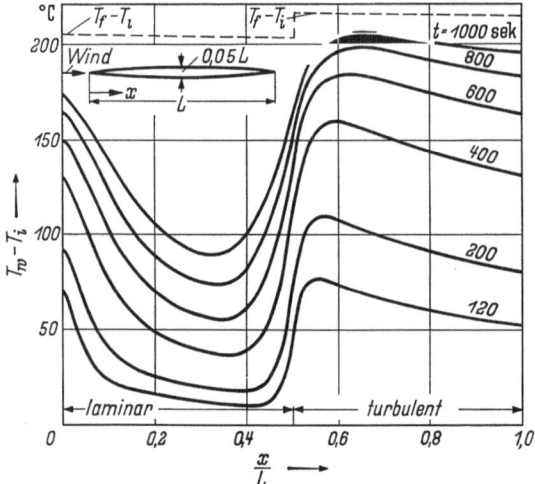

Abb. 5. Temperaturverteilung in einer 10 mm dicken Duraluminplatte, die die Wand eines 5% dicken symmetrischen Flügels mit Parabelbogenprofil bildet. Vorne laminare, hinten turbulente Grenzschicht. Umschlagpunkt bei 50% Flügeltiefe. x Abstand von der Vorderkante, L Flügeltiefe. Einfluß der Wandtemperaturverteilung auf die Wärmeübergangszahl vernachlässigt. Kein Wärmeübergang durch Strahlung. Plötzliche Beschleunigung von $M = 0,9$ auf $2,5$ in 20 km Höhe

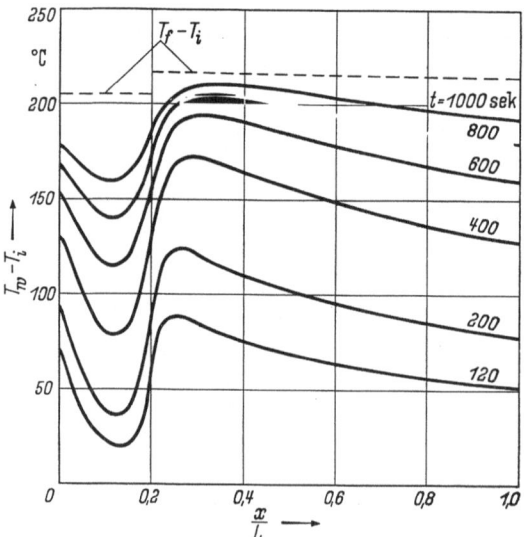

Abb. 6. Umschlagspunkt bei 20% Flügeltiefe, im übrigen wie bei Abb. 5

an der ebenen Platte. Die starken Temperaturänderungen beim Umschlagspunkt erklären sich daraus, daß die turbulente Wärmeübergangszahl hier ungefähr 10mal so groß wie die laminare ist. Beachtlich sind die
langen Zeiten, die zum Temperaturausgleich erforderlich sind, also bis
die durch T_f gekennzeichnete stationäre Temperaturverteilung erreicht
ist. Die Temperaturdifferenzen zwischen einer Stelle kurz vor und kurz
nach dem Umschlagspunkt sind um so größer, je weiter hinten der Umschlagspunkt liegt. Dies ist besonders für aeronautische Anwendungen

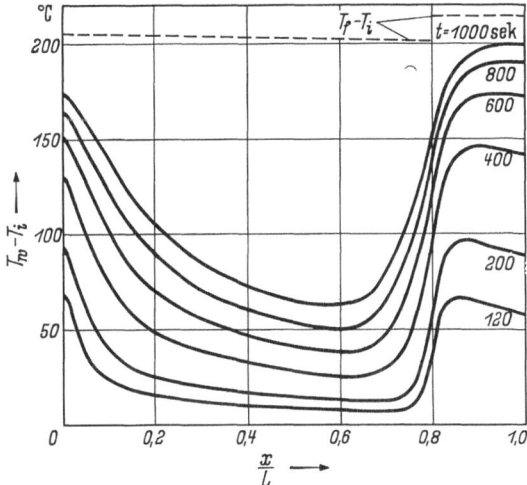

Abb. 7. Umschlagspunkt bei 80% Flügeltiefe, im übrigen wie bei Abb. 5.

interessant, bei denen man sich für die durch Temperaturunterschiede
hervorgerufenen Wärmespannungen interessiert.

Bisher wurde der Einfluß der Wandtemperaturverteilung auf die
Wärmeübergangszahl vernachlässigt. An Keilkörpern wurden bei inkompressibler laminarer Grenzschicht und bei Wandtemperaturen der Form
$T_w - T_f = C\,x^\beta$ (C, β Konstante, x Abstand von der Vorderkante) be-
deutende Änderungen in den Wärmeübergangszahlen, verglichen mit
denen bei konstanter Wandtemperatur, gefunden. Zum Beispiel ergaben
sich an einer ebenen Platte bei $\beta = 1$ um 65% größere Werte als bei
konstanter Wandtemperatur, $\beta = 0$ [14]. Wir wollen nun diesen Einfluß
bei dem oben behandelten Flügel mit einem Umschlagspunkt bei 50%
Flügeltiefe näherungsweise untersuchen, wobei die Wärmeübergangszahlen nach der Methode von Lighthill [15] und von Rubesin [16] für
laminare bzw. turbulente Strömung berechnet wurden.

Die so erhaltenen neuen Temperaturen unterscheiden sich nach Abb. 8
nicht viel von den alten, erfordern aber höheren Rechenaufwand. Am

meisten merkbar ist der Unterschied natürlich an den Stellen mit den
größten lokalen Temperaturdifferenzen. In vielen Fällen, vor allem bei
technischen Anwendungen, wird es also genügen, wenn man mit Wärmeübergangszahlen für konstante Wandtemperatur rechnet.

Als letztes Beispiel behandeln wir einen Fall mit Wärmestrahlung. Wir

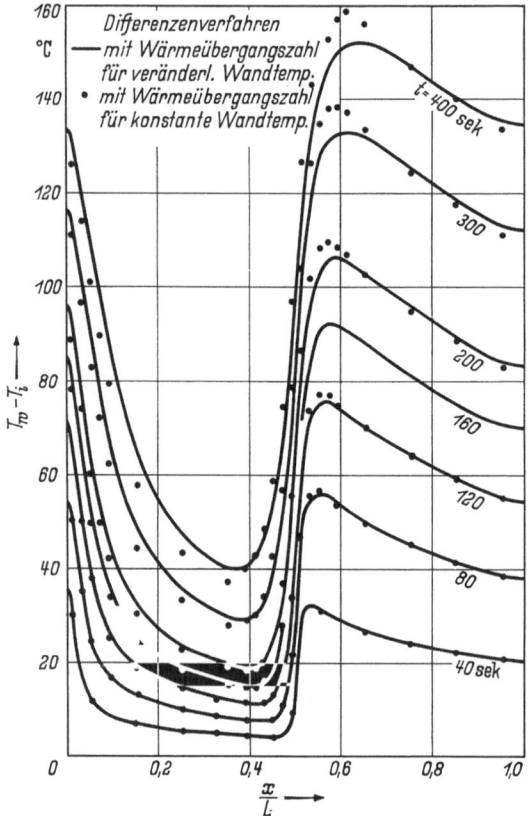

Abb. 8. Wie bei Abb. 5, Wärmeübergangszahl mit und ohne Einfluß der Wandtemperaturverteilung

nehmen wieder denselben Flügel an, aber denken uns eine plötzliche Be-
schleunigung von $M = 0,9$ auf $M = 4$ in 30,5 km Höhe durchgeführt.
Der Emissionskoeffizient der Flugzeugoberfläche sei $\varepsilon = 0,8$. Laminare
Grenzschichtströmung wurde nun über den ganzen Flügel angenommen.
Bei einer Reynoldszahl des Flügels von $4,2 \cdot 10^6$ kann dies unter günsti-
gen Umständen (glatte Oberfläche und Vermeidung von Störungen durch
Umströmung der Vorderkante) eintreten [17, 18]. Die Abschnittsteilun-
gen sind aus Abb. 3a und b zu sehen, die bei den Differenzenrechnungen

von $t = 0$ bis 643 s bzw. $t = 643$ bis 2568 s verwendet wurden. Zum vorderen Teil in Abb. 3a und b gehören die Zeitschritte $(\varDelta t)_{\mathrm{I}} = 26{,}8$ bzw. 68,75 s und im hinteren $(\varDelta t)_{\mathrm{II}} = 107{,}2$ bzw. 275 s. Nach Abb. 9 ist natürlich der Einfluß der Wärmestrahlung im stationären Zustand am größten, er macht sich bei der Erwärmung der Platte längere Zeit nicht geltend, dann aber in kurzer Zeit recht kräftig. Der Gleichgewichtszu-

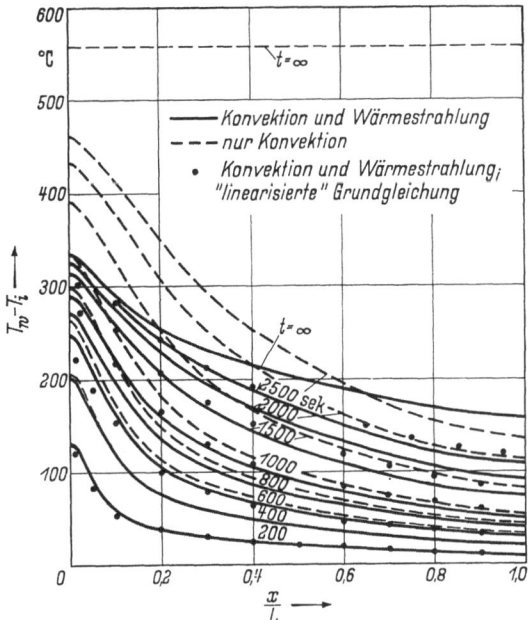

Abb. 9. Flügel wie in Abb. 5. Laminare Grenzschicht. Wärmeübergang durch Strahlung berücksichtigt. Emissionskoeffizient $\varepsilon = 0{,}8$. Einfluß der Wandtemperaturverteilung auf den konvektiven Wärmeübergang vernachlässigt. Plötzliche Beschleunigung von $M = 0{,}9$ auf $M = 4$ in 30,5 km Höhe

stand wird bei den höheren Temperaturen mit Strahlung viel rascher erreicht als ohne diese.

Es soll nun untersucht werden, ob sich das Strahlungsglied in der Wärmeleitungsgleichung derart linearisieren läßt, daß man dieses näherungsweise durch entsprechend abgeänderte konvektive Glieder berücksichtigen kann. Gelingt dies, so könnte man z. B. die in Abb. 1 dargestellten Lösungen auch bei Wärmestrahlung anwenden. Im allgemeinen besteht die hier verwendete Methode darin, daß die nichtlineare Wärmeleitungsgleichung bei einer oder zwei Temperaturen mit dem Näherungsansatz übereinstimmt, man aber im übrigen Abweichungen zuläßt. Es soll hier nur eine der vielen Möglichkeiten untersucht werden: Wir begnügen uns mit Übereinstimmung im stationären Endzustand. Dies ist insofern sinnvoll, als der Strahlungseinfluß zuletzt, wenn die Temperatur am größten ist, auch am größten ist. Wir vereinfachen das Problem noch dadurch, daß wir im stationären Endzustand auch von der Wärmeleitung in der Platte absehen. Für die entsprechende Wandtemperatur T_g erhalten wir nach Streichung der Ableitungen nach x und t in Gl. (2):

$$0 = \alpha \, (T_f - T_g) - \varepsilon \, C_s \, (T_g{}^4 - T_1{}^4). \tag{12}$$

Die Glieder in Gl. (2), die vom Wärmeübergang durch Konvektion und Wärmestrahlung herrühren, ersetzen wir durch folgenden linearen Ausdruck:

$$\alpha_e \, (T_g - T_w), \tag{13}$$

wobei nach hier nicht wiedergegebenen Zwischenrechnungen

$$\alpha_e = \alpha + \frac{\varepsilon \, C_s}{T_g - T_1} \, (T_g{}^4 - T_1{}^4) \, . \tag{14}$$

Rechnungen mit dem linearisierten Ansatz ergeben nach Abb. 9 für viele Zwecke brauchbare Näherungswerte. Vermutlich wäre die Übereinstimmung noch besser, wenn man die Endtemperatur T_g unter Berücksichtigung des Wärmeleitungsgliedes in Gl. (2) nach üblichen Differenzenverfahren berechnet hätte.

Literatur

[1] Parker, H. M.: Transient Temperature Distributions in Simple Conducting Bodies Steadily Heated Through a Laminar Boundary Layer. NACA TN 3058 (1953).

[2] Emmons, H. W.: The Non-steady Aerodynamic Heating of a Plate. „50 Jahre Grenzschichtforschung", eine Gedenkschrift, herausgegeben von H. Görtler und W. Tollmien (1955), S. 385.

[3] Heaps, N. S.: Transient Thermal Stress in a Flat Plate Due to Non-uniform Heat Transfer Across One Surface. Aeronautical Research Council Current Paper 299 (1956).

[4] Tideman, M.: On the Temperature Distribution in Thin Flat Plates with Laminar Supersonic Boundary Layers. Svenska Aeroplan A. B. Tech. Note Nr. 39 (1958).

[5] O'Brien, G. G., M. A. Hyman and S. Kaplan: A Study of the Numerical Solution of Partial Differential Equations. Journ. of Math. and Phys. **XXIX**, 223 (1951).

[6] Dusinberre, G. M.: Numerical Analysis of Heat Flow. New York 1949.

[7] Schuh, H.: Differenzenverfahren zum Berechnen von Temperaturausgleichsvorgängen bei eindimensionaler Wärmeströmung in einfachen und zusammengesetzten Körpern. VDI-Forschungsheft 459 (1957).

[8] Monaghan, R. J.: On the Behaviour of Boundary Layers at Supersonic Speeds. Institute of the Aeron. Sciences. Preprint No. 557 (1955).

[9] Eckert, E. R. G.: Survey on Heat Transfer at High Speeds. Wright Air Development Center Technical Report 54—70 (1954).

[10] RUBESIN, M. W., and H. A. JOHNSON: A Critical Review of Skin Friction and Heat Transfer Solutions of the Laminar Boundary Layers on a Flat Plate. Trans. Amer. Soc. Mech. Engineers. **71**, 385 (1949), No. 4.

[11] SOMMER, S. C., and B. J. SHORT: Free Flight Measurements of Turbulent-Boundary-Layer Skin Friction in the Presence of Severe Aerodynamic Heating at Mach Numbers from 2.8 to 7.0. NACA TN 3391 (1955).

[12] WILSON, R. W.: Turbulent Boundary-Layer Characteristics at Supersonic Speeds—Theory and Experiment. Jour. Aero. Sci. **17**, 585 (1950).

[13] VAN DRIEST, E. R.: Investigation of Laminar Boundary Layer in Compressible Fluids Using the Crocco Method. NACA TN 2597 (1952).

[14] CHAPMAN, R., and M. W. RUBESIN: Temperature and Velocity Profiles in the Compressible Laminar Boundary Layer with Arbitrary Distribution of Surface Temperature. Journ. of the Aeronautical Sciences, **16**, 547 (1949), No. 9.

[15] LIGHTHILL, M. I.: Contributions to the Theory of Heat Transfer Through a Laminar Boundary Layer. Proc. Roy. Soc., A **202**, 359 (1950).

[16] RUBESIN, M. W.: The Effect of Arbitrary Surface-Temperature Along a Flat Plate on the Convective Heat Transfer in an Incompressible Turbulent Boundary Layer. NACA TN 2345 (1951).

[17] STERNBERG, J.: A Free Flight Investigation of the Possibility of High Reynolds Number Supersonic Laminar Boundary Layers. Journ. of the Aeronautical Sciences **19**, 721 (1952), No. 11.

[18] DUNNING, R. W., and E. F. ULMANN: Effects of Sweep and Angle of Attack on Boundary-Layer Transition on Wings at Mach-Number 4.40. NACA TN 3473 (1955).

Cross-stresses in air, demonstrated by means of a centripetal vacuum pump[1]

By .

B. Popper and M. Reiner[2]

Israel Institute of Technology, Haifa

REINER (1945) has shown that when a viscous fluid is sheared in laminar flow, the resulting stress may have not only the customary tangential components "which arise from the lack of slipperiness of the parts of the liquid" as assumed by NEWTON in his Principia and taken over in the Navier-Stokes equations of classical hydrodynamics, but that there will be present, in general, also "cross-stresses," i.e. normal stresses in the direction of the velocity of flow and of the velocity gradient. REINER (1948) has also shown that such cross-stresses will be present in the simple shear of an elastic material, and he has shown (1956) that this is the case even if the strain is infinitesimal. It therefore stands to reason that cross-stresses must be present in the flow of elastico-viscous materials. REINER (1957) has shown that cross-forces are present in the flow of air and has interpreted them as resulting from elastico-viscous cross-stresses. POPPER and REINER (1956, 1957) have used this phenomenon for the design of a centripetal air pressure pump, but TAYLOR (1957) has shown that such cross-forces can result from (A) a slight error in the perpendicularity of the plane of either the stator or the rotor to the axis of rotation, or (B) from vibrations of the rotor in the direction of its axis, without departing from the Navier-Stokes equations, provided the compressibility of the air is taken into account. In order to provide additional information which may lead to a definite theory of cross-stresses in air, a centripetal vacuum pump suitable for the observation of cross-stresses in the laminar flow of air was therefore built by us.

The essential feature of the vacuum pump is the same as of the pressure pump, namely two parallel circular metal plates, one stationary, the other rotating opposite it with a very narrow gap between both.

---

[1] The research reported in this document has been sponsored in part by the Air Research and Development Command, United States Air Force, under Contract AF 61 (052) — 10, through the European Office ARDC.

[2] Vorgetragen von M. REINER.

The figure shows a cross section of the instrument. The stationary plate S is underneath the rotating one R. Plate R—the rotor—is driven by the main shaft M. However, R is not fixed to M, it can rotate freely around the ball B, and it can also be displaced along the axis of rotation. It rotates together with the shaft about the axis of M. This is achieved by means of the pin P, which protrudes from B, and enters into two

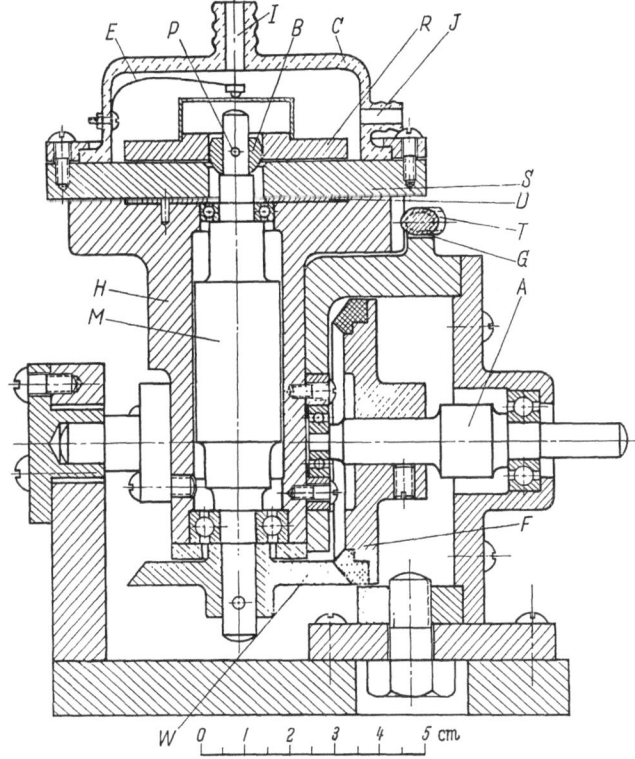

short radial slots in R. A bevel shaped wheel W, on shaft M, is rotated by the friction wheel F with a bevel shaped rubber rim. Wheel F is driven by an electric motor through shaft A.

During rotation, a cross-pressure in the direction of the velocity gradient is produced which maintains a gap between the plates, supporting the rotor. An airbearing is thus formed which, in contradistinction to the usual airbearings, does not require the application of an outside pressure. Rotation is started when the main shaft M, together with its housing H, is tilted to a horizontal position, so that the weight of the rotor does not press the plates together before the cross-pressure is generated.

A hand wheel is fixed to the toothed pivot T, which is engaged with a stationary gear segment G, holding pivot T and housing H in position. The pivot T has seven helical teeth and must be turned seven full revolutions to tilt H. It is thus possible to bank H slowly enough so that the free suspended revolving rotor R will not precess around an axis perpendicular to those of the shafts.

When the axis of the rotor is changed by means of the hand wheel, from the horizontal to the vertical position, the weight component (P) of the rotor which tends to bring the plates together is gradually changed from zero to a maximum, the weight (W) of the rotor.

When the instrument is in action, centripetal pumping results which sucks the air from the cup C. The cup is connected to two manometers I and J, one at the top, the other at the side, which register the suction or pressure, as the case may be. The air escapes through a slot (not shown in the figures) in disc D to an outlet in the housing (not shown in the figures) until a steady state is reached.

The rotor is insulated electrically from the stator. The plates thus form a capacitor with a capacity as a function of the distance between them. This property was used to find the distance D, by measuring the capacity. A special electrical switch is provided, by means of which the sense of rotation can be reversed.

This instrument was demonstrated at the Symposium. A description of the experiments with theoretical analysis will appear in the Proceedings of the Royal Society of London.

References

REINER, M. (1945): Amer. J. Math. **67**, 350.
REINER, M. (1948): Amer. J. Math. **70**, 433.
REINER, M. (1956): in Grammel (Editor) Deformation and Flow of Solids, Springer, Berlin.
REINER, M. (1957): Proc. Roy. Soc. (London). A **240**, 173.
POPPER, B., and M. REINER (1956): Brit. J. Appl. Phys. **7**, 452; (1957): ibid. **8**, 493.
TAYLOR, G. I. and P. G. SAFFMAN (1957): J. Aeron. Sci. **24**, 553.

Boundary layers in a rarefied gas[1]

By

S. A. Schaaf and L. Talbot

University of California, Berkeley

Theoretical and experimental results are presented relative to the fundamental phenomena associated with laminar boundary layers in the "slip flow" range of densities. This corresponds to values of the ratio of the mean free path λ to the boundary layer thickness δ in the range of $0.01 < \lambda/\delta < 1.0$.

It is shown that two effects are of importance in this range: first, that of the slip boundary condition at the surface; and second, that of the interaction of the boundary layer with the essentially inviscid outer flow. Theoretical results are derived which show that the effect of the slip boundary condition is to decrease the thickness of the boundary layer, and to reduce the local skin friction coefficient; however, these effects are of second order in the mean free path when the inviscid pressure gradient is zero—as for example with a flat plate, or a cone or wedge in supersonic flow. No direct experiments have yet been carried out for a case involving a boundary layer with first order slip effect.

A number of experiments, by the authors, as well as other groups, have been carried out to determine the pressure distribution on cones and flat plates in supersonic flow in the slip flow range. These show chiefly the interaction effect; i.e. the boundary layer displaces the inviscid flow outwards, increasing the surface pressure above the inviscid value. This effect is very pronounced in the slip flow range, although it has no direct connection with the slip boundary condition. It is shown, in fact, that there is no first order effect of slip on the induced pressure— even for the inviscid pressure gradient case. The most extensive tests have been carried out at BERKELEY with flat plate at $M = 6$ and are presented here. These measurements determined the induced pressure distribution on flat plates with very sharp leading edges. The induced pressure attained values of the order of five times the free stream pressure. A theoretical analysis, based on conventional boundary layer assump-

---

[1] Der Bericht wurde von S. A. SCHAAF erstattet. Da die Verfasser ihre Ergebnisse im Rahmen einer anderweitig erscheinenden Darstellung zu veröffentlichen beabsichtigen, kann hier nur die im Tagungsmaterial enthaltene Inhaltsankündigung wiedergegeben werden. — D. Hrsg.

tions, is presented which is successful in predicting the measured pressure distributions, except in the immediate vicinity (within about fifty mean free paths) of the leading edge.

Aus der Diskussion

G. E. Gadd (Teddington): Is it true that the leading edge thickness does not seem to make very much difference to the pressure distribution?

S. A. Schaaf (Berkeley): No, it made quite a difference as long as it was reasonably thick, but if the Reynolds number at the leading edge is less than 100, say, then it does not make much more difference. This has been confirmed at Princeton and at Caltech.

Analyse théorique du champ aérodynamique dans le voisinage immédiat d'un plan illimité, basée sur la théorie cinétique des gaz[1]

Par

Michał Łunc et **Jan Lubonski**

Polska Akademia Nauk, Warszawa

On sait que chaque écoulement peut être déterminé entièrement par le moyen du champ de la fonction de distribution des vitesses moléculaires. Cette fonction est la solution de l'équation intégro-différentielle aux différences de L. BOLTZMANN, restée insoluble jusqu'à présent. Nous allons montrer comment peut-on trouver la valeur approchée de la fonction de distribution, sans avoir à résoudre formellement l'équation de BOLTZMANN, dans le cas du mouvement stationnaire d'un gaz, au voisinage d'une paroi plane illimitée. On en déduira alors les champs des vitesses, de la densité et de la température. On se limitera au mouvement, où toutes les grandeurs cherchées ne dépendent que d'une seule variable indépendante — la distance z de la paroi.

L'idée principale consiste à classer toutes les molécules, présentes simultanément à l'intérieur d'un certain élement de volume, d'après leur «origine», c.-à-d. d'après le lieu où chacune d'elles avait subi son dernier choc. Les molécules, ayant l'origine commune, auront aussi des propriétés cinématiques semblables. Cette méthode «démographique» s'apparente à celle qui permit à MAXWELL de lier le libre parcours moyen des molécules avec la viscosité du gaz.

On admettra que la fonction de distribution des molécules à peine nées, se trouvant donc tout-près de leur commune origine, est du type maxwellien. En d'autres termes — aussi bien la paroi solide que le milieux gazeux sont parfaitement diffusants. Après chaque collision les molécules «oublient» leur passé mécanique. Ceci élargit l'hypothèse connue de KNUDSEN, applicable aux chocs entre les molécules et la paroi, à toute espèce des chocs intermoléculaires. On sait que la réflexion diffuse se vérifie bien dans la plupart des expériences.

La seconde hypothèse simplificatrice consistera à admettre que le libre parcours moyen l des molécules est constant — indépendant aussi bien du lieu que de la vitesse.

---

[1] Vorgetragen von M. ŁUNC.

Désignons par $\vec{c}$ la vitesse des molécules, par $d_3\vec{c}$ l'élement de volume de l'espace des vitesses. On désignera par c_3 la composante de vitesse normale à la paroi et par c_1 la composante parallèle à la paroi et à la vitesse moyenne (macroscopique). Le point courant de l'espace gazeux sera désigné par $\vec{r}$ ou par $\vec{r}'$. La distance du point $\vec{r}$ au plan par z. On introduira encore les variables sans dimensions $\zeta = z/l$ et $\zeta' = z'/l$.

Supposons maintenant qu'autour de chaque point, situé à la distance z' de notre plan naissent, par unité de temps et par unité de volume,

$$G\left(\vec{c}, z'\right) d_3\vec{c}$$

molécules, dont les vitesses sont comprises à l'intérieur de l'élement de volume $d_3\vec{c}$. L'hypothèse de «maxwellicité» de la fonction de distribution des vitesses des molécules naissantes nous conduit à écrire

$$G\left(\vec{c}, z'\right) = \frac{c\, v\left(z'\right)}{\pi^{3/2} \cdot l \cdot \left[\omega\left(z'\right)\right]^3} \exp\left\{-\left[\frac{\vec{c} - \vec{a}\left(z'\right)}{\omega\left(z'\right)}\right]^2\right\}. \tag{1}$$

Admettons, aussi, qu'en chaque point de la paroi naissent, par unité de temps et par unité de surface

$$H\left(\vec{c}\right) d_3\vec{c}$$

molécules de vitesse comprise dans $d_3\vec{c}$. On supposera que

$$H\left(\vec{c}\right) = \frac{c\, v_s}{\pi^{3/2}\, \omega_s^3} \exp\left(-\frac{c^2}{\omega_s^2}\right). \tag{2}$$

Les fonctions $v\left(z'\right)$, $\omega\left(z'\right)$ et $\vec{a}\left(z'\right)$ possèdent, respectivement, les dimensions physiques de la densité numérique, du module de vitesse et de la vitesse, sans se rattacher aux grandeurs moléculaires réelles. Sans renoncer à la généralité, on peut poser $\vec{a}\left(0\right) = 0$. Toutes les grandeurs prises en 0 seront notées à l'aide d'un indice $_0$ inférieur.

Introduisons maintenant la probabilité $p\left(\vec{c}, \vec{r}, \vec{r}'\right)$ pour qu'une molécule de vitesse $\vec{c}$, née en $\vec{r}'$ puisse parvenir en $\vec{r}$ avant une nouvelle collision. Bien entendu, la vitesse $\vec{c}$ est supposée parallèle à $\vec{r} - \vec{r}'$ et l'on a

$$\left|\vec{r} - \vec{r}'\right| = \left|\frac{(z - z')\, c}{c_3}\right|.$$

Nous aurons, puisque $l = \mathrm{const}$,

$$p\left(\vec{c}, \vec{r}, \vec{r}'\right) = \exp\left[-\frac{\left|\vec{r} - \vec{r}'\right|}{l}\right] = \exp\left[-\left|\frac{(z - z')\, c}{l\, c_3}\right|\right] = \exp\left[-\left|\frac{(\zeta - \zeta')\, c}{c_3}\right|\right]. \tag{3}$$

La fonction de distribution en un point quelconque $\vec{r}$ sera alors donnée par l'une des deux formules alternatives:

1. lorsque $c_3 < 0$

$$f\left(\vec{c}, z\right) = f_- = \int p\left(\vec{c}, \vec{r}, \vec{r}'\right) \frac{G\left(\vec{c}, z'\right)}{c}\, ds, \tag{4}$$

où ds est un élement de longueur le log d'une demi-droite qui part de $\vec{r}$ vers l'infini, parallèlement à $-\vec{c}$;

2. lorsque $c_3 > 0$

$$f(\vec{c}, z) = f_+ = \frac{1}{c} p(\vec{c}, \vec{r}, \vec{r}_{\Sigma}) H(\vec{c}) + \frac{1}{c} \int p(\vec{c}, \vec{r}, \vec{r}) G(\vec{c}, z') ds, \qquad (5)$$

où $\vec{r}_{\Sigma}$ est le point d'intersection de la droite issue de $\vec{r}$, parallèle à $\vec{c}$, avec le plan et l'intégrale étant prise le long du segment $[\vec{r}_{\Sigma}, \vec{r}]$.

Puisque

$$ds = \frac{c\,dz'}{|c_3|} = \frac{lc\,d\zeta'}{|c_3|},$$

on aura

$$f_- = -\frac{c}{\pi^{3/2}\,c_3} \int_{\zeta'=\zeta}^{\infty} \frac{\nu(l\,\zeta')}{[\omega(l\,\zeta')]^3} \exp\left\{-\left[\frac{\vec{c}-\vec{a}(l\,\zeta')}{\omega(l\,\zeta')}\right]^2 + \frac{(\zeta'-\zeta)c}{c_3}\right\} d\zeta' \qquad (6)$$

et

$$f_+ = \frac{\nu_s}{\pi^{3/2}\,\omega_s^3} \exp\left(-\frac{c^2}{\omega_s^2} - \frac{\zeta\,c}{c_3}\right) +$$

$$+ \frac{c}{\pi^{3/2}\,c_3} \int_{\zeta'=0}^{\zeta} \frac{\nu(l\,\zeta')}{[\omega(l\,\zeta')]^3} \exp\left\{-\left[\frac{\vec{c}-\vec{a}(l\,\zeta')}{\omega(l\,\zeta')}\right]^2 + \frac{(\zeta'-\zeta)c}{c_3}\right\} d\zeta'. \qquad (7)$$

La fonction sous le signe de l'intégrale pourrait être intégrée au cas où $G(\vec{c}, z)$ est développable en série, suivant z, dans un domaine suffisamment étendu. Supposons donc que l'on puisse écrire

$$G(\vec{c}, z) = G(\vec{c}, 0) + z\left(\frac{\partial G}{\partial z}\right)_0 + \cdots = G(\vec{c}, 0) \sum_{m=0}^{\infty} \frac{z^m}{l^m} A_m(\vec{c}), \qquad (8)$$

où

$$A_m(\vec{c}) = \frac{l^m}{m!\,G(\vec{c}, 0)}\left(\frac{\partial^m G}{\partial z^m}\right)_0 = \frac{1}{m!\,G(\vec{c}, 0)}\left(\frac{\partial^m G}{\partial \zeta^m}\right)_0. \qquad (9)$$

Si l'on substitue alors (8) et (9) dans (6) et (7), et si l'on effectue l'intégration suivant ζ', on obtiendra

$$f_-(\vec{c}, \zeta) = \frac{l}{c} G(\vec{c}, 0) \sum_{m=0}^{\infty}\left(-\frac{c}{c_3}\right) \exp\left(-\frac{\zeta\,c}{c_3}\right) A_m(\vec{c}) \int_{\zeta'=\zeta}^{\infty} \zeta'^m \exp\left(\frac{\zeta'c}{c_3}\right) d\zeta' =$$

$$= \frac{l}{c} G(\vec{c}, 0) \sum_{k=0}^{\infty} \frac{\zeta^k}{k!} \sum_{m=k}^{\infty}\left(-\frac{c_3}{c}\right)^{m-k} m!\,A_m(\vec{c}) = \qquad (10)$$

$$= \frac{\nu_0}{\pi^{3/2}\,\omega_0^3} \exp\left(-\frac{c^2}{\omega_0^2}\right) \sum_{k=0}^{\infty} \frac{\zeta^k}{k!} \sum_{m=k}^{\infty} m!\,A_m(\vec{c})\left(-\frac{c_3}{c}\right)^{m-k}$$

et

$$f_+(c,\zeta) = \frac{v_0}{\pi^{3/2}\,\omega_0^3}\exp\left(-\frac{c^2}{\omega_0^2}\right)\sum_{k=0}^{\infty}\frac{\zeta^k}{k!}\sum_{m=k}^{\infty}m!\,A_m\left(\vec{c}\right)\left(-\frac{c_3}{c}\right)^{m-k} +$$

$$+ \exp\left(-\frac{\zeta c}{c_3}\right)\left[\frac{v_s}{\pi^{3/2}\,\omega_s^3}\exp\left(-\frac{c^2}{\omega_s^2}\right) - \right. \tag{11}$$

$$\left. - \frac{v_0}{\pi^{3/2}\,\omega_0^3}\exp\left(-\frac{c^2}{\omega_0^2}\right)\sum_{m=0}^{\infty}m!\,A_m\left(\vec{c}\right)\left(-\frac{c_3}{c}\right)^{m}\right].$$

On admet que les séries (10) et (11) sont convergentes. Les deux formules (10) et (11) peuvent être réunies en une seule, dans laquelle on écrira au lieu des $A_m\left(\vec{c}\right)$ les expressions

$$m!\,A_m\left(\vec{c}\right) = B_m\left(\vec{c}\right) = \sum_{i,\,j\,\geqslant\,0}^{i+j\,\leqslant\,m} B_{m,\,i,\,j}\,\frac{c_1{}^i\,c^{2j}}{\omega_0{}^{i+2j}}, \tag{12}$$

que l'on obtient en différentiant (1).

De cette manière on obtiendra pour la fonction de distribution l'expression

$$f\left(\vec{c},\zeta\right) = \frac{v_0\exp\left(-\frac{c^2}{\omega_0^2}\right)}{\pi^{3/2}\,\omega_0^3}\sum_{k=0}^{\infty}\frac{\zeta^k}{k!}\sum_{m=k}^{\infty}\sum_{i,\,j\,\geqslant\,0}^{i+j\leqslant m} B_{m,\,i,\,j}\,\frac{(-1)^{m-k}\,c_1{}^i\,c_3{}^{m-k}\,c^{2j-m+k}}{\omega^{i_0+2j}} +$$

$$+ \left\{\exp\left(-\frac{\zeta c}{c_3}\right)\left[\frac{v_s\exp\left(-\frac{c^2}{\omega_s^2}\right)}{\pi^{3/2}\,\omega_s^3} - \right.\right. \tag{13}$$

$$\left.\left. - \frac{v_0\exp\left(-\frac{c^2}{\omega_0^2}\right)}{\pi^{3/2}\,\omega_0^3}\sum_{m=0}^{\infty}\sum_{i,\,j\,\geqslant\,0}^{i+j\,\leqslant\,m} B_{m,\,i,\,j}\,\frac{(-1)^m\,c_1{}^i\,c_3{}^m\,c^{2j-m}}{\omega_0{}^{i+2j}}\right]\right\}c_3 > 0.$$

Les coefficients $B_{m,\,i,\,j}$, définis par (12), sont des polynômes en $\omega_0',\ldots,$ $\omega_0^{(m)},\,v_0',\ldots,\,v_0^{(m)},\,a_0',\ldots,\,a_0^{(m)}$. En particulier, comme celà résulte de (9), $B_{000} = 1$.

La densité du gaz

$$n(\zeta) = \int f\left(\vec{c},\zeta\right)d_3\vec{c} \tag{14}$$

et la vitesse macroscopique

$$v(\zeta) = \frac{1}{n(\zeta)}\int c_1\,f\left(\vec{c},\zeta\right)d_3\vec{c} \tag{15}$$

doivent posséder, pour des raisons d'ordre physique, des dérivées finies d'ordre 1 et 2, lorsque $\zeta \to 0$ (c.-à-d. près de la paroi). Quant à la température

$$T(\zeta) = \frac{m}{3\,k\,n(\zeta)}\int\left\{c^2 - 2\,c_1\,v(\zeta) + [v(\zeta)]^2\right\}f\left(\vec{c},\zeta\right)d_3\vec{c} \tag{16}$$

sa dérivée première, tout au moins, doit être limitée sur la paroi. Dans (16) m désigne la masse des molécules et k est la constante de BOLTZMANN. De ces conditions on peut aisément déduire que l'on doit avoir

$$\nu_s = \nu_0 \tag{17}$$

$$\omega_s = \omega_0 \tag{18}$$

$$B_{100} = B_{101} = B_{110} = 0. \tag{19}$$

Les conditions (19) entraînent l'annulation de tous les coefficients $B_{2,\,i,\,j}$ et $B_{3,\,i,\,j}$ à l'exception, peut-être, de B_{200}, B_{201}, B_{210}, B_{300}, B_{301}, B_{310}. Pour les déterminer nous ferons usage des équations générales du mouvement des fluides. Ce sont:

1. l'équation de continuité

$$\int c_3\, f\left(\overrightarrow{c}, \zeta\right) d_3\, \overrightarrow{c} = 0 , \tag{20}$$

2. l'équation de la conservation de l'énergie totale

$$\int c_3\, c^2\, f\left(\overrightarrow{c}, \zeta\right) d_3\, \overrightarrow{c} = 0 \qquad \text{et} \tag{21}$$

3. l'équation de la constance du flux de la quantité de mouvement, exprimant l'absence des forces extérieures agissantes sur le gaz, en dehors de celles qui sont appliquées le long des parois

$$\frac{d}{d\zeta} \int c_3{}^2\, f\left(\overrightarrow{c}, \zeta\right) d_3\, \overrightarrow{c} = 0 . \tag{22}$$

Les équations (20), (21), et (22) doivent être satisfaites pour toutes les valeurs de la variable indépendante ζ. Si nous nous limitons à l'identification au zéro des coefficients des termes en ζ^0 et ζ^1, on aura 6 équations linéaires que l'on peut écrire en une seule formule

$$\sum_{m=1}^{\infty} \sum_{\substack{l,\,j \geqslant 0}}^{2l+j \leqslant m} B_{m,\,2l,\,j} \frac{\Gamma\left(j + l + \dfrac{3}{2} + \dfrac{s}{2}\right) \cdot (2l-1)\,!\,!}{(m+t)(m+t+2)\ldots(m+t+2l)} = 0 , \tag{23}$$

où $s = 1, 2, 3$ et $t = 1, 2$ et $(2l-1)\,!\,! = 1\cdot 3\cdot 5 \ldots (2l-1)$.

Si l'on se limite aux coefficients $B_{m,i,j}$ avec $m \leqslant 3$, alors tous les coefficients $B_{m,\,0,\,j}$ doivent être nuls. Ainsi nous restent uniquement les 2 coefficients non-nuls B_{210} et B_{310}. Nous allons les calculer, en se servant des conditions aux limites.

A cet effet on effectuera les intégrations (14), (15), (16) en se servant des valeurs calculées des coefficients $B_{m,i,j}$. D'autre part, il serait utile de calculer également les dérivées de ces fonctions par rapport à ζ. Parmi d'autres grandeurs physiques que l'on peut utiliser dans les équations aux limites se trouvent la force tangentielle par unité de surface de la paroi

$$F = m \int c_1 \, c_3 \, f\left(\vec{c}, \, 0\right) d_3 \, \vec{c} =$$

$$= -\frac{2 \, m \, v_0 \, \omega^2{}_0}{\sqrt{\pi}} \left(\frac{1}{24} \, B_{210} + \frac{1}{35} \, B_{310}\right) \tag{24}$$

et le flux de chaleur Q par unité de paroi

$$Q = F \, v_0 = -\frac{m \, v_0 \, \omega_0{}^3}{2 \, \sqrt{\pi}} \left(\frac{1}{24} \, B_{210} + \frac{1}{35} \, B_{310}\right)\left(\frac{1}{5} \, B_{210} + \frac{1}{8} \, B_{310}\right). \tag{25}$$

L'on voit immédiatement que les coefficients B_{210} et B_{310} peuvent être déterminés soit des valeurs de la vitesse à la paroi et de sa dérivée, soit de l'une de ces deux grandeurs et de l'une de deux grandeurs F et Q.

Si l'on admet l'existence de la relation de Maxwell entre le saut de la vitesse à la paroi et son gradient

$$v_0 = \left(\frac{d}{d \, \zeta} \, v\right)_0 = v_0{}', \tag{26}$$

la connaissance de $v_0{}'$ suffit pour la solution du problème.

Dans ce dernier cas on trouvera

$$F = -\frac{24}{11 \, \sqrt{\pi}} \, m \, v_0 \, \omega_0 \, v_0{}' \tag{27}$$

et, si l'on introduit le coefficient de viscosité μ, on aura

$$F = -\frac{24}{11} \, \mu \left(\frac{d \, v}{d \, z}\right)_0 \approx -2{,}18 \, \mu \left(\frac{d \, v}{d \, z}\right)_0. \tag{28}$$

La température s'exprimera par la relation

$$T \, (z) = T_s - \frac{m \, v^2}{3 \, k} = T_s + \left(1 + \frac{2 \, z}{l}\right) \triangle T, \tag{29}$$

où $\triangle T$ est le saut de la température à la paroi, donné par

$$\triangle T = -\frac{m \, v_0{}'^2}{3 \, k}. \tag{30}$$

Enfin la densité reste constante, aux termes d'ordre 2 en z, près.

Aus der Diskussion

G. Temple (Oxford): The demographic method used by Mr. Łunc has the great advantage that it evades the solution of the Boltzmann equation. As is well known, this equation is very difficult to solve, and Mr. Truesdell has made some very trenchant criticisms of the distinguished authors who have attempted to solve it by series and iterations. However, the demographic method suffers from two disadvantages:—

1. The mean free path l is assumed to be constant and independant of the velocities of the colliding molecules. This simplification may be necessary in order to make the equations tractable, but it seems to be too drastic for accurate calculation.

2. Only one distribution function is used, whereas it seems that we should distinguish between the distribution function of the molecules which are incident in a volume element and the distribution function of the molecules which emerge from this volume element after collision.

M. ŁUNC (Warszawa): I agree with your first remark. The mean free path is not only a function of density but also of speed. However, this is a very complicated function. We have tried to introduce something more realistic than our strong simplification but, in all cases tried out, we have failed in trying to deal with the function assumed. Of course, with computing machines it may be possible but without these, only with brains, we failed.

Your second remark is also extremely true. However, that which may appear best to do from the theoretical standpoint proves to be practically extremely complicated.

Diskussionsveranstaltung zur V. Sitzung

Beitrag zum Thema

Grenzschichten in verdünnten Gasen[1]

Von H. J. Kaeppeler und A. Zaddach[2]

Forschungsinstitut für Physik der Strahltriebwerke, Stuttgart

Dieser Diskussionsbeitrag soll eine kurze Ausführung über unsere Arbeiten über Transporterscheinungen in sehr heißen Gasen bei relativ geringer Dichte, also großen freien Weglängen, darstellen. Es wurde von uns berücksichtigt, daß ein solches mehrkomponentiges Gassystem Reaktionen unter den einzelnen Komponenten sowie weitreichende intermolekulare Wechselwirkungskräfte aufweisen kann. Insbesondere anwendbar sind diese Überlegungen auf mehr oder weniger teilionisierte Gase, also Plasmen.

Die praktische Anwendungsmöglichkeit solcher Überlegungen tritt neben solchen der reinen und angewandten Plasmaphysik auf bei Hyperschallerwärmung von Flugkörpern in relativ großen Höhen, etwa bei Machzahlen oberhalb $M = 15$.

Ein Teil der Arbeiten wurde in Zusammenarbeit mit G. BAUMANN [1, 2] durchgeführt, neuere Überlegungen, insbesondere in bezug auf Transporterscheinungen in verdünnten Gasen bei sehr hohen Temperaturen, entstanden in Zusammenarbeit mit A. ZADDACH und A. KOLLER [3].

Bei diesen Untersuchungen handelt es sich um den Versuch der Ableitung möglichst allgemeiner makroskopischer Erhaltungssätze für mehrkomponentige Gasgemische mit inneren Freiheitsgraden und beliebigen Reaktionen sowie beliebigen intermolekularen Wechselwirkungskräften, mit Hilfe stochastischer Methoden.

Es sei $f = f^{(N)}$ die Verteilungsfunktion eines geschlossenen Systems von N Teilchen, welches in $k = 1, 2 \ldots \nu$ gleiche Komponenten unterteilbar sei. Für $f^{(N)}$ gilt dann die LIOUVILLE-Gleichung

$$\frac{D f^{(N)}}{D t} = \frac{\partial f^{(N)}}{\partial t} + [H, f^{(N)}] = 0, \tag{1}$$

---

[1] Diese Untersuchungen wurden als Teilarbeiten im Rahmen von Aufträgen des European Office, Air Research and Development Command durchgeführt.

[2] Vorgetragen von H. J. KAEPPELER.

wobei D/Dt ein generalisierter Stokes-Operator, $[H, \ ]$ die Poisson-Klammern und H eine generalisierte Hamilton-Funktion des Gesamtsystems sei. Der Übergang zwischen inneren Freiheitsgraden sowie chemische Reaktionen werden als Diffusion in einem sogenannten „inneren" Phasenraum beschrieben. Es wird somit

$$[H, \ ] = \left(\frac{\partial H}{\partial p_k} \cdot \frac{\partial}{\partial r_k}\right) - \left(\frac{\partial H}{\partial r_k} \cdot \frac{\partial}{\partial p_k}\right) + \Sigma_i \left\{\left(\frac{\partial H}{\partial \eta_i} \cdot \frac{\partial}{\partial q_i}\right) - \left(\frac{\partial H}{\partial q_i} \cdot \frac{\partial}{\partial \eta_i}\right)\right\}. \quad (2)$$

Der Erwartungswert einer dynamischen Variablen α sei

$$< \alpha > = < \alpha, f > = \int \alpha f \, dp \, dr \, d\eta \, dq. \quad (3)$$

Es gelte ferner die Normierung

$$\int f \, dp \, dr \, d\eta \, dq = 1. \quad (4)$$

Die zeitliche Änderung des Erwartungswertes von α ist dann mit Gl. (1),

$$\frac{\partial}{\partial t} <\alpha> = < \frac{\partial \alpha}{\partial t}, f > + < \alpha, \frac{\partial f}{\partial t} >$$

$$= < \frac{\partial \alpha}{\partial t}, f > - < \alpha, [H, f] > \quad (5)$$

$$= < \frac{\partial \alpha}{\partial t}, f > + < [H, \alpha] f > .$$

Dies ist die generalisierte Transportgleichung. Setzt man für α summationsinvariante Größen ein, (a) die *Gesamt*massendichte, (b) die *Gesamt*impulsdichte, (c) die *Gesamt*energiedichte, dann folgen die makroskopischen Erhaltungssätze für das *Gesamt*system [1]. Man sieht aus den Gleichungen, daß zunächst noch die Navier-Stokesschen Gleichungen resultieren. Entwickelt man aber die Bewegungsgleichungen für eine einzelne Komponente [3], dann resultieren nicht mehr die Navier-Stokesschen Gleichungen, wie auch BRITTIN gezeigt hat [4].

Für hinreichend verdünnte Gase kann man die BOLTZMANN-Gleichung als Näherung für die LIOUVILLE-Gleichung verwenden. Lösung derselben ist erforderlich für eine numerische Auswertung der makroskopischen Gleichungen. Die ENSKOGsche Lösung der BOLTZMANN-Gleichung liefert uns eine beschränkte Klasse von Lösungen, bei denen der Zustand des Systems durch die makroskopischen Variablen Dichte, Temperatur und mittlere Massengeschwindigkeit ausreichend beschrieben wird. Dies ist generell, insbesondere bei merklichen Abweichungen vom Gleichgewicht, nicht der Fall.

Die von GRAD entwickelte Methode [5] erlaubt beliebige Lösungen der BOLTZMANN-Gleichung. Es werden die sogenannten „Geschwindigkeitsmomente" zur Beschreibung des Systems herangezogen. Diese sind definiert durch

$$S_{ijk}\ldots = \int V_{ijk}\ldots f \, d\vec{v}; \qquad V_{ijk}\ldots = V_i V_j V_k \cdots; \quad (6)$$

$$V_i = v_i - \overline{v_i}; \qquad \overline{v_i} = \int v_i f \, dv_i.$$

Diesen Geschwindigkeitsmomenten kann man eine physikalische Bedeutung zuordnen. Es ist z. B.

$$P_{jk} = m \int V_j V_k \, d\vec{v} \quad (7)$$

der Drucktensor und

$$q_j = \frac{m}{2} \int V^2 V_j \, d\vec{v} \quad (8)$$

der Wärmestromvektor.

Für den Fall eines mehrkomponentigen Gases hat A. ZADDACH [3] eine Lösung durchgeführt. Die Differentialgleichungen für die Einzelkomponente (unter Verwendung des Index v für die Komponenten) sind

$$\frac{\partial P_{jk,v}}{\partial t} + \frac{2}{5} \sum_i \frac{\partial q_{i,v}}{\partial r_i} \delta_{jk} + \frac{2}{5} \frac{\partial q_{j,v}}{\partial r_k} + \frac{2}{5} \frac{\partial q_{k,v}}{\partial r_j} +$$

$$+ \sum_i \left\{ \frac{\partial \bar{v}_{j,v}}{\partial r_i} P_{ik,v} + \frac{\partial \bar{v}_{k,v}}{\partial r_i} P_{ij,v} + \frac{\partial}{\partial r_i} (\bar{v}_{i,v} \cdot P_{jk,v}) \right\} \qquad (9)$$

$$= m_v \sum_s \int V_{j,v} V_{k,v} A_{v,s} d\vec{v}_v$$

$$\frac{\partial q_{j,v}}{\partial t} + \sum_i \left\{ -\frac{3}{2} \frac{p_v}{\varrho_v} \frac{\partial P_{ij,v}}{\partial r_i} - \frac{1}{\varrho_v} \sum_k \frac{\partial P_{ik,v}}{\partial r_i} P_{jk,v} + \frac{7}{5} \frac{\partial \bar{v}_{i,v}}{\partial r_i} q_{j,v} + \right.$$

$$\left. + \frac{7}{5} \frac{\partial \bar{v}_{j,v}}{\partial r_i} q_{i,v} + \frac{2}{5} \frac{\partial \bar{v}_{i,v}}{\partial r_j} q_{i,v} + \frac{\partial q_{j,v}}{\partial r_i} + \frac{7}{2} \frac{\partial}{\partial r_i} (R T_v P_{ij,v}) \right\} -$$

$$- \frac{\partial}{\partial r_j} (\dot{r}_v R T_v) = \frac{m_v}{2} \sum_s \int X_{j,v} V_v{}^2 A_{v,s} d\vec{v}_v - \qquad (10)$$

$$- \frac{1}{n_v} \sum_k P_{jk,v} \sum_{s \neq v} \int v_{i,v} A_{v,s} d\vec{v}_v - \frac{3}{2} \frac{p_v}{n_v} \sum_{s \neq v} \int v_{j,v} A_{v,s} d\vec{v}_v .$$

Die Integrale wurden analytisch ausgewertet für den Fall MAXWELLscher Moleküle. Das Gleichungssystem ist anwendbar für sehr extreme Verhältnisse großer freier Weglängen und für starke Gradienten, wobei jede einzelne Komponente durch eine verschiedene Temperatur T_v gekennzeichnet sein kann.

Literatur:

[1] KAEPPELER, H. J., und G. BAUMANN: Mittl. Forsch. Inst. Phys. Strahlantr. Stuttgart, No. **8** (1956).
[2] KAEPPELER, H. J.: Z. Physik **148**, 425—434 (1957).
[3] KAEPPELER, H. J.: Mittl. Forsch. Inst. Phys. Strahlantr. Stuttgart, No. **14** (1958).
[4] BRITTIN, W. E.: Phys. Review **106**, 843—849 (1957).
[5] GRAD, H.: Comm. Pure Appl. Math. **2**, 331—372 (1949).

Contribution to the Subject
Thermal Boundary Layers

By **M. G. Scherberg,** Wright Air Development Center, Wright-Patterson Air Force Base, Ohio

Some experimental techniques related to free convection thermal boundary layers are the subject of this contribution to the discussions. It was the aim of experimentation by E. SOEHNGEN of our Aerodynamics Branch, and his students to explore the limits of validity of conventional boundary layer and heat transfer correlation theory in the range of Prandtl numbers up to one million. The experimental approach consisted of free convection tests on heated vertical plates and cylinders under transient and steady state conditions. Test fluids were air, hydrocarbon polymers (large temperature-viscosity dependency), silicon polymers (low temperature-viscosity dependency) and engine oils at — 70° F. Boundary layer velocity profiles on the cylinder were obtained from the fluid motion transport of a line of mushroom spores (a white powder, 4 μ size) suspended in the fluid. The Prandtl numbers here were up to one million and the Grashof-Prandtl numbers range from 10^4 to 10^8. In the plate tests a 25 μ wire suspended across the plate (heated to 50° F above ambient temperature) near the bottom edge and in about the middle of the free convection boundary layer. This wire was to introduce electrically produced heat pulse disturbances into the boundary layer. The motion was observed and photographically recorded from interferimetric fringe lines. The heat pulses were of .1 watt sec energy and of 4 mil sec duration. They were used over a range of frequencies up to the Reynolds number-critical of the boundary layer. Motion pictures showed very elegantly the flow of the boundary layer and the effects of single and multiple pulses. The following results were obtained. Within the range of the Grashof-Prandtl numbers 10^4 to 10^8 the $^1/_4$ power law holds up to Prandtl numbers of 10^6. The constant $Nu(GrPr)^{-1/4}$ was found to be 20% lower than indicated by the theory of SCHUH, OSTRACH and others. On the flat plate experiments the boundary layer "natural frequency" was found to be a function of the Reynolds number based on boundary layer thickness.

Interactions between shock waves and boundary layers

By

G. E. Gadd

National Physical Laboratory, Teddington

Interactions between shock waves and boundary layers are of frequent practical occurrence, but often the configuration of the region of inter-action is very complex, and this makes it difficult to understand the processes involved. To throw light on the matter, therefore, simplified types of interaction, mostly with what is intended to be two-dimensional flow, have been studied by many workers. Cases that have received the greatest attention are (a) the interaction between the boundary layer on a flat plate and a shock wave generated by a wedge held in the main-stream above the plate (fig. 1) and [1] to [6], (b) the case where the surface on which the boundary layer is formed has an abrupt change of

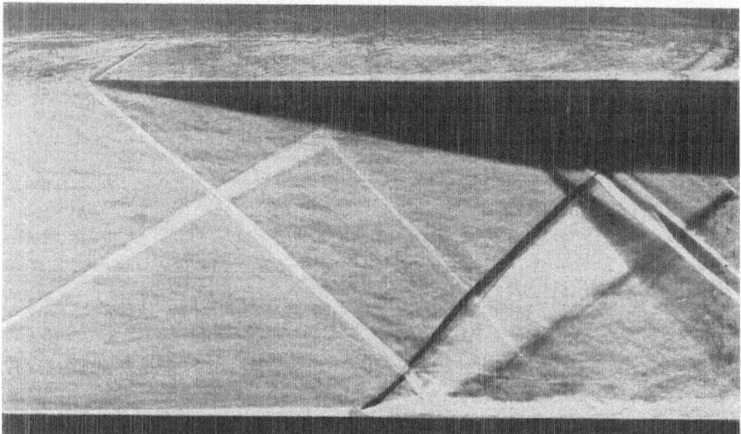

Fig. 1. External shock. $M = 2$, turbulent layer

slope, the surface upstream of the corner being level and downstream being inclined upwards, though still flat (fig. 2) and [1, 4, 5, 7], and (c) the case of a step on a flat wall (fig. 3) and [5, 8]. If the shock in case (a) is sufficiently strong, boundary-layer separation will take place well ahead of the point where the shock strikes the boundary layer. Similarly if the change of surface slope in case (b) is sufficiently large, or the step in case (c) sufficiently high, separation takes place well ahead of the

corner or step. In these circumstances the flow in the immediate vicinity
of separation is virtually independent of the particular agency—external
shock, corner, or step—provoking separation. Thus, for example, for
given upstream conditions, the pressure distribution at the wall in the
immediate vicinity of separation will, as in fig. 4 taken from [8], be the
same if separation is provoked by an external shock as it will be if
separation is provoked by a step, provided that the step is suitably
arranged to cause separation at the same position as with the external

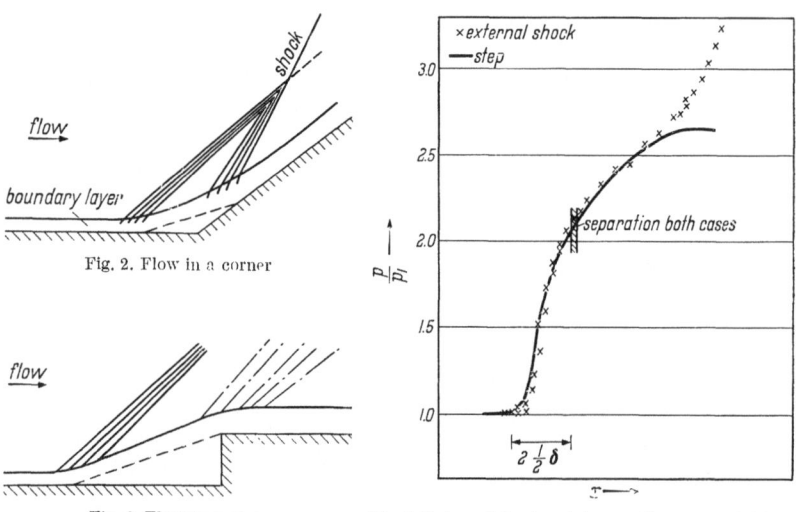

Fig. 2. Flow in a corner

Fig. 3. Flow up a step

Fig. 4. External shock and step results compared. $M = 3$,
turbulent layer. (From Kepler and Bogdonoff [8])

shock, and provided also that in both cases the regions of separated flow
are large in extent. The reason for this lack of dependence of the conditions near separation on the agency which causes separation is as
follows. In the vicinity of separation there are adverse pressure gradients
which thicken the stream tubes of low velocity in the boundary layer.
This thickening of the inner part of the boundary layer deflects the
external flow from its original direction parallel to the wall and hence
generates a band of compression waves. Clearly the compression and
the thickening of the layer must adjust themselves to be in equilibrium.
This equilibrium process will be insensitive to downstream conditions,
provided that for some distance downstream of separation there are no
externally-generated shocks or waves incident upon the boundary layer,
and no steps or abrupt changes of slope or curvature in the wall. The flow
under these conditions may be described for convenience as a "free
interaction", to use Chapman's term [5]. The simplest type of free-
interaction flow is two-dimensional, with the separation line straight

and perpendicular to the flow. Even here, however, the flow is affected by several variables. These are 1. the state (laminar or turbulent) of the upstream boundary layer, 2. Reynolds number, 3. free-stream Mach number, 4. free-stream turbulence level, 5. heat transfer at the wall, and 6. curvature of the wall, the flow being parallel to the lines of maximum curvature to satisfy the assumed two-dimensional requirement.

At separation, the pressure gradients are usually relatively steep, becoming much less steep downstream. This is true both when the flow

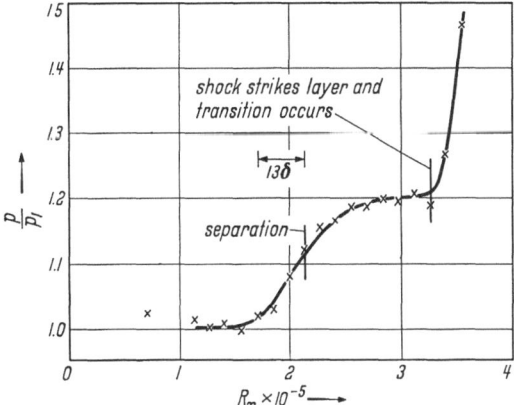

Fig. 5. Laminar separation produced by external shock. $M = 2$. (From Hakkinen and Trilling [6])

remains laminar and also when it is turbulent throughout the region considered (see figs. 4 and 5). The two cases are indeed qualitatively similar although quantitatively they differ greatly, the pressure co-efficients associated with turbulent separations being much greater than those for laminar separations. Also a laminar layer gives rise to a band of compression waves spread over several boundary layer thicknesses in the region of the separation point (fig. 6), so that the change of pressure across the boundary layer is, at moderate Mach numbers, relatively small. By contrast (fig. 1), the compressions generated by a turbulent separation have more the character of a shock wave, and close to the separation point the pressure is far from being constant across the boundary layer. It is because he neglects the variation in pressure across a turbulent boundary layer near separation that Crocco [9] is led to the conclusion that there is a fundamental difference between the physical processes underlying turbulent separation on the one hand and laminar separation on the other. On the assumption that the pressure is constant across the layer, he shows that the turbulent boundary layer which has developed under constant pressure on a flat plate will be reduced in thickness by a small pressure increase. The reason for this is that for all but the lowest supersonic Mach numbers the greater part of the upstream turbulent layer is supersonic, and supersonic stream tubes tend to get thinner on encountering rises of pressure. If the boundary layer were to become thinner, the external-flow streamlines would have to turn in towards the wall, and this would be incompatible with a rise of pressure.

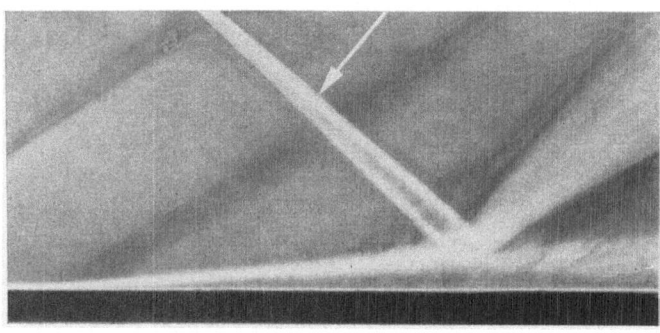

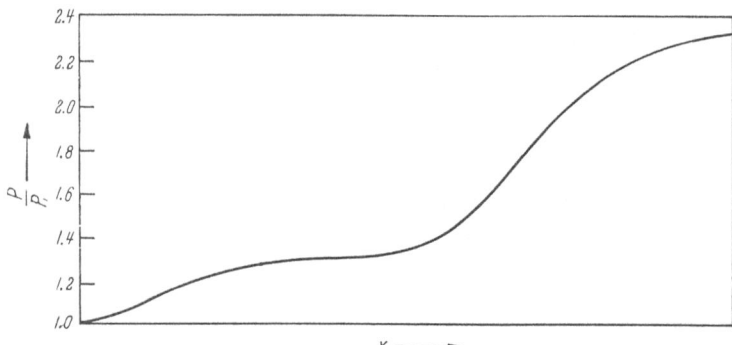

Fig. 6. Flow pattern and pressure distribution to same longitudinal scale. $M = 2$. Laminar at separation, turning turbulent downstream

However, in reality the slowest-moving stream tubes adjacent to the wall expand when they meet a pressure rise, and in the outer part of the boundary layer the pressure tends to be constant along Mach lines inclined at an oblique angle to the wall. Thus there are compression waves emanating from well inside the boundary layer. These waves deflect both the external flow and the flow in the outer part of the layer away from the wall. Hence, regarded from this viewpoint, the difference between laminar and turbulent separation is only one of degree, not of fundamental physical principle.

Recently, reliable experimental information [5, 6] has become available for laminar separations from a flat wall under zero heat-transfer conditions. For free-stream Mach numbers M in the range 1.1 to 3.5 and Reynolds numbers R_x (based on free-stream conditions and the distance x from the leading edge) in the range 10^4 to 10^6, it has been established that the pressure coefficient at separation in a two-dimensional free interaction is approximately proportional to $(M^2-1)^{-1/4} R_x^{-1/4}$. Further experiments,

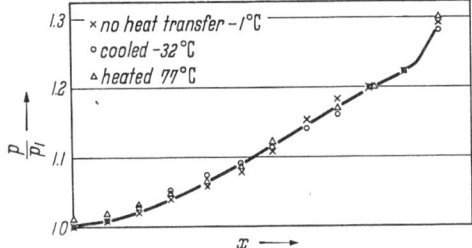

Fig. 7. Laminar separations with various wall temperatures. $M = 3$

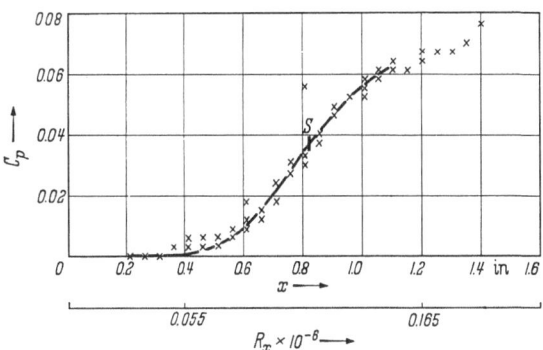

Fig. 8. Pressure coefficient distribution—comparison between theoretical curve and experimental points. $M = 3$, zero heat transfer. S marks theoretical position of separation point

[10] and fig. 7, have shown that moderate degrees of heat transfer appear to have little effect on the overall pressure distribution through the separated region, though the pressure coefficient at the actual separation point may perhaps vary: more experimental work is needed to settle the matter, however. Convex surface curvature, such as is normally present on a two-dimensional aerofoil, is found [11] to reduce the pressure coefficient at separation. Various approximate theories [5, 6] and [12, 13], some of which are extremely simple, give quite good agreement in most respects with experiment. (See, for example, fig. 8, a comparison with the theory of [13].) However, the theories do not seem to cope very satisfactorily with heat-transfer effects.

For turbulent layers in two-dimensional flow there is again a good deal of experimental information available covering the same sorts of conditions as for laminar layers [3, 4, 5] and [7, 8]. Variations in the Reynolds number R_x in the range of about 10^6 to 2.10^7 appear to have only quite small effects on the pressure coefficients at separation. Heat transfer and surface-curvature effects are also fairly small in the ranges investigated experimentally [10, 11] and [14]. The main important effect is that of free-stream Mach number, the pressure coefficient at separation becoming smaller at higher Mach numbers, approximately like $(M^2 - 1)^{-1/3}$. There are various semi-empirical theories which predict the pressure coefficient at the separation point or at related points reasonably well [15] to [19].

The practical utility of information concerning conditions at separation is enhanced by the fact that the pressure gradients tend to flatten off downstream of separation, so that there is often a "plateau pressure" related to the pressure at separation (c.f. figs. 4 and 5). This nearly-constant plateau pressure may prevail over a relatively large region, and knowledge of it is sometimes therefore of use in the estimation of the forces acting on a body, if the extent of the plateau region can be estimated. If the flow is laminar at separation and turns turbulent downstream in the separated region, the pressure immediately downstream of transition will usually rise steeply above the laminar plateau pressure (fig. 6). The distance between transition and separation can also have a large effect on the position of separation [4] and [5]. Hence, the overall pressure distribution is very sensitive to transition position, which is partly governed by the free-stream turbulence level. Accordingly, for cases with mixed laminar and turbulent flow, (cases which are more frequently encountered than the pure laminar type), it is difficult to estimate what the pressure distribution will be in free flight, for example, even for the simple case of a spoiler or step, a configuration for which in purely laminar or purely turbulent flow the upstream effect can be estimated reliably. For other types of separation, even when the flow is turbulent throughout the region of interaction, small differences in experimental arrangement may cause large discrepancies in the experimental results, as is discussed further below. Because of these considerations, it is often difficult to estimate the extent of regions of separation in practice, where the flow pattern may be complicated: however, the data obtained with simple configurations are useful qualitatively in, for example, emphasizing the critical importance of transition position. The data for separation in free interactions are also useful, although it must be borne in mind that separation certainly does not always occur whenever the boundary layer is subjected to a pressure increase greater than the pressure rise up to a free-interaction separation point. This is because free-interaction conditions only apply after extensive separation has occurred, and the separation point has moved well upstream of the agency used to impose the pressure rise. The conditions leading to the first occurrence of separation depend on the nature of the agency used to produce the pressure rise (see [4] and [20]). Thus, consider, for example, the flow in a corner as in fig. 2. Here, especially at high Mach numbers, a large part of the overall pressure rise occurs downstream of the corner, whether separation is present or not. Hence, since separation, if it occurs, must be upstream of the corner, the overall pressure rise must greatly exceed that at a free-interaction separation point before any detectable extent of separated flow can be formed at the higher Mach numbers. With other configurations separation may occur when the overall pressure rise is less than the pressure rise to a free-interaction separation point. Thus with the flow up a step, as in fig. 3, separation probably occurs however small the step height, but the overall pressure increase produced by the step will presumably fall smoothly to zero as the ratio of step height to boundary-layer thickness is reduced to zero. However, it is safe to conclude in any practical case which is approximately two-dimensional that separated flow, if present, will be limited to a small region if the overall pressure rise through the region of interaction is less than that for free-interaction separation. This gives a useful design guide for the avoidance of serious separation effects even though in many cases, as discussed above, it may be unduly conservative.

It was mentioned above that there are certain discrepancies in experimental measurements of the extent of separated turbulent-flow regions. The measurements concern interactions between externally-generated oblique shocks, as in fig. 1, and turbulent layers. For this configuration with shocks strong enough to cause appreciable separation, the only detailed experiments involving pressure measurements that have been discussed in published reports are those described in [3, 4], and [20]. In [3] the test boundary layer was that formed on the tunnel wall, whereas in [4] a flat plate was used to produce the test layer, the tunnel-wall boundary layer being ducted underneath. As can be seen from fig. 9, the results differ greatly. It does not seem possible to attribute the difference to the difference in the Reynolds number R_δ based on boundary-layer thickness δ, because the flat-plate experiments of [4] showed no trend of variation with Reynolds number over a wide range of R_δ. Further, the boundary layer upstream of the region of interaction appears to be of the same shape on the tunnel wall as on the flat plate. BOGDONOFF has suggested that the cause is a slight departure from two-dimensional conditions in the flat plate experiments. This he attributes to side-wall boundary layer effects, effects which, he suggests, are less significant when the test boundary layer is relatively thick. The grounds for his suggestion are that he has found in unpublished experiments that the large upstream effects obtained with flat-plate boundary layers can be greatly reduced if boundary-layer fences are fitted. This certainly seems persuasive evidence, and it is intended to try and repeat these experiments at the N.P.L. Nevertheless, it seems hard to reconcile BOGDONOFF'S suggestion with the agreement, reported in [4], between the main set of results obtained in a small tunnel and the check data obtained in a much larger tunnel. For the latter case the overall length of the

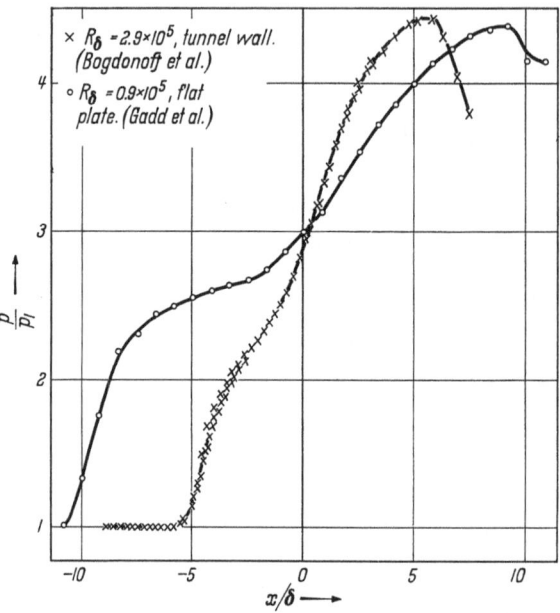

Fig. 9. Interactions between turbulent boundary layers and shock waves generated by 12° wedge held above wall. Plane of shock intersects wall at point 0. $M = 3$

separated region was quite small compared with the tunnel width, the ratio being much smaller than with most of the data from the small tunnel; however, all the results were in reasonable agreement. It seems difficult to imagine that any effects due to departures from two-dimensional flow would be as large for the large-tunnel tests as for the others. Hence, it is felt at the N.P.L. that BOGDONOFF'S suggestion needs further investigation, and may perhaps not prove to be correct. Be that as it may, the fact remains that what are apparently only minor differences in experimental arrangement can produce large differences in the flow with the configuration of fig. 1. Results with separation produced by a step, as in fig. 3, are more easily repeatable, provided that transition does not occur within the separated region. With a step the maximum height of the separated region is the height of the step, and is thus fixed in advance, but for an oblique shock as in fig. 1 the height and extent of the separated region are not fixed, but must adjust themselves to permit the boundary layer to reattach in equilibrium with its self-induced pressure gradients. There is thus, as it were, an extra "degree of freedom" for the external shock case as compared with the step, and this is probably associated with the greater difficulty in obtaining repeatable experimental results.

Experimental discrepancies which may conceivably be related to those discussed above occur in tests on interactions between nearly-normal shocks and turbulent layers, interactions such as are encountered in transonic flow past aerofoils. Tests at the N.P.L. [21] on a wide variety of two-dimensional aerofoil models have suggested that separation usually occurs when the Mach number at the edge of the boundary layer just upstream of the shock exceeds about 1.23. However, in the recent experiments of MICHEL [22], separation did not occur till considerably higher upstream Mach numbers were reached. The test configuration used by MICHEL was that of a circular arc "bump" on a tunnel wall, so that the test boundary layer had developed along the tunnel wall. The pioneer experiments of ACKERET, FELDMANN and ROTT [23] investigated both a bump on the tunnel wall and also a curved plate underneath which the tunnel wall boundary layer was ducted. Fig. 10a shows one of the pressure distributions obtained on the curved plate in [23], together with the corresponding boundary-layer profiles. Fig. 10b shows a pressure distribution and set of profiles taken from [22], where the boundary layer was that on the tunnel wall. The boundary-layer thicknesses happened to be roughly the same in the two cases (the tunnels being of different sizes), and the pressure distributions have been drawn to similar longitudinal scales. The curvature of the bump in [22] relative to the thickness of the boundary layer was much sharper than that of the plate in [23]; this accounts for the much steeper favourable pressure gradient upstream of the shock in fig. 10b. It seems likely that this favourable gradient is the reason why the upstream boundary layer is much "fuller" in shape in fig. 10b than in fig. 10a. In turn, this difference in upstream profile shape may account for the fact that the boundary layer in fig. 10a comes much closer to separation than in fig. 10b. It is true that the profile at the most downstream station, station E, in fig. 10b is close to the separation condition; this, however, is associated with the fact that the bump on the tunnel wall has its trailing edge just downstream of E, so that the wall has a corner there, approaching which the pressure rises steeply, tending to cause separation. Hence, the pressure distribution at the wall in fig. 10b shows gradients which fall off after the shock and then begin to steepen again downstream, whilst the gradients in fig. 10a fall off progressively more and more downstream of the shock. However, if attention is confined to the region of interaction between the boundary layer and shock, it certainly seems that there is much less tendency for separation to occur with the tunnel-wall boundary layer than with the boundary layer on the curved plate, despite the fact that for this latter case the upstream Mach number was slightly lower.

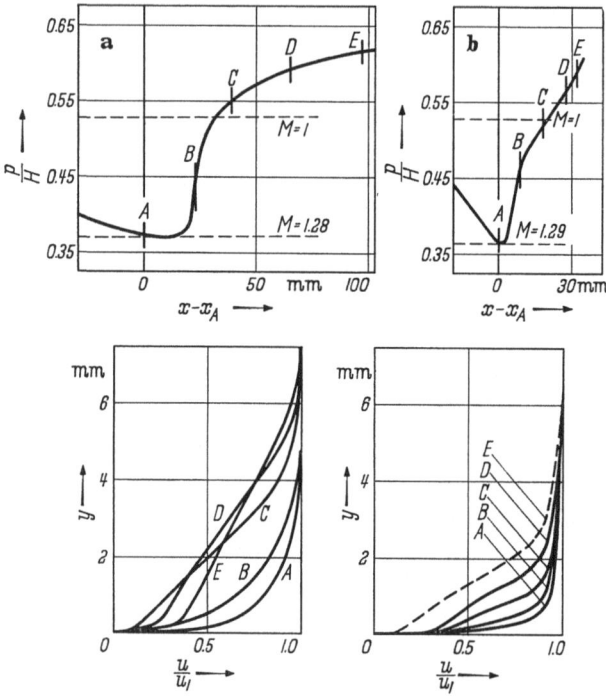

Fig. 10. Pressure distributions and boundary layer profiles for interactions of the transonic aerofoil type between turbulent layers and nearly normal shocks. (a) Curved plate with tunnel-wall boundary layer ducted underneath. (ACKERET et. al. [23]) (b) "Bump" on tunnel wall. (MICHEL [22])

A similar reduced tendency for separation is also evident in those tests of [23] which were done on a bump on the tunnel wall. Possibly the cause of all this is related to the cause of the discrepancy which is illustrated in fig. 9. However, there, where the walls were flat in the test region, the boundary layer profiles just upstream of the regions of interaction were approximately the same shape in the two cases, whereas they differ considerably in fig. 10. This difference in profile shape seems perhaps more likely to be the cause of the differences in the interactions, rather than any departure from two-dimensional conditions or whatever the cause of the discrepancy shown in fig. 9 may be.

Side-wall boundary layer effects may, however, be important in wind tunnel tests of configurations resembling the transonic flow past an aerofoil, even if they are not responsible for the discrepancies between results obtained with a bump on the tunnel wall and those obtained with a curved plate or aerofoil model. The shock wave in such a test frequently extends over a large fraction of the distance between the model and the top wall of the tunnel. The side-wall boundary layers must thicken on passing through the shock, so that the effective area of the tunnel is reduced downstream. This must tend to reduce the downstream pressures, and if the effect is a significant one, the whole flow pattern may indirectly be appreciably affected. It does not follow that such effects would become negligible if the Mach number were reduced, because, although the thickening of the boundary layers must be small with weak shocks at Mach numbers near 1, the pressure becomes very sensitive to small changes of stream-tube area at such Mach numbers. It is intended to investigate the matter at the N.P.L. by using suction to reduce the thickness of the side-wall boundary layers in the transonic test of an aerofoil. A fundamental programme to investigate interactions between weak normal shocks and turbulent boundary layers is also in hand. This should be helpful in interpreting the transonic flow over thin aerofoil sections, such as are now coming into use. With such sections at low angles of incidence weak nearly-normal shocks with upstream Mach numbers near to 1 often occur. These shocks are not strong enough to cause boundary-layer separation. Nevertheless, it is found experimentally [24] that the abrupt increase of pressure on the aerofoil underneath the shock is considerably less than the full theoretical normal-shock pressure rise corresponding to the Mach number just upstream of the shock. Partly this may be due to the finite nature of the supersonic region (c.f. [23]), but with thin, fairly flat, sections the rate of change of the Mach number upstream of the shock with distance away from the surface is small, and hence in the absence of boundary-layer effects one would expect something close to the full normal-shock pressure rise to be achieved. The boundary layer on the aerofoil underneath the shock would be expected to "spread out" the pressure distribution over a distance proportional to boundary layer thickness, though not to affect the ultimate downstream pressure. However, an approximate theory [25] suggests that the pressure gradients at the downstream end of the region of interaction should be relatively gentle, so that the pressure distribution has a "tail". This, coupled with the possible side-wall boundary layer effects mentioned above, might account for the reduced downstream pressures observed experimentally. To see whether this is so or not it is desirable to do experiments in which boundary layer effects are studied in isolation from any effects due to non-uniformity in the supersonic external flow. Thus it is planned to investigate a normal shock in a tunnel of circular cross section with a slightly-supersonic uniform mainstream. The pressure downstream of the shock here will depend on the

ratio of the boundary layer thickness to the tunnel radius, but if this
ratio is sufficiently small, the overall pressure rise at the wall should be
approximately equal to the theoretical inviscid value. It should then be
possible to determine how much the boundary layer "spreads out" the
pressure rise, and what the shape of the pressure distribution is. An
alternative indirect experiment which should yield the same information
(though it needs checking by the direct experiment) has been performed,
using the arrangement shown in fig. 11. An oblique shock is generated by

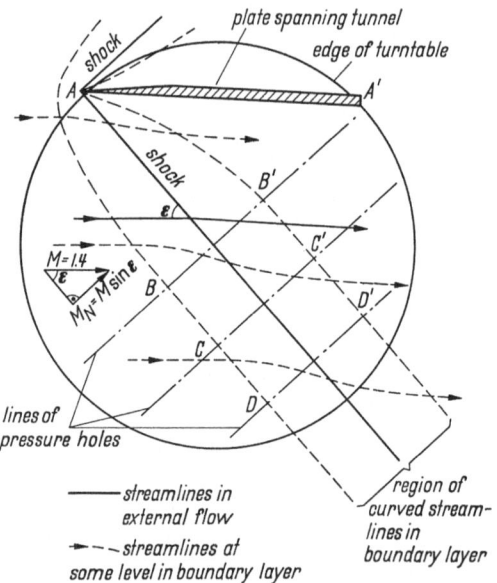

Fig. 11. Experiment on interaction between side-wall boundary layer and weak oblique shock, related
to normal shock in stream of Mach number $M_N = M \sin \varepsilon$

a flat plate spanning the tunnel and mounted between turntables so that
the incidence of the plate and strength of the shock can be varied. This
shock, which is arranged always to be weak, interacts with the side-wall
boundary layers, and the pressure distributions along lines BB^1, CC^1,
DD^1, (fig. 11), which are approximately normal to the shock, can be
measured. In the region of interaction the streamlines within the
boundary layer do not remain parallel to those in the external flow. This
is primarily because the velocity components normal to the shock suffer
less proportional reduction at the edge of the layer than they do inside
the layer, where the velocities are lower. Upstream of the region of interaction the boundary-layer profiles are approximately the same everywhere along the shock, because the effective origin of the layer is a long
way upstream. For weak shocks it seems likely that within the region of

interaction also, at a sufficient distance from the shock apex A (fig. 11), a quasi-two-dimensional state will be achieved, such that flow conditions do not vary along the shock. This state of affairs, however, cannot apply near to A, because near to the plate AA¹ even the streamlines within the boundary layer must be approximately parallel to AA¹. Also if the shock were strong enough to cause separation of the side-wall boundary layer, it seems probable that two-dimensional conditions could not occur anywhere, because the streamlines very close to the wall would bend round parallel to the shock on its upstream side, and would not cross underneath it. Hence, the zone of interaction would presumably, due to the piling up of separated fluid, widen indefinitely with increasing distance from the shock apex A. If, however, with weak shocks, in the absence of separation, a region of quasi-two-dimensional flow does occur, then the pressure distributions normal to the shock in this region should be much the same as for the interaction between a boundary layer of the same upstream thickness and an infinite normal shock at the normal-component Mach number M_N. For denote by x, y and z and u, v and w, the directions and velocity components parallel to the wall normal to the shock, parallel to the wall parallel to the shock, and normal to the wall, respectively. Then in the full boundary layer momentum equations for the u and w components, all the terms explicitly involving v disappear if the flow is quasi-two-dimensional, because this, by definition, means that gradients with respect to y are zero. Thus, if the density and viscosity were constant, the flow in planes normal to the shock would be entirely independent of the flow in planes parallel to the shock. Hence, the pressure distribution in the normal direction would be the same as for a normal shock at the component Mach number M_N. In reality, density and viscosity depend on temperature and are indirectly affected by the terms in v; however, since the free stream Mach number in the present experiments is fairly low, 1.4, the situation should not differ greatly from that with a normal shock provided that a quasi-two-dimensional flow pattern is, in fact, achieved.

This condition seems to be approximately met, since the pressure distributions along the lines BB¹, CC¹ and DD¹ are roughly the same. Three typical pressure distributions are shown in fig. 12. There is a considerable experimental scatter because of stray disturbances in the tunnel. These disturbances, arising from small roughnesses at the edges of tunnel windows, etc., are small for ordinary purposes, but are relatively important in the present experiments. It is, however, difficult to obviate them entirely. It will be observed from fig. 12 that the amount by which the pressure distributions are "spread out" does not diminish as the normal-component Mach number M_N approaches 1, although the thickening of the boundary layer diminishes with the reduced pressure rise. The

explanation put forward in [25] of this effect is that the maximum
deflection of the external-flow streamlines from their original free-stream
direction parallel to the wall decreases even more markedly as the Mach

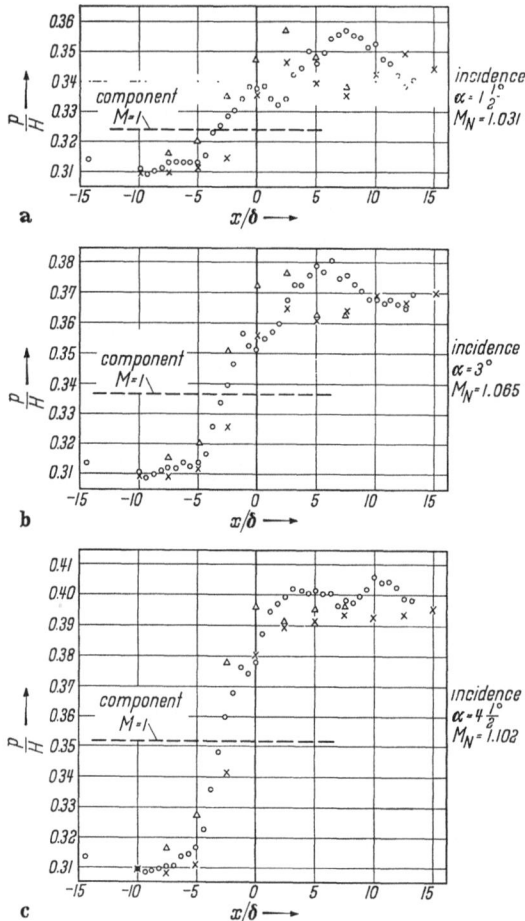

Fig. 12. Pressure distributions with arrangement of Fig. 11
× on BB', ○ on CC', △ on DD'

number approaches 1 than does the increase in boundary-layer thickness.
The spreading out of the distribution is proportional to the ratio of the
increase in boundary-layer thickness to the deflection of the external
flow, and hence it tends, if anything, to increase as M_N approaches 1.
The experimental scatter makes it difficult to determine the precise shape
of the pressure distributions, but, in accordance with [25], the average
pressure gradients do seem to be less steep at the downstream end of the

region of interaction. It is hoped that the direct experiments planned for a tunnel of circular cross section will give the shape of the pressure distributions more accurately.

Acknowledgment. This paper is published by permission of the Director of the National Physical Laboratory.

References

[1] LIEPMANN, H. W., A. ROSHKO and S. DHAWAN: On reflection of shock waves from boundary layers. N.A.C.A. Report 1100.

[2] BARRY, F. W., S. H. SHAPIRO and E. P. NEUMANN: The interaction of shock waves with boundary layers on a flat surface. J. Aero. Sci. **18**, 229 (1951).

[3] BOGDONOFF, S. M., C. E. KEPLER and E. SANLORENZO: A study of shock wave turbulent boundary layer interaction at M = 3. Princeton University Aeronautical Engineering Dept. Report 222, July 1953.

[4] HOLDER, D. W., and G. E. GADD: The interaction between shock waves and boundary layers and its relation to base pressure in supersonic flow. Paper No. 8 presented at the Symposium on Boundary Layer Effects in Aerodynamics, at the National Physical Laboratory, March/April, 1955. London, H.M.S.O.

[5] CHAPMAN, D. R., D. M. KUEHN and H. K. LARSON: Investigation of separated flows in supersonic and subsonic streams with emphasis on the effect of transition. N.A.C.A. T.N. 3869, March, 1957.

[6] HAKKINEN, R. J., and L. TRILLING: On the interaction of an oblique shock wave with a laminar boundary layer. Paper presented at the 9th International Congress of Applied Mechanics, at Brussels, September, 1956.

[7] DROUGGE, G.: Experimental investigation of the influence of strong adverse pressure gradients on turbulent boundary layers at supersonic speeds. Paper presented at the 8th International Congress on Theoretical and Applied Mechanics, at Istanbul, 1952.

[8] KEPLER, C. E., and S. M. BOGDONOFF: Interaction of a turbulent boundary layer with a step at M = 3. Princeton University Aeronautical Engineering Dept. Report 238, September, 1953.

[9] CROCCO, L.: Considerations on the shock-boundary layer interaction. Proceedings of the Conference on High Speed Aeronautics, Polytechnic Institute of Brooklyn, January 20th—22nd, 1955.

[10] GADD, G. E.: An experimental investigation of heat transfer effects on boundary layer separation in supersonic flow. Journal of Fluid Mechanics, **2**, 105 (1957).

[11] GADD, G. E.: The effects of convex surface curvature on boundary layer separation in supersonic flow. British A.R.C. C.P. No. 289.

[12] BRAY, K. N. C.: The application of the theory of Crocco and Lees to shock-boundary layer interactions. Princeton University Aeronautical Engineering Dept. Report 322, July, 1955.

[13] GADD, G. E.: A theoretical investigation of laminar separation in supersonic flow. To be published in J. Aero. Sci.

[14] Czarnecki, K. R., and A. R. Sinclair: A note on the effect of heat transfer on peak pressure rise associated with separation of turbulent boundary layer on a body of revolution (NACA RM-10) at a Mach number of 1.61. N.A.C.A. T.N. 3997, April, 1957.

[15] Tyler, R. D., and A. H. Shapiro: Pressure rise required for separation in interaction between turbulent boundary-layer and shock wave. J. Aero. Sci., 20, 858 (1953).

[16] Crocco, L., and R. F. Probstein: The peak pressure across an oblique shock emerging from a turbulent boundary-layer over a plane surface. Princeton University Aeronautical Engineering Dept. Report 254, 1954.

[17] Gadd, G. E.: Interactions between wholly laminar or wholly turbulent boundary layers and shock waves strong enough to cause separation. J. Aero. Sci., 20, 729 (1953).

[18] Mager, A.: Prediction of shock-induced turbulent boundary-layer separation. J. Aero. Sci., 22, 201 (1955).

[19] Schuh, H.: On determining turbulent boundary-layer separation in incompressible and compressible flow. J. Aero. Sci., 22, 343 (1955).

[20] Bogdonoff, S. M.: Some experimental studies of the separation of supersonic turbulent boundary layers. Paper presented at the meeting of the Heat Transfer and Fluid Mechanics Institute, at the University of California, June, 1955.

[21] Pearcey, H. H.: Some effects of shock-induced separation of turbulent boundary layers in transonic flow past aerofoils. Paper No. 9 presented at the Symposium on Boundary Layer Effects in Aerodynamics, at the National Physical Laboratory, March/April, 1955. London, H.M.S.O.

[22] Michel, R.: Etude experimentale de la couche limite turbulente et de son interaction avec l'onde de choc sur un demi-profil en ecoulement transsonique. ONERA Note Technique 11/1675. A. Oct., 1956.

[23] Ackeret, J., F. Feldmann and N. Rott: Untersuchungen an Verdichtungsstößen und Grenzschichten in schnell bewegten Gasen. Institut für Aerodynamik E.T.H. Zürich, Nr. 10, 1946.

[24] Holder, D. W.: Experiments with a two-dimensional aerofoil designed to be free from turbulent boundary-layer separation at small angles of incidence for all Mach numbers. To be published.

[25] Gadd, G. E.: The interaction between a weak normal shock wave and a turbulent boundary layer. British A.R.C. 19,352 (1957).

Aus der Diskussion

R. Michel (Châtillon-sous-Bagneux): Des différences importantes ont été relevées entre les critères de décollement turbulent de Gadd-Pearcy et des expériences sur un profil circulaire effectuées à l'O.N.E.R.A. en transsonique.

Pour un nombre de Mach en amont du choc égal à 1,4, il n'a pas été trouvé de décollement. Celui-ci ne se produit que loin en aval, sous l'action du gradient de pression positif qui prolonge celui dû à l'onde de choc.

Le calcul de la couche limite, effectué en introduisant dans l'équation des quantités de mouvement les pressions expérimentales à la paroi, est en assez bon accord avec l'expérience à la traversée de l'onde de choc.

En aval, lorsque la couche limite se rapproche des conditions de décollement, l'expérience conduit à des épaisseurs de couche limite nettement supérieures à celles données par le calcul.

Cette différence est à rapprocher de différences semblables relevées par différents auteurs en incompressible, pour des couches limites turbulentes sous gradient de pression intense. Elle semble poser à nouveau le problème de la validité des hypothéses classiques de la couche limite à l'approche du décollement.

On peut se demander s'il ne serait pas utile d'étudier déjà ce problème pour une recompression continue afin de mieux comprendre le cas de l'onde de choc, pour lequel on observera d'ailleurs que la recompression est elle-même pratiquement toujours continue.

Remark added by Mr. G. E. GADD (N. P. L., Teddington): The question of the different conditions for separation is discussed in the written paper above, in a section omitted from the spoken version given at Freiburg.

Influence of the boundary layer on airfoil pressures at supersonic speeds[1]

By

Carlo Ferrari

Politecnico di Torino, Torino

Introduction

The problem to be investigated analytically in this study is not the usual determination of the moderate alteration in pressures associated with an apparent thickening of an airfoil section when enveloped by an unseparated boundary layer, but rather an examination is made of the regenerative influence exerted far upstream, at supersonic speeds, between the boundary layer formed over the aft end of an airfoil and the shock-wave interference system which is set up. The trailing-edge shock when first formed tends to introduce a pressure increment along the subsonic part of the boundary layer so that a pressure rise $\varDelta p$ is propagated upstream. Even though this increment is small of itself when the flow is laminar, nevertheless the rate of change of this pressure increase (grad $\varDelta p$) is large enough to cause the boundary layer to detach itself from the surface of the airfoil. By so separating from the solid wall contour, the fluid flowing in the detached boundary layer acquires sufficient kinetic energy to overcome the remaining pressure differential produced by the shock in the potential-governed streamlined flow of the external stream.

What is sought, therefore, is not merely a means of determining the pressure adjustments which are produced over the forward part of the airfoil as a result of the apparent thickening that is associated with the progressive growth of the boundary layer, but knowledge is also required as to what size of pressure increment, $\varDelta p$, will produce a detachment, such as is known from optical studies to take place when the flow is laminar. After obtaining the upstream flow characteristics, it is then possible to determine the detailed nature of the detached jet, and the position of the separation point then becomes resolved by examining the conditions which must be fulfilled downstream. It is recognized that the wake-like flow off the dorsal surface, constituting the jet, and the viscously

---

[1] This work was carried out for the Applied Physics Laboratory of Johns Hopkins University under sponsorship of the U.S. Navy Bureau of Ordnance, Contract NOrd-7386.

retarded vein of fluid which comes off the ventral side of the airfoil will commingle at a point of confluence located relatively far behind the trailing edge. By making some plausible assumptions as to the geometry of this flow and introducing the simplifying assumption that the pressure is equal at the point of confluence of these layers coming off the top and bottom faces of the airfoil, a sufficiency of conditions is established for determining the detachment point and the associated pressure jump.

If the boundary layer is turbulent, the structure of the flow about the airfoil is altered much less violently as a result of boundary-layer-shock-wave interference, because the ability of the turbulent layer to force itself "uphill" against the shock-produced pressure differential is much more pronounced; for this reason only the case of a laminar boundary layer is considered here, although the analysis is applicable for any shape of profile whatsoever.

1. General equations and boundary conditions for the flow ahead of the separation point

Under the usual set of hypotheses pertaining to the boundary layer in a compressible fluid the following set of governing equations apply:

momentum equation

$$
\left.
\begin{aligned}
\varrho u \frac{\partial u}{\partial x} + \varrho v \frac{\partial u}{\partial y} &= -\frac{\partial p}{\partial x} + \frac{\partial}{\partial y}\left(\mu \frac{\partial u}{\partial y}\right) \\
\frac{\partial p}{\partial y} &= 0 \, ;
\end{aligned}
\right\}
\tag{1}
$$

continuity equation

$$
\frac{\partial}{\partial x}(\varrho u) + \frac{\partial}{\partial y}(\varrho v) = 0 \, ;
\tag{2}
$$

energy integral

$$
\frac{T}{T_e} = 1 + \frac{\gamma - 1}{2} M_e^2 \left(1 - \frac{u^2}{u_e^2}\right)
\tag{3}
$$

where the energy equation is the particular one which is valid under the assumption that the wall is at the adiabatic compression temperature and that the Prandtl number is unity.

The notation is almost obvious, inasmuch as u and v represent the velocity components which are directed along the x and y axes, respectively, where the x-axis is taken to lie along the surface of the airfoil section, but the curvature is assumed to be so small that the axis may be imagined as being straight for all intents and purposes. The subscript e is used to denote that the value of the quantity to which it is affixed is the one attained at the outer edge of the boundary layer.

It is also sufficient for present purposes to assume that the viscosity dependence upon the temperature is a simple proportionality; i. e.,

$$\frac{\mu}{T} = C \frac{\mu_\infty}{T_\infty} \tag{4}$$

where the subscript ∞ denotes conditions pertaining to the free-stream flow.

If $u^{(i)}$ is used to denote the velocity that is produced at a generic point P on the airfoil when the flow is assumed to be non-viscous, the usual second order BUSEMANN formula will suffice to give its value as a function of the free-stream Mach number and local angle that the element of surface makes with the free-stream. The velocity, u_e, at the outer edge of the boundary layer at the location P may then be sought as a perturbation adjustment to $u^{(i)}$, i. e., it is expected that

$$u_e = u^{(i)} \left\{ 1 - \varepsilon \cdot F \left[\varepsilon^n R_s{}^m \frac{X^{(i)}}{X_s^{(i)}} \right] \right\} \tag{5}$$

where n and m are positive numbers which are not specified a priori and where R_s is the special Reynolds number which applies to the detachment point, S, on the dorsal surface of the airfoil, and is defined as

$$R_s = \frac{u_s^{(i)} X_s^{(i)}}{\nu_s^{(i)}} \tag{6}$$

where $\nu_s^{(i)}$ is the kinematic viscosity coefficient evaluated under the conditions of pressure and temperature which pertain to the detachment-point location when the boundary layer is absent. In this eq. (6) the reduced streamwise coordinate $X^{(i)}$ is obtained from the integral expression

$$X^{(i)} = C \int_0^x \frac{c_e^{(i)}}{c_s^{(i)}} \cdot \frac{p_e^{(i)}}{p_s^{(i)}} \cdot d x = C \int_0^x \left(\frac{c_e}{c_s^{(i)}} \right)^{\frac{3\gamma-1}{\gamma-1}} \cdot d x \tag{7}$$

where $c_e^{(i)}$ is the velocity of sound at the generic point P also for the case where the flow is inviscid. The corresponding symbols $c_s^{(i)}$ and $X_s^{(i)}$ mean that these quantities are to be evaluated at the separation point, S.

It should be noted that ε^n is taken to be a small quantity in relation to unity, but it is large with respect to $R_s{}^{-m}$. The angular deflection factor F is not known beforehand, except that it is a function of the argument indicated between the parentheses in eq. (5), and it must become equal to unity when $X^{(i)} = X_s^{(i)}$.

In like manner, it will then hold true for the velocity component directed along the y-axis (considered to be normal to the surface of the airfoil) that, neglecting squared terms, the value at the outer boundary is obtainable from the relation

$$\frac{p - p^{(i)}}{\varrho^{(i)} u^{(i)2}} = \varepsilon F = \frac{1}{\sqrt{M_e^{(i)2} - 1}} \cdot \frac{v_e}{u^{(i)}}$$

where the superscript (i) denotes, as usual, that the values obtained by neglecting the viscosity are meant. Then, explicitly, the normal component of velocity is

$$v_e = \frac{u^{(i)}}{u_\infty} \cdot \varepsilon \cdot F \cdot u_\infty \sqrt{M_e^{(i)2} - 1} \qquad (8)$$

where u_∞ is the velocity in the undisturbed free-stream.

It is obvious that because of the constancy of the pressure across the boundary layer, as indicated in the second part of eq. (1), one may recast the momentum equation into a simplified expression for the pressure gradient, namely

$$-\frac{\partial p}{\partial x} = -\frac{dp}{dx} = \varrho_e \cdot u_e \cdot \frac{du_e}{dx}. \qquad (9)$$

2. Stewartson's transformation

By referring back to the equation of continuity it is apparent that a stream function $\varPsi$ may be set up in terms of the definitions

$$\varrho u = \varrho_s^{(i)} \sqrt{\nu_s^{(i)}} \cdot \frac{\partial \varPsi}{\partial y} \quad \text{and} \quad \varrho v = -\varrho_s^{(i)} \sqrt{\nu_s^{(i)}} \frac{\partial \varPsi}{\partial x} \qquad (10)$$

and if one then subjects the x and y coordinates to STEWARTSON'S transformation[1] in which

$$dY = \frac{c_e}{c_s^{(i)}} \cdot \frac{\varrho}{\varrho_s^{(i)}} \cdot \frac{dy}{\sqrt{\nu_s^{(i)}}} \quad \text{and} \quad dX = \frac{c_e}{c_s^{(i)}} \cdot \frac{p_e}{p_s^{(i)}} \cdot C \cdot dx$$

so that

$$Y = \frac{1}{\sqrt{\nu_s^{(i)}}} \cdot \frac{c_e}{c_s^{(i)}} \int_0^y \frac{\varrho}{\varrho_s^{(i)}} \, dy \quad \text{and} \quad X = C \int_0^x \left(\frac{c_e}{c_s^{(i)}} \right)^{\frac{3\gamma - 1}{\gamma - 1}} \cdot dx \qquad (11)$$

it follows that the first momentum relationship given in eq. (1) will reduce to

$$\varPsi_Y \varPsi_{XY} - \varPsi_X \varPsi_{YY} = c_s^{(i)2} M_e \frac{dM_e}{dX} + \varPsi_{YYY} \qquad (12)$$

where

$$M_e = \frac{u_e}{c_e}. \qquad (13)$$

To reduce the apparent order of the differential equation it is convenient to make the substitutions

$$U = \varPsi_Y \quad \text{and} \quad V = -\varPsi_X \qquad (14)$$

so that then

$$U \frac{\partial U}{\partial X} + V \frac{\partial U}{\partial Y} = c_s^{(i)2} M_e \frac{dM_e}{dX} + \frac{\partial^2 U}{\partial Y^2}. \qquad (12')$$

It is immediately seen by substitution from eqs. (10), (11), (14), and
(5) that

$$U_e = \frac{c_s^{(i)}}{c_e} \cdot u_e = \frac{c_s^{(i)}}{c_e} u^{(i)} (1 - \varepsilon F) \tag{15}$$

while if one employs the energy integral to obtain the value of $c_e^{(i)}/c_e$ it
will be seen that the reduced velocity at the outer edge of the boundary
layer is

$$U_e = c_s^{(i)} M_e^{(i)} \left[1 - \varepsilon \left(1 + \frac{\gamma - 1}{2} M_e^{(i)2} \right) F \right] = c_s^{(i)} M_e^{(i)} (1 + F^*) \tag{15'}$$

where $M_e^{(i)} = u^{(i)}/c_e^{(i)}$.

Since the Mach number for the viscous case is given by

$$M_e = U_e/c_s^{(i)}$$

it follows that the first term on the right hand side of eq. (12) may be
recast as

$$c_s^{(i)2} M_e \frac{d M_e}{d X} = U_e \frac{d U_e}{d X} \simeq c_s^{(i)2} \left[M_e^{(i)} \cdot \frac{d M_e^{(i)}}{d X} + \frac{d}{d X} (M_e^{(i)2} F^*) \right]$$

where

$$F^* = - \varepsilon \left(1 + \frac{\gamma - 1}{2} M_e^{(i)2} \right) F \tag{16}$$

and where the approximation consists of neglecting terms of the order of
magnitude of ε with respect to terms of the order of magnitude unity.

Let another transformation of coordinates and a separation of solutions
be effected through use of the relations:

$$\left. \begin{array}{l} \Psi (X, Y) = \Phi (X, \eta) \sqrt{u_s^{(i)} X} \\ Y = \eta \sqrt{X/u_s^{(i)}} \quad \text{and} \quad X = X_s \cdot e^t \\ \Phi (t, \eta) = f (t, \eta) + g (t, \eta) \end{array} \right\} \tag{17}$$

and

and let a further simplification in notation be made by taking

$$m^{(i)} = M_e^{(i)}/M_s^{(i)}$$

so that then eq. (12) becomes converted into a separable set of partial
differential equations.

The one involving only $f (t, \eta)$ is:

$$\frac{\partial f}{\partial \eta} \cdot \frac{\partial^2 f}{\partial \eta \partial t} - \frac{\partial^2 f}{\partial \eta^2} \cdot \frac{\partial f}{\partial t} = m^{(i)} \cdot \frac{d m^{(i)}}{d t} + \frac{\partial^3 f}{\partial \eta^3} + \frac{1}{2} \cdot f \cdot \frac{\partial^2 f}{\partial \eta^2} \tag{18}$$

with the boundary conditions that

$$\left. \begin{array}{ll} f = \dfrac{\partial f}{\partial \eta} = 0 & \text{for} \quad \eta = 0 \\[2ex] \dfrac{\partial f}{\partial \eta} = m^{(i)} & \text{for} \quad \eta = \infty \end{array} \right\} \tag{18'}$$

and

while the one involving $g\,(t,\eta)$ is:

$$\frac{\partial f}{\partial \eta}\cdot\frac{\partial^2 g}{\partial \eta\,\partial t}+\frac{\partial g}{\partial \eta}\cdot\frac{\partial^2 f}{\partial \eta\,\partial t}-\frac{\partial^2 f}{\partial \eta^2}\cdot\frac{\partial g}{\partial t}-\frac{\partial^2 g}{\partial \eta^2}\cdot\frac{\partial f}{\partial t}=\frac{d}{d\,t}\,(m^{(i)2}\,F^*)+$$
$$+\frac{f}{2}\cdot\frac{\partial^2 g}{\partial \eta^2}+\frac{g}{2}\cdot\frac{\partial^2 f}{\partial \eta^2}+\frac{\partial^3 g}{\partial \eta^3}+\left[-\frac{\partial g}{\partial \eta}\cdot\frac{\partial^2 g}{\partial \eta\,\partial t}+\frac{\partial^2 g}{\partial \eta^2}\cdot\frac{\partial g}{\partial t}+\frac{g}{2}\frac{\partial^2 g}{\partial \eta^2}\right] \tag{19}$$

with the boundary conditions that

$$g=\frac{\partial g}{\partial \eta}=0 \quad \text{for} \quad \eta=0$$

and $\qquad\qquad\dfrac{\partial g}{\partial \eta}=m^{(i)}\,F^* \quad \text{for} \quad \eta=\infty$ $\qquad\qquad$ (19')

3. Linearization of the equation involving the g function

By checking through the terms appearing in eqs. (19) and (19') it may be seen that

$$\frac{\partial g}{\partial \eta}=O\,(F^*)\equiv O\,(\varepsilon) \quad \text{and then} \quad \frac{\partial^2 g}{\partial \eta\,\partial t}\equiv O\left(\frac{d\,F^*}{d\,t}\right)\equiv O\,(\varepsilon^{1\,+\,n}\,R_s{}^m)$$

as well as that

$$\frac{\partial^3 g}{\partial \eta^3}\equiv O\left(\frac{d\,F^*}{d\,t}\right)\equiv O\,(\varepsilon^1\,+\,^n\,R_s{}^m)\,,$$

so it clearly follows that the effect of applying the operator $\dfrac{\partial}{\partial t}$ is to augment the quantity operated upon in the ratio $\varepsilon^n\,R_s{}^m$ to 1, while the effect of the operator $\dfrac{\partial}{\partial \eta}$ is to augment the quantity operated upon in the ratio $\varepsilon^{n/2}\,R_s{}^{m/2}$ to 1.

Consequently, it may be deduced that

$$g\equiv O\left(\varepsilon^{1-\frac{n}{2}}\,R_s{}^{-\frac{m}{2}}\right) \quad \text{and} \quad \frac{\partial^2 g}{\partial \eta^2}\equiv O\left(\varepsilon^{1+\frac{n}{2}}\,R_s{}^{\frac{m}{2}}\right).$$

If one now makes the stipulation that $\varepsilon\,(\varepsilon^n\,R_s{}^m)^{1/2}$ is to be quite small with respect to unity (and the significance of this assumption will be made more clear in Article 7) then the last three terms, set off in brackets, on the right hand side of eq. (19) may be dropped and the equation becomes linear in g. On the other hand the function f appearing in eq. (18) defines the behavior of the flow within the boundary layer for the case where the pressure distribution at the external edge is assigned and the fluid is incompressible; consequently, it may be taken for granted that this function f is known for any profile in question.

4. Determination of the f and g functions

Inasmuch as one wishes to know the function g merely to determine the external pressure distribution, it is suitable to use the KÁRMÁN-POHLHAUSEN integral procedure which requires that g satisfy the linearized eq. (19) just on the average. Thus the sought solution may be obtained by first integrating eq. (19) across the boundary layer, and after some mathematical manipulation it will be found that g is then governed by the integral relations:

$$2 \frac{d}{dt} \int_0^\infty \left[\frac{\partial f}{\partial \eta} \cdot \frac{\partial g}{\partial \eta} - \left(\frac{\partial f}{\partial \eta} \right)_\infty \left(\frac{\partial g}{\partial \eta} \right)_\infty \right] d\eta + \left(\frac{\partial g}{\partial \eta} \right)_\infty \cdot \frac{d}{dt} \int_0^\infty \left[\left(\frac{\partial f}{\partial \eta} \right)_\infty - \frac{\partial f}{\partial \eta} \right] d\eta +$$

$$+ \left(\frac{\partial f}{\partial \eta} \right)_\infty \cdot \frac{d}{dt} \int_0^\infty \left[\left(\frac{\partial g}{\partial \eta} \right)_\infty - \frac{\partial g}{\partial \eta} \right] d\eta = - \left(\frac{\partial^2 g}{\partial \eta^2} \right)_0 + \frac{1}{2} \int_0^\infty \left(f \cdot \frac{\partial^2 g}{\partial \eta^2} + g \cdot \frac{\partial^2 f}{\partial \eta^2} \right) d\eta .$$

$$\text{(20)}$$

Following the lead given by TIMMAN in [2], the expression for f may be assumed to have the form of

$$\frac{\partial f}{\partial \eta} = f_\eta = m^{(i)} - \int_{\eta^*}^\infty e^{-\eta^{*2}} (a_0 + b_0 \eta^* + c_0 \eta^{*2}) \, d\eta^* \qquad (21)$$

where $\eta^* = \eta / \delta^{(i)}$
in which $\delta^{(i)}$ represents a certain function of t, which will be specifically defined at a later stage of the development. The quantities a_0, b_0, and c_0 are functions of t, which will be determined in such a way as to satisfy the stated conditions imposed on the function f at the surface of the profile, but the exponential term will automatically take care of the conditions met at $\eta = \infty$.

By repeated differentiation of f_η, therefore, a set of determining relations for a_0, b_0, and c_0 are found, and this procedure results, therefore, in the solution

$$f_\eta = m^{(i)} H_0 + m^{(i)} \cdot \frac{d m^{(i)}}{dt} \cdot \frac{\delta^{(i)2}}{2} \cdot H_1 \qquad (22)$$

where

$$H_0 = \text{erf.} \, \eta^* - \frac{2}{3\sqrt{\pi}} \eta^* \cdot e^{-\eta^{*2}} \quad \text{and} \quad H_1 = e^{-\eta^{*2}} \left(1 - \frac{2}{3\sqrt{\pi}} \eta^* \right) - \text{erfc.} \, \eta^* . \quad (23)$$

Coming back to consideration of the precise definition of $\delta^{(i)}$ it is convenient to define first a coefficient of friction which is the one corresponding to the location with abscissa X and computed for the profile in question when immersed in incompressible flow for which the pressure distribution is given by

$$\frac{-d p^{(i)}}{d X} = U_e^{(i)} \cdot \frac{d U_e^{(i)}}{d X} \cdot \varrho_s^{(i)}$$

where $U_e^{(i)}$ is determined by inserting the value zero for ε in eq. (15′); this friction coefficient is to be represented by the notation:

$$c_f^{(i)} = \frac{\tau^{(i)}}{\dfrac{1}{2}\,\varrho_s^{(i)}\,u_s^{(i)2}}\,.$$

Now since $\quad f_{\eta\eta}\,(0,t) = a_0/\delta^{(i)} = \sqrt{R_X^{(i)}}\cdot\dfrac{c_f^{(i)}}{2}\,, \quad$ then it follows that

$$\delta^{(i)} = \frac{2}{\sqrt{R_X^{(i)}}\cdot c_f^{(i)}}\cdot\frac{4}{3\sqrt{\pi}}\,m^{(i)}\left(1+\frac{\delta^{(i)2}}{2}\cdot\frac{d\,m^{(i)}}{d\,t}\right) \tag{24}$$

where

$$R_X^{(i)} = \frac{u_s^{(i)}\,X}{\nu_s^{(i)}}\,.$$

By reference to the BLASIUS solution for a flat plate it is known that $(f_{\eta\eta})_0 = 0.332$; so that for such a simple profile it is found that $\delta^{(i)}_{\text{flat plate}} = 2.279$.

In an entirely analogous manner to what was done in the case of the f function, it may now be assumed that the g function will appear in the form:

$$g_\eta = m^{(i)}\,F^* - \int_{\eta^*}^{\infty} e^{-\eta^{*2}}\,(a + b\,\eta^* + c\,\eta^{*2} + d\,\eta^{*3})\,d\eta^*. \tag{25}$$

By invoking the boundary conditions at the surface where $\eta = 0$, and by carrying out the integrations which are involved, it will be found that the sought solution is

$$g_\eta = m^{(i)}\,F^*\,H_0' + \delta^{(i)2}\cdot\frac{d}{d\,t}\,(m^{(i)2}\,F^*)\cdot H_1' + a\cdot H_2' \tag{25'}$$

where

$$\left.\begin{array}{l}
H_0' = 1 - e^{-\eta^{*2}}\,(1+\eta^{*2}) \quad\text{and}\quad H_1' = -\dfrac{1}{2}\,\eta^{*2}\,e^{-\eta^{*2}} \\[2mm]
\text{while}\quad H_2' = -\dfrac{1}{2}\left[e^{-\eta^{*2}}\left(\eta^* - \dfrac{3\sqrt{\pi}}{2} - \dfrac{3\sqrt{\pi}}{2}\,\eta^{*2}\right) + \dfrac{3\sqrt{\pi}}{2}\cdot\text{erfc. }\eta^*\right].
\end{array}\right\} \tag{26}$$

The unknown function "a" standing in this expression for g_η may be determined by stipulating that the condition derived as eq. (20) must be satisfied. When the expressions for f_η and g_η just obtained above are substituted into eq. (20) it turns out that the determining equation for "a" becomes

$$\frac{d}{d\,t}\,(a\,\delta^{(i)}) + \mathfrak{G}_0\cdot a\,\delta^{(i)} = \mathfrak{G}_1\cdot m^{(i)}\,F^* + \mathfrak{G}_2\cdot\frac{d}{d\,t}\,(m^{(i)2}\,F^*) + \mathfrak{G}_3\cdot\frac{d^2}{d\,t^2}\,(m^{(i)2}\,F^*) \tag{27}$$

in which

$$\mathfrak{G}_0 = \frac{\dfrac{1}{\delta^{(i)2}} + \mathfrak{Z}_4 + \mathfrak{Z}_8}{\mathfrak{Z}_5}$$

$$\mathfrak{G}_1 = \frac{\mathfrak{Z}_i - \mathfrak{Z}_1}{\mathfrak{Z}_5}; \quad \mathfrak{G}_2 = \frac{\mathfrak{Z}_7 - \mathfrak{Z}_2}{\mathfrak{Z}_5}$$

and $\quad \mathfrak{G}_3 = \mathfrak{Z}_3 / \mathfrak{Z}_5$

$$(28)$$

where $\quad \mathfrak{Z}_3 = 0.09403 \, m^{(i)} \, \delta^{(i)3} + 0.02914 \, \delta^{(i)3} \, m^{(i)} \, \dfrac{d \, m^{(i)}}{d \, t} \cdot \dfrac{\delta^{(i)2}}{2}$

and $\quad \mathfrak{Z}_5 = 0.2306 \, m^{(i)} + 0.2679 \, m^{(i)} \cdot \dfrac{d \, m^{(i)}}{d \, t} \cdot \dfrac{\delta^{(i)2}}{2}$

$$(29)$$

and where the other $\mathfrak{Z}$'s are functions of (t) which are not required to be further specified here, because they will not have any effect on the course of the subsequent derivations. It may easily be checked at this point that for t lying in the range $-\infty \le t \le 0$, then it is true that $\mathfrak{G}_0 (t) > 0$ and also $\mathfrak{G}_3 (t) > 0$.

Inasmuch as the generalized thickness of the boundary layer is certainly finite, it seems natural to suppose that "a" must remain finite also as $t \to -\infty$; if this condition is agreed upon, then it turns out that the solution to eq. (27) may be written down as

$$a \, \delta^{(i)} = \mathfrak{G}_3 \, m^{(i)2} \, \frac{d \, F^*}{d \, t} + \cdots$$

$$= - m^{(i)2} \, \mathfrak{G}_3 \left(1 + \frac{\gamma - 1}{2} \, M_e^{(i)2} \right) \cdot \varepsilon^{1+n} \, R_s{}^m \cdot \frac{d \, F}{d \, X^*} + \cdots \qquad (30)$$

where $\quad X^* = \varepsilon^n \, R_s{}^m \cdot \dfrac{X^{(i)}}{X_s^{(i)}}$

and where the terms dropped out of the sequence are only of the order of magnitude of ε or less.

5. Conditions at the point of separation

At the separation point where $t = 0$ and $x = x_s$, it must hold true that

$$g_{\eta\eta} (0,0) + f_{\eta\eta} (0,0) = 0$$

and thus it is necessary that for this point one should have

$$\mathfrak{G}_3 (0) \left(1 + \frac{\gamma - 1}{2} \, M_s^{(i)2} \right) \varepsilon^{1+n} \, R_s{}^m \left(\frac{d \, F}{d \, X^*} \right)_{X^* = X_s{}^*} =$$

$$= \frac{4}{3 \sqrt{\pi}} \, \delta_s^{(i)} \left[1 + \frac{\delta_s^{(i)2}}{2} \cdot \left(\frac{d \, m^{(i)}}{d \, t} \right)_{x = x_s} \right]. \qquad (31)$$

The terms on the left hand side of eq. (31) which multiply $\varepsilon^{1+n} \, R_s{}^m$ are of the order of magnitude of unity, and inasmuch as the order of magnitude of the terms on the right hand side is also unity, it follows that $\varepsilon = O\left(R_s^{-\frac{m}{1+n}}\right)$. Making use of this deduction it will be permissible to say that

$$\varepsilon = E_0\,R_s^{-\frac{m}{1+n}} \quad \text{where} \quad E_0 \equiv O\,(1).$$

Then eq. (31) may be recast as

$$\mathfrak{G}_3\,(0)\left(1 + \frac{\gamma - 1}{2}\,M_{s^{(i)2}}\right)E_0^{1+n}\left(\frac{d\,F}{d\,X^*}\right)_{X^* = X_s^{\,*}} =$$
$$= \frac{4}{3\,\sqrt{\pi}}\,\delta_s^{(i)}\left[1 + \frac{\delta_s^{(i)2}}{2}\left(\frac{d\,m^{(i)}}{d\,t}\right)_{t=0}\right]. \tag{32}$$

This relationship thus constitutes the condition which must be satisfied at the separation point between the generalized thickness, $\delta^{(i)}$, and the change in the angular deflection factor F.

6. The displacement thickness of the boundary layer

In order to obtain the pressure distribution existing along the airfoil profile and thus in turn to obtain the value of F, it is necessary to find the component of velocity taken in the direction of the y-axis at locations along the external edge of the boundary layer; this may be accomplished on the basis of the knowledge about the behavior of the u-component in the boundary layer as considered in the previous articles.

Now it is well known, see [3] for example, that

$$\frac{v_e}{u} \cong \frac{d\,\delta^*}{d\,x} - \frac{\delta - \delta^*}{\varrho^{(i)}\,u^{(i)}} \cdot \frac{d}{d\,x}\,(\varrho_e\,u_e) \tag{33}$$

to the accuracy of terms of the order of magnitude of ε, and where δ is the boundary layer's physical thickness while δ^* represents the displacement thickness which is conventionally defined as

$$\delta^* = \int_0^\infty\left(1 - \frac{\varrho\,u}{\varrho_e\,u_e}\right)d\,y\,. \tag{34}$$

It may, therefore, be seen that

$$\delta^* = \lim_{y\to\infty}\int_0^y\left(d\,y - \frac{\varrho\,u}{\varrho_e\,u_e}\,d\,y\right) = \lim_{y\to\infty}\left(y - \frac{\varrho_s^{(i)}\,\sqrt{\nu_s^{(i)}}}{\varrho_e\,u_e} \cdot \Psi\right). \tag{34'}$$

On the other hand, the STEWARTSON'S transformation gives that

$$y = \frac{c_s^{(i)}}{c_e} \cdot \varrho_s^{(i)}\,\sqrt{\nu_s^{(i)}}\int_0^Y\frac{d\,Y}{\varrho} =$$
$$= \frac{c_s^{(i)}}{c_e}\,\sqrt{\nu_s^{(i)}} \cdot \frac{p_s^{(i)}}{p_e} \cdot \frac{T_e}{T_s^{(i)}}\left[Y + \frac{\gamma - 1}{2} \cdot \frac{u_e^2}{c_e^2}\int_0^Y\left(1 - \frac{U^2}{U_e^2}\right)d\,Y\right]. \tag{35}$$

After a great deal of mathematical manipulation and replacement of terms in these eqs. (34') and (35) by expressions obtained earlier it will be found that the displacement thickness may be written as

$$\frac{\delta^*}{X_s^{(i)}} = \frac{c_s^{(i)}}{c_e} \cdot \frac{p_s^{(i)}}{p_e} (R_s^{(i)})^{-1/2} \left(\frac{X^{(i)}}{X_s^{(i)}}\right)^{1/2} \left[\frac{1 + \dfrac{\gamma-1}{2} M_s^{(i)2}}{1 + \dfrac{\gamma-1}{2} M_e^{(i)2}}\right] \cdot$$

$$\cdot \left\{ m^{(i)} \delta^{(i)3} \left(\mathfrak{R}_1 + 2 \cdot \frac{\gamma-1}{2} M_e^{(i)2} \mathfrak{R}_3\right) \frac{X^{(i)}}{X_s^{(i)}} \varepsilon^{1+n} R_s^m \left(1 + \frac{\gamma-1}{2} M_e^{(i)2}\right) \frac{dF}{dX^*} + \right.$$

$$\left. + \mathfrak{L}_1 + \frac{\gamma-1}{2} M_e^{(i)2} \mathfrak{L} \right\}$$

(36)

which is correct at least to terms of the order of magnitude of ε. The terms $\mathfrak{R}_1$ and $\mathfrak{R}_3$ appearing herein are defined as

$$\mathfrak{R}_1 = 0{,}7672 \frac{\mathfrak{G}_3}{\delta^{(i)3}} - 0{,}2216$$

and

$$\mathfrak{R}_3 = 0{,}498 \frac{\mathfrak{G}_3}{\delta^{(i)3}} - 0{,}1578 + \frac{d\,m^{(i)}}{d\,t} \cdot \frac{\delta^{(i)2}}{2}\left(0{,}056 \frac{\mathfrak{G}_3}{\delta^{(i)3}} - 0{,}1457\right)$$

(37)

while the quantities $\mathfrak{L}$ and $\mathfrak{L}_1$ are functions of t which will not be written out explicitly here inasmuch as they will not enter into the subsequent derivations. For future reference it may be worth noting, however, that $\mathfrak{L}(t) \equiv O(1)$, and likewise $\dfrac{d\,\mathfrak{L}(t)}{d\,t} \equiv O(1)$, and in addition it is also true that $\mathfrak{L}_1(t) \equiv O(1)$.

7. Determination of the angular deflection factor, F

By reference to eq. (36) and recalling the comments anent the order of magnitude of ε made in connection with eq. (32) the following approximation analysis may be made. The order of magnitude of $\delta^*/X_s^{(i)}$ is equivalent to $R_s^{-1/2}$, and so is that of δ itself. Moreover, the order of magnitude of $d\,(\varrho_e u_e)/d\,x$ is equivalent to $\varepsilon^{1+n} R_s^m$, or, that is to say, of the order of magnitude of unity. Consequently it may be stated that

$$\frac{\delta - \delta^*}{\varrho^{(i)} u^{(i)}} \cdot \frac{d\,(\varrho_e u_e)}{d\,x} \equiv O(R_s^{-1/2}).$$

(38)

On the other hand, it may be easily checked that

$$\frac{d\,\delta^*}{d\,x} \equiv O(\varepsilon^{1+2n} R_s^{2m-1/2}) \equiv O(R_s^{-1/2}/\varepsilon)$$

(39)

so that this latter term dominates the one given in eq. (38). It is clear, therefore, that in first approximation it is legitimate to say

$$\frac{v_e}{u^{(i)}} \cong \frac{d\,\delta^*}{d\,x}\,. \qquad (40)$$

Without introducing any greater degree of inexactness then, it may also be noted that an explicit expression for this derivative is

$$\frac{d\,\delta^*}{d\,x} \cong C\,\frac{c_s^{(i)}}{c_e^{(i)}} \cdot \frac{p_s^{(i)}}{p_e^{(i)}}\left(1 + \frac{\gamma-1}{2}\,M_s^{(i)2}\right)\frac{c_e^{(i)}}{c_s^{(i)}} \cdot \frac{p_e^{(i)}}{p_s^{(i)}}\,m^{(i)}\,\delta^{(i)3}\,.$$

$$\cdot\left(\Re_1 + 2 \cdot \frac{\gamma-1}{2}\,M_e^{(i)2}\,\Re_3\right)\varepsilon^{1+2\,n}\,R_s^{2\,m-1/2}\,.$$

$$\frac{d}{d\,X^*} \cdot \left[X^{*3/2}\,(\varepsilon^n\,R_s^m)^{-3/2}\,\frac{d\,F}{d\,X^*}\right] \qquad (41)$$

so that the elimination of v_e between eqs. (8) and (41) results in the relation

$$\varepsilon\,F\,(M_e^{(i)2}-1)^{1/2} = C\left(1 + \frac{\gamma-1}{2}\,M_s^{(i)2}\right)m^{(i)}\,\delta^{(i)3}\,.$$

$$\cdot\left(\Re_1 + 2 \cdot \frac{\gamma-1}{2}\,M_e^{(i)2}\,\Re_3\right)\cdot$$

$$\cdot\,\varepsilon^{1+2\,n}\,R_s^{2\,m-1/2}\,\cdot\,\frac{d}{d\,X^*}\left[X^{*3/2}\,(\varepsilon^n\,R_s^m)^{-3/2}\,\frac{d\,F}{d\,X^*}\right]. \qquad (42)$$

Now inasmuch as the quantity $X^{*3/2}\,(\varepsilon^n\,R_s^m)^{-3/2}$, which is equivalent to $(X/X_s)^{3/2}$, is of the order of magnitude of unity, it may be seen by reference to eq. (42) that one must admit that

$$\varepsilon^{2\,n}\,R_s^{2\,m-1/2} \equiv O\,(1) \qquad (43)$$

and consequently it follows that

$$\varepsilon \equiv O\left(R_s^{-\frac{(2\,m-1/2)}{2\,n}}\right). \qquad (43')$$

By comparison with what was stated earlier in eq. (32), it may then be deduced that

$$\frac{2\,m-1/2}{2\,n} = \frac{m}{1+n} \qquad (44)$$

or simply that $m = \dfrac{n+1}{4}\,.$ \qquad (45)

It then follows that

$$\varepsilon^n\,R_s^m \equiv O\,(R_s^{1/4}) \;\Big\}$$
$$\varepsilon = E_0\,R_s^{-1/4} \;\Big\} \qquad (46)$$

or

At this juncture it is worth mentioning that the result just elicited may be used to gain a more precise appreciation of the kind of approximation which was resorted to when the non-linear terms were dropped in the derivation of eq. (20). In fact, it is now possible to estimate the value of the term

$$\varepsilon \cdot \varepsilon^{n/2} \cdot R_s{}^{m/2} \equiv O\left(\varepsilon^{1/2}\right)$$

so that it is plainly evident that the linearization previously resorted to was quite legitimate.

Continuing with the primary objective in mind, the elucidation of the nature of the F function, it may now be shown that eq. (42) is convertible to the form

$$\frac{d}{d X^*} X^{*3/2} \cdot \frac{d F}{d X^*} = \mathfrak{A} F \tag{47}$$

where

$$\mathfrak{A} = \frac{R_s{}^{3/8}}{C E_0{}^{n/2}} \cdot \frac{(M_e^{(i)2} - 1)}{\left(1 + \dfrac{\gamma - 1}{2} M_s^{(i)2}\right) m^{(i)} \delta^{(i)3} \left(\mathfrak{R}_1 + 2 \cdot \dfrac{\gamma - 1}{2} M_e^{(i)2} \mathfrak{R}_3\right)} \cdot \tag{47'}$$

This equation may be made less cumbersome by subjecting it to the transformation defined as

$$d \mathfrak{z} = - \frac{d X^*}{X^{*3/2}} \tag{48}$$

which merely means that $\mathfrak{z} = 2 X^{*-1/2}$
so that now

$$\frac{d^2 F}{d \mathfrak{z}^2} = 8 \mathfrak{A} \cdot \mathfrak{z}^{-3} \cdot F. \tag{49}$$

If it is further agreed that $\mathfrak{A}_s$ is to stand for the value that $\mathfrak{A}$ takes on when $X^* = X_s^*$ (and thus when $t = 0$) a more standard form for eq. (49) may be concocted, namely,

$$\mathfrak{z}^3 \frac{d^2 F}{d \mathfrak{z}^2} = 8 \mathfrak{A}_s \left[F - \left(1 - \frac{\mathfrak{A}}{\mathfrak{A}_s}\right) F \right] \tag{49'}$$

which may be solved by applying the method of Fubini [4].

When such a solution is carried out, it will be found that at large values of the argument $\sqrt{16 \mathfrak{A}_s} \cdot X^{*1/4}$ the general integral for eq. (49') will be expressible as

$$F = B_1 \cdot \sqrt{\mathfrak{z}} \cdot I_1\left(2 \sqrt{8 \mathfrak{A}_s} \cdot \mathfrak{z}^{-1/2}\right) + B_2 \sqrt{\mathfrak{z}} K_1\left(2 \sqrt{8 \mathfrak{A}_s} \cdot \mathfrak{z}^{-1/2}\right) \tag{50}$$

where I_1 and K_1 are the modified Bessel functions of the first and second kinds, respectively.

Since $\mathfrak{z} = \infty$ when $X^* = 0$, and provided it is assumed that F must remain finite for $X = 0$, it may be presumed that it must follow that

$B_2 = 0$. Upon imposition of the boundary condition that F must equal unity when $x = x_s$, or that is to say, for $X^* = X_s^*$, it is found that

$$F = \left(\frac{X_s}{X}\right)^{1/4} \cdot \frac{I_1(\sqrt{16\,\mathfrak{A}_s}\,X^{*1/4})}{I_1(\sqrt{16\,\mathfrak{A}_s}\,X_s^{*1/4})} \,. \tag{51}$$

For large values of the argument it is well known that the modified Bessel function of the first kind may be approximated by a single term, giving

$$F = \left(\frac{X_s}{X}\right)^{3/8} \cdot e^{-\sqrt{16\,\mathfrak{A}_s}\cdot X_s^{*1/4}\left[1-\left(\frac{X}{X_s}\right)^{1/4}\right]} \,. \tag{51'}$$

8. Determination of the factor E_0

Returning to the condition that must be satisfied at the separation point, as embodied in eq. (31'), the value of E_0 appearing therein may be determined by replacing the derivative of F which enters this relationship by the formula for $\left(\dfrac{dF}{dX^*}\right)_{X^* = X_s^*}$ immediately obtainable from eq. (51') above. When this is done, it turns out that

$$\mathfrak{G}_3(0)\left(1 + \frac{\gamma-1}{2}\,M_s^{(i)2}\right)^{1/2}(M_s^{(i)2} - 1)^{1/4}\,E_0 =$$

$$= C^{1/2}\,\delta_s^{(i)3/2}\left(\mathfrak{R}_1 + 2\cdot\frac{\gamma-1}{2}\,M_s^{(i)2}\,\mathfrak{R}_3\right)^{1/2}\cdot\frac{4\,\delta_s^{(i)}}{3\sqrt{\pi}}\left[1 + \frac{\delta_s^{(i)2}}{2}\left(\frac{d\,m^{(i)}}{d\,t}\right)_{t=0}\right] \tag{52}$$

and this expression thus allows one to evaluate the factor E_0.

For the simple situation wherein the profile shape is assumed to be that of a flat plate set at an angle of attack of α radians, it may be readily computed that

$$E_0 = \frac{1.7839}{4.8276}\cdot\frac{C^{1/2}}{(M_s^{(i)2} - 1)^{1/4}} \,. \tag{52'}$$

Once having obtained the value of E_0, it is then possible to obtain directly the value of the pressure which exists at the point of detachment, because for this location

$$\frac{\Delta p}{p_s^{(i)}} = \frac{p_s - p_s^{(i)}}{p_s^{(i)}} = \gamma\,M_s^{(i)2}\cdot\varepsilon \,. \tag{53}$$

9. Behavior of the separated flow—Fundamental equations and hypotheses

By means of the analysis given in the preceding articles, it is now known what the velocity distribution will be through the boundary layer at the beginning of the separated jet, provided one takes the position S of the point of detachment of this layer to be known, i.e., provided the value of X_s/L is given. In fact, by referring back to eqs. (14) and (17), the velocity distribution in question may be defined by the relation

$$\frac{U}{u_s^{(i)}} = \left(\frac{\partial f}{\partial \eta} + \frac{\partial g}{\partial \eta}\right)_{t=0}. \tag{54}$$

The direction that the velocity vector aligned with the streamline at the outer edge of the boundary layer makes with the free-stream velocity $\vec{U}_\infty$ may be denoted by the angle

$$\varDelta = \alpha + \beta_s - \frac{(v_e)_s}{u_s^{(i)}}$$

where α is the angle of attack of the airfoil chord with respect to the free-stream velocity vector $\vec{U}_\infty$, and where β_s is the local angle that the tangent to the airfoil contour makes with the chordline. This angle is known, provided the detachment point is specified.

It would appear to be feasible to undertake the study of the behavior of the flow existing downstream of the separation point then, by application of the concept that this fluid motion will have the characteristics of a jet. The nature of this jet will be such that the velocity distribution existing across it at the initial cross-section, corresponding to the separation point, will be identical with the distribution of velocities just defined in eq. (54). It seems natural to postulate that the zeroeth streamline, $\varPsi = 0$, will detach itself from the surface of the airfoil at the point S and will continue on in the direction of the velocity which exists at the outer edge of the boundary layer at the point S. The tangent to this particular streamline at the point S may be selected as a convenient new axis of abscissae, X, while the origin of coordinates may be selected to fall at the separation point itself.

Inasmuch as the pressure distribution along the edge of the separated jet will be prescribed by a known law which will be presented a little further on, the distribution of velocities, U_e, which will exist along this jet boundary may, consequently, be considered to behave in a precisely defined manner.

In order to obtain the velocity profile in the jet, therefore, one may proceed firstly by applying to eq. (12) the von Mises transformation, so that the momentum equation now appears as

$$\frac{\partial U^2}{\partial X} = U \frac{\partial^2 U}{\partial \varPsi^2} + U_e \frac{d U_e}{d X}. \tag{55}$$

Let a new transformation of coordinates be effected by setting

$$U = u_s^{(i)} U^*; \quad X = X_s \cdot X^{**}; \quad \text{and} \quad Y = \eta^{**}\sqrt{X_s/u_s^{(i)}}$$

and make the substitutions

$$\varPsi = \varPhi(X^{**}, \eta^{**})\sqrt{X_s u_s^{(i)}}; \quad U_e^* = U_e/u_s^{(i)}; \quad \text{and} \quad Z = U_e^{*2} - U^{*2}$$

so that then eq. (55) becomes converted to

$$\frac{\partial Z}{\partial X^{**}} = \frac{\sqrt{U_e^{*2} - Z}}{2} \cdot \frac{\partial^2 Z}{\partial \Phi^2} \tag{56}$$

and the equivalent boundary conditions may be stated now as

$$Z = Z_0(\Phi) \quad \text{for} \quad X^{**} = 0 \tag{57}$$

where $Z_0(\Phi)$ represents a known function of Φ and also

$$\left.\begin{array}{ll} Z_0 = 0 & \text{for } \Phi = +\infty \\ Z = U_e^{*2} & \text{for } \Phi = \Phi_0(X^{**}) \end{array}\right\} \tag{57'}$$

while

where $\Phi = \Phi_0(X^{**})$ denotes the line which separates the (Φ, X^{**})-plane into a nether region (which is not disturbed by the jet) from the outer region occupied by the jet. This line of division is not known beforehand, but a suitable assumption will be made later on concerning its location and the justification for its selection will then be presented.

It may be observed at this juncture that along the zeroeth streamline, $\Psi = 0$, the value of U^* is only going to be zero at the point of detachment S. Except for a small region near this separation point, the value of U^* will not actually differ greatly from the quantity U_e^*, which is the value of the velocity which exists at the edge of the jet. On the other hand, it must be recognized that the suppositions now being made are not really going to be valid for describing the flow close to this region in the neighborhood of S.

Thus, it may be taken for granted for present purposes that it is permissible to simplify the coefficient of the $\partial^2 Z / \partial \Phi^2$ term appearing in eq. (56) by setting $U_e^* = U^*$. This method of attack will be justifiable provided one does not approach too close to the aforementioned zone close to S, and provided it is merely the flow in the region external to the streamline $\Phi = 0$ which is of interest. It will be useless to try to improve the degree of approximation being made concerning the behavior of the flow in this intractable region near S, so long as one does not relinquish any of the special assumptions now made nor allow alterations in the hypotheses underlying the boundary-layer type flow under consideration.

Consequently, on the basis of the agreements now proposed, one may write in place of eq. (56) the following simplified version

$$\frac{\partial Z}{\partial \xi} = \frac{1}{2} \frac{\partial^2 Z}{\partial \Phi^2} \tag{58}$$

where the transformation

$$\xi = \int_0^{X^{**}} U_e^* \, dX^{**} \tag{58'}$$

has also been introduced.

10. Solution to the integral equation, eq. (58), governing the velocity distribution in the wake

In order to obtain the integral solution which satisfies the equation of motion just derived as eq. (58) and which obeys the stipulation to be invoked at the boundaries, as expressed by eqs. (57) and (57′), it may be observed, first of all, that the following integral may be considered as part of the sought answer, inasmuch as it will obey the differential equation and will likewise satisfy eq. (57) and the first part of eq. (57′); this suggested partial solution is

$$Z_1(\xi,\,\Phi) = \frac{1}{\sqrt{\pi}} \int\limits_{-\infty}^{\infty} e^{-\beta^2} Z_0 [\Phi - \sqrt{2\xi} \cdot \beta] \, d\beta. \tag{59}$$

Note that on the streamline $\Phi = \Phi_0(\xi)$ the value of this partial solution is particularized to read

$$Z_1[\xi;\,\Phi_0(\xi)] = \frac{1}{\sqrt{\pi}} \int\limits_{-\infty}^{\infty} e^{-\beta^2} Z_0[\Phi_0(\xi) - \sqrt{2\xi} \cdot \beta] \, d\beta. \tag{60}$$

The assumption may now be introduced that $|\Phi_0(\xi)| >> 1$, so that it is possible to impose the condition that

$$Z_1[\xi;\,\Phi_0(\xi)] = (U_e{}^{*2})_s \tag{61}$$

where the notation $(U_e{}^{*2})_s$ signifies that the value of $U_e{}^{*2}$ which applies at the separation point S is the one meant. The stipulation may be considered fully justifiable for large enough values of ξ, but it is also taken to be true for $\xi = 0$.

Now in order to concoct a solution which will satisfy all the stipulations, i.e., in order to ensure that the requirement set down as the second part of eq. (57′) is fulfilled, it will be found necessary to add to the Z_1 integral just defined, another function having the following characteristics:

This additional integral, call it Z_2, is to satisfy likewise the differential eq. (58) but it is to vanish for $\xi = 0$ and for $\Phi = \infty$, while when $\Phi = \Phi_0(\xi)$ the value of Z_2 must be given by

$$Z_2[\xi;\,\Phi_0(\xi)] = U_e{}^{*2} - (U_e{}^{*2})_s. \tag{62}$$

Consequently, the sought second part of the solution is to have the form

$$Z_2 = \frac{1}{\sqrt{\pi}} \int\limits_{0}^{\xi} \mu(\xi') \cdot \frac{\Phi - \Phi_0(\xi')}{(\xi - \xi')^{3/2}} \cdot e^{-1/2\,[\dot\Phi - \Phi_0(\xi')]^2 / (\xi - \xi')} \cdot d\xi'$$

where $\mu(\xi')$ is a function which is not specified beforehand, but it will be determined as the solution to the following nonhomogeneous Volterra integral equation of the second kind [5]

$$2\,\mu\,(\xi) + \frac{1}{\sqrt{\pi}} \int_0^{\xi} \mu\,(\xi') \cdot -\frac{\varPhi_0\,(\xi) - \varPhi_0\,(\xi')}{(\xi - \xi')^{3/2}}\, e^{-\,1/2\,[\varPhi_0\,(\xi) - \varPhi_0\,(\xi')]^2\,/\,(\xi - \xi')} \cdot d\,\xi'$$
$$= U_e{}^{*2} - (U_e{}^{*2})_s.$$

The sought complete solution to eq. (58) which obeys all the boundary conditions is thus
$$Z = Z_1 + Z_2.$$

11. Kind of velocity distribution premised along edge of boundary layer

The solution for the behavior of the separated flow depends on how the velocity, $U_e{}^*$, varies along the edge of the boundary layer. In order to gain some insight into this question, it is worth noting the following features which characterize the flow as it passes downstream of the separation point S.

The pressure distribution along the airfoil is going to remain practically constant downstream of S, and likewise it will thus be constant

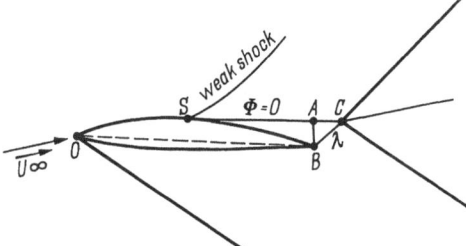

Fig. 1. Geometric relationships subsisting between the airfoil, the separated jet, the zeroeth streamlines, and the corresponding shocks about an airfoil from which the laminar boundary layer has separated

along and within the detached jet; this pressure may be taken as the one which is attained at the separation point where $p = p_s$. Now consider the thin stratum of air which is scouring along the under surface of the airfoil, and in particular observe that the streamline, call it λ, which constitutes the outer edge of this lower-surface boundary-layer will have to expand when it gets to the trailing edge of this surface (denoted by the letter B in fig. 1) in order to attain the pressure p_s.

This lower-surface streamline, λ, will intersect the zeroeth streamline coming off the upper surface of the airfoil at the point C, as depicted in fig. 1 also. For present intents and purposes this point C may be considered as the intersection of the streamline λ with the streamline which marks the outer edge of the detached jet issuing from the upper surface. Again, in way of approximation, it may be supposed that this outer boundary of the detached jet is a straight line, having a direction which is aligned with the jet boundary where it leaves the airfoil at the point S.

At the point C there will be a compression shock formed in that part of the flow sweeping past the bottom surface of the airfoil. Quite analogously, another compression shock will be formed, extending away from the detached jet, out into the field of flow that sweeps over the upper side of the airfoil. Now, because of the relatively small thickness of the jet, it seems that it would be permissible to assume that the two layers of flow from off the ventral and dorsal sides of the airfoil will be in actual contact at the point C, and, consequently, the pressure and velocity existing immediately downstream of these shocks may be calculated on the basis of this idealization. The common value for the pressure existing downstream of the shocks is thus to be denoted by p_c, which is the value of the pressure producted at the point C. This pressure may be considered as known, therefore, provided p_s is known.

On the other hand, it may be appreciated that the velocity must come to stagnation as it travels downstream along the zeroeth streamline and reaches the point C, because the velocity is subsonic on this $\Phi = 0$ streamline upstream of the point C, and is subjected to a change of direction at C, which aligns it nearly parallel to the free-stream. Of course, further substantiation for the arresting of the velocity to stagnation conditions at point C may be found by considering the unacceptable implication of an alternate schematization. Observe first of all that the fluid which courses along below the zeroeth streamline and constituting the jet, as well as the vein of retarded flow which comes from the ventral surface of the airfoil, cannot become commingled in the potential flow downstream of C. If these two viscous wakes joint at C, they can only make a sharp obtuse-angled reversal, turning in towards each other and merging together as a retrograde wake, directed back towards the airfoil. In this way the sort of circulatory backwater-type of flow regularly observed in the region BSC is set up. If it were to be presumed, consequently, that the flow did not become subsonic at C on the zeroeth stream-line, then at least a part of the jet wake would be supersonic, interior to the zeroeth streamline, and since it would have to be subjected to the sharp change in direction just mentioned, the subsequent accelerated pattern of velocities would be completely incompatible with the commonly observed behavior of the retarded retrograde flow in the back-water region.

Even though the flow is agreed to be subsonic on $\Phi = 0$ near C, it is still necessary to explain how the pressure can increase from p_s to p_c as one progresses downstream along the exterior edge of the jet; this change may be ascribed to the viscous reactions occurring in this region. To give some precision to a discussion of this situation it is convenient to proceed as follows. Let a normal be drawn to the $\Phi = 0$ streamline and passing through B. This normal will intersect the streamline $\Phi = 0$ at the point which may be labeled A. Now on the stretch of the zeroeth streamline from S to A the pressure may be assumed to remain constant, to conform with the straight segment of the jet boundary being postulated. From location A to location C, however, there will be a rise of pressure experienced; i.e., it goes from the p_s value at A to the p_c value at C, where the velocity has come to rest by following along the streamline $\Phi = 0$. Thus, in order to justify this latter assumption, one must have recourse to the idea that the viscous stresses developed in the wedge-shaped region $AC : BC$ are the agency through which the pressure change is brought about. This concept of the local flow near point C thus controverts the assumption of a straight zeroeth streamline upon which the picture of the local pressure variation near C was built up, but such a contradiction will actually have very little influence on the main pattern of the flow, which may now be treated in analytic fashion on the basis of the understandings thus arrived at.

12. Determination of the length SC of the separated jet

In order to find the distance downstream that the point C lies below the separation point S, it is first necessary to find the value of the velocity at the outer edge of the boundary layer (now the jet) which corresponds to the point C; let this value of the velocity be denoted by $(U_e{}^*)_c$. Consequently, on the basis of the deductions made in article 10, it may now be stated that

$$(U_e{}^{*2})_c = Z_1\,(\xi_c, 0) + Z_2\,(\xi_c, 0) =$$

$$= \frac{1}{\sqrt{\pi}} \int\limits_{-\infty}^{\infty} e^{-\beta^2} \cdot Z_0\,(-\sqrt{2\,\xi_c} \cdot \beta)\, d\beta + Z_2\,(\xi_c, 0)\,. \tag{64}$$

The value of $(U_e{}^{*2})_c$ is to be obtained under the hypothesis that, as premised in article 11, the pressure is the same at the point C whether it is considered as occurring on the part of the streamtube coming off of the upper surface or that coming off of the lower surface of the airfoil; thus the pressure at the juncture of these two veins of boundary layer air is to be assigned the common value, p_c. With this stipulation, it may then be assumed that

$$\frac{(U_e{}^{*2})_c}{(U_e{}^{*2})_s} = \frac{1 - \dfrac{2}{M_s{}^2}\left[\left(\dfrac{p_c}{p_s^{(i)}}\right)^{\frac{\gamma-1}{\gamma}} - 1\right]}{1 + (\gamma - 1)\left[\left(\dfrac{p_c}{p_s^{(i)}}\right)^{\frac{\gamma-1}{\gamma}} - 1\right]}\,. \tag{65}$$

It is now worth remarking that the contribution to the $(U_e{}^*)_c$ velocity which is indicated by the $Z_2\,(\xi_c, 0)$ term is going to be very small because the increment in pressure which takes place from A to C is going to occur principally in the neighborhood of the point C, while the pressure is considered to remain constant all along the straight line from S to A, as pointed out previously. Thus it would appear to be legitimate to neglect the Z_2 $(\xi_c, 0)$ term entirely, and consequently one obtains the following two relationships for the velocity:

$$
(U_e^{*2})_c = (U_e^{*2})_s \cdot \frac{1 - \dfrac{2}{M_s^2}\left[\left(\dfrac{p_c}{p_s^{(i)}}\right)^{\frac{\gamma-1}{\gamma}} - 1\right]}{1 + (\gamma - 1)\left[\left(\dfrac{p_c}{p_s^{(i)}}\right)^{\frac{\gamma-1}{\gamma}} - 1\right]} =
$$

$$
= \frac{1}{\sqrt{\pi}} \int\limits_{-\infty}^{\infty} e^{-\beta^2} \cdot Z_0 \left(-\sqrt{2\,\xi_c} \cdot \beta\right) d\beta .
$$

(66)

From this relationship, consequently, one may determine ξ_c, and this is tantamount to finding X_c, inasmuch as $\xi_c \cong X_c/X_s$.

13. Determination of the location of the separation point

The location of the separation point, S, may be discovered by carrying through the following procedure which is admittedly not exactly precise but at least of tolerable accuracy. From what has been done so far, therefore, the following two graphs may be plotted:

a) Thanks to eq. (66) just derived, one may set down a series of values for the abscissae of the point C, as a function of selected abscissa values corresponding to point S, where the abscissae are measured along the chord of the airfoil section, starting from an origin at the point B.

b) Thanks to a geometric construction based on the expansion process which takes place at point B, another set of abscissae may be derived for the point C, as a function of selected abscissae values corresponding to point S and measured in the same coordinate system. This construction requires that C be obtained as the point of intersection of two rays, one issuing from the point S and constituting the X-axis as the straight edge of the jet, and the other issuing from the point B. The ray coming from B may be imagined to represent the streamline which constitutes the Prandtl-Meyer expansion boundary for the stratum of fluid which has coursed along below the airfoil and has experienced a sharp turn going about the corner B in order to attain the pressure p_s, the common value for the separated region.

Thus from the intersection of these two plots (a) and (b), one obtains the sought separation-point location for a given angle of attack, α, and for a given Mach number of the free-stream, M_∞.

14. Numerical application

The method developed in the foregoing articles has been applied to the very elementary case of a flat-plate airfoil set at an angle of attack α with the respect to the free-stream velocity vector $\overrightarrow{U}_\infty$. The angles of attack selected for purpose of illustration happen to be $\alpha = 5°$, $10°$, and

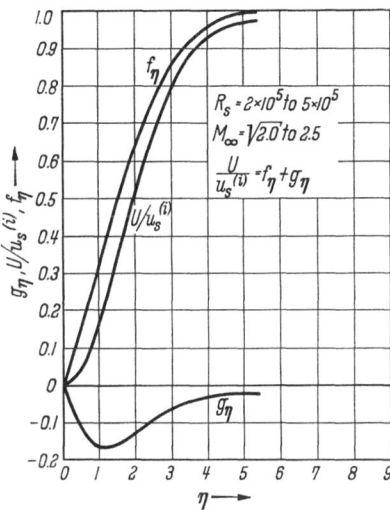

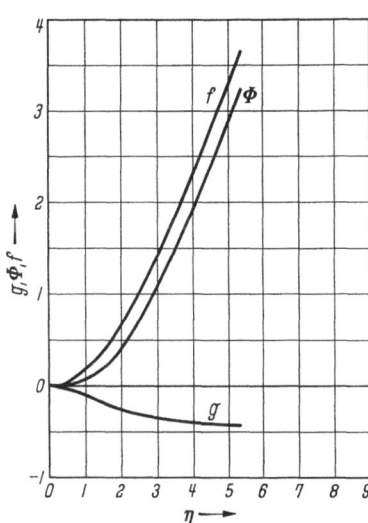

Fig. 2. Reduced-velocity contributions at the separation point together with total-velocity ratio in the boundary layer

f_η = velocity due to external pressure distribution with no interference

g_η = velocity due to the growth of boundary layer with interference from shock wave

$U/u_s(i) = f_\eta + g_\eta$

Fig. 3. Reduced stream-function contributions from external pressure distribution and from shock wave interference together with total Φ-value existing across the boundary layer at the separation point

f = stream function due to external pressure distribution with no interference

g = stream function due to growth of boundary layer with interference from shock wave

$\Phi = f + g$

$15°$. For each one of these three angles of attack the analysis has been carried out for three Mach numbers and for a pair of Reynolds numbers; these selected values are:

$$M_\infty = \sqrt{2},\ 2,\ \text{and}\ 2.5$$

$$\text{while} \quad R_s = 2 \cdot 10^5 \text{ and } 5 \cdot 10^5.$$

The values of the velocities pertaining to the separation-point location that correspond to the derivatives of the two contributions to the reduced stream function that are due to the external pressures (f_η) and due to the growth of the boundary layer itself (g_η), are plotted in fig. 2; the total velocity ratio obtained from these contributions, $U/u_s(i)$, is also shown on this curve. It is found that the differences between these velocities is so small that no noticeable effect can be discerned in the result regardless of which set of starting values of M_∞, R_s, or α is employed in the computations; the single set of curves presented in fig. 2 is thus universally valid for the range of starting values listed above.

Upon integration of these velocity contributions, one obtains the streamfunction components f and g, and the streamfunction Φ itself, since $f + g = \Phi$. This computation

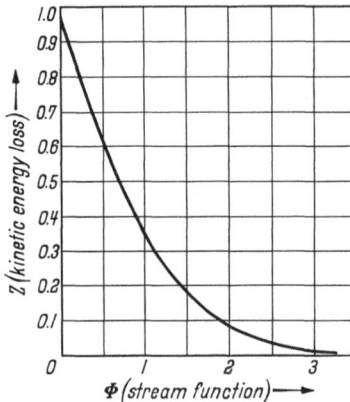

produces results such as illustrated in fig. 3. By reference to these curves, then, one may go on to obtain the value of the kinetic-energy-loss function at the separation point, namely Z_0 (Φ). This dependence of Z_0 on the streamfunction is displayed in fig. 4.

By making use of the values for the kinetic-energy-loss function just given in fig. 4, the corresponding values for the reduced velocity on the zeroeth streamline and applying to the point of confluence of the coelescing retarded-

Fig. 4. Reduced kinetic energy loss at the separation point as a function of the reduced stream function

flow layers from the upper and lower profile faces may be computed by aid of eq. (64); these results are presented in fig. 5, as a function of the distance that C lies away from the separation point. The contribution of the term Z_2 to the values of $(U_e^{*2})_c$ obtained from eq. (64) have been neglected, of course, in this calculation.

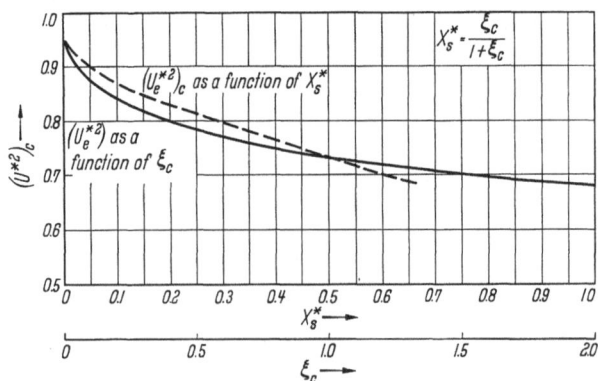

Fig. 5. Reduced velocity on the zeroeth streamline and located at point of intersection of the coalescing upper and lower boundary layers versus distance downstream from the separation point

By substitution in eq. (52′) for E_0, and subsequent evaluation of ε by means of eq. (46), the pressure increment due to the shock wave interference at the point of detachment may then be obtained from eq. (53);

these pertinent values of $\Delta p/p_s^{(i)}$ are given in fig. 6. Finally, now that a working curve of $(U_e^{*2})_c$ vs. X_s^* is available as fig. 5, it is a simple matter to enter this diagram on the ordinate axis with a value of $(U_e^{*2})_c$ obtained from use of a standard gas-dynamics table giving pressures and velocities for a Prandtl-Meyer expansion when the pressures from fig. 6 are known. Upon carrying out this step, the derived theoretical curve for the abscissa distance of the point of detachment, measured upstream from the trailing edge of the airfoil, can be constructed as shown in fig. 7,

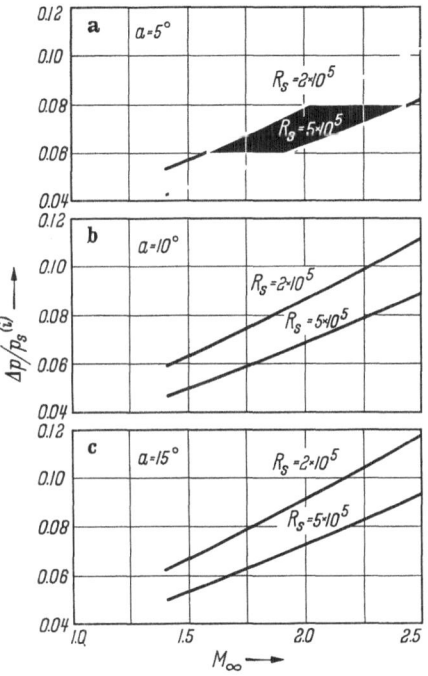

for the case of $M_\infty = 2.0$. The other selected Mach numbers will not actually show much of an alteration to these results because $(U_e^*)_c$ appears to be rather insensitive to Mach number.

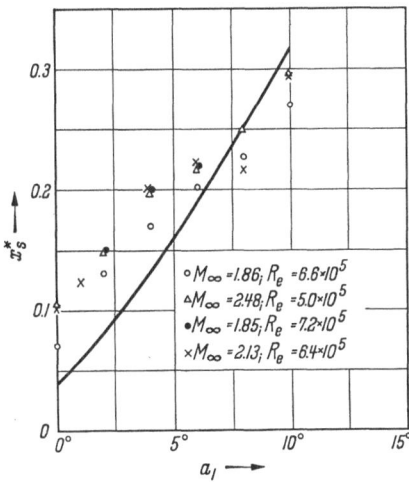

Fig. 6. Pressure increments produced on upper surface of the profile by action of the interference from the local shock wave that initiates the separation of the laminar boundary layer

Fig. 7. Location of the separation point measured from the after end of the profile as a function of angle of attack. Computed results are based on eq. (66). Experimental results pertain to a ten-per-cent-thick circular-arc profile

15. Comparison of the calculated results with experimental findings

The computed results as now obtained and plotted in fig. 7 are not going to be directly comparable to any available experimental data on airfoils with appreciable thickness. In addition, it must be recognized also that in the case of a flat plate there even arise in real flows certain mixing phenomena not accounted for in the analytic idealization resorted to here. Thus, along the rear end of the plate in the vicinity of the jet there exists a retrograde flow along the surface, Σ, of the plate and the mixing process set up between this flow and the jet which has been detached from Σ further upstream introduces a quite different set of physical conditions into the actual picture of the flow from what has been premised in the foregoing treatment.

Despite these dissimilarities, however, it will be instructive to make a comparison between the computed and experimental results, if only to obtain a gross chek on the general trends indicated by the calculations.

Naturally, in order to obtain a rigorous check it would be required that the velocity profile on the actually tested model reproduce in detail the velocities expected of a flat plate, while in particular it must be acknowledged that if a wedge section were tested, the expansion which takes place around the corner at the mid-chord position will have a rather profound local effect in altering the flow from what is assumed in deriving the BLASIUS solution. Nevertheless, it may not be too difficult to imagine that as one moves downstream from the shoulder point on the wedge the boundary layer should tend to re-assume characteristics more nearly comparable to those for a flat plate.

Thus, in order to make a rough comparison of separation distances it is merely necessary to make a suitable adjustment to the angle of attack in order to bring about a meaningful confrontation between the measured results on a wedge and the calculated data. It may be presumed, therefore, that when the angle of inclination of the faces of a double-wedge profile is $5°$ to the chordline (so that the thickness is 8.7%), then the separation distances obtained for a flat plate (when at a given angle of attack α) should be equivalent approximately to the corresponding distances which would be measured on the double-wedge airfoil when it is set at the angle $\alpha - 5°$.

It would also appear that results for a 10% thick circular-arc profile might also be compared to the computed results, after a similar adjustment in angle is introduced. It is recalled that for a circular-arc profile the leading (or trailing) edge half angle is twice that of the double-wedge profile with equivalent thickness ratio, so that the average angle of inclination for the arc's surface over which separation may occur is somewhere nearly the same as the constant angle of the flat face on the rear of a double-wedge profile of the same thickness ratio.

For purposes of a gross check, therefore, it appears legitimate to take the results for a 10% thick circular-arc profile to be approximately equivalent to those which would pertain to a 8.7% thick symmetric double-wedge section, and in turn both of these practical shapes may have a modicum of relevancy for comparison with computed flat-plate separation characteristics. Consequently, in fig. 7 the abscissa values are listed as $\alpha_1 = \alpha - 5°$ where α_1 is the angle of attack (measured with respect to the chordline) for the experimental airfoil and is the shifted angle of attack for the computed flat-plate results. The experimental results for the 10% thick circular-arc profile have been excerpted from [6].

The correspondence between theory and experiment made in this fashion appears to be quite good, because the differences between the computed and measured results are really rather small and the discrepancy which is evident is of the correct sign to be explainable on the basis of the difference between the idealized profile and the thick section actually employed.

A further comparison between the predictions of the theory and the data obtained from wind-tunnel measurements may also be made on the basis of the pressure jumps experienced by the boundary layer flow which are sufficient to produce separation observed; such a confrontation is presented in fig. 8. In this case, too, it is seen that the comparison of

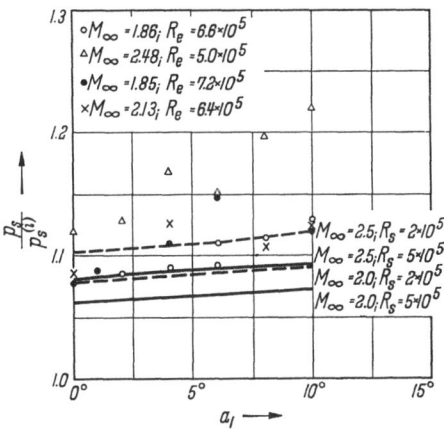

Fig. 8. Comparison of theoretical and experimental results for the pressure jump ratio experienced at the separation point caused by interference from the weak shock wave at that location. Experimental results pertain to a ten-per-cent-thick circular-arc profile

theory and experiment is also quite acceptable. It may seem that the experimental results for a free-stream Mach number of 2.48 deviate quite far from the predictions, but it seems likely that such a large discrepancy between the data at this Mach number and that shown at the other lower Mach numbers may indicate that some doubt could be reserved about their reality, inasmuch as it is difficult to imagine that all this difference could suddenly be brought about by a simple progressive Mach number change. Thus, if an exception is made of the Mach number 2.48 data, the correspondence between theory and experiment is vastly improved.

References

[1] STEWARTSON, K.: On the Interaction Between Shock Waves and Boundary Layers, Proceedings of the Cambridge Philosophical Society, **47**, 545—553, (1951).

[2] TIMMAN, R.: A One Parameter Method for the Calculation of Laminar Boundary Layers, Report No. F-35, Reports and Transactions, Vol. XV, Nationaal Luchtvaartlaboratorium, Amsterdam, 1949.

[3] YOUNG, A. D.: The Calculation of the Profile Drag of Aerofoils and Bodies of Revolution at Supersonic Speeds, Report No. 73, The College of Aeronautics, Cranfield, England, April, 1953.

[4] FUBINI, G.: Studi Asintotici per Alcune Equazioni Differenziali, Rendiconti della Reale Accademia Nazionale dei Lincei, Classe di Scienze Fisiche, Matematiche, e Naturali, **26**, 253—259 (1937), No. 6.

[5] GOURSAT, E.: «Cours d'Analyse Mathématique», Vol. III, Fifth Edition, Gauthier-Villars, Paris, France (1942), pp. 316—319.

[6] ZIENKIWICZ, H. K.: An Investigation of Boundary Layer Effects on Two Dimensional Supersonic Aerofoils, Report No. 49, The College of Aeronautics, Cranfield, England, December, 1951.

The wall boundary layer behind a moving shock wave

By

Harold Mirels

Lewis Flight Propulsion Laboratory, National Advisory Committee for Aeronautics
Cleveland, Ohio

1. Introduction

If a shock wave advances into a stationary fluid bounded by a wall, a boundary layer is established along the wall behind the wave. Some characteristics of this boundary layer are investigated herein. The problem is reduced to a steady state by using a coordinate system wherein the shock is stationary (fig. 1). In this coordinate system the fluid and wall both approach the shock with the same velocity, u_w. The shock reduces the fluid velocity to u_e while the wall velocity is unaffected, resulting in a boundary layer for $x > 0$. The laminar case has been treated in [1] to [8] and the turbulent case has been treated in [3]. The purposes of the present paper are to present modifications of the

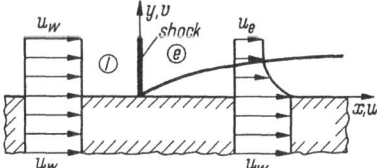

Fig. 1. Coordinate system and velocity boundary layer

solutions of [2] and [3] which make them more applicable for the strong wave case, to investigate further the wall surface temperature variation and to discuss some experimental transition results.

2. Laminar boundary layer

The laminar boundary layer equations for zero pressure gradient and uniform wall surface temperature are applicable [3]. A stream function exists such that

$$\partial \psi / \partial y = \varrho u / \varrho_w \qquad \partial \psi / \partial x = - \varrho v / \varrho_w \tag{1}$$

Introducing new independent variables

$$\xi = x \qquad \eta = \sqrt{u_e / 2 \, v_w \, x} \int_0^y (\varrho / \varrho_w) \, d y \tag{2}$$

and assuming $\psi = \sqrt{2 u_e \xi \, v_w} f(\eta)$ permits the boundary layer equations to be expressed as

$$[C f'']' + f f'' = 0 \tag{3a}$$

$$\left[\frac{1}{\sigma} C g'\right]' + f g' + \frac{u_e^2}{2 H_e}\left[2 C\left(1-\frac{1}{\sigma}\right) f' f''\right]' = 0 \tag{3b}$$

$$f(0) = 0 \qquad f'(0) = U \qquad f'(\infty) = 1$$

$$g(0) = H_w/H_e \qquad g(\infty) = 1 \tag{3c}$$

where $C = \varrho\,\mu/\varrho_w\,\mu_w$, $h =$ static enthalpy, $H =$ stagnation enthalpy $= h + (u^2/2)$, $g = H/H_e$, $U = u_w/u_e$, and $\sigma = \mu c_p/k$. No assumption is made regarding the variation of fluid properties. Note that $f' = u/u_e$. Eqs. (3) are identical with those for the flow over a semi-infinite flat plate except for the nonzero wall velocity herein. The wall shear and heat transfer are

$$\tau_w/u_e\,f''(0) = -\,q_w\,\sigma_w/h'(0) = \sqrt{u_e\,\varrho_w\,\mu_w/2\,x} \tag{4}$$

If $\sigma = 1$, eq. (3b) can be integrated, yielding

$$g - g(0) = [(f' - U)/(1 - U)]\,(1 - g(0)) \tag{5}$$

which is a form of the Crocco relation between stagnation enthalpy and velocity.

For U only slightly greater than one (i.e., weak shocks) eqs. (3) reduce to RAYLEIGH's problem for the impulsive start of an infinite plate (with t replaced by x/u_w) and the solution is well known [2]. If C is a constant, eq. (3a) is independent of eq. (3b). The latter case was integrated numerically in [2] and [3] for $C = 1$; $\sigma = 1, 0.72$; $c_p =$ constant; $U = 1.5, 2.0, 3.0, 4.0, 5.0$, and 6.0. The following interpolation formulas, presented in [3], agree with these numerical results to within 1 percent:

$$\frac{-f''(0)}{U-1} = \frac{h'(0)}{h_r - h_w}\,\sigma - (0.48 + 0.022\,U) = 0.489\,\sqrt{1 + 1.665\,U} \tag{6a}$$

$$h_r/h_e = 1 + (U - 1)^2\,[u_e^2/2\,h_e]\,\sigma^{0.39 - 0.023\,U} \tag{6b}$$

When considering strong shocks, the temperature variation normal to the wall is large and it is desirable to use the proper variation of C and σ. As a first step in this direction, solutions employing the proper variation of C, but with $\sigma = 1$, have been obtained for air at an initial temperature and pressure of $522°$ R and 0.001 atmospheres, respectively, and shock Mach numbers in the range $4 \leq M_s \leq 14$. Conditions across the shock were obtained using thermodynamic charts for air at high temperatures. The wall surface temperature was assumed to remain at $522°$ R for all x. This assumption will be justified in a later section. The expressions

$$C = \frac{1.5481}{\sqrt{h/h_w}} - \frac{0.5481}{h/h_w} + 0.0028\left(\frac{h}{h_w} - 1\right) - 5.74 \cdot 10^{-5}\left(\frac{h}{h_w} - 1\right)^2 \tag{7a}$$

$$\frac{h}{h_w} = \left[1 - \frac{u_e^2}{2 H_e}\,(f')^2\right]\bigg/\left[1 - \frac{u_e^2}{2 H_e}\,U^2\right] \tag{7b}$$

were used to integrate eq. (3a). (Eq. (7a) is an empirical relation which agrees within 1 percent with exact values for C for the pressures and temperatures under consideration and eq. (7b) follows from $g = 1$. The numerical results are listed in table 1. Included are values obtained by an integration of eq. (3a) using $C = 1$. The latter agree within 1 percent with eqs. (6a), indicating that eqs. (6a) may be considered as an accurate representation of the $C = 1$ solution for values of U beyond 6 (the highest value of U considered in [3]).

Table 1. *Laminar boundary layer behind strong shock*
$[T_w = T_1 = 522°$ R; $p_1 = 0.001$ atmospheres; $\sigma = 1]$

M_s	$\dfrac{u_e{}^2}{2\,H_e}$	U	$C =$ eq. (7a)		$C = 1$	
			$-f''(0)$	$\dfrac{h'(0)}{h_w}$	$-f''(0)$	$\dfrac{h'(0)}{h_w}$
4	0.032072	4.875	5.1548	6.7788	5.7213	7.5237
6	.022920	6.190	7.1007	16.543	8.5278	19.867
8	.014279	8.060	10.146	32.264	13.091	41.628
10	.009318	10.11	13.673	54.133	18.774	74.330
12	.008090	10.93	14.650	77.270	21.229	111.97
14	.007597	11.33	14.748	102.45	22.462	156.04

Several authors have proposed that the solution for variable C could be estimated by multiplying the corresponding $C = 1$ solution by $C_e = \varrho_e \mu_e / \varrho_w \mu_w$ to a suitable exponent. The exponent 0.29 results in the equations

$$\frac{-f''(0)}{U-1} = \frac{h'(0)}{h_r - h_w} = 0.489 \sqrt{1 + 1.665\,U}\,(C_e)^{0.29}$$

which correlate the data of table 1 to within 3 percent.

The effect of taking σ at a constant value other than one, and using the proper variation of C, has not been determined. The result might be estimated using eqs. (6) as a guide. The resulting estimate is

$$\frac{-f''(0)}{U-1} = \frac{h'(0)}{h_r - h_w}\,\sigma^{-(0.48 + 0.022\,U)} = 0.489 \sqrt{1 + 1.665\,U}\,(C_e)^{0.29} \tag{8}$$

where h_r is found from eq. (6b).

3. Turbulent boundary layer

An estimate of the turbulent boundary layer development can be made by extending the empirical integral relations developed for semi-infinite flat plates.

Assume the boundary layer profile, relative to the wall, has a seventh power variation. Then

$$(u_w - u)/(u_w - u_e) = \zeta^{1/7} \qquad 0 \leq \zeta \leq 1 \tag{9}$$

where $\zeta = y/\delta$, δ being the boundary layer thickness from a POHLHAUSEN viewpoint. By definition of displacement and momentum thickness

$$\frac{\delta^*}{\delta} = \int_0^1 \left(1 - \frac{\varrho\,u}{\varrho_e\,u_e}\right) d\zeta \qquad \frac{\theta}{\delta} = \int_0^1 \frac{\varrho\,u}{\varrho_e\,u_e}\left(1 - \frac{u}{u_e}\right) d\zeta \tag{10}$$

To permit integration of eq. (10) it is further assumed that

$$(\varrho_e/\varrho) \approx (h/h_e) \approx (h_w/h_e) [1 + b\,\zeta^{1/7} - c\,\zeta^{2/7}] \tag{11}$$

where $b = (h_r/h_w) - 1$ and $c = (h_e/h_w)[(h_r/h_e) - 1]$. Eq. (11) is exact for a perfect gas having $\sigma = 1$ but must be regarded as a simplifying assumption for the case of strong shocks in air. Integration of eq. (10) then yields

$$\frac{\delta^*}{\delta} = 1 - 7\,\frac{h_e}{h_w}\left\{\frac{U}{7} + \left(1 - \frac{h_e}{h_w}U\right)I_7 + c\,U\,(I_8 - I_7)\right\}$$

$$\frac{\theta}{\delta} = -7\,\frac{h_e}{h_w}\,(U-1)\left\{U\left(\frac{1}{7} - \frac{h_e}{h_w}I_7\right) + (I_8 - I_7)\left[(c+1)\,U - 1\right]\right\} \tag{12}$$

where $I_N = \int_0^1 z^N dz/(1 + bz - cz^2)$. The reciprocals of I_7 and $(I_7 - I_8)$ are tabulated in table 2. Reciprocals are used to permit linear interpolation, except for $c/(b+1)$ near 1 [3]. Table 2 can be used, with the identity $(N+1)[I_N + bI_{N+1} - cI_{N+2}] = 1$ to evaluate I_N for N other than 7 or 8.

For uniform wall surface temperature δ^*/δ and θ/δ are independent of x and the momentum integral equation can be written

$$\tau_w/\varrho_e u_e{}^2 = (\theta/\delta)\,(d\delta/dx) \tag{13}$$

To permit integration of eq. (13) it is assumed that the incompressible BLASIUS relation between turbulent boundary layer thickness and wall shear [9] can be extended to the present case if velocities relative to the wall are used and if static properties are based on a mean enthalpy. Then

$$\tau_w/[\varrho_m(u_w - u_e)^2] = 0.0225\,[\nu_m/\delta\,(u_w - u_e)]^{1/4} \tag{14}$$

Integration of eq. (13) yields

$$\tau_w/\varrho_e u_e{}^2 = 0.0460\,(\theta/\delta)\,[\varphi\,(1 - U)\,\delta/\theta]^{4/5}\,(U - 1)^{3/5}\,(\nu_e/u_e x)^{1/5} \tag{15}$$

where $\varphi = (\mu_m/\mu_e)^{1/4}\,(\varrho_m/\varrho_e)^{3/4}$. Using a modified Reynolds analogy, the heat transfer to the wall can be estimated from

$$q_w = [(h_r - h_w)\,\tau_w]/[(u_w - u_e)\,\sigma_m{}^{2/3}] \tag{16}$$

with $h_r = h_e + [(U - 1)^2\,u_e{}^2\,\sigma_m{}^{1/3}/2]$. The choice of the proper mean reference condition has not been resolved. One estimate might be based on [10] which proposes that fluid properties be based on a mean enthalpy defined by $h_m = 0.5\,(h_w + h_e) + 0.22\,(h_r - h_e)$.

The above theory must be viewed as a first attempt to construct a solution for the turbulent boundary layer behind a shock. Experiments are required to establish the degree of accuracy of the method and to provide a basis for improvements.

Table 2. Evaluation of $I_N = \int_0^{\,} \dfrac{z^i\,w_i}{1+bz-cz^2}$

$\dfrac{c}{b+1}$	b												
	0	2	4	6	8	10	15	20	25	30	35	50	100
	Reciprocal of I_7												
0	8.0000	22.094	36.119	50.132	64.139	78.144	113.15	148.15	183.16	218.16	253.15	358.16	708.27
.1	7.3577	20.223	33.022	45.807	58.588	71.365	103.31	135.24	167.18	199.11	231.05	326.85	646.18
.2	6.7102	18.343	29.910	41.464	53.013	64.560	93.422	122.28	151.14	179.99	208.85	295.42	583.97
.3	6.0562	16.450	26.780	37.097	47.409	57.719	83.489	109.26	135.02	160.79	186.55	263.84	521.48
.4	5.3938	14.541	23.626	32.699	41.767	50.833	73.493	96.149	118.81	141.46	164.11	232.08	458.61
.5	4.7205	12.609	20.440	28.258	36.072	43.883	63.407	82.929	102.45	121.97	141.49	200.04	395.23
.6	4.0317	10.647	17.207	23.756	30.301	36.843	53.195	69.545	85.893	102.24	118.59	167.63	331.09
.7	3.3200	8.6368	13.904	19.160	24.412	29.663	42.785	55.905	69.023	82.142	95.260	134.61	265.78
.75	2.9513	7.6046	12.211	16.807	21.400	25.991	37.465	48.937	60.408	71.878	83.347	117.76	232.45
.8	2.5699	6.5450	10.477	14.400	18.319	22.237	32.028	41.817	51.605	61.393	71.180	100.54	198.41
.85	2.1698	5.4447	8.6809	11.909	15.134	18.358	26.414	34.469	42.523	50.576	58.629	82.787	163.31
.9	1.7390	4.2764	6.7811	9.2791	11.775	14.269	20.502	26.733	32.964	39.195	45.425	64.115	126.41
.95	1.2448	2.9656	4.6622	6.3539	8.0438	9.7328	13.953	18.173	22.392	26.610	30.829	43.484	85.666
.975	.93665	2.1715	3.3885	4.6019	5.8139	7.0253	10.052	13.079	16.105	19.130	22.156	31.233	61.487
1.0	0	0	0	0	0	0	0	0	0	0	0	0	0
	Reciprocal of $I_7 - I_8$												
0	72.000	185.42	297.85	410.06	522.18	634.26	914.39	1194.4	1474.5	1754.5	2033.9	2874.6	5680.4
.1	67.262	171.91	275.59	379.07	482.45	585.80	844.09	1102.3	1360.5	1618.7	1876.9	2651.5	5233.2
.2	62.467	158.29	253.18	347.87	442.48	537.04	773.38	1009.7	1245.9	1482.2	1718.4	2427.1	4789.3
.3	57.604	144.54	230.57	316.41	402.17	487.89	702.12	916.30	1130.5	1344.6	1558.7	2201.1	4342.4
.4	52.648	130.60	207.70	284.61	361.44	438.24	630.17	822.04	1013.9	1205.7	1397.6	1973.1	3891.3
.5	47.581	116.43	184.48	252.35	320.14	387.90	557.28	726.58	895.87	1065.2	1234.4	1742.2	3434.8
.6	42.359	101.94	160.78	219.46	278.09	336.67	483.08	629.44	775.78	922.11	1068.4	1507.4	2970.5
.7	36.916	86.973	136.38	185.64	234.85	284.03	406.91	529.77	652.59	775.41	898.23	1266.7	2494.7
.75	34.075	79.231	123.79	168.21	212.58	256.92	367.73	478.50	589.25	700.00	810.74	1142.9	2250.3
.8	31.121	71.236	110.81	150.27	189.67	229.06	327.46	425.84	524.20	622.56	720.91	1015.9	1999.4
.85	28.006	62.879	97.285	131.58	165.84	200.08	285.63	371.15	456.65	542.16	627.65	884.13	1739.0
.9	24.643	53.951	82.880	111.72	140.53	169.32	241.26	313.18	385.09	456.99	528.89	744.57	1463.5
.95	20.813	43.926	66.777	89.568	112.34	135.09	191.95	248.80	305.64	362.47	419.31	589.79	1158.1
.975	18.512	37.979	57.267	76.514	95.744	114.97	163.00	211.03	259.05	307.06	355.08	499.12	972.39
1.0	0	0	0	0	0	0	0	0	0	0	0	0	0

4. Wall surface temperature

For $x < 0$, the fluid and wall are both at the initial temperature, T_1. However, for $x > 0$ the fluid temperature has been increased by the shock and heat is conducted into the wall. The resulting surface temperature, T_w, will now be considered.

The temperature distribution in the wall is defined by the steady state conduction equation

$$u_w \, \partial T / \partial x = \alpha \, [\partial^2 T / \partial x^2 + \partial^2 T / \partial y^2] \tag{17}$$

where $\alpha = k / \varrho \, c_p$. Boundary layer approximations may be applied to eq.

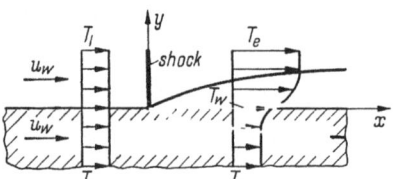

Fig. 2. Thermal boundary layers

(17) (e.g., [3] and [4]) resulting in

$$u_w \, \partial T / \partial x = \alpha \, \partial^2 T / \partial y^2 \tag{18}$$

which is the classical one-dimensional heat conduction equation with t replaced by x / u_w. Thus, there is a thermal boundary layer in the wall which is directly analogous to that in the fluid (fig. 2). The well known solution for the wall temperature distribution, in terms of the heat transfer at $y = 0$, is

$$T - T_1 = \frac{-1}{\sqrt{\pi \, \varrho \, c_p \, k \, u_w}} \int_0^x \frac{q_w(\xi)}{(x - \xi)^{1/2}} \, e^{- y^2 u_w / 4 \, \alpha \, (x - \xi)} \, d\xi \tag{19}$$

where ξ is an integration variable for x and $q_w \equiv q_w(\xi)$ is the heat transfer in the $+y$ direction. If $q_w(\xi)$ can be expressed in the form

$$q_w(\xi) = - Q / \xi^n \tag{20}$$

where Q is independent of ξ, then substituting into eq. (19) and setting $y = 0$ gives

$$T_w - T_1 = Q \, x^{(1/2) - n} \, B(1 - n, 1/2; 1) / \sqrt{\pi \, \varrho \, c_p \, k \, u_w} \tag{21}$$

(The incomplete Beta function is denoted herein by

$$B(p, q; Z) = \int_0^Z z^{p-1} (1 - z)^{q-1} \, dz \, .$$

Eq. (21) can be solved for T_w.

For laminar boundary layers eq. (20) applies with $n = 1/2$, $Q = Q_L$, giving

$$T_w - T_1 = Q_L \sqrt{\pi / \varrho \, c_p \, k \, u_w} \tag{22}$$

An expression for Q_L can be obtained from eqs. (4) and (6). Iteration is sometimes required to find T_w since Q_L depends on T_w. However, in most cases the wall value of $\varrho \, c_p \, k$ is much greater than that of the fluid, resulting in values of $T_w - T_1$ which are small compared with $T_r - T_w$.

In these cases, the departure of T_w from T_1 does not significantly affect the heat transfer and Q_L can be evaluated at $T_w = T_1$. Note that T_w is independent of x, which is consistent with the assumption to this effect in the laminar boundary layer analysis. The temperature distribution within the wall is

$$(T - T_w)/(T_1 - T_w) = \text{erf}\left[-y\sqrt{u_w}\,/2\sqrt{\alpha x}\,\right] \qquad (23)$$

If eq. (20), with $n = 1/5$, is assumed to apply for turbulent boundary layers, eq. (21) becomes

$$T_w - T_1 = 2.299\, Q_T\, x^{3/10}\,/\,\sqrt{\pi\,\varrho\,\bar{c}_p\,\bar{k}\,\bar{u}_w} \qquad (24)$$

where Q_T is evaluated from eqs. (15) and (16). Eq. (24) shows T_w to vary with x, contradicting the uniform wall surface temperature assumption in the turbulent boundary layer analysis and in eq. (20). However, in most cases $T_w - T_1$ is sufficiently small so that the heat transfer is relatively unaffected. Eq. (24) is then valid with Q_T evaluated from eqs. (15) and (16) using $T_w = T_1$.

Cases wherein the boundary layer is first laminar and then becomes turbulent might be treated, by assuming q_w to have the form $q_w = -\,Q_L/\xi^{1/2}$ for $0 \le \xi < x_t$ and $q_w = -\,Q_T/(\xi - x_{\text{ef}})^{1/5}$ for $\xi > x_t$ where x_t is the transition point, x_{ef} is the effective origin of the turbulent boundary layer (i.e., the turbulent boundary layer for $x > x_t$ is assumed to be equivalent to one originating at $x = x_{\text{ef}}$) and the quantities Q_L and Q_T are constants evaluated for $T_w = T_1$. Substituting into eq. (19) and integrating yields, for $x > x_t$ and $y = 0$

$$\sqrt{\pi\,\varrho\,\bar{c}_p\,\bar{k}\,\bar{u}_w}\,(T_w - T_1) = 2\,Q_L \tan^{-1}\left(\frac{x}{x_t} - 1\right)^{-1/2} +$$

$$+\,Q_T\,(x - x_{\text{ef}})^{3/10}\left[2.299 - B\left(4/5, 1/2;\frac{x_t - x_{\text{ef}}}{x - x_{\text{ef}}}\right)\right] \qquad (25)$$

To facilitate evaluation of eq. (25) the incomplete Beta function is tabulated in table 3. The value of x_t must be obtained from a knowledge of transition Reynolds number (this is discussed later). The value of x_{ef} must also be known. If it is assumed that the momentum thickness, or equivalently, the thermal energy decrement is continuous across x_t, then x_{ef} can be found from

$$(x_t - x_{\text{ef}})^{4/5} = 1.6\,Q_L\,x_t^{1/2}/Q_T \qquad (26)$$

Eq. (26) is obtained by integrating $q_w = -\,Q_L/\xi^{1/2}$ for $0 \le \xi \le x_t$ and $q_w = -\,Q_T/(\xi - x_{\text{ef}})^{1/5}$ for $x_{\text{ef}} \le \xi \le x_t$ and equating the results.

Table 3

Z	0	0.05	0.1	0.2	0.3	0.4	0.5	0.6	0.7	0.8	0.9	0.95	1.0
B $(4/5,1/2;Z)$	0	0.115	0.203	0.362	0.515	0.668	0.826	0.994	1.179	1.392	1.663	1.851	2.299

It may be concluded that the wall surface temperature rises discontinuously across the shock, is uniform behind the shock until transition occurs and, thereafter, starts to increase with distance. Eq. (25) indicates a rate of increase, after the transition point, which is ultimately proportional to $(x - x_{ef})^{3/10}$.

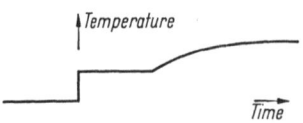

Fig. 3. Typical oscillograph trace of wall surface temperature variation as indicated by thin film resistance thermometer

Experimental observations of the heat transfer from shock induced boundary layers have been made using thin film resistance thermometers mounted on the wall of a shock tube [5, 11, 12 and 13]. A typical oscillograph trace of the response of a thin film resistance thermometer to shock wave passage is given in fig. 3. The vertical ordinate represents the film temperature while the abscissa can be viewed as representing either time or distance behind the shock (i.e., $x = u_w t$). The variation agrees with the discussion of the previous paragraph.

5. Transition

A suitable Reynolds number must be defined if experimental shock tube transition data are to be correlated. If the characteristic velocity is taken to be $u_w - u_e$ (the velocity of the free stream relative to the wall), and the characteristic distance is taken to be $x_t (U - 1)$ (the distance a particle of the free stream has moved relative to the wall before it reaches the transition point), the transition Reynolds number becomes

$$(Re)_t = u_e (U - 1)^2 \, x_t / \nu_e \qquad (27)$$

where x_t is the distance between the shock and the transition point and ν is arbitrarily referred to free stream conditions. For $u_w = 0$ eq. (27) reduces to the familiar form used for semi-infinite flat plates. Eq. (27) may be viewed as a generalization of the latter to account for nonzero u_w.

Experimental studies of transition on shock tube walls are currently being conducted at University of Toronto Institute of Aerophysics (UTIA), Cornell Aeronautical Laboratory (CAL), and Lehigh University (LU). In each case the end of the constant temperature "plateau," as indicated by a wall resistance thermometer, is taken to be the transition point. At UTIA, observations were made in a 3-inch by 3-inch and a 2-inch by 7-inch shock tube for $M_s = 1.6, 1.9$, and $50 \text{ mm} \leq p_1 \leq 760 \text{ mm}$ Hg. These results, presented in [14], indicate transition Reynolds numbers varying from $0.2 \cdot 10^6$ to $0.9 \cdot 10^6$. The variation with M_s and p_1 was not always systematic and also differed between the 2 inch, 3 inch, and 7 inch walls. These results are of the same order of magnitude as the preliminary data of LU [15] which indicated values of $(Re)_t$ around $0.75 \cdot 10^6$ for M_s up to 2.2. Values of this magnitude might have been anticipated from the incompressible semi-infinite flat plate value of about $0.5 \cdot 10^6$.

No systematic data are currently available on transition behind strong shocks. Some information can be obtained from [16] which contains schlieren photographs of strong shocks and the flow behind the waves. Transition clearly appears as an abrupt change in the texture of flow. Using real gas properties, transition Reynolds numbers of from $6.8 \cdot 10^6$ for $M_s = 6.8$ to $57 \cdot 10^6$ for $M_s = 9.7$ are indicated. The free stream Mach number relative to the wall was about 2.4 and the ratio of wall-to-free-stream temperature was about 0.1 for these tests. The data are somewhat in doubt since a combustible mixture was used to drive the shock tube and the resulting shock attenuation rate and flow non-uniformities may have been excessive. At any rate, a large increase in $(Re)_t$ is observed with increase in M_s. These results are consistent with stability analysis of the boundary layer on semi-infinite flat plates in supersonic flow. The latter indicate that large amounts of heat transfer to the wall (as occurs behind strong shocks in a shock tube), combined with supersonic flow in the free stream (relative to the wall), tend to have a very stabilizing effect on laminar boundary layers.

The author is indebted to Drs. R. HARTUNIAN, J. G. HALL, and A. J. CHABAI for having made available unpublished transition data from CAL, UTIA, and LU, respectively.

6. Concluding remarks

Experimental observations of the wall boundary layer in a high pressure shock tube can provide fundamental information (and verification of theory) concerning the development of laminar and turbulent boundary layers in undissociated or partly dissociated gases. A resistance thermometer gives a continuous record, with time, of the heat transfer from the boundary layer. This is equivalent to the variation with distance behind the shock. Hence, a single thermometer gives the entire heat transfer characteristics of the boundary layer including laminar, transition and turbulent values. Correlation between theory and experiment should ultimately be applicable to the boundary layer on a semi-infinite flat plate since the latter is a special case (i.e., zero wall velocity) of the shock induced boundary layer.

Basic information relating to transition can also be obtained since transition is easily and accurately detected. In particular, the stabilizing effects of supersonic velocities relative to the wall and large amounts of cooling can be noted. This, coupled with a theoretical study of the stability of the shock induced boundary layer, should again yield information applicable to semi-infinite flat plates.

The application of shock tube boundary layer theory to the calculation of nonuniformities and attenuation in shock tubes is discussed in [17] to [19].

19*

Notation

Subscripts:

e conditions in free stream external to boundary layer
L laminar flow
m mean value
r value corresponding to insulated wall
T turbulent flow
t transition
w value at wall surface ($y = 0$)
1   conditions upstream of shock wave

Superscripts:

$(^-)$ property of wall material

References

[1] HOLLYER, R. N., Jr.: A Study of Attenuation in the Shock Tube. Eng. Res. Inst., Univ. of Michigan, July 1, 1953. Contract N6-ONR-232-TO IV.

[2] MIRELS, H.: Laminar Boundary Layer Behind Shock Advancing into Stationary Fluid. NACA TN 3401, 1955.

[3] MIRELS, H.: Boundary Layer Behind Shock or Thin Expansion Wave Moving into Stationary Fluid. NACA TN 3712, 1956.

[4] ROTT, N., and R. HARTUNIAN: On the Heat Transfer to the Walls of a Shock Tube. Grad. School of Aero. Eng., Cornell Univ., Nov. 1955. Contract AF 33-(038)-21406.

[5] BERSHADER, D., and J. ALLPORT: On the Laminar Boundary Layer Induced by a Traveling Shock Wave. Princeton Univ., Dept. of Physics, Tech. Rept. II-22 (May 1956).

[6] COHEN, N. B.: A Power Series Solution for the Unsteady Laminar Boundary-Layer Flow in an Expansion Wave of Finite Width Moving Through a Gas Initially at Rest. NACA TN 3943, 1957.

[7] TRIMPI, R. L., and N. B. COHEN: An Integral Solution to the Flat-Plate Laminar Boundary Layer Flow Existing Inside and After Expansion Waves and After Shock Waves Moving into Quiescent Fluid with Particular Application to the Complete Shock Tube Flow. NACA TN 3944, 1957.

[8] BROMBERG, R.: Use of the Shock Tube Wall Boundary Layer in Heat Transfer Studies. Jet Propulsion. Sept. 1956.

[9] SCHLICHTING, H.: Boundary Layer Theory. McGraw Hill Book Co., Inc. (1955), p. 406.

[10] ECKERT, E. R. G.: Engineering Relations for Friction and Heat Transfer to Surfaces in High Velocity Flow. Jour. Aero. Sci., 22 (1955), No. 8.

[11] CHABAI, A. J., and R. J. EMRICH: Measurement of Wall Temperature and Heat Flow in a Shock Tube. Jour. Appl. Phys. 26, 779—780 (1955), No. 6.

[12] VIDAL, R.: Model Instrumentation Techniques for Heat Transfer and Force Measurements in a Hypersonic Shock Tunnel. Cornell Aero. Lab. Rept. AD-917-A-1 WADC-TN-56-315. Feb. 1956.

[13] RABINOWICZ, J., M. E. JESSEY and C. A. BARTSCH: Resistance Thermometer for Heat Transfer Measurements in a Shock Tube. GALCIT Hypersonic Res. Proj. Memo. No. 33, July 1956.

[14] BOYER, A.: UTIA Tech. Rept. No. 15—To be published.

[15] CHABAI, A. J.: Private communication. June 1957.

[16] ANON.: Memorandum for the record. Cornell Aero. Lab. November 3, 1954.

[17] TRIMPI, R. L., and N. B. COHEN: A Theory for Predicting the Flow of Real Gases in Shock Tubes with Experimental Verification. NACA TN 3375, 1955.

[18] MIRELS, H.: Attenuation in a Shock Tube Due to Unsteady Boundary-Layer Action. NACA TN 3278, 1956.

[19] MIRELS, H., and W. H. BRAUN: Nonuniformities in Shock-Tube Flow Due to Unsteady Boundary-Layer Action. NACA TN 4021, 1957.

Aus der Diskussion

F. K. G. ODQVIST (Stockholm): I would like to call your attention to papers by ROSENTHAL of UCLA, first printed in the 1940: ies which, although concerned with quite a different problem, may have some bearing upon a detail in Dr. MIRELS' exposition. It is the problem of the temperature distribution in an infinite plate in the neighborhood of a heat source of constant intensity, moving at constant speed. ROSENTHAL's solution was developed for applications to continuous seam welding and is an exact solution of the two-dimensional equation of heat conduction. Thus it does not make use of Dr. MIRELS' simplification to neglect the temperature drop in the direction of motion.

Reference:

ROSENTHAL, D.: a. R. SCHMERBER; Weld. J., April 1938.

ROSENTHAL, D.: Weld. J., May 1941.

L. N. PERSEN (Trondheim): Having, a few years ago, had the opportunity to examine some experimental results on the attenuation of shock-waves, and having been able to establish an approximate analytic expression for it, I would be interested in knowing whether or not you have been able to get a reliable analytic expression for the attenuation. I want to add that my own expression has not a reliable theoretical background and, even though it matched the experimental results, it has therefore not been published.

H. MIRELS (Cleveland): References 17, 18 and 19 of the present paper deal with attenuation and nonuniformities in a shock tube. Analytical expressions have been obtained which are applicable when the boundary layer is thin relative to the tube diameter (references 18 and 19). However, the case where the wall boundary layer tends to span the tube cross section requires further study.

Diskussionsveranstaltung zur VI. Sitzung

Contribution sur le sujet

Équations hydrodynamiques

Par **D. Massignon,** Centre d'Etudes Nucléaires, Saclay (Seine et Oise)

Je voudrais faire quelques remarques sur la validité de l'équation de Navier-Stokes dans les régions d'un fluide où existent d'importantes fluctuations de la densité de masse, par exemple au voisinage d'un front d'ondes de choc dans une couche limite, ou dans un fluide au voisinage de son point critique de condensation.

L'équation de Navier-Stokes peut être déduite de l'équation de Liouville en Mécanique statistique de Gibbs: dans ce cas la vitesse locale $\bar{u}$ (R, t) au point R du fluide et à l'instant t est, comme l'ont montré par exemple Kirkwood et Irving [1], définie par le rapport

$$\bar{u}\,(R, t) = \frac{\bar{J}\,(R, t)}{\bar{\varrho}\,(R, t)} \qquad (1)$$

où $\bar{J}$ et $\bar{\varrho}$ sont les valeurs moyennes sur l'espace des phases Γ du système mécanique des molécules du fluide, de fonctions explicites des positions q et des impulsions p de ces molécules et d'un point R du fluide: par exemple

$$\bar{J}\,(R, t) = \int\limits_{\Gamma} J\,(q,\,p,\,R)\,f_N\,(q,\,p,\,t)\,dq\,dp \qquad (2)$$

où $f_N\,(q,\,p,\,t)$ est la densité de répartition dans l'espace des phases de la Mécanique statistique. L'équation de Navier s'obtient alors en formant l'équation de transport d'impulsion vérifiée par $\bar{J}$, obtenue en prenant la valeur moyenne (2) dans Γ du premier membre de $\dot{J}\,(q,\,p,\,R) - [J\,(q,\,p,\,R),\,H] = 0$, où $[J,\,H]$ est le crochet de Poisson de J et de l'Hamiltonien H: il suffit alors d'introduire la viscosité dans le courant d'impulsion.

Cette démonstration ne concerne que le cas où $\bar{u}$ est la vitesse *moyenne* locale. Si la densité de masse $\varrho\,(q, R)$ n'a que de faible fluctuations (*fluides incompressibles*), et si la vitesse locale a d'importantes fluctuations (turbulence), $\varrho \simeq \bar{\varrho}$ et $u \neq \bar{u}$. On peut encore définir [2], en Mécanique statistique de Gibbs, cette vitesse locale *fluctuante* par

$$u\,(R) = \frac{J\,(q,\,p,\,R)}{\bar{\varrho}\,(R)} \qquad (3)$$

et on montre alors que l'équation de Navier-Stokes est encore vérifiée par cette grandeur aléatoire (3) à partir de l'équation de transport $\dot{J} - [J,\,H] = 0$, et non de sa valeur moyenne.

Mais si la densité de masse et le courant de masse ont, à la fois d'importantes fluctuations, $\varrho \neq \bar{\varrho}$ et $u \neq \bar{u}$ (*fluides compressibles*), il n'est plus possible d'établir l'équation de Navier-Stokes par la Mécanique statistique. De ce point de vue théorique, le mouvement d'un fluide devrait etre décrit, dans ce cas très particulier, par le courant de masse J et par son équation de transport, et non par la vitesse locale u et par l'équation de Navier-Stokes.

Une remarque analogue peut être faite à propos du tenseur de corrélation des vitesses $\overline{u\,(R)\,u\,(R')}$. Ce tenseur ne peut être calculé ni en théorie cinétique des gaz, ni en Mécanique statistique de Gibbs. Mais le tenseur de corrélation $\overline{J\,(R)\,J\,(R')}$ peut être définie par (3) à partir de $J\,(q,\,p,\,R)$ et ce tenseur a une valeur finie [2]. Si la densité de masse n'a que de faibles fluctuations, $\overline{u\,(R)\,u\,(R')}$ peut être défini par $\overline{J\,(R)\,J\,(R')}\,/\,\bar{\varrho}\,(R)\,\bar{\varrho}\,(R')$. Mais si ses fluctuations sont importantes, on ne peut plus définir $\overline{u\,(R)\,u\,(R')}$ et il faut se servir du tenseur de corrélation des courants de masse.

Références:

[1] Kirkwood, J. G., und J. H. Irving: J. Chem. Phys. **18**, 1950, 817.

[2] Massignon, D.: Mécanique statistique des fluides. Fluctuations et propriétés locales, Dunod, Paris, 1957.

Contribution to the Subject
Equations of Hydrodynamics

By **A. Iberall**, Rand Development Corporation, Cleveland, Ohio

It would appear that the equations of hydrodynamics for accoustics, and for laminar and turbulent flow fields can be reduced, making no more assumptions than that the medium is continuous and elastic, and that field gradients are small. The following dimensionless set of equations then results for boundary value problems.

In Cartesian tensor notation

for the mean flow (0 subscripts; 00 subscripts denote constants)

$$\nabla^2 \, \nabla \times R_0 = \nabla \times A_0 \qquad \text{vector potential}$$
$$\nabla \cdot R_0 = 0 \qquad \text{scalar potential}$$
$$\nabla^2 \, T_0 = - \, \sigma_{00} \, \Phi_0 \qquad \text{temperature}$$
$$\nabla^3 \, P_0 = - \, \nabla \cdot A_0 \qquad \text{pressure}$$

for the fluctuating flow (1 subscripts)

$$\left[\nabla^2 - \frac{\partial}{\partial \tau} \right] \nabla \times R_1 = \nabla \times A_1$$

$$\left[\; \right] \left[\nabla \cdot R_1 = -n \left[\frac{\partial}{\partial \tau} - \frac{\gamma_{00}}{\sigma_{00}} \, \nabla^2 \right] \nabla \cdot \frac{\partial A_1}{\partial \tau} + n \left[\frac{\partial}{\partial \tau} - \frac{\nabla^2}{\sigma_{00}} \right] \nabla^2 B_1 - n \, \nabla^2 \cdot \frac{\partial \, \Phi_1}{\partial \tau} \right.$$

$$\left[\; \right] \left[\; T_1 = (\gamma_{00} - 1) \, \nabla \cdot \frac{\partial A_1}{\partial \tau} - (\gamma_{00} - 1) n \left[\frac{\partial}{\partial \tau} - \frac{4}{3} \, \nabla^2 \right] \frac{\partial B_1}{\partial \tau} + \left[\gamma_{00} \, n \, \frac{\partial^2}{\partial \tau^2} - \nabla^2 \right] \Phi_1 \right.$$

$$\left[\; \right] \left[\; P_1 = \left[\frac{\partial}{\partial \tau} - \frac{\nabla^2}{\sigma_{00}} \right] \nabla \cdot A_1 - n \left[\frac{\partial}{\partial \tau} - \frac{\nabla^2}{\sigma_{00}} \right] \left[\frac{\partial}{\partial \tau} - \frac{4}{3} \, \nabla^2 \right] B_1 + \right.$$

$$+ \, n \left[\frac{\partial}{\partial \tau} - \frac{4}{3} \, \nabla^2 \right] \frac{\partial \, \Phi_1}{\partial \tau}$$

$$A = R \cdot \nabla \, R \qquad \Phi = \frac{\gamma_{00} - 1}{2 \, \alpha_{00} \, \Theta} \, [R_{i,j} + R_{j,i}]^2 + R \cdot \nabla \, [(\gamma_{00} - 1) P - T]$$

$$B = R \cdot \nabla \, [\gamma_{00} \, P - T] \qquad n = \left(\frac{\nu}{c \, D} \right)^2_{00} \qquad \left[\; \right] \equiv \left[\frac{\partial}{\partial \tau} - \frac{\nabla^2}{\sigma_{00}} \right] \left[n \, \frac{\partial^2}{\partial \tau^2} - \nabla^2 \right]$$

The parameters used are:

D = characteristic dimension in field
α_{00} = thermal coefficient of expansion
c_{00} = Laplacian velocity of sound
t = $D^2 \, \tau / \nu_{00}$ (t is real time)
x^i = $D y^i$ (x^i is real dimension)
Θ = $n \, T / \alpha_{00}$ (Θ is real temperature)
p = $\mu_{00} \, \nu_{00} \, P / D^2$ (p is real pressure)
V_i = $\nu_{00} \, R_i / D$ (V_i is real velocity)

$[UV]_0 = U_0 \, V_0 + \widehat{U_1 V_1}$
$[UV]_1 = U_0 \, V_1 + U_1 \, V_0$

The continuum assumption is that $\dfrac{\nu}{c \, D}$ and $\dfrac{\nu \omega}{c^2}$ are small.

C. C. LIN (Cambridge, Mass.): There was a paper by KOVASZNAY, two years back, on the same subject, and the linear part of the theory here seems to be in complete agreement with KOVASZNAY's.

VII. SITZUNG

Vorsitzender: J. Ackeret

On the separation of the unsteady laminar boundary layer[1]

By

Franklin K. Moore

Cornell Aeronautical Laboratory, Inc., Buffalo, New York

1. Introduction

When a boundary layer is subjected to unsteady conditions, it strongly tends, if it is thin, to be quasi-steady; that is, to be described at each instant by the appropriate steady equations of motion. In many cases of practical interest a thin boundary layer is nearly quasi-steady, but, owing to some imposed unsteadiness, its departures from quasi-steadiness require estimation. Such is the case during rotating stall in an axial flow compressor and stalling flutter of an airfoil. In these problems, any essential unsteadiness in the boundary layer tends to produce a hysteresis in the relation between lift and angle of attack, thus affecting the energy balance of the flow oscillation. In this connection, of course, the unsteady features of separation must be accounted for, at least approximately, in any attempt to predict consequences of unsteadiness of the airfoil boundary layer. It is the particular problem of separation to which the present paper is addressed.

For the unsteady (but nearly quasi-steady) incompressible boundary layer of an airfoil, it is possible by approximate methods to calculate solutions of the boundary-layer equations[2]

$$\psi_{yt} + \psi_y \, \psi_{xy} - \psi_x \, \psi_{yy} = u_{e_t} + u_e \, u_{e_x} + \nu \, \psi_{yyy} \left.\begin{array}{c} \\ \\ \end{array}\right\} \quad (1)$$
$$\psi_y \, (x, \infty, t) = u_e \, (x, t) \, ; \quad \psi_y \, (x, 0, t) = \psi_x \, (x, 0, t) = 0$$

in the integrated form

$$(u_e \, \delta^*)_t + (2 \, \theta + \delta^*) \, u_e \, u_{e_x} + u_e{}^2 \, \theta_x = \nu \, (u_y)_w \quad (2)$$

Assuming that the contribution of the unsteady terms in these equations is small, one may adopt a perturbation procedure in which the solution is considered to differ only slightly from that which would be obtained neglecting the unsteady terms entirely. A preliminary investigation of this type for an airfoil boundary layer has been conducted by the writer [1].

---

[1] The present paper is based on analyses conducted under the sponsorship of the U.S. Air Force through the Office of Scientific Research and through the Aeronautical Research Laboratory of the Wright Air Development Center.

[2] Symbols are defined in the Appendix.

A serious difficulty, however, attends an unsteady boundary-layer calculation of the foregoing type when the prediction of the effects of unsteadiness on the position of the separation point is required. Specifically, one must entertain doubts that the familiar criterion for separation, namely, the vanishing of shear stress at the surface, is valid in an unsteady approximation.

Restricting consideration to the vicinity of the separation point of laminar boundary layer, it will be shown that the unsteady problem near separation may properly be split into two sub-problems. In one of these the quasi-steady separation point is fixed on the surface as would be the case, for example, when a sphere accelerates through a fluid in a rectilinear path. Owing to a slight unsteadiness, this type of separation point may be expected to shift slightly. This shift is illustrated by solutions of an unsteady variant of the familiar FALKNER-SKAN problem [2] of the laminar boundary layer developing over a downwardly-inclined ramp. Results of that study will be briefly reviewed here. The second problem of importance is, in effect, that of identifying the proper definition of separation to be applied when the separation point is in motion over the surface as would be true for an airfoil undergoing oscillation in pitch. Detailed consideration will be given to this second problem, and the implications of the results for approximate boundary-layer calculation will be cited.

Special acknowledgment is due Dr. RICHARD A. HARTUNIAN of the Cornell Aeronautical Laboratory for his contributions to the analysis upon which this paper is based. In particular, he conducted the calculations for the FALKNER-SKAN problem with moving wall presented herein.

2. The two sub-problems of unsteady separation

We assume that the laminar incompressible boundary layer governed by eq. (1) is nearly quasi-steady; i.e., only terms of first order in unsteadiness need be considered. Further, it is assumed that the quasi-steady position of the separation point is in motion upstream with a small velocity ε relative to the surface. Accordingly, a coordinate system is selected in which the quasi-steady separation point is stationary:

$$x' \equiv x + \varepsilon t; \; y' \equiv y; \; t' \equiv t; \; \psi' \equiv \psi + \varepsilon y; \; u_e' \equiv u_e + \varepsilon \qquad (3)$$

Next, the stream function is expressed as the sum of a quasi-steady function and a perturbation proportional to the velocity of movement of the separation point ε, assumed small.

$$\psi' \equiv \Phi + \varepsilon \varphi \qquad (4)$$

Substituting eqs. (3) and (4) into eq. (1) yields the following system of differential equations and boundary conditions:

$$\left. \begin{aligned} \Phi_{y'}\,\Phi_{x'y'} - \Phi_{x'}\,\Phi_{y'y'} &= u_{e'}\,u_{e_{x'}}' + \nu\,\Phi_{y'y'y'} \\ \Phi_{y'}\,(x',\infty,t') &= u_{e'};\;\; \Phi_{y'}\,(x',0,t') = \varepsilon;\;\; \Phi_{x'}\,(x',0,t') = 0 \end{aligned} \right\} \quad (5)$$

$$\left. \begin{aligned} \Phi_{y'}\,\varphi_{x'y'} + \Phi_{x'y'}\,\varphi_{y'} - \Phi_{x'}\,\varphi_{y'y'} - \Phi_{y'y'}\,\varphi_{x'} &= \frac{1}{\varepsilon}\,(u'_e - \Phi_{y'})_t + \nu\varphi_{y'y'y'} \\ \varphi_{y'}\,(x',\infty,t') = \varphi_{y'}\,(x',0,t') &= \varphi_{x'}\,(x',0,t') = 0 \end{aligned} \right\} \quad (6)$$

Evidently, eqs. (5) represent a quasi-steady boundary layer on a surface which is in slow motion downstream with a speed ε at the separation point. The position of actual separation of the boundary layer may, of course, be considered fixed in view of the definition of ε. Eqs. (6) govern the unsteady perturbation which is subject to homogeneous boundary conditions. The coefficients in eqs. (6) may, if ε is small, be approximated by the quasi-steady solution for a stationary wall.

The brief development described above shows that near the position of separation the essential problems to be understood and solved when the boundary layer is nearly quasi-steady are, first, the unsteady boundary-layer perturbation for a fixed separation point (eqs. (6)), and second, the quasi-steady boundary layer over a surface which, relative to a separation point assumed fixed, is in motion (eqs. (5)). In the two sections to follow, these basic problems will be considered in greater detail.

3. Unsteady boundary layer, stationary separation point

For the type of problem described by eqs. (6), one may expect that the effect of unsteadiness is simply to cause a shift in the position of separation. If this is true, one would not expect it to be necessary to make any fundamental change in the usual definition of separation; i.e., the separation point may be regarded as the point where the wall shear vanishes, in the first unsteady approximation.

The best indication that the quasi-steady definition of separation may be carried over for this type of problem may be had by considering the singularity at separation. GOLDSTEIN [3] has shown that, in steady flow, a singular solution of the boundary-layer equations is possible when the shear and velocity both vanish at the wall surface. Of course, the full Navier-Stokes equations do not show such a singularity [4]. However, the existence of a singular boundary-layer solution is no doubt a reliable indication of separation, insofar as the boundary-layer equations are able to describe it.

We now proceed to show that the singularity of the type found by GOLDSTEIN is also possible in the unsteady approximation at the point on the surface where the shear vanishes. GOLDSTEIN's approach could, of course, be used here. However, a somewhat different, though less rigorous, approach is adopted here in the interest of brevity, and for the sake of a certain insight which the present approach provides. Eqs. (1) may be integrated directly to yield

$$u^2 - u_s^2 = 2\,x\,(u_{es}\,u_{es_x} + u_{es_t} + v\,u_{syy}) + 2\int_0^x (-u_t + u_y \int_0^x u_x\,dy)\,dx \qquad (7)$$

where x is measured from the separation point. An equivalent equation

$$(u^2 - u_s^2)_{\psi=\text{const.}} = 2\,x\,(u_{es}\,u_{es_x} + u_{es_t} + v\,u_{syy}) \qquad (8)$$

is obtained by integration of the boundary-layer momentum equation written under the von Mises transformation. In eq. (8), the squares of velocity are compared along a given streamline rather than for constant y, as is the case in eq. (7). In both equations it is assumed that the viscous stress terms are essentially regular in x at the separation point. Both equations may be regarded as determining the velocity profile u upstream of separation where the profile u_s is postulated to be known.

Clearly, both eqs. (7) and (8) suggest that when the velocity u_s vanishes a necessary condition exists for the occurrence of a singularity in u as a function of x. If we define the velocity profile precisely at separation by the series

$$u_s = b_2\,\frac{y^2}{2!} + b_3\,\frac{y^3}{3!} + ---- \qquad (9)$$

and postulate that the velocity profile nearby is

$$u = a_1\,y + a_2\,\frac{y^2}{2!} + ---- \qquad (10)$$

and we insert these expansions into eq. (7), collection of the coefficients of the various powers of y yields the following relations:

$$\left.\begin{aligned}
v\,b_2 &= -u_{es_x} - u_{es_t} \\
v\,b_3 &= a_{1t} \\
a_1 &= [2\,(b_{2t} - v\,b_4)]^{1/2}\,(-x)^{1/2} \\
a_2 &= b_2
\end{aligned}\right\} \qquad (11)$$

This result is essentially equivalent to that obtained by GOLDSTEIN [3], and indicates that the surface shear approaches zero with a half-order singularity in distance in unsteady as well as steady flow, provided that the separation point may be regarded as stationary.

Of course, the foregoing discussion of the separation singularity is not concerned with the actual location of the separation point. In order to study the shift of the separation point caused by unsteadiness, it is interesting to consider an unsteady variant of the FALKNER-SKAN problem in which the boundary layer is assumed to develop from the leading edge of

a ramp inclined downward in such a degree that the velocity distribution
outside the boundary layer is given by

$$u_e = A(t)\, x^m \tag{12}$$

where A is a function of time rather than the usual constant. In the quasi-
steady approximation, at a particular negative value of m for which
$\beta \equiv 2m/(m+1) = -0.1988$, solution of the boundary-layer equations
yields a velocity profile having vanishing shear at the surface. Actually,
the steady shear vanishes with a square-root singularity in β, as illustrated
in fig. 1a.

Owing to unsteadiness, we might expect that this singular curve will
shift (as in fig. 1a) in such a way that if A is an increasing function of

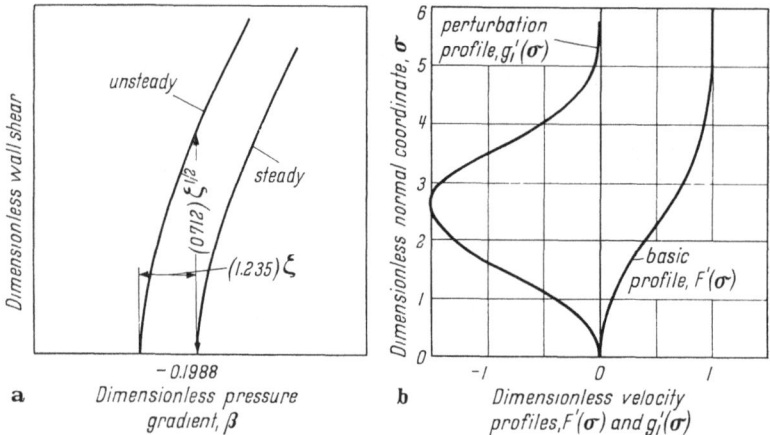

Fig. 1. Results of linear unsteady perturbation of separating Falkner-Skan wedge flow
a) Wall shear near separation; b) Velocity profiles at separation

time, a more negative β corresponds to the vanishing of shear. In order to
determine this shifted separation point, the stream function is expanded
about its quasi-steady value as follows:

$$\psi = \sqrt{\frac{2\nu A}{m+1}}\; x^{\frac{m+1}{2}} \left[F(\sigma) + \frac{x^{1-m} A'}{A^2}\, g_1(\sigma) + ----- \right] \tag{13}$$

where σ is the similarity variable customary for this problem:

$$\sigma \equiv \sqrt{\frac{A(m+1)}{2\nu}}\; y\, x^{\frac{m-1}{2}} \tag{14}$$

and where the quantity $x^{1-m} A'/A^2$ is the parameter governing the effect
of the unsteadiness. It may be shown that this parameter is, in effect,
the ratio of the time required for change to diffuse across the boundary
layer compared with a characteristic time based on the rate of change

of $A(t)$. For nearly quasi-steady flow, this quantity must be assumed small. In addition to the expansion given in eq. (13), the fact of a singularity at separation requires also that β be expanded:

$$\beta = -0.1988 + \frac{x^{1-m} A'}{A^2} \gamma + - - - \tag{15}$$

where γ is an unknown numerical coefficient to be determined by the condition that the perturbation shear vanish; i.e., $g_1''(0) = 0$.

The details of this analysis are available elsewhere [5], and therefore only the results will be summarized. Inserting eqs. (13) and (15) into the appropriate differential equation yields the profile functions and the value of γ corresponding to zero shear. As fig. 1a indicates, the shift in the separation value of β is toward negative β with $\gamma = 1.235$. It is obvious from the geometry of the singularity that measuring shear directly at the quasi-steady value of β for separation yields a shift of order $(x^{1-m} A'/A^2)^{1/2}$.

It is interesting that the FALKNER-SKAN problem shows a half-order singularity of the same sort that is predicted by GOLDSTEIN [3] and by eqs. (11). Further, the analysis just described indicates that, owing to unsteadiness, an acceleration of the outer flow velocity shifts the separation singularity toward more adverse pressure gradient, in an amount proportional to the appropriate unsteadiness parameter. The velocity profiles obtained at separation are illustrated in fig. 1b. The basic and the perturbation velocity profiles have zero shear at the surface. The perturbation profile shows that a velocity defect occurs at the shifted separation point in proportion to the unsteadiness parameter.

4. Steady boundary layer, moving separation point

Turning now to the problem posed in eq. (5), we see that the essential complications introduced by a slight movement of the separation point over the surface are also found in the steady boundary layer on a slowly moving wall. It is necessary to distinguish between two consequences of the moving wall: First, the motion of the wall will affect the development of the boundary layer as it approaches separation. For example, one might expect that, if the wall were moving downstream, separation would be delayed. Second, the basic question arises as to the proper definition of separation to be applied when the wall is in motion. Inasmuch as the first effect may presumably be calculated by approximate methods, it is the second effect with which we will chiefly be concerned.

Fig. 2a illustrates the streamline pattern generally accepted to apply in the vicinity of separation when the wall is stationary. A bubble separated from the rest of the boundary layer appears at the point where the surface shear vanishes. If the wall is in motion downstream, as shown in

fig. 2 b, we may imagine that at some point in the boundary layer, zero
shear will be achieved at the wall, as illustrated by the velocity profile
at the left. It seems unlikely that this velocity profile represents
a separation condition, however, because at every point in the boundary layer, there is a nonvanishing velocity in the downstream direction.

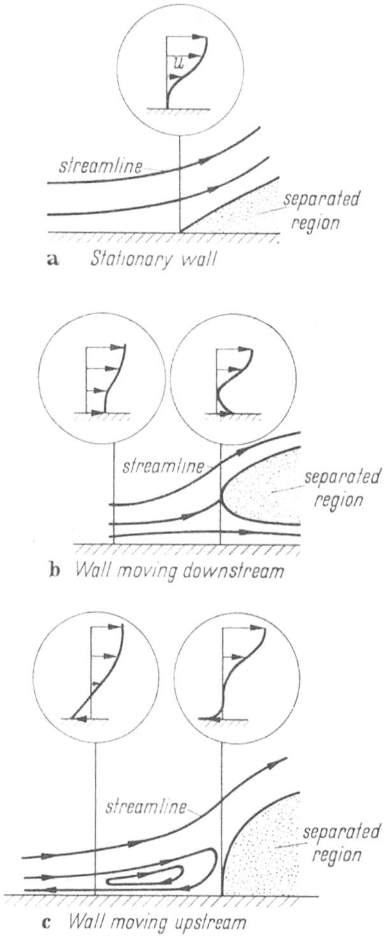

If a bubble is to be found within the boundary layer, the profile velocity must vanish at the front of the bubble. Further, if, as one would expect, the front of the bubble constitutes a vertical streamline "shoulder" in the boundary layer, the profile velocity should have vanishing shear there. Therefore, one is led to formulate the criterion that, for a slowly moving wall, separation occurs when, at some point in the boundary layer, the profile velocity and shear simultaneously vanish[1]. If the wall is moving downstream, this criterion would be met when the adverse pressure gradient causes the velocity profile to osculate the vertical axis, as illustrated by the velocity profile sketched on the right in fig. 2 b. The streamline pattern associated with the velocity profiles described above would apparently show a bubble-like separated region beginning at the point where shear and velocity vanish, and the head of this bubble should presumably be identified as the separation point.

Fig. 2. Expected streamlines and velocity profiles at separation for stationary wall and moving wall

Fig. 2 c illustrates a problem for which speculation is difficult. When the wall is moving upstream, at some point well upstream of separation, the velocity profile would be as illustrated in the left-hand sketch, crossing the vertical axis toward negative values of velocity. In consider-

---

[1] Prof. Sears and Rott at Cornell University have formulated this criterion also (private communication).

ing the subsequent progress of the boundary layer downstream from this point, it should be realized that the retarding action of the pressure gradient is to diminish the square of the velocity; i.e., whether the velocity is positive or negative, its magnitude will be diminished by the action of the pressure gradient. Of course, the pressure gradient effect is modified by the gradient of shear acting on the fluid particles as well; however, it would seem that as the boundary layer proceeds downstream that portion of the velocity profile which crosses the axis should, in effect, be rotated toward congruence with the vertical axis, owing to the effect of pressure gradient. This process may be imagined to continue until the velocity profile achieves the shape as illustrated in the right-hand sketch in fig. 2c, in which an extensive region of vanishing velocity appears in the profile, together with a rather abrupt change in velocity near the base of the boundary layer in order that the required condition of wall motion be met. If this picture is correct, then the streamline pattern in the boundary layer would be as shown in fig. 2c. A separated region would begin at the point where the velocity profile has fully rotated to the vertical axis.

Another possibility for the development of a separation point is that, before the segment of the profile crossing the axis rotates to the axis, a part of the profile at larger distances from the wall is retarded sufficiently to osculate the vertical axis in the same manner as shown in fig. 2b. If this should occur, the streamline pattern of fig. 2c would be slightly altered, though its overall features would be similar.

Upstream of the separation point there must be an upstream motion in the boundary layer. Following a streamline from a point near the surface, it is apparent that the streamline must reverse direction and proceed back downstream, as in fig. 2c. It is clear from eq. (8) that this effect must occur. As we have already noted, if the velocity anywhere in the boundary layer should vanish, then the velocity field may be singular there. Thus, in the vicinity of the cross-over point of the velocity profile, the profile velocity has a half-order singularity in x. Such a singularity implies that some streamline performs a loop at each value of x, the head of the loop occurring where the velocity vanishes. Now, when the streamline performs this reversal and proceeds back downstream, continuity requires that the streamline close on itself; i.e., the streamline loops must be closed, and embedded in the thin boundary layer upstream of separation. Of course, in a part of the boundary layer farther away from the wall, streamlines never participate in the upstream movement, and therefore would be directed entirely in the downstream direction, as illustrated by the outermost streamline in fig. 2c. Apparently, the loops just described should not be related to separation, because they are consistent with the idea of a thin boundary layer. Indeed, streamline loops must occur all the way forward to the beginning point of the boundary layer.

It is now necessary to inquire whether the separation profiles suggested in figs. 2b and c correspond to a possible singularity in the boundary layer of the same sort as that found by Goldstein for a stationary surface. Before proceeding with an analytical discussion of this question, we should note that Goldstein's analysis does not actually prove that a singularity must occur at separation, but simply notes that such a possibility exists. Therefore, having in mind the rather complicated problem of the moving wall, it is important to decide on physical grounds whether a singularity is, in fact, to be expected at separation. First, we have already noted that the effect of adverse pressure gradient is to diminish the magnitude of the velocity; adverse pressure gradient itself will not produce reverse flow. The action of the transverse shear stresses contemplated in the boundary-layer equations will be to drag fluid particles downstream, if, as one would expect, the profile curvature is positive near separation. Thus, the boundary-layer equations embody no mechanism which can provide reverse flow, and accordingly, if velocity and shear come to zero, they must do so with a singularity in x.

Actually, reverse flow does occur (with no singularity) at separation, presumably as a consequence of the viscous normal stress $\nu\, u_{xx}$ neglected in deriving the equations of a thin boundary layer. One might say that singular vanishing of velocity and shear would be the best approximation to the occurrence of reverse flow possible, under the boundary-layer assumptions.

Thus, singular approaches to the separation profiles of fig. 2 are expected, and may be studied by reference to eq. (7). The unsteady terms are dropped out, and in the final term, the integration is carried over y from $-h$ to y, where $y = 0$ is the separation point within the boundary layer, and $y = -h$ is the surface upon which the vertical velocity v must vanish and u must attain the prescribed wall velocity. First, considering the case of the downstream-moving wall, we postulate that the velocity profile at separation is as shown in fig. 2b. Measuring y from the point of osculation, we write

$$u_s = b_2 \frac{y^2}{2!} + b_3 \frac{y^3}{3!} + ---- \qquad (16)$$

$$u = a_0 + a_1\, y + a_2 \frac{y^2}{2!} + ---- \qquad (17)$$

Eq. (17) contemplates the possibility that both the shear and velocity are singular at $y = 0$. In carrying out the integration on the last term of eq. (7), it is assumed that the Taylor series development can be extended to $y = -h$, on the grounds that if ε is small, presumably h is also small.

In fact, if the streamline curvature is of unit order, h must be approximately $(2\varepsilon/b_2)^{1/2}$. Inserting eqs. (16) and (17) into eq. (7) modified as described, and equating powers of y yields the following relations:

$$a_0{}^2 = 2\,x\,(u_e\,u_{e_x} + v\,b_2) - 2\int_0^x v\,(x,0)\,a_1\,dx \tag{18}$$

$$2\,a_0\,a_1 = 2\,x\,v\,b_3 + 2\int_0^x [a_1\,a_0{}' - v\,(x,0)\,a_3]\,dx \tag{19}$$

$$2\,(a_1{}^2 + a_0\,a_2) = 2\,x\,v\,b_4 + 2\int_0^x [a_1\,a_1{}' + 2\,a_2\,a_0{}' - v\,(x,0)\,a_2]\,dx \tag{20}$$

where, as may be shown,

$$-\,v\,(x,0) = a_0{}'\,h - a_1{}'\,\frac{h^2}{2!} + - - - \tag{21}$$

As previously,

$$v\,b_2 = v\,a_2 = -\,u_e\,u_{e_x}; \quad v\,b_3 = v\,a_3 = 0 \tag{22}$$

Then, if a_0 and a_1 are to have singularities according to some power of $-\,x$, consistency of eqs. (18), (19), and (20) requires that

$$\left.\begin{array}{l} a_0 = (k^2/2\,b_2)\,(-\,x)^{1/2} \\ a_1 = k\,(-\,x)^{1/4} \\ v\,(x,0) = (k^3/8\,b_2{}^2)\,(-\,x)^{-1/4} \end{array}\right\} \tag{23}$$

to lowest order in x, where k is an arbitrary constant, in effect. We see from eqs. (23) that the velocity approaches zero through positive values with a half-order singularity in distance, while shear and vertical velocity have one-fourth-order singularities. If, as one would expect, the vertical velocity approaches positive, rather than negative, infinity as separation is approached, then the shear approaches zero through positive values, consistent with one's expectation that the point of zero shear should rise off the wall as separation is approached.

It remains to find the dependence of the coefficient k on ε, in order to estimate the distance (in x) between the position of vanishing shear and the head of the separation bubble. It is obvious that if the foregoing development were carried to higher order in x that results equivalent to eqs. (11) would arise; i.e., a contribution to a_1 of order $(-\,x)^{1/2}$ would be found:

$$a_1 = (-\,2\,v\,b_4)^{1/2}\,(-\,x)^{1/2} \tag{24}$$

where, presumably, b_4 has no dependence on ε. Now, since we know that $v\,(x,0)$ must be of order $x^{-1/4}$, we may turn to eq. (21), and, assuming that h is approximately $(2\,\varepsilon/b_2)^{1/2}$, cancellation of terms of order $x^{-1/2}$ in eq. (21) would require that

$$k = 0\,[\varepsilon^{1/4}] \tag{25}$$

The required one-fourth-order singularity of v would enter eq. (21) from a higher-order solution of a_0. The conclusion expressed in eq. (25) must be regarded as tentative. The analysis would have to be carried out to several higher orders in $x^{1/4}$ before the relation between k and ε could satisfactorily be determined.

If k is of order $\varepsilon^{1/4}$, one may infer from eqs. (23) that the distance upstream at which the velocity at y becomes of order ε (and also that the shear becomes of order $\varepsilon^{1/2}$) is of order ε. Thus, we may conclude that a movement at the wall in the downstream direction with a velocity of order ε produces a separation point downstream by a distance of order ε from the point where shear vanishes at the wall.

In the case of the wall surface moving upstream, the results are not so clear-cut. However, it may be reasonable to expect that when the wall is moving upstream, the singularity of the boundary-layer equations occurs when the velocity profile flattens against the y-axis to the degree that all derivatives of the profile approach zero. If such is the case, then as separation is approached, the profile of boundary-layer velocity is decelerated purely by action of the pressure gradient; i.e.,

$$u = (- u_e u_{e_x})^{1/2} (- x)^{1/2} \tag{26}$$

and a singularity must be postulated at the foot of the boundary-layer profile in order to achieve the required wall velocity. A suitable singular solution may be described as follows:

$$\left.\begin{array}{c} u = - \varepsilon G' \left(y \sqrt{\dfrac{\varepsilon}{-\nu x}} \right) \\[2mm] G G'' + 2 G''' = 0 \,; \ G'(0) = 1 \,; \ G(0) = G'(\infty) = 0 \end{array}\right\} \tag{27}$$

As in the case of the wall moving downstream, eq. (27) would indicate that the abrupt adjustment to separation takes place over distances of order ε, because in the singular solution postulated, when x becomes of order ε, y must become of the order of the boundary-layer thickness.

It should be emphasized that the foregoing discussion of the case of the wall moving upstream is quite speculative, owing to the rather complete uncertainty as to the proper assumption for the shape of the final separation profile.

Finally, the information which may be obtained from FALKNER-SKAN flow when the surface is considered to be in motion should be considered. Calculations have been carried out for the wall moving downstream with a velocity $\varepsilon\, x^m$. Fig. 3 illustrates some of the results of these calculations. It was necessary in this analysis to extend the FALKNER-SKAN curves beyond the singular point. It is perhaps of interest to notice that the lower branches of all the curves, including the one for zero wall velocity, tend to curve back upward as β is increased. Exact calculations were carried out for $\varepsilon = 1/3$ and $1/10$.

As one would expect, the occurrence of zero shear is delayed to more negative values of β if the wall velocity is moving in the streamwise direction. This, however, is not the chief point of interest here. Apparently, the minimum possible values of β occur along the line AD, which extends into regions of negative shear. Points along the line AD do not, however, correspond to the osculation profile discussed earlier (fig. 2b). Fig. 4 illustrates the fact that the profile at point D, while having a negative

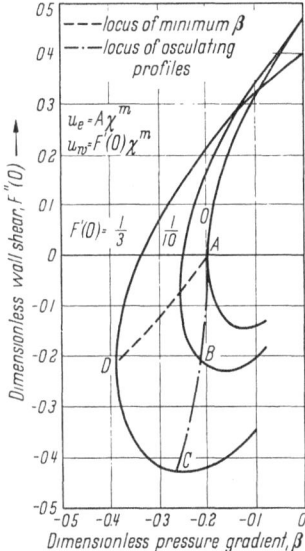

Fig. 3. Dependence of wall shear stress on surface velocity, for wedge flow near separation

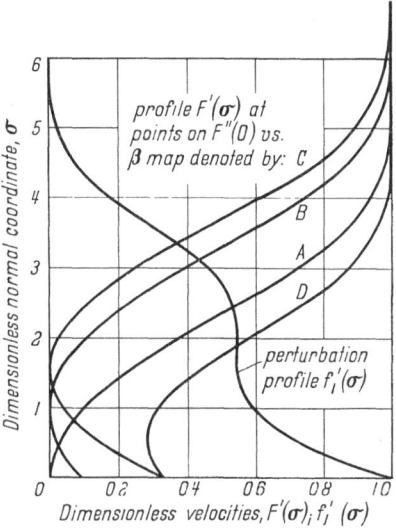

Fig. 4. Velocity profiles for wedge flow near separation

shear, does not approach the y-axis at all closely. One must proceed farther down these curves to much more negative values of shear before the osculation condition is reached at point C for $\varepsilon = 1/3$ and at point B for $\varepsilon = 1/10$. These latter points appear to differ only slightly from the separation value of β for zero wall velocity. Thus, while the FALKNER-SKAN flow with a moving surface does exhibit the type of profile postulated here to be indicative of separation, the occurrence of the osculation does not coincide, as one might perhaps expect, with the minimum value of β which one is accustomed to regard as signifying separation of FALKNER-SKAN flow. It may be, of course, that the FALKNER-SKAN treatment described here is significantly complicated by the assumption that the wall velocity varies with x, made in order to preserve similarity.

A perturbation analysis was also carried out for small ε, yielding a perturbation profile as shown in fig. 4. The perturbation analysis was carried out to determine not only the shift of the point of zero shear, but also the location of the value of shear for minimum β. The perturbation velocity profile corresponding to minimum β is illustrated in fig. 4. Addition of this profile, properly weighted, to the basic profile predicts quite closely the features of the exact solution for $\varepsilon = 1/3$ and $1/10$.

5. Remarks on approximate analysis

In the approximate method of THWAITES, a semi-empirical relation between shear and pressure gradient is proposed for steady flow which embodies, in effect, the sort of singularity described by GOLDSTEIN [3].

——*Profile at stationary separation point*

– – –*Profile at moving separation point*

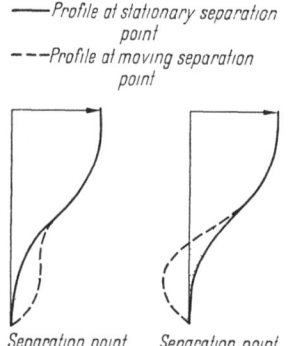

Separation point   Separation point
moving downstream   moving upstream

Fig. 5. Comparison of expected velocity profiles at separation points when stationary and when in motion

For the two types of problems arising in the unsteady case, THWAITES' relation could be revised to take account of unsteady changes in profile shape, perhaps by the aid of the FALKNER-SKAN results described herein. In the case of a stationary quasi-steady separation point, separation would be determined by the occurrence by the same type of profile shape as is customary in steady flow and would be denoted by vanishing shear.

When the quasi-steady separation point is in motion, particular care would be required for formulating a THWAITES relation near separation. In particular, the relation would have to be arranged to be singular, in the case of the separation point moving upstream, at a negative value of shear. When the wall moves upstream the singularity should be represented at positive shear.

Returning to the coordinate system fixed in the surface, it is apparent that an approximate analysis must contemplate changes in profile shape such as are shown in fig. 5, and must include singular behavior when the proper velocities are reached in these disturbed profiles.

6. Conclusions

It has been shown that the general motion near a separation point in an unsteady incompressible laminar boundary layer which is nearly quasi-steady may be resolved into two problems, one of which is that of finding unsteady corrections to the boundary-layer solution when the quasi-steady separation point is stationary. In such a case, the familiar definition of separation as the place where wall shear vanishes is correct. The second problem is in essence that of a steady boundary layer separating over moving wall. Here, the chief problem is to achieve a proper definition of separation.

It is shown that a singularity of the type first described by GOLDSTEIN may exist in the boundary-layer solution when the wall shear vanishes for the first problem. The analysis is made using an integrated form of the momentum equation which yields the singularity quite directly. It is shown by treatment of the FALKNER-SKAN problem in which the magnitude of the outer velocity field is allowed to vary with time that the separation point shifts, owing to unsteadiness, in the direction of more adverse pressure gradient, if the velocity field is increasing with time.

In connection with the second problem, the question of steady separation over a moving wall is discussed in some detail. If the wall is moving downstream, corresponding to a separation point moving upstream, it is expected that the separation bubble will be lifted off the surface and its leading edge will be signified by the simultaneous vanishing of velocity and shear at some point in the boundary layer away from the surface. It is concluded that a criterion of simultaneous vanishing of shear and velocity is the proper generalization of the usual definition of steady separation, for the case of a separation point moving slowly along a surface. When applied to the case of a wall moving upstream, it appears that the velocity profile, where it crosses through zero, should decay to zero over an extensive portion of the boundary layer, yielding a bubble with a shoulder which may be identified as a separation point. Upstream of this shoulder may be expected a long, thin bubble fully embedded within the boundary layer and to be regarded as part of the boundary layer.

Consideration of the singularity at separation shows that, for the downstream moving wall, a singularity is in fact possible when the velocity profile has simultaneously zero velocity and shear at a point above the surface. Upstream of this point, the velocity increases according to the square-root of distance, and the shear increases according to the fourth-root of distance. The distance intervening between the first appearance of vanishing shear at the surface and the leading edge of the bubble is expected to be of the order of the wall velocity.

When the surface is moving upstream there is uncertainty concerning the proper definition of velocity profile precisely at the separation point. However, it seems possible that for this problem the entire velocity profile collapses over the lower part of the boundary layer, with the result that a singularity appears at the foot of the boundary layer, across which the adjustment to the given wall velocity is achieved. In this problem also, the degeneration of the velocity profile is expected to take place over distances of the order of the wall velocity.

Results are presented of a FALKNER-SKAN analysis extended to include the effect of a moving wall. It is found that the position of zero shear is shifted in proportion to wall velocity, that the minimum possible β

corresponds to negative shear but not, however, to the expected osculating profile. Rather, the osculating profile occurs on the lower branch of the skin-friction versus pressure-gradient curve.

It is concluded that further research is required in order to formulate a suitable approximate boundary-layer theory to include the effects described in the present study, especially in view of the necessity of including the proper singular behavior at separation.

References

[1] Moore, F. K.: Lift Hysteresis at Stall as an Unsteady Boundary-Layer Phenomenon. NACA TN 3571 (1955).

[2] Falkner, V. M., and Miss S. W. Skan: Phil. Mag. **12**, 865 (1931).

[3] Goldstein, S.: Quart. Jour. Mech. Appl. Math. **1**, 43—69 (1948).

[4] Dean, W. R.: Note on the Motion of Liquid Near a Position of Separation. Proc. Camb. Phil. Soc., Pt. 2, **46**, 293—306 (1950).

[5] Moore, F. K.: The Unsteady Laminar Boundary Layer of a Wedge, and a Related Three-Dimensional Problem. Heat Transfer and Fluid Mech. Inst. Stanford University Press (1957).

Appendix—Notation

$A(t)$ coefficient of Falkner-Skan flow (eq. (12))

a_n Taylor coefficients of profile near separation

b_n Taylor coefficients of profile at separation

F basic Falkner-Skan profile function (eq. (13))

G singular solution at separation (eq. (27))

g unsteady perturbation of Falkner-Skan profile function (eq. (13))

h vertical distance from surface to point of osculation

k constant in singular solution (eq. (23))

m exponent of x, Falkner-Skan flow (eq. (12))

t time

u velocity component parallel to surface

v normal component of velocity

x coordinate along surface

y coordinate normal to surface

β dimensionless pressure gradient $(= 2\,m/(m+1))$

γ numerical coefficient (eq. (15))

δ^* displacement thickness $\left(\equiv \int\limits_0^\infty (1-u/u_e)\,dy\right)$

ε velocity of movement of separation point

θ momentum thickness $\left(\equiv \int\limits_0^\infty (u/u_e)\,(1-u/u_e)\,dy\right)$

ν kinematic viscosity

σ similarity variable (eq. (14))

Φ, φ stream functions (eq. (4))

ψ stream function $(u = \psi_y;\ v = -\psi_x)$

Primes denote ordinary differentiation, and, in eq. (3), a coordinate transformation

Subscripts denote partial differentiation, also:

e evaluation at outer edge of boundary layer
s evaluation at the separation value of x
w evaluation at the surface

Aus der Diskussion

Dem Vortrage folgte eine sehr ausgedehnte Diskussion über den Begriff „Ablösung" in instationären Strömungen. Insbesondere vertraten ihre Standpunkte bzw. wiesen in instruktiven Beispielen auf die Schwierigkeiten hin die Herren M. B. GLAUERT, C. C. LIN, G. I. TAYLOR und H. GÖRTLER. Auf der einen Seite kam der mehr formale Standpunkt zum Ausdruck, „Ablösung" als einen vom Bezugssystem abhängigen Begriff zu definieren, auf der anderen Seite stand der vom physikalischen Vorgang her vertretene Standpunkt, dann von „Ablösung" zu sprechen, wenn eine Rückströmung aus Gebieten, die in Hauptströmungsrichtung weit stromabwärts liegen, erfolgt. Begreiflicherweise wurde eine Einigung über eine zweckmäßige Definition der „Ablösung" nicht erzielt. Es sei daher hier nur die abschließende Bemerkung des Vorsitzenden der Sitzung, J. ACKERET, wiedergegeben: Mir scheint, daß mit dem Begriff der Ablösung ein neues Symposium bestritten werden könnte. — D. Hrsg.

Lösungen für instationäre Grenzschichtströmungen mit Hilfe von Integraltransformationen

Von

Leif N. Persen

Norges Tekniske Høgskole, Trondheim

1. Einleitung

An Hand zweier Beispiele sollen hier einige Vorteile der Anwendung von Integraltransformationen auf instationäre Grenzschichtströmungen gezeigt werden. Es wird sich herausstellen, daß in vielen Fällen die Methode der Integraltransformationen leichter zum Ziel führt als die gewöhnlichen klassischen Methoden. Ferner wird gezeigt, daß diese Methode sehr oft einen einfachen Weg zu angenäherten Ausdrücken angibt.

2. Beispiel 1

Die zu behandelnde Strömung sei folgendermaßen gegeben. Eine unendlich ausgedehnte Platte, über der sich eine ruhende Flüssigkeit befindet, wird zur Zeit $t = 0$ plötzlich aus der Ruhelage heraus in ihrer eigenen Ebene mit der Geschwindigkeit $U_0 + u_0 \sin \omega t$ bewegt. Man hat es also mit einem Anfahrvorgang zu tun, und man wünscht die Entwicklung der Geschwindigkeitsprofile mit der Zeit zu verfolgen. Die folgende Frage kann jetzt gestellt werden: Gibt es eine Lösung der Navier-Stokesschen Gleichungen von einer solchen Form, daß

$$v = w = 0, \quad u = u(y, t), \quad p = \text{konstant}$$

wo u die Geschwindigkeitskomponente in der Bewegungsrichtung der Platte ist? In diesem Falle werden sich nämlich die Navier-Stokesschen Gleichungen auf die Wärmeleitungsgleichung reduzieren, und man erhält:

$$\frac{\partial u}{\partial t} = \nu \frac{\partial^2 u}{\partial y^2} \tag{2.1}$$

mit den Grenzgleichungen:

$$u(0, t) = U_0 + u_0 \sin \omega t$$
$$u(\infty, t) = 0 \tag{2.2}$$
$$u(y, 0) = 0$$

Wird die Laplacetransformierte von u mit $\bar{u}$ bezeichnet, so erhält man von Gl. (2.1):

$$\bar{u}'' - \frac{s}{\nu}\,\bar{u} = 0 \tag{2.3}$$

mit den Grenzbedingungen

$$\bar{u}(0, s) = \frac{U_0}{s} + u_0\,\frac{\omega}{s^2 + \omega^2} \tag{2.4}$$

$$\bar{u}(\infty, s) = 0\,.$$

Die diese Grenzbedingungen befriedigende Lösung der Gl. (2.3) läßt sich sofort folgendermaßen angeben:

$$\bar{u}(y, s) = \left(\frac{U_0}{s} + u_0\,\frac{\omega}{s^2 + \omega^2}\right) e^{-\sqrt{\frac{s}{\nu}}\,y}\,. \tag{2.5}$$

Hier ist s der durch die Laplacetransformation eingeführte Parameter. Mit Hilfe von Tabellen über Laplacetransformationen [1] erhält man aus Gl. (2.5) sofort folgende Lösung in geschlossener Form:

$$u(y, t) = U_0\,\mathrm{Erfc}\left[\frac{1}{2}\,y\,(\nu\,t)^{-1/2}\right] +$$

$$+ \frac{u_0}{2\sqrt{\pi}} \int\limits_0^t y\,\nu^{-1/2}\,\xi^{-3/2}\,e^{-\frac{y^2}{4\nu\xi}}\,\sin[\omega\,(t - \xi)]\,d\xi\,. \tag{2.6}$$

Bei der numerischen Auswertung dieser leicht angebbaren Lösung stellt sich aber heraus, daß sie bereits für mäßig große Werte von t unangenehm wird. Für eine Umformung steht aber die Umkehrformel der Laplacetransformation zu Verfügung

$$u(y, t) = \frac{1}{2\,\pi\,i} \int\limits_{c - i\infty}^{c + i\infty} \bar{u}(y, s)\,e^{st}\,ds \qquad Re\,c > Re\,i\,\omega = 0\,. \tag{2.7}$$

Durch Einsetzen von Gl. (2.5) in (2.7) sieht man leicht, daß $u(y, t)$ aus zwei Beiträgen zusammengesetzt ist, erstens dem Beitrag von den Polen $s = \pm\,i\,\omega$ und vom Verzweigungspunkt $s = 0$, zweitens dem Beitrag von der Integration längs des Verzweigungsschnittes. Der erste gibt also den periodischen Endzustand an, und der zweite enthält die Anfangsstörungen. Hiermit ist die Lösung in eine solche Form gebracht, daß die physikalischen Vorgänge in ihren Einzelheiten verfolgt werden können. Zum Beispiel kann die Frage beantwortet werden, nach wie vielen Perioden der Endzustand erreicht wird, und es zeigt sich, daß dies schon nach einer Periode mit guter Näherung der Fall ist (für den Fall $U_0 = 0$). Dies wurde beim Vortrag in einem kurzen Film näher gezeigt, wovon Abb. 1 und 2 genommen sind.

3. Beispiel 2

Die hier zu behandelnde Strömung sei folgendermaßen gegeben. Eine unendlich ausgedehnte Halbebene in einer ruhenden Flüssigkeit wird zur Zeit $t = 0$ plötzlich aus der Ruhe heraus mit der Geschwindigkeit

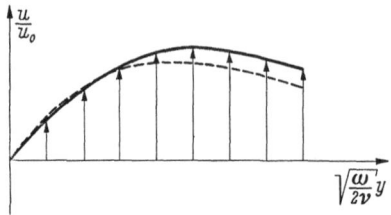

Abb. 1. Vergleich des Geschwindigkeitsprofils nach einer halben Periode (ausgezogen) mit dem voll entwickelten Profil (gestrichelt)

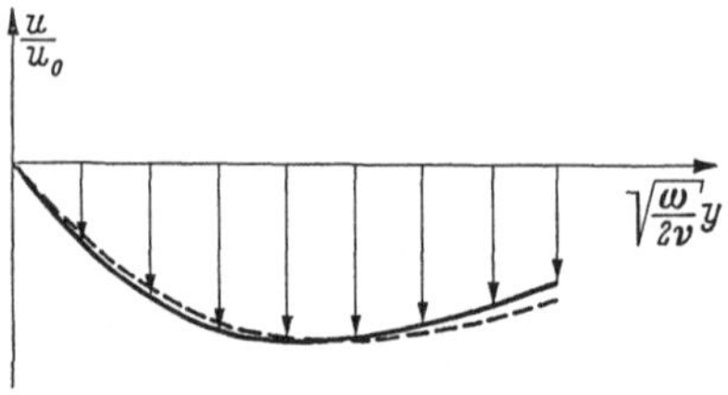

Abb. 2. Vergleich des Geschwindigkeitsprofils nach einer Periode (ausgezogen) mit dem voll entwickelten Profil (gestrichelt). Die Übereinstimmung ist schon sehr gut

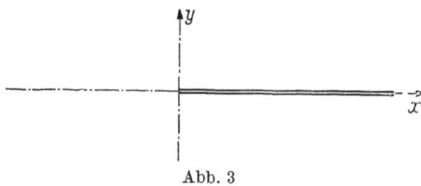

Abb. 3

$W_0 + w_0 \sin \omega t$ parallel zur z-Achse bewegt (Abb. 3). Man hat es also wieder mit einem Anfahrvorgang zu tun. Auch in diesem Falle wird eine Lösung gesucht, für die

$$u = v = 0, \quad w = w (x, y, t), \quad p = \text{konstant}$$

sind, und wo w die Geschwindigkeitskomponente in der Bewegungsrichtung der Platte ist. Die Navier-Stokesschen Gleichungen reduzieren sich dann auf folgende Gleichung

$$\frac{\partial w}{\partial t} = \nu \left\{ \frac{\partial^2 w}{\partial x^2} + \frac{\partial^2 w}{\partial y^2} \right\} \tag{3.1}$$

mit den nicht-analytischen Grenzbedingungen

$$\left. \begin{array}{ll} w (x, y, 0) = 0 & \\ w (x, \infty, t) = 0 & \\ w (x, 0, t) = W_0 + w_0 \sin \omega t & \text{für } x > 0 \\ \dfrac{\partial}{\partial y} [w (x, 0, t)] = 0 & \text{für } x < 0 \end{array} \right\} \tag{3.2}$$

Durch Anwendung der Laplacetransformation entsteht folgende Gleichung für $\overline{w}$

$$s \overline{w} = \nu \left\{ \frac{\partial^2 \overline{w}}{\partial x^2} + \frac{\partial^2 \overline{w}}{\partial y^2} \right\} \tag{3.3}$$

mit den Grenzbedingungen

$$\overline{w}\,(x, \infty, s) = 0$$

$$\overline{w}\,(x, 0, s) = \frac{W}{s} + w_0\,\frac{\omega}{s^2 + \omega^2} \qquad \text{für } x > 0$$

$$\frac{\partial}{\partial y}\,[\overline{w}\,(x, 0, s)] = 0 \qquad \text{für } x < 0$$

$$(3.4)$$

Auf Grund des nichtanalytischen Charakters der Grenzbedingungen wird die Lösung der Gl. (3.3) am besten mit Hilfe der Fouriertransformation gesucht. Wird die Fouriertransformation folgendermaßen definiert:

$$f^*\,(p, y, s) = \int\limits_{-\infty}^{+\infty} f\,(x, y, s)\,e^{-\,i\,p\,x}\,d\,x \qquad (3.5)$$

und werden außerdem folgende Voraussetzungen gemacht:

$$\frac{\partial}{\partial x}\,[w\,(\pm \infty, y, t)] = 0$$

$$w\,(+\infty, y, t) \text{ beschränkt}$$

$$w\,(-\infty, y, t) = 0$$

$$(3.6)$$

läßt sich durch Anwendung dieser Transformation auf Gl. (3.3) in einfacher Weise die Differentialgleichung für $\overline{w}^*$ ableiten.

$$\frac{\partial^2\,\overline{w}^*}{\partial y^2} - \left(p^2 + \frac{s}{\nu}\right)\overline{w}^* = 0\,. \qquad (3.7)$$

Die Lösung dieser Gleichung läßt sich sofort hinschreiben

$$\overline{w}^*\,(p, y, s) = C_1\,(p, s)\,e^{-\,y\,\sqrt{p^2 + \frac{s}{\nu}}}\,. \qquad (3.8)$$

Diese Lösung befriedigt die erste Bedingung (3.4), und die noch zu bestimmende Konstante C_1, die von p und s abhängig ist, muß so bestimmt werden, daß die nichtanalytischen Bedingungen (3.4) längs der x-Achse befriedigt werden. Zu diesem Zweck werden die Hilfsfunktionen f_1 und f_2 eingeführt, wo

$$\overline{w}\,(x, 0, s) = f_1\,(x, s) + f_2\,(x, s) \qquad (3.9)$$

und ferner

$$f_1\,(x, s) = \begin{cases} \dfrac{W}{s} + w_0\,\dfrac{\omega}{s^2 + \omega^2} & \text{für } x > 0 \\[2mm] 0 & \text{für } x < 0 \end{cases}$$

$$f_2\,(x, s) = \begin{cases} 0 & \text{für } x > 0 \\[2mm] \text{unbekannt} & \text{für } x < 0 \end{cases}$$

$$(3.10)$$

Durch Anwendung der Fouriertransformation (3.5) auf (3.9) und (3.10) erhält man

$$\overline{w}^* (p, 0, s) = f_1^* + f_2^*$$

wo

$$f_1^* = + \frac{1}{i\,p} \left(\frac{W_0}{s} + w_0 \frac{\omega}{s^2 + \omega^2} \right)$$

$$f_2^* = \int\limits_{-\infty}^{0} f_2\, e^{-i\,p\,x}\, d\,x$$

(3.11)

sind. Wird ferner zur Abkürzung die Funktion $h\,(p, s)$ eingeführt, wo

$$h\,(p, s) = \int\limits_{0}^{\infty} \frac{\partial}{\partial y}\, [\overline{w}\,(x, 0, s)]\, e^{-i\,p\,x}\, d\,x$$

(3.12)

ist, so sieht man sofort ein, daß f_2^* eine analytische Funktion von p in der ganzen oberen Halbebene der komplexen p-Ebene sein muß, während $h\,(p, s)$ eine analytische Funktion von p in der ganzen unteren Halbebene der p-Ebene ist. Dies wird im weiteren durch die Indizes $\oplus$ bzw. $\ominus$ gekennzeichnet. Man muß beachten, daß der Punkt $p = 0$ als zur oberen Halbebene gehörend betrachtet werden muß. Aus den Gln. (3.8), (3.11), (3.12) läßt sich nun die Funktion $C_1\,(p, s)$ eliminieren, wodurch die folgende Gleichung entsteht

$$\frac{1}{i\,p} \left[\frac{W_0}{s} + w_0 \frac{\omega}{s^2 + \omega^2} \right] + f_2^*{}_{\oplus} = - \frac{1}{\sqrt{p^2 + \dfrac{s}{\nu}}}\, h_{\ominus}\,(p, s)\,.$$

(3.13)

Diese Gleichung enthält zwei unbekannte Funktionen, $f_2^*{}_{\oplus}$ und $h_{\ominus}$, und behält ihre Gültigkeit in einem kleinen Bereich unterhalb der reellen Achse der p-Ebene. Die Aufspaltung dieser Gleichung gelingt nun durch eine Betrachtung über das Verhalten der auftretenden Funktion in der p-Ebene. Durch eine einfache Umformung von Gl. (3.13) entsteht

$$\left[\frac{W_0}{s} + w_0 \frac{\omega}{s^2 + \omega^2} \right] \left\{ \frac{1}{i\,p} \sqrt{p + \sqrt{-s/\nu}} - \frac{1}{i\,p} \sqrt{\sqrt{-s/\nu}} \right\} -$$

$$- f_2^*{}_{\oplus} \sqrt{p + \sqrt{-s/\nu}} = - \frac{1}{\sqrt{p - \sqrt{-s/\nu}}}\, h_{\ominus}\,(p, s) -$$

(3.14)

$$- \left[\frac{W_0}{s} + w_0 \frac{\omega}{s^2 + \omega^2} \right] \frac{\sqrt{\sqrt{-s/\nu}}}{i\,p}\,.$$

Hier läßt sich leicht nachprüfen, daß die linke Seite der obigen Gleichung eine analytische Funktion von p in der oberen Halbebene der p-Ebene ist, während die rechte Seite eine analytische Funktion von p in der unteren Halbebene ist. Da die beiden Seiten analytische Fortsetzungen voneinander sein müssen, sind beide gleich einer Konstanten, die in diesem Falle gleich Null ist. Dadurch wird sowohl $h_{\ominus}\,(p, s)$ als $f_2^*{}_{\oplus}\,(p, s)$

bestimmt, wonach die Funktion $C_1(p, s)$ bestimmt werden kann. Schließlich erhält man also

$$\overline{w}^*(p, y, s) = \left[\frac{W_0}{s} + w_0 \frac{\omega}{s^2 + \omega^2}\right] \frac{\sqrt{\sqrt{-s/\nu}}\, e^{-y\sqrt{p^2 + \frac{s}{\nu}}}}{i\,p\,\sqrt{p + \sqrt{-s/\nu}}}. \qquad (3.15)$$

Mit Hilfe der Umkehrformel der Fourier- und Laplacetransformationen läßt sich nun von Gl. (3.15) ausgehend der Ausdruck für das Geschwindigkeitsprofil ableiten. Darauf soll hier verzichtet werden, um einen anderen Vorteil dieser Methode vorzubringen. Wenn zum Beispiel der Reibungseinfluß gefunden werden soll, braucht man nicht den Umweg über das Geschwindigkeitsprofil zu gehen, weil die Funktion $h_\ominus(p, s)$ bekannt ist und somit eine direkte Ermittlung der Geschwindigkeitsgradienten an der Platte ermöglicht. Die Umkehrformel der Fouriertransformation liefert sofort

$$\frac{\partial}{\partial y}[\overline{w}(x, 0, s)] =$$

$$-\frac{1}{2\pi} \int_{-\infty}^{+\infty} \frac{\sqrt{\sqrt{-s/\nu}}\left[\frac{W_0}{s} + w_0 \frac{\omega}{s^2 + \omega^2}\right] \sqrt{p - \sqrt{-s/\nu}}\, e^{i\,p\,x}}{i\,p}\, d\,p \qquad (3.16)$$

wonach

$$\frac{\partial}{\partial y}\overline{w}[(x, 0, s)] = \begin{cases} -\left[\frac{W_0}{s} + w_0 \frac{\omega}{s^2 + \omega^2}\right]\left\{\left(\frac{s}{\nu}\right)^{1/2} - \right. \\ \left. -\left(\frac{s}{\nu}\right)^{1/4} \frac{1}{2\sqrt{\pi}} \int_{x}^{\infty} \xi^{-3/2}\, e^{-\xi\sqrt{s/\nu}}\, d\,\xi\right\} \qquad & x > 0 \\[2mm] 0 & x < 0. \end{cases}$$

Schließlich erhält man mit Hilfe der Umkehrformel der Laplacetransformation die endgültige Lösung:

$$\frac{\partial}{\partial y}[w(x, 0, t)] = -w_0 \sqrt{\frac{\omega}{\nu}} \sin\left(\omega t + \frac{\pi}{4}\right) +$$

$$+ \frac{W_0}{2\pi}\left(\frac{\omega}{\nu}\right)^{1/4} \int_{x}^{\infty} \xi^{-3/2} \sin\left(\omega t + \frac{\pi}{8} - \xi\sqrt{\frac{\omega}{2\nu}}\right) e^{-\xi\sqrt{\frac{\omega}{2\nu}}}\, d\,\xi +$$

$$+ \frac{1}{\pi} \int_{0}^{\infty}\left[-\frac{W_0}{\zeta} + w_0 \frac{\omega}{\zeta^2 + \omega^2}\right]\sqrt{\frac{\zeta}{\nu}}\, e^{-\zeta t}\, d\,\zeta + \qquad (3.18)$$

$$+ \frac{1}{\pi} \int_{0}^{\infty}\left[-\frac{W_0}{\zeta} + w_0 \frac{\omega}{\zeta^2 + \omega^2}\right]\left(\frac{\zeta}{\nu}\right)^{1/4} g(x, \zeta)\, e^{-\zeta t}\, d\,\zeta$$

mit

$$g(x, \zeta) = \frac{1}{2\sqrt{\pi}} \int_{x}^{\infty} \xi^{-3/2} \sin\left(\xi\sqrt{\frac{\zeta}{\nu}} - \frac{\pi}{4}\right) d\,\xi.$$

4. Bemerkungen zu den Beispielen

Die hier durchgerechneten Beispiele sind so gewählt, daß durch sie einige Vorteile der Methode der Integraltransformationen vorgebracht werden konnten. Es läßt sich aber eine Reihe von instationären Strömungen dieser Art angeben, bei der sich zylindrische Körper in der Richtung ihrer Erzeugenden in vorgeschriebener Weise bewegen, oder wo Rotationskörper um ihre Achsen in beliebiger Weise rotieren, die sich sämtlich mit Hilfe der angegebenen Methode leicht lösen lassen. Solche Strömungen sind damit nicht nur physikalisch verwandt, sondern zeichnen sich auch durch die einheitliche mathematische Behandlung aus. Die große Tragweite dieser Methode liegt darin, daß bei der Berechnung die Ergebnisse der ganzen Funktionstheorie zur Verfügung stehen. Bei der Integration in der komplexen Ebene entsteht oft die Möglichkeit, gute Näherungslösungen in relativ einfacher Weise anzugeben. Zum Beispiel braucht man in einigen Fällen nur die Beiträge von einigen von den vielen Singularitäten zu berücksichtigen (vgl. G. F. Carrier [2]), oder man wird zu schnell konvergierenden Rekursionsverfahren geleitet (vgl. Leif N. Persen [3]).

Literatur

[1] Bateman Manuscript Project, California Institute of Technology "Tables of Integral Transforms" Vol. 1 and 2 McGraw Hill (1954).
[2] Carrier, G. F.: "Integral Equation Boundary Layer Problems", 50 Jahre Grenzschichtforschung. Braunschweig: Vieweg (1955), S. 13.
[3] Persen, L. N.: "On the theory of the oscillating disk viscometer", Proc. IX Intern. Congr. Appl. Mech., Brussels (1956) (im Druck).

Aus der Diskussion

J. Kampé de Fériet (Lille): Une remarque sur un point de détail: dans une Communication au Congrès International des Mathématiciens, Zürich 1932, j'ai déterminé tous les mouvements plans d'un fluide visqueux incompressible où les termes d'inertie s'annulent identiquement.

Literatur:

J. Kampé de Fériet: Sur quelques cas d'intégration des équations du mouvement plan d'un fluide visqueux incompressible. — 3e Congrès Intern. Mec. Appliquée Stockholm. 1 — 1931 — p. 334-338.
   Détermination des mouvements plans d'un fluide visqueux incompressible ou le tourbillon est constant le long des lignes de courant. Verh. Internat. Mathematiker Kongresses Zürich. 2 — 1932 — p. 298-300.

Periodische Absaugegrenzschichten

Von

W. Wuest

Aerodynamische Versuchsanstalt, Göttingen

Übersicht

Bei periodisch längs einer unendlich ausgedehnten ebenen Wand verteilter Absaugegeschwindigkeit nimmt auch die Grenzschichtströmung
einen rein periodischen Charakter an. Es lassen sich dabei die Fälle
unterscheiden, daß die Absaugegeschwindigkeit einfach periodisch in
Strömungsrichtung, quer zur Strömungsrichtung oder in einem gewissen
Winkel hierzu oder aber doppeltperiodisch verteilt ist. Spezielle Sonder-
fälle solcher Verteilungen entsprechen einer unendlichen Folge paralleler
Schlitze gleichen Abstandes sowie Lochreihen. Es wird zunächst an der
reibungslosen Strömung untersucht, wie der Einflußbereich der periodischen Störungen des Geschwindigkeits- und Druckfeldes nach außen
abklingt. Danach wird auch die Grenzschichtströmung selbst, teilweise
auf der Grundlage einer Störung des (asymptotischen) homogenen Ab-
saugeprofils in linearisierter Näherung, berechnet.

1. Einführung

Durch Absaugen der wandnahen Grenzschicht kann die Laminar-
haltung und das Ablöseverhalten günstig beeinflußt werden und daher
auch der Strömungswiderstand verringert werden. Man benötigt hierzu
halbdurchlässige Wände, die man entweder durch Verwendung von
Werkstoffen erhält, die einen faserigen oder körnigen Aufbau haben, oder
aber dadurch, daß man eine ursprünglich glatte und undurchlässige
Wand mit Perforationen versieht. Von besonderer technischer Bedeutung
sind im ersten Fall Sintermetallbleche, bei denen der Abstand der
Durchlaßkanäle durch die Korngröße bestimmt ist, im anderen Fall aber
geschlitzte oder gelochte Bleche.

In allen genannten Fällen wird also nicht die von der Theorie gewöhnlich vorausgesetzte gleichmäßige Verteilung der Absaugegeschwindigkeit
erreicht. Vielmehr wird durch eine große Anzahl feiner Kanäle abgesaugt,
die regelmäßig oder nur im Mittel geordnet an der Wandfläche vorhanden
sein können. Dementsprechend werden auch in die Grenzschicht Störungen hineingetragen, die von periodischer oder regelloser Struktur
sind und vom Verhältnis des mittleren Abstandes der Durchlaßkanäle

zur Grenzschichtdicke abhängen. Es ist naheliegend, diese Einflüsse am idealisierten Modell der rein periodischen Absaugegrenzschichten zu untersuchen. Diese können bei einer konstanten Außenströmung auftreten, wenn die Absaugegeschwindigkeit periodisch verteilt ist, oder umgekehrt bei gleichmäßig verteilter Absaugung, wenn die Außenströmung periodisch gestört ist. Auch eine Überlagerung beider Fälle ist möglich. Im folgenden wollen wir aber nur den Fall der konstanten Außenströmung an einer unendlich ausgedehnten ebenen Wand be-

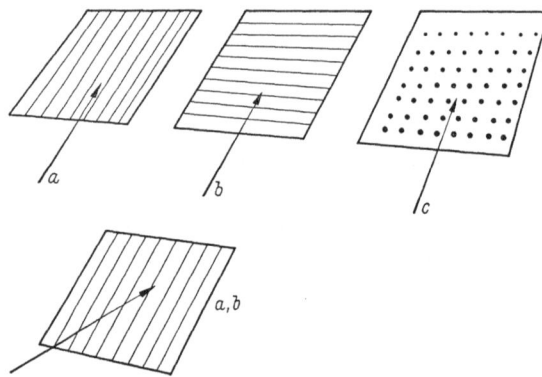

Abb. 1. Beispiele periodischer Absaugegrenzschichten. a) Querperiodisch (Senkenlinien in Strömungsrichtung); b) Längsperiodisch (Senkenlinien quer zur Strömungsrichtung); c) Doppeltperiodisch (regelmäßige Lochanordnungen); a, b) Überlagerung von a) und b) bei Schräganblasung von Senkenlinien

trachten, längs der die Absaugegeschwindigkeit in einer periodischen Weise verteilt ist. Es sollen dabei folgende Fälle unterschieden werden:

a) Verteilung der Absaugegeschwindigkeit periodisch quer zur Strömungsrichtung, konstant in Strömungsrichtung (Abb. 1 a).

b) Verteilung der Absaugegeschwindigkeit periodisch in Strömungsrichtung, konstant quer dazu (Abb. 1 b).

c) Verteilung der Absaugegeschwindigkeit periodisch in beiden Richtungen (Abb. 1 c).

Der in Abb. 1 ebenfalls dargestellte Fall der schrägangeströmten Schlitze ist nicht in echter Weise doppeltperiodisch, sondern kann als lineare Überlagerung der Fälle a) und b) aufgefaßt werden.

Im folgenden bezeichnen wir die Koordinaten in Strömungsrichtung mit x, diejenige senkrecht zur Wand mit y und die dritte Koordinate quer zur Strömungsrichtung mit z. Entsprechend seien auch die Geschwindigkeitskomponenten u, v, w festgelegt. U sei die als konstant angenommene Geschwindigkeit der Außenströmung, $v_\infty = \mathrm{const}$ die gleichmäßig verteilte Absaugegeschwindigkeit in genügendem Abstand von der Wand.

2. Querperiodische Absaugegrenzschichten

(Absaugegeschwindigkeit an der Wand konstant in Strömungsrichtung, periodisch quer dazu).

In diesem Fall sind alle drei Geschwindigkeitskomponenten unabhängig von x. In den Navier-Stokesschen Differentialgleichungen

$$u\,\frac{\partial u}{\partial x} + v\,\frac{\partial u}{\partial y} + w\,\frac{\partial u}{\partial z} = -\frac{1}{\varrho}\,\frac{\partial p}{\partial x} + \nu\left(\frac{\partial^2 u}{\partial x^2} + \frac{\partial^2 u}{\partial y^2} + \frac{\partial^2 u}{\partial z^2}\right) \tag{1}$$

$$u\,\frac{\partial v}{\partial x} + v\,\frac{\partial v}{\partial y} + w\,\frac{\partial v}{\partial z} = -\frac{1}{\varrho}\,\frac{\partial p}{\partial y} + \nu\left(\frac{\partial^2 v}{\partial x^2} + \frac{\partial^2 v}{\partial y^2} + \frac{\partial^2 v}{\partial z^2}\right) \tag{2}$$

$$u\,\frac{\partial w}{\partial x} + v\,\frac{\partial w}{\partial y} + w\,\frac{\partial w}{\partial z} = -\frac{1}{\varrho}\,\frac{\partial p}{\partial z} + \nu\left(\frac{\partial^2 w}{\partial x^2} + \frac{\partial^2 w}{\partial y^2} + \frac{\partial^2 w}{\partial z^2}\right) \tag{3}$$

$$\frac{\partial u}{\partial x} + \frac{\partial v}{\partial y} + \frac{\partial w}{\partial z} = 0 \tag{4}$$

fallen daher die unterstrichenen Glieder weg. Die letzten drei Gleichungen enthalten die u-Komponente nicht, und die Querströmung (v, w)

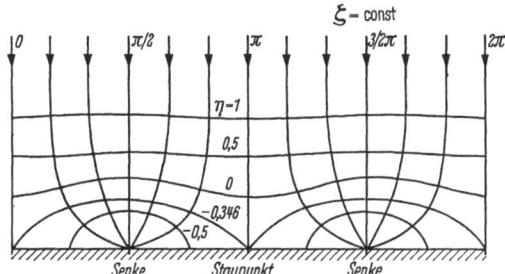

Abb. 2. Sonderfall einer querperiodischen Absaugegrenzschicht, bei der die zähe Strömung mit einer Potentialströmung identisch ist. Als Wandlinie kann irgendeine Linie $\eta = $ const gewählt werden. Die Verteilung der Absaugegeschwindigkeit ist in bestimmter Weise vorgegeben. Auf den Linien $\eta = $ const ist auch $u = $ const

kann also vorab unabhängig von u berechnet werden. Nachträglich kann dann die u-Komponente aus einer linearen Differentialgleichung ermittelt werden. Für spezielle Fälle, in denen die Querströmung identisch mit der Potentialströmung ist, kann die Lösung in expliziter Form angegeben werden. Sie sind von mir in einer früheren Untersuchung [1] behandelt worden. Sie entsprechen allerdings nicht der periodischen Absaugung an einer ebenen Wand, sondern an einer welligen oder gekerbten Wand, oder einer unendlichen Folge von längsangeströmten ovalen Zylindern mit Absaugung (Abb. 2).

Die u-Komponente ist in diesen Fällen gegeben durch:

$$\frac{u}{U} = 1 - e^{-\frac{F}{\nu}\,(\eta - \eta_0)} \tag{5}$$

wobei η die in Abb. 2 dargestellte krummlinige Koordinate und η_0 die Wandkontur darstellt, während F eine Konstante ist, die bis auf einen Maßstabsfaktor gleich v_∞ ist.

Interessanter ist jedoch in diesem Zusammenhang der Fall, daß an einer ebenen Wand in Strömungsrichtung in gleichen Abständen Liniensenken angeordnet sind. Bei Nichtberücksichtigung der Flüssigkeitsreibung ist das Potential der Querströmung hier gegeben durch:

$$\Phi = \frac{2\,v_\infty\,l}{\pi}\;ln\;\sin\frac{\pi\,\zeta}{2\,l} \tag{6}$$

wobei $\zeta = z + i\,y$ und $2\,l$ der Abstand zweier Liniensenken ist. Die Geschwindigkeitskomponenten der Querströmung sind gegeben durch:

$$v = v_\infty\,\frac{\sin\pi\,z/l}{\mathrm{Cos}\,\pi\,y/l - \cos\pi\,z/l} \tag{7}$$

$$w = v_\infty\,\frac{\mathrm{Sin}\,\pi\,y/l}{\mathrm{Cos}\,\pi\,y/l - \cos\pi\,z/l}\;. \tag{8}$$

Das Stromliniennetz Abb. 2 zeigt in der Mitte einen Staupunkt, und in der Umgebung dieses Staupunktes gehen für kleine $\bar{z} = z - z_{st}$ die Gl. (7) und (8) näherungsweise über in

$$v \approx v_\infty\,\frac{\pi}{2\,l}\,\bar{z} \tag{7a}$$

$$w \approx v_\infty\,\frac{\pi}{2\,l}\,y\;, \tag{8a}$$

was ganz der ebenen Staupunktströmung entspricht, bei der die Wirkung der Wandreibung als strenge Lösung der Navier-Stokesschen Gleichungen bekannt ist. Die Übertragung auf den vorliegenden Fall setzt natürlich voraus, daß der Grenzschichteinfluß in einem Bereich abklingen muß, in dem die Näherungen (7a) und (8a) noch genau genug sind. Eine rohe Abschätzung zeigt, daß hierzu die mit der mittleren Absaugegeschwindigkeit v_∞ und dem Abstand $2\,l$ der Liniensenken gebildete Reynoldszahl $-\,v_\infty\,2\,l/\nu > 100$ sein muß. In Abb. 3 sind für den als untere Grenze angenommenen Fall $-\,v_\infty\,2\,l/\nu = 100$ die Grenzschichtbereiche eingetragen. Entwickelt man andererseits die Formeln (7) und (8) in der Umgebung von $y = z = 0$ in eine Reihe, so findet man mit $y^2 + z^2 = r^2$ erwartungsgemäß eine reine Radialströmung

$$v_r = \sqrt{v^2 + w^2} = -\,2\,v_\infty\,l/r\,\pi\;. \tag{9}$$

Der Einfluß der Wandreibung auf eine derartige Radialströmung ist von mir in einer früheren Arbeit [2] untersucht worden. Dem Vorgehen von JEFFERY (1915) und HAMEL (1916) entsprechend ergeben sich dabei als strenge Lösungen elliptische Funktionen. Bei dem hier in Betracht kommenden Re-Zahlbereich geht die strenge Lösung näherungsweise in die von K. Pohlhausen (1921) abgeleitete Grenzschichtlösung über:

$$v_r = \frac{A}{r} \left\{ -3 \, \mathrm{Tg}^2 \left[\sqrt{\frac{A}{2\nu}} \left(\frac{\pi}{2} - \vartheta\right) + 1{,}146 \right] + 2 \right\}. \tag{10}$$

Hierbei ist A dadurch bestimmt, daß

$$\int_0^{\pi/2} v_r \, r \, d\vartheta = v_\infty l$$

ist.

Nachdem die Komponenten v und w der Querströmung bestimmt sind, kann die u-Komponente aus einer linearen Differentialgleichung er-

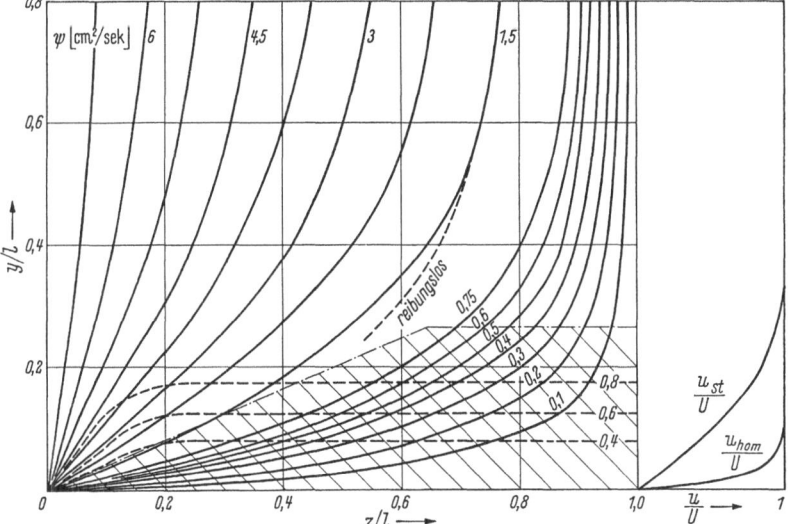

Abb. 3. Stromlinienbild einer querperiodischen Absaugegrenzschicht für $Re = (v_\infty \cdot 2\,l)/\nu = 100$. Bei $\nu = 0{,}15$ cm²/s (Luft) ist also $\psi_0 = v_\infty \cdot l = 7{,}5$ cm²/s. Der Grenzschichtbereich ist schraffiert, Linien $u = $ const sind gestrichelt eingetragen. Rechts u-Profil in der Nähe des Staupunktes (u_{st}) und gleichwertiges homogenes Absaugeprofil (u_{hom})

mittelt werden. Die Grenzschicht der u-Komponente wird im allgemeinen dicker sein als diejenige der Querströmung. Außerhalb der Grenzschicht der Querströmung ist die u-Grenzschicht durch die Formel (5) gegeben, wobei η die in Abb. 2 dargestellte krummlinige Koordinate ist. Im Bereich der v-w-Grenzschicht ändert sich auch die u-Grenzschicht gegenüber der Lösung (5). Sowohl in der Umgebung des Staupunktes als auch in der Umgebung der Senkenlinie kann man die u-Grenzschicht aus einer gewöhnlichen linearen Differentialgleichung bestimmen und an die äußere Lösung anschließen. Abb. 3 zeigt für den Fall $-2 v_\infty \, l/\nu = 100$ den Verlauf der u-Grenzschicht in Form von Linien konstanter Geschwindigkeit.

21*

3. Längsperiodische Absaugegrenzschichten

(Absaugegeschwindigkeit an der Wand konstant quer zur Strömungsrichtung, periodisch in Strömungsrichtung).

In diesem Fall fällt die Querkomponente w weg, es handelt sich also
um ein ebenes Problem. Geschlossene Lösungen können für die reibungs-
freie Strömung bei regelmäßig aufeinanderfolgenden Senkenlinien leicht
angegeben werden. Abgesehen von der Vertauschung der z- und x-Ko-
ordinate und dem Hinzutreten eines konstanten Gliedes entsprechen sie

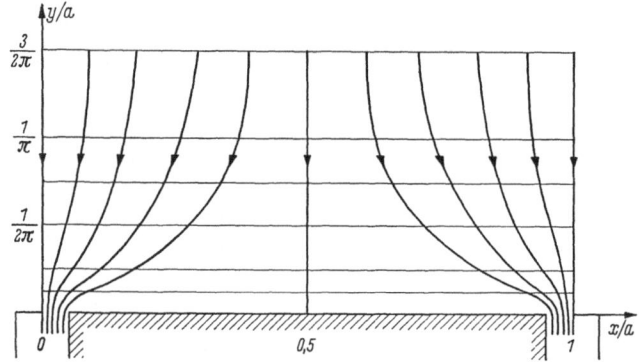

Abb. 4. Stromlinienbild einer reibungslosen Absaugeströmung allein (Verhältnis von Schlitzbreite zu
Schlitzabstand $\lambda = 0{,}1$)

ganz der bereits in den Gln. (6) bis (8) angeschriebenen Potentialströmung. Im Hinblick auf allgemeinere periodische Verteilungen ziehen
wir es aber vor, die Geschwindigkeitskomponenten in Form von Fourier-
entwicklungen anzuschreiben:

$$\frac{u}{U} = 1 + c_Q \frac{l}{\pi} \sum_{n=0}^{\infty} \left[a_n'(y) \cos \frac{n\pi x}{l} + b_n'(y) \sin \frac{n\pi x}{l} \right]$$

$$\frac{v}{U} = - n\, c_Q \sum_{n=1}^{\infty} \left[a_n(y) \sin \frac{n\pi x}{l} + b_n(y) \cos \frac{n\pi x}{l} \right] \tag{11}$$

$$a_n = A_n\, e^{-\frac{n\pi y}{l}} \quad \text{und} \quad b_n = B_n\, e^{-\frac{n\pi y}{l}} \quad \text{und} \quad c_Q = \frac{v_\infty}{U}$$

und die Striche Ableitungen nach y bezeichnen.

Für Schlitze endlicher Breite, also für eine Verteilung der Absaugegeschwindigkeit entsprechend einer Rechteckkurve werden die Konstanten

$$A_n = 0, \qquad B_n = \frac{2 \sin n\lambda\pi}{n^2\, \lambda\, \pi}$$

wobei $\lambda = $ Schlitzbreite/Schlitzabstand ist. Abb. 4 zeigt das Stromlinienbild, das sich für die reibungslose Absaugeströmung allein ergibt.

Die Darstellung läßt erkennen, daß in einem Abstand etwa gleich dem halben Schlitzabstand die diskontinuierliche Absaugung in eine annähernd gleichmäßige Absaugung übergegangen ist.

Zur Berücksichtigung der Wandreibung gehen wir von den Navier-Stokesschen Differentialgleichungen aus, die wir nach Elimination des Druckes in der Form schreiben:

$$u \frac{\partial \omega}{\partial x} + v \frac{\partial \omega}{\partial y} = \nu \, \varDelta \, \omega \qquad (12)$$

wobei

$$2 \, \omega = \frac{\partial v}{\partial x} - \frac{\partial u}{\partial y} .$$

Bei Einführen der Stromfunktion ψ geht (12) über in:

$$\frac{\partial \psi}{\partial y} \frac{\partial \varDelta \psi}{\partial x} - \frac{\partial \psi}{\partial x} \frac{\partial \varDelta \psi}{\partial y} = \nu \, \varDelta \, \varDelta \, \psi . \qquad (13)$$

Macht man für die Stromfunktion ψ den Fourieransatz:

$$\psi = U \, y + v_\infty \left[x + \sum_{n=0}^{\infty} \left\{ a_n (y) \cos \frac{n \, \pi \, x}{l} + b_n (y) \sin \frac{n \, \pi \, x}{l} \right\} \right] \qquad (14)$$

und führt dies in (13) ein, so erhält man für jedes Fourierglied eine gewöhnliche Differentialgleichung, die aber alle anderen Fourierglieder in Produktform enthält. Eine direkte Berechnung dieser nichtlinearen Gleichungssysteme ist ziemlich aussichtslos. Im folgenden wird daher ein Näherungsverfahren vorgeschlagen.

Wenn man die Absaugegeschwindigkeit gleichmäßig über die Wand verteilte, würde sich das asymptotische Absaugeprofil einstellen, das gegeben ist durch:

$$\frac{u_0}{U} = 1 - e^{\frac{v_\infty \, y}{\nu}} \qquad \omega_0 = + \frac{v_\infty}{\nu} \cdot e^{\frac{v_\infty \, y}{\nu}} .$$

Es empfiehlt sich, das Geschwindigkeitsfeld in diesen homogenen Anteil und einen periodischen Anteil aufzuteilen in der Form:

$$u = u_0 + u_1, \qquad v = v_\infty + v_1, \qquad \omega = \omega_0 + \omega_1$$

wobei also u_1, v_1, ω_1 periodische Funktionen in x sind, die durch eine Fourierreihe darstellbar sind. Da der homogene Teil für sich die Differentialgleichung (12) befriedigt, erhält man dann

$$(u_0 + \underline{u_1}) \frac{\partial \omega_1}{\partial x} + v_1 \left(\frac{\partial \omega_0}{\partial y} + \underline{\frac{\partial \omega_1}{\partial y}} \right) + v_\infty \frac{\partial \omega_1}{\partial y} = \nu \, \varDelta \, \omega_1 . \qquad (15)$$

Bei der hier allein interessierenden schwachen Absaugung kann man die unterstrichenen Glieder vernachlässigen, die Gleichung also linearisieren. Wie weit diese Vernachlässigungen zulässig sind, erkennt man am sichersten dadurch, daß man mit den Näherungslösungen iterativ in die vollständigen Gleichungen eingeht.

Die linearisierte Differentialgleichung (15) entspricht aber der von J. PRETSCH [3] behandelten Störungsgleichung für das asymptotische Absaugeprofil mit dem Unterschied, daß hier die Randbedingung für die v-Komponente eine andere Form hat. Mehrere Beispiele sind berechnet worden. Abb. 5a und 5b zeigen für den Fall einer im Verhältnis zum Schlitzabstand sehr dünnen Grenzschicht den Verlauf der ersten drei Fourierglieder, und man erkennt, daß das erste besonders stark in Erscheinung tritt. Die ebenfalls berechneten 4. bis 7. Glieder sind wegen ihrer Kleinheit in der Zeichnung weggelassen. Die Ergebnisse unterscheiden sich grundlegend von der reibungslosen Theorie. Das Maximum von b_1 ist etwa viermal so groß, wie nach der reibungslosen Theorie zu erwarten wäre. Zeichnet man ein Stromlinienbild der Zusatzströmung (Abb. 6), so treten zwei Wirbel in Erscheinung, und in der Mitte zwischen den Schlitzen strömt sogar die Grenzschicht nach außen. Das Auftreten der Wirbel kann man sich an Hand der Gl. (15) etwa in folgender Weise klarmachen: Beschränkt man sich auf die wesentlichen reibungsfreien Glieder der linken Seite, so ist

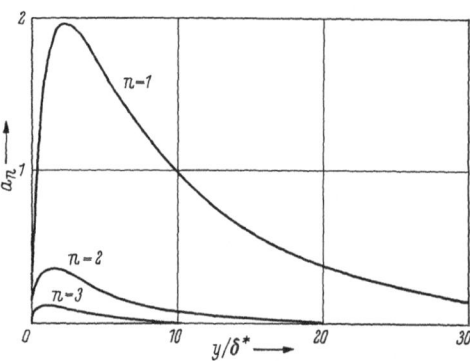

Abb. 5a. Verlauf der Fourier-Koeffizienten a_n für $Re_\delta{}^* = U\, \delta^*/\nu = 10^4$, $c_Q = 10^{-4}$

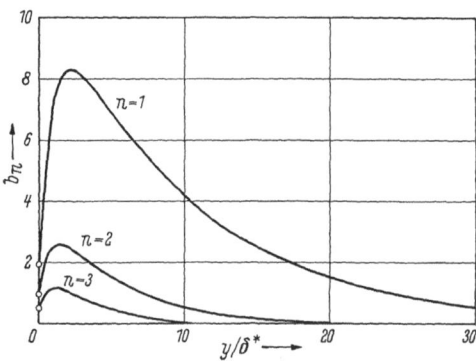

Abb. 5b. Verlauf der Fourier-Koeffizienten b_n für $Re_\delta{}^* = U\, \delta^*/\nu = 10^4$, $c_Q = 10^{-4}$

$$\frac{\partial\,\omega_1}{\partial\,x} \approx \frac{1}{u_0}\,\frac{d^2\,u_0}{d\,y^2}\,v_1.$$

Da die zweite Ableitung des Grundprofils negativ ist, ergibt sich also für einen bestimmten Wandabstand die in Abb. 7 schematisch dargestellte Verteilung der Wirbelstärke. Für die Wirbelbildung ist es also wesentlich, daß eine periodische Strömung normal zur Wand in einer Grenzschicht

mit nicht verschwindender zweiter Ableitung des Geschwindigkeits-
profils vorhanden ist. Wenn die Grenzschichtdicke größer als der
Schlitzabstand ist, klingen die Störungen bereits in einem wandnahen
Bereich ab, in dem das Geschwindigkeitsprofil durch eine Gerade an-

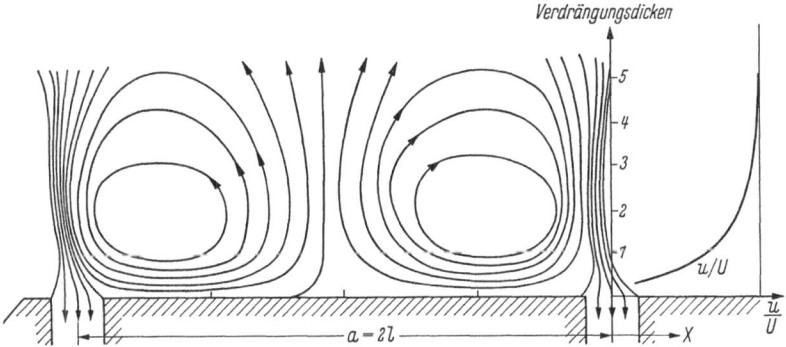

Abb. 6. Stromlinienbild der Zusatzströmung bei Absaugung durch Schlitze unter Berücksichtigung
der Reibung und der Wandgrenzschicht. Die Gesamtströmung ergibt sich durch vektorielle Addition
der rechts dargestellten Grenzschichtströmung. $(a/\delta^* = 20\,\pi,\ Re = U\,\delta^*/\nu = 10^4,\ c_Q = 10^{-4})$

genähert werden kann. Tatsächlich
verschwindet auch die Wirbelbildung
bei dieser Idealisierung völlig.

Man kann die aus der linearisierten
Gleichung gewonnenen Näherungslösungen dazu benutzen, um iterativ
aus der vollständigen Differentialgleichung eine genauere Lösung ab-
zuleiten. Für das konstante Glied ist
diese zweite Näherung auch numerisch berechnet worden in der Form:

$$\frac{u}{U} = 1 - e^{-\eta} + c_Q{}^2\,Re^*\,\varDelta\,a_0{}'(\eta) \quad (16)$$

mit $\eta = -\dfrac{v_\infty\,y}{\nu}$ und $Re^* = \dfrac{U\,l}{\pi\,\nu}$.

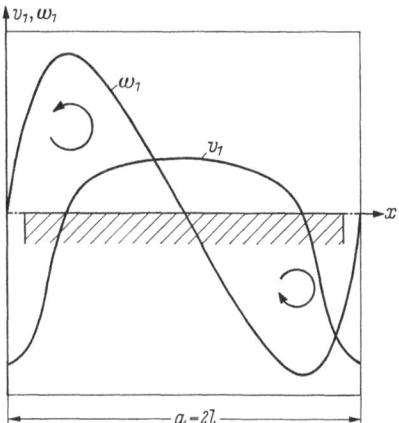

Abb. 7. Schematische Darstellung des Zusam-
menhanges zwischen Wirbelstärke und Geschwindigkeit normal zur Wand in einer
Grenzschicht mit konvex gekrümmtem Geschwindigkeitsprofil

Das letzte Glied ist demnach der
Fehler der ersten Näherung gegenüber der zweiten. Im ungünstigsten
der berechneten Beispiele betrug der maximale Fehler 2%. Auch für die
übrigen Fourierglieder sind Formeln zur iterativen Berechnung einer
höheren Näherung abgeleitet worden, aus denen man erkennt, daß der Fehler der ersten linearisierten Näherung dann klein ist, wenn ε klein genug ist

$$\left(\varepsilon = \sqrt[3]{Re^*\,c_Q{}^2}\,\right).$$

4. Doppeltperiodische Absaugegrenzschichten

Doppeltperiodische Absaugegrenzschichten, wie sie durch aufeinander-folgende Lochreihen hervorgerufen werden, sind für die praktische An-wendung von besonderer Bedeutung. Einer Berechnung sind sie aber nur schwer zugänglich, da hier alle drei Geschwindigkeitskomponenten auftreten, die auch von allen drei Koordinaten abhängen.

Einen gewissen Einblick in das Abklingen der Fourierglieder gibt aber auch schon die reibungslose Lösung. Wegen der doppeltperiodischen Randbedingungen wählen wir als Ansatz für die Geschwindigkeit v quer zur Wand

$$\frac{v}{U} = A \sum_m \sum_n c_{mn} \cos \frac{2\pi m x}{a} \cdot \cos \frac{2\pi n z}{b} e^{-\alpha_{mn} y} \qquad (17)$$

a ist hierbei der Lochabstand in x-Richtung, b in z-Richtung, A und c_{mn} sind noch zu bestimmende Konstante.

Wegen rot $\mathfrak{v} = 0$ und div $\mathfrak{v} = 0$ erhält man daraus für die u und w Komponenten:

$$\frac{u}{U} = 1 + \frac{2\pi}{a} A \sum_m \sum_n c_{mn} \frac{m}{\alpha_{mn}} \sin \frac{2 m \pi x}{a} \cos \frac{2 n \pi z}{b} e^{-\alpha_{mn} y} \qquad (18)$$

$$\frac{w}{U} = \frac{2\pi}{b} A \sum_m \sum_n c_{mn} \frac{n}{\alpha_{mn}} \cos \frac{2 m \pi x}{a} \sin \frac{2 n \pi z}{b} e^{-\alpha_{mn} y} . \qquad (19)$$

Aus der Kontinuitätsgleichung folgt dann

$$\alpha_{mn} = \sqrt{\left(\frac{2\pi m}{a}\right)^2 + \left(\frac{2\pi n}{b}\right)^2} . \qquad (20)$$

Diese Formel läßt erkennen, daß für das Abklingen der kleinere der Abstände a und b maßgebend ist. Für regelmäßig verteilte Punktsenken ist dann $A = -c_Q$ und $c_{mn} = 2$ für alle m, $n = 0$ bzw. 1 für $m = n = 0$.

Zur Berechnung der Reibungsströmung kann man wieder in ähnlicher Weise wie vorher eine Linearisierung vornehmen, wobei man von dem asymptotischen Absaugeprofil als Grundprofil ausgeht, das gleich u_0 gesetzt wird. Es ist dann

$$\left.\begin{aligned} u &= u_0 + u_1 (x, y, z) \\ v &= v_\infty + v_1 (x, y, z) \\ w &= w_1 (x, y, z) \end{aligned}\right\} \qquad (21)$$

wobei dann u_1, v_1, w_1 ebenso wie v_∞ als kleine Größen behandelt werden, deren Produkte in den Gleichungen vernachlässigt werden. Nach Elimi-nation des Druckes durch Einführen der Rotation

$$\left.\begin{aligned} 2\omega_1 &= \frac{\partial v_1}{\partial x} - \frac{\partial u_1}{\partial y}, \quad 2\omega_0 = -\frac{d u_0}{d y} \\ 2\chi_1 &= \frac{\partial w_1}{\partial y} - \frac{\partial v_1}{\partial z} \\ 2\zeta_1 &= \frac{\partial u_1}{\partial z} - \frac{\partial w_1}{\partial x} \end{aligned}\right\} \qquad (22)$$

lauten dann die linearisierten Navier-Stokesschen Differentialgleichungen:

$$\left.\begin{aligned}
u_0\,\frac{\partial\,\omega_1}{\partial\,x} + v_1\,\frac{d\,\omega_0}{d\,y} - \omega_0\,\frac{\partial\,\omega_1}{\partial\,z} &= \nu\,\varDelta\,\omega_1 \\[2mm]
u_0\,\frac{\partial\,\chi_1}{\partial\,x} - \zeta_1\,\frac{d\,u_0}{d\,y} - \omega_0\,\frac{\partial\,\omega_1}{\partial\,z} &= \nu\,\varDelta\,\chi_1 \\[2mm]
u_0\,\frac{\partial\,\zeta_1}{\partial\,x} - \omega_0\,\frac{\partial\,v_1}{\partial\,z} &= \nu\,\varDelta\,\zeta_1
\end{aligned}\right\} \qquad (23)$$

Die weitere Rechnung hat dann offenbar so vorzugehen, daß die reibungslosen Ansätze (17) bis (19) in der folgenden Weise verallgemeinert werden:

$$\left.\begin{aligned}
\frac{u_1}{U} &= f_1(y)\,\sin\frac{2\,m\,\pi\,x}{a}\,\cos\frac{2\,n\,\pi\,z}{b} + f_2(y)\,\cos\frac{2\,m\,\pi\,x}{a}\,\sin\frac{2\,n\,\pi\,z}{b} \\[2mm]
\frac{v_1}{U} &= g_1(y)\,\cos\frac{2\,m\,\pi\,x}{a}\,\cos\frac{2\,n\,\pi\,z}{b} + g_2(y)\,\sin\frac{2\,m\,\pi\,x}{a}\,\sin\frac{2\,n\,\pi\,z}{b} \\[2mm]
\frac{w_1}{U} &= h_1(y)\,\sin\frac{2\,m\,\pi\,x}{a}\,\cos\frac{2\,n\,\pi\,z}{b} + h_2(y)\,\cos\frac{2\,m\,\pi\,x}{a}\,\sin\frac{2\,n\,\pi\,z}{b}
\end{aligned}\right\} \qquad (24)$$

Durch Einsetzen dieser Ansätze in die Differentialgleichungen (23) und in die Kontinuitätsgleichung erhält man dann gewöhnliche Differentialgleichungen zur Bestimmung der Funktionen $f_1(y)$ usw. Für große Werte von y müssen diese Funktionen verschwinden, ebenso wie u_1 und w_1 an der Wand, während hier v_1 durch die Absaugung vorgegebene Werte hat.

Wenn das Geschwindigkeitsfeld periodischer Absaugegrenzschichten vollständig bestimmt ist, wird es eine Aufgabe weiterer Untersuchungen sein, den Einfluß dieser stehenden Wellen auf die Stabilität des laminaren Geschwindigkeitsprofils zu untersuchen.

Literatur

[1] WUEST, W.: Asymptotische Absaugegrenzschichten an längs angeströmten zylindrischen Körpern. Ing. Arch. **23**, 198—208 (1955).

[2] WUEST, W.: Strömung durch Schlitz- und Lochblenden bei kleinen Reynoldszahlen. Ing. Arch. **22**, 357—367 (1954).

[3] PRETSCH, J.: Umschlagbeginn und Absaugung. Jahrbuch der Deutschen Luftfahrtforschung (1942), S. 1—7.

Aus der Diskussion

L. N. PERSEN (Trondheim): Sind die Wirbel, die wir gesehen haben, etwa dieselben wie die GÖRTLER-Wirbel, und wirken sie sich entsprechend auf die Stabilität aus?

W. WUEST (Göttingen): Nein, sie haben damit unmittelbar nichts zu tun. Die GÖRTLER-Wirbel sind Längswirbel, während diese in Richtung senkrecht dazu liegen.

Einige Strömungen der idealen Flüssigkeit mit entstehender Ablösung und ihre Betrachtung vom Standpunkt der Grenzschichttheorie

Von

A. A. Nikolskij

Akademie der Wissenschaften, Moskau

In einer im Jahre 1924 veröffentlichten Arbeit über „Die Entstehung von Wirbeln in idealen Flüssigkeiten" [1] hat LUDWIG PRANDTL auf die Möglichkeit und die Zweckmäßigkeit hingewiesen, die Entstehung und weitere Entwicklung spiraliger Wirbelflächen γ an umströmten Kanten zu untersuchen, insbesondere die ähnlichen Strömungen dieser Art. Untersuchungen dieser besonderen Bewegungsform, die im folgenden als „zweite Bewegungsform" bezeichnet werden soll, sind gegenwärtig notwendig geworden, da gerade diese Bewegungsform die auf dünne Tragflügel kleiner Streckung bei Anströmung unter einem Anstellwinkel wirkenden Kräfte in erheblichem Maße bestimmt. Ebenso hängt der Durchgang von Stoßwellen um Hindernisse mit Kanten davon ab.

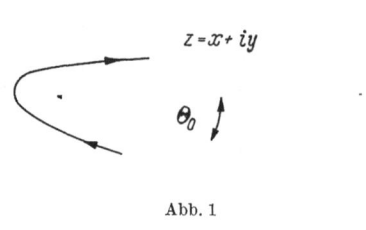

Abb. 1

Wenn ein ebener Körper mit Kanten zur Zeit $t = 0$ in Ruhe ist und sich zur Zeit $t > 0$ mit der Geschwindigkeit $v = v_0(t)$ bewegt, so entsteht für

$$\frac{1}{l} \int_0^t v_0(t)\, dt \ \ll 1$$

an den Kanten des Körpers die „zweite Bewegungsform", die in einem mit der umströmten Kante fest verbundenen Koordinatensystem ein komplexes Potential besitzt, dessen Hauptglied

$$W_1 = \varphi_1 + i\, \psi_1 = c\, l^{1-n}\, v_0(t)\, z^n \tag{1}$$

ist. l ist dabei eine für die Abmessungen des Körpers charakteristische Länge, und es gilt:

$$n = \frac{\pi}{2\,\pi - \theta_0}$$

mit dem Kantenwinkel θ_0; c ist eine dimensionslose Konstante und $z = x + i\,y$ eine komplexe Variable (s. Abb. 1).

Wir betrachten die folgende allgemeine Form des komplexen Ausgangspotentials der stetigen Strömung um einen unendlich ausgedehnten Keil:

$$W_1 = f_1(t)\, z^n + f_2(t)\, z^{2n}.$$

Wenn

$$W_1 = \varkappa_1\, t^m\, z^n + \varkappa_2\, t^{\frac{2m(1-n)-n}{2-n}}\, z^{2n} \tag{2}$$

mit $\varkappa_1 = \text{const}$, $\varkappa_2 = \text{const}$, $m = \text{const}$, $n = \text{const}$ gilt, dann ist die Strömung für $t > 0$ einschließlich der entwickelten Wirbelfläche γ „ähnlich". Die gesamte Zirkulation Γ_1 der Wirbel, aus denen die Fläche γ besteht, ändert sich mit der Zeit nach dem Gesetz:

$$\Gamma_1 = \varkappa_1^{\frac{2}{2-n}}\, t^{\frac{2m+n}{2-n}}\, g_1$$

wobei g_1 eine dimensionslose reelle Konstante ist. Wenn $m = -n/2$ ist, dann hängt Γ_1 nicht von t ab. Es sei zunächst $m > -n/2$ und strebe dann gegen diesen Wert: $m \to -n/2$. Bei diesem Prozeß wird eine immer größere Zahl der äußeren Windungen γ frei von Zirkulation, und in der Grenze $m = -n/2$ ist schließlich die ganze Spirale frei von Wirbeln. Das gesamte Wirbelsystem hat sich zu einem diskreten Wirbel zusammengezogen, der im Zentrum der Spirale γ liegt.

Wir betrachten die Strömung für $m = -n/2$, wenn sich das ganze Wirbelsystem zu einem diskreten Wirbel im Punkte $z = z_1(t)$ zusammengezogen hat, der die mit der Zeit konstante Zirkulation Γ_1 hat. Wir setzen:

$$\varkappa_1^{\frac{n}{n-2}}\, t^{-\frac{n}{2}}\, z_1^{\,n} = R_1\, e^{i\,\vartheta_1}\,; \qquad g_1 = \varkappa_1^{\frac{2}{n-2}}\, \Gamma_1$$

wobei R_1, g_1, ϑ_1 vorläufig unbestimmte reelle Konstante sind. Unter der Voraussetzung der Abwesenheit äußerer Kräfte im Punkte $z = z_1(t)$ und bei Forderung einer endlichen Geschwindigkeit für $z = 0$, also an der Spitze des Keils, erhalten wir die Gleichungen:

$$\frac{1}{2}\, R_1^{\frac{2}{n}} = n\, R_1 \cos\vartheta_1 + 2\, n\, \beta\, R_1^2 \cos 2\,\vartheta_1 - \frac{g_1}{4\,\pi}\, n\, \frac{\cos\vartheta_1}{\sin\vartheta_1}$$

$$n\, R_1 \sin\vartheta_1 + 2\, n\, \beta\, R_1^2 \sin 2\,\vartheta_1 = \frac{g_1}{4\,\pi} \tag{3}$$

$$\pi\, R_1 = g_1 \sin\vartheta_1\,.$$

Dabei ist

$$\beta = \varkappa_1^{\frac{2(n-1)}{2-n}}\, \varkappa_2$$

ein dimensionsloser Parameter. Für $\beta = 0$ ($\varkappa_2 = 0$) läßt sich das System (3) leicht lösen und ergibt:

$$R_1 = \left\{ (1-n)\, \sqrt{n}\, \sqrt{4\,n-1} \right\}^{\frac{n}{2-n}}\,; \qquad g_1 = 2\,\pi\, \sqrt{n}\, R_1\,; \qquad \sin\vartheta_1 = \frac{1}{2\,\sqrt{n}}\,. \tag{4}$$

Im Falle der Umströmung einer unendlich ausgedehnten Halbplatte ($n = 1/2$) hat man als Lösung des Gleichungssystems (3):

$$R_1 = \frac{1}{\sqrt[3]{4 \cos \vartheta_1}} \; ; \qquad g_1 = \frac{\pi}{\sqrt[3]{4 \cos \vartheta_1} \, \sin \vartheta_1} \; ; \qquad \beta = \frac{\cos 2\,\vartheta_1 \sqrt[3]{4 \cos \vartheta_1}}{4 \sin \vartheta_1 \, \sin 2\,\vartheta_1} \, . \qquad (5)$$

Bei ähnlicher Bewegung, die durch das komplexe Ausgangspotential (2) hervorgerufen wird, bewegt sich der Bildpunkt eines betrachteten Teilchens der z-Ebene in den Ebenen:

$$\mu = \varkappa_1^{\frac{n}{n-2}} \, t^{\frac{n(m+1)}{n-2}} \, z^n \; ; \qquad \lambda = \varkappa_1^{\frac{1}{n-2}} \, t^{\frac{m+1}{n-2}} \, z$$

längs einer gewissen Kurve, der „Bahn" der ähnlichen Bewegung. Die Gleichung der Bahn hängt nicht von der Zeit ab. Eine Kurve der z-Ebene, die einer Bahn in der μ- oder in der λ-Ebene entspricht, besteht für beliebige Zeiten t stets aus den gleichen Flüssigkeitsteilchen. Das Abbild der Wirbelspirale γ in den Ebenen μ und λ ist eine Kurve, die von zwei verschiedenen, sich vom Körper ablösenden Bahnen der ähnlichen Bewegung eingeschlossen wird. In dem betrachteten, ausgearteten Fall, für den $m = -n/2$ gilt, hat die Differentialgleichung der ähnlichen Bewegung in der μ-Ebene die Form:

$$\frac{d \ln \bar{\mu}}{d \ln \mu} = \frac{\dfrac{1}{2} - n\,\mu \mid \mu \mid^{-\frac{2}{n}} \left(1 + 2\,\beta\,\mu - \dfrac{1}{2\,\pi\,i} \dfrac{g_1}{\mu - \mu_1} + \dfrac{1}{2\,\pi\,i} \dfrac{g_1}{\mu - \mu_1} \right)}{\dfrac{1}{2} - n\,\bar{\mu} \mid \mu \mid^{-\frac{2}{n}} \left(1 + 2\,\beta\,\bar{\mu} + \dfrac{1}{2\,\pi\,i} \dfrac{g_1}{\overline{\mu - \mu_1}} - \dfrac{1}{2\,\pi\,i} \dfrac{g_1}{\overline{\mu - \mu_1}} \right)} \qquad (6)$$

mit

$$\mu_1 = R_1 \, e^{i\,\vartheta_1} .$$

R_1, ϑ_1, g_1 werden aus (3) ermittelt. Es existiert eine Bahn γ_1, die vom Körper — also von der festen Wand — ausgeht und sich auf den Wirbelpunkt $\mu = R_1 e^{i\vartheta_1}$ wickelt. Die Gestalt dieser Bahn ist die Grenzform des Bildes der Wirbelspirale γ in der μ-Ebene für $m \to -n/2$, aber die Kurve γ_1 ist nicht mehr eine tangentiale Trennkurve.

Im Falle $n = 1/2$; $m = 1/4$ wird das Vorzeichen der Geschwindigkeit im Punkte $z = 0$ durch das Vorzeichen von $\beta R_1 - \cos \vartheta_1$ bestimmt. Der Wert von β, für den $\beta R_1 = \cos \vartheta_1$ ist, soll mit β_k bezeichnet werden. Im Falle $\beta < \beta_k$ ist dann der Punkt $z = 0$ ($\mu = 0$) der Ablösepunkt der Strömung vom Körper (bzw. des Fortgehens der Bahn γ_1 vom Körper), wie es Abb. 2 zeigt. Im Falle $\beta > \beta_k$ ist der Punkt $z = 0$ (bzw. $\mu = 0$) zwar der Eingangspunkt der Strömung, aber nicht der Ablösepunkt, an dem sich die Bahn γ_1 vom Körper entfernt. Wie Abb. 3 zeigt, liegt der Ablösepunkt C, an dem die Kurve γ_1 den Körper verläßt, bei $x > 0$. Für $\beta \to \infty$ folgt $\vartheta_1 \to 0$, $g_1 \to \infty$, $R_1 \to \dfrac{1}{\sqrt[3]{4}}$. In der z-Ebene lokalisiert sich die zweite Bewegungsform dann um den beweglichen Punkt

$$y = 0 ; \qquad x = 4^{-\frac{2}{3}} \varkappa_1^{\frac{2}{3}} \sqrt{t}$$

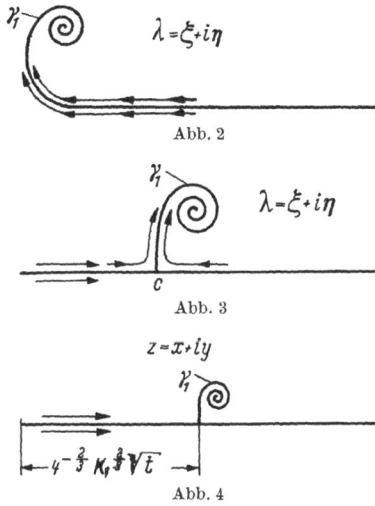

Abb. 2

Abb. 3

wie es Abb. 4 zeigt.

Wir bezeichnen mit $[a]$ die Dimension einer beliebigen physikalischen Größe a, mit L die Dimension einer Länge, mit T die Dimension der Zeit. Mit ν soll die kinematische Zähigkeit bezeichnet werden; dann gilt:

$$[\nu] = \frac{L^2}{T} .$$

Im betrachteten Fall mit $m = -n/2$ bekommen wir

$$[\varkappa_1] = \frac{L^{2-n}}{T^{\frac{2-n}{2}}}$$

und deshalb $\quad [\nu] = \varkappa_1^{\frac{2}{2-n}}$.

Abb. 4

Im wesentlichen bringt daher die Berücksichtigung des Zähigkeitskoeffizienten ν keine neue Dimensionskonstante herein. Man kann deshalb die Strömung der zähen Flüssigkeit in entsprechender Form ansetzen wie die oben behandelte Strömung der idealen Flüssigkeit mit diskreten Wirbeln. Die Dimension von ν stimmt mit der Dimension der Zirkulation der Wirbel überein. Man kann feststellen, daß die bekannte Lösung für das Zerfließen eines Wirbels für $t > 0$ der für $t = 0$ einen diskreten Wirbel mit der Zirkulation bildet, gerade wegen $[\nu] = [\varGamma]$ ähnlich ist.

In den dimensionslosen Größen

$$\xi = \varkappa_1^{\frac{1}{n-2}} t^{-\frac{1}{2}} x ; \qquad \eta = \varkappa_1^{\frac{1}{n-2}} t^{-\frac{1}{2}} y ; \tag{7}$$

$$U = \varkappa_1^{\frac{1}{n-2}} t^{\frac{1}{2}} u ; \quad V = \varkappa_1^{\frac{1}{n-2}} t^{\frac{1}{2}} v ; \quad P = \varkappa_1^{\frac{2}{n-2}} t \frac{p}{\varrho} ; \quad \varepsilon = \nu \varkappa_1^{\frac{2}{n-2}} = \mathrm{const}$$

bei denen u und v die Komponenten der Geschwindigkeit, p den Druck und ϱ die Dichte der Flüssigkeit bedeuten, bekommt die Bewegungsgleichung für die zähe Flüssigkeit die Form:

$$-\frac{1}{2} U + \left(U - \frac{1}{2}\xi \right) \frac{\partial U}{\partial \xi} + \left(V - \frac{1}{2}\eta \right) \frac{\partial U}{\partial \eta} = -\frac{\partial P}{\partial \xi} + \varepsilon \left(\frac{\partial^2 U}{\partial \xi^2} + \frac{\partial^2 U}{\partial \eta^2} \right)$$

$$-\frac{1}{2} V + \left(U - \frac{1}{2}\xi \right) \frac{\partial V}{\partial \xi} + \left(V - \frac{1}{2}\eta \right) \frac{\partial V}{\partial \eta} = -\frac{\partial P}{\partial \eta} + \varepsilon \left(\frac{\partial^2 V}{\partial \xi^2} + \frac{\partial^2 V}{\partial \eta^2} \right) \tag{8}$$

$$\frac{\partial U}{\partial \xi} + \frac{\partial V}{\partial \eta} = 0 .$$

Bei kleinen Werten von ε vereinfacht sich die Aufgabe entsprechend und führt zur Lösung eines ähnlichen Grenzschichtproblems, bei dem die Randbedingungen an der äußeren Grenze von der für die ideale Flüssigkeit gefundenen Lösung übernommen werden. Für die Halbplatte ($n = 1/2$) erhält man bei $\beta < \beta_k$ offensichtlich ein hinreichend klares Bild der Strömung: die Grenzschicht besteht aus einem an der Wand anliegenden Teil und einer freien Schicht, die sich längs der Spiralbahn γ_1 erstreckt. Die freie Schicht ist als Ablösungsprodukt der an der Wand gebildeten Grenzschicht aufzufassen. Der diskrete Wirbel wird in der zähen Strömung durch eine stetige Wirbelverteilung ersetzt, in deren Mitte die Strömung etwa der bekannten Strömung für einen zerfließenden Wirbel entspricht. Bei der Umströmung von Kanten mit $n > 1/2$ sowie bei der Umströmung von Platten im Falle $\beta > \beta_k$ ist der Mechanismus der Strömung weniger übersichtlich, da die Geschwindigkeit des Abfließens der Teilchen vom Körper im Ablösepunkt (Punkt, an dem sich die Kurve γ_1 vom Körper entfernt) verschwindet. Die Untersuchung dieser Aufgaben für zähe Flüssigkeiten wäre zweifellos für die allgemeine Theorie der Strömungen mit Ablösung sehr interessant.

Literatur

[1] PRANDTL, L.: Über die Entstehung von Wirbeln in der idealen Flüssigkeit, mit Anwendung auf die Tragflügeltheorie und andere Aufgaben. Vorträge aus dem Gebiete der Hydro- und Aerodynamik (Innsbruck 1922), hrsg. v. TH. V. KÁRMÁN u. T. LEVI-CIVITA. Berlin: Springer (1924), S. 18—33.

Grenzschicht-Theorie
der homogenen und Zweikomponenten-Flüssigkeit mit zwei Geschwindigkeiten

Von

H. A. Rachmatulin

Moskauer Staatsuniversität und Akademie der Wissenschaften der Usbek. SSR

In letzter Zeit entwickelt sich die Methode der Grenzschichttheorie zu einer der universellen Methoden der Mechanik der kontinuierlichen Massen, wobei sie Anwendung bei Problemen findet, die sich von denen der Bewegung der zähen Flüssigkeit unterscheiden.

In der Sowjetunion haben die Methoden der Grenzschichttheorie in letzter Zeit bei folgenden Forschungen Anwendung gefunden:

1. starke, nicht punktförmige Detonation (G. G. TSCHERNY);

2. instationäre Filterung (P. J. KOTSCHIN, G. I. BARENBLATT);

3. instationäre Wärmeübertragung (J. A. DEMJANOW).

Mit den vorliegenden Ausführungen ist beabsichtigt, die Aufmerksamkeit von Forschern noch auf eine Reihe von Anwendungsgebieten der Grenzschichttheorie zu lenken.

Dabei faßt der Verfasser die Anwendung der Methode der Grenzschichttheorie auf die sogenannte Zweigeschwindigkeits-Hydro-Gasdynamik ins Auge.

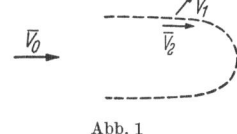

Abb. 1

Um zu erklären, worum es sich handelt, werden die konkreten Probleme angeführt, bei denen die Zweigeschwindigkeits-Hydrodynamik Anwendung findet.

1. Umströmung der porösen Oberfläche einer homogenen Flüssigkeit:

In diesem Falle kann man die Aufgabe so stellen, daß die Strömung einer idealen Flüssigkeit ermittelt wird, von der ein Teil mit der mittleren Dichte ϱ_1 und mit der Geschwindigkeit $\overline{V}_1$ durch die Fläche strömt und ein anderer Teil mit der mittleren Dichte ϱ_2 und der Geschwindigkeit $\overline{V}_2$ die letztere umströmt (vgl. Abb. 1).

Hierbei ist $\varrho_1 + \varrho_2 = \varrho$ die wirkliche Dichte der Flüssigkeit. Offensichtlich sind die Grenzbedingungen:

$$k \, V_{1n} = \varDelta \, p$$

$$V_{2n} = 0$$

$$x^2 + y^2 + z^2 \to \infty: \ \overline{V}_1 \to \overline{V}_2 \to \overline{V}_0 \qquad (1.1)$$

wobei V_{1n}, V_{2n} die normalen Geschwindigkeitskomponenten bedeuten. Die oben angegebenen Bedingungen stellen rein kinetische dar. Dynamisch hängt die Möglichkeit des Zweigeschwindigkeitszustandes von der richtigen Wahl der Kräftehypothese ab. Anscheinend sind entweder besondere Massenkräfte einzuführen, oder die Hypothese der Möglichkeit von zwei Drücken in der Flüssigkeit ist anzunehmen.

Augenscheinlich wird ein solcher Zweigeschwindigkeitszustand in Wirklichkeit nur in dem an die poröse Oberfläche grenzenden Gebiet bestehen, obwohl derselbe theoretisch überall möglich ist.

Demgemäß gelangen wir zum Problem der Grenzschicht einer idealen Flüssigkeit (eine analoge Aufgabe kann auch für eine zähe Flüssigkeit gestellt werden). Im vorliegenden Falle hat man nicht das klassische Problem der Grenzschicht, wobei die Strömung außerhalb der Schicht bekannt ist, sondern das Problem der Wechselwirkung der Grenzschicht mit der Außenschicht (analog dem Falle großer Überschallgeschwindigkeiten; beiläufig sei bemerkt, daß die erste zu lösende Aufgabe in der Strömung eines Überschallstromes um eine poröse Platte besteht).

2. Das Problem der porösen Abkühlung, ein Problem, das sich im Grunde genommen nicht vom ersten unterscheidet, nur ist in diesem Falle die Bedingung (1.1) durch die entsprechende zu ersetzen.

3. Das Problem der Grenzschichtabsaugung.

4. Das Problem von mit kleinsten Makroteilchen gesättigten Flüssigkeiten und Gasen [1].

Letzteres Problem stellt im wesentlichen eine Verallgemeinerung des Problems der Filterungstheorie dar.

1. Ableitung der Differentialgleichungen der Grenzschicht einer Zweikomponentenflüssigkeit

Anscheinend kann man mit kleinsten Makroteilchen gesättigte Flüssigkeiten und Gase berechtigterweise als Gemische mehrerer kontinuierlicher Massen betrachten.

Offensichtlich sind für jede der an der Bewegung teilnehmenden Massen die Theorien der Spannung und der Deformation gerechtfertigt.

Folglich erhält man:

$$\sigma_{ix} = -p + 2\,\mu_i\,\frac{\partial u_i}{\partial x} - \frac{2}{3}\,\mu_i\,div\,\overline{V}_i$$

$$\sigma_{iy} = -p + 2\,\mu_i\,\frac{\partial v_i}{\partial y} - \frac{2}{3}\,\mu_i\,div\,\overline{V}_i$$

$$\sigma_{iz} = -p + 2\,\mu_i\,\frac{\partial w_i}{\partial z} - \frac{2}{3}\,\mu_i\,div\,\overline{V}_i$$

$$\tau_{ixy} = \mu_i\left(\frac{\partial u_i}{\partial y} + \frac{\partial v_i}{\partial x}\right)$$

$$\tau_{iyz} = \mu_i\left(\frac{\partial v_i}{\partial z} + \frac{\partial w_i}{\partial y}\right)$$

$$\tau_{ixz} = \mu_i\left(\frac{\partial u_i}{\partial z} + \frac{\partial w_i}{\partial x}\right)$$

wobei $\qquad i = 1,2 \qquad$ ist. $\hfill (2.1)$

Die Bezeichnungen $f_1 = \dfrac{\varrho_1}{\varrho_{1i}}$; $f_2 = \dfrac{\varrho_2}{\varrho_{2i}}$ werden eingeführt, wobei ϱ_1, ϱ_2 die mittlere und ϱ_{1i}, ϱ_{2i} die wirkliche Dichte entsprechend der ersten und der zweiten Masse bedeuten.

Bei den angenommenen Bezeichnungen erhalten die Geschwindigkeits-gleichungen folgende Form:

$$\varrho_i\,\frac{d\,v_i}{d\,t} = -f_i\,\frac{\partial p}{\partial x} + \frac{\partial}{\partial x}\,f_i\,\mu_i\left(2\,\frac{\partial u_i}{\partial x} - \frac{2}{3}\,div\,\overline{V}_i\right) +$$

$$+ \frac{\partial}{\partial y}\,f_i\,\mu_i\left(\frac{\partial u_i}{\partial y} + \frac{\partial v_i}{\partial x}\right) + k\,(u_2 - u_1)$$

$$\varrho_i\,\frac{d\,v_i}{d\,t} = -f_i\,\frac{\partial p}{\partial y} + \frac{\partial}{\partial y}\,f_i\,\mu_i\left(2\,\frac{\partial v_i}{\partial y} - \frac{2}{3}\,div\,\overline{V}_i\right) +$$

$$+ \frac{\partial}{\partial x}\,f_i\,\mu_i\left(\frac{\partial u_i}{\partial y} + \frac{\partial v_i}{\partial x}\right) + k\,(v_2 - v_1)$$

$$\frac{\partial\,(\varrho_i u_i)}{\partial x} + \frac{\partial\,(\varrho_i v_i)}{\partial y} = 0$$

$$\varSigma f_i = 1; \quad i = 1,2 \,. \hfill (2.2)$$

In diesen sieben Gleichungen erweisen sich folgende sieben Größen als unbekannt: u_1, u_2, v_1, v_2, ϱ_1, ϱ_2, p.

Wenn in diesen Gleichungen ϱ_1, ϱ_2 als konstant und die Bedingung $\varSigma f_i = 1$ als befriedigt angenommen wird, so erhält man sechs Gleichungen mit fünf Unbekannten.

Folglich kann man im Falle der Bewegung einer Zweikomponenten-masse, bei der jede Komponente nicht kompressibel ist, die mittleren Dichten ϱ_1, ϱ_2 nicht als konstant annehmen.

Trotzdem ist eine solche Annahme für die Bewegung in der Grenzschicht zulässig und führt nicht zur Überbestimmtheit der Aufgabe, da
zwei Gleichungen aus der Betrachtung ausscheiden.

In Wirklichkeit erhält man, wenn man ϱ_1 und ϱ_2 als konstant annimmt
und die gewöhnlichen Bewertungen für den Fall der Bewegung in der
Grenzschicht durchführt:

$$\varrho_1 \frac{d u_1}{d t} = -f_1 \frac{\partial p}{\partial x} + f_1 \mu_1 \frac{\partial^2 u_1}{\partial y^2} + k (u_2 - u_1); \quad \frac{\partial u_1}{\partial x} + \frac{\partial v_1}{\partial y} = 0$$

$$\varrho_2 \frac{d u_2}{d t} + - f_2 \frac{\partial p}{\partial x} + f_2 \mu_2 \frac{\partial^2 u_2}{\partial y^2} + k (u_1 - u_2); \quad \frac{\partial u_2}{\partial x} + \frac{\partial v_2}{\partial y} = 0. \qquad (2.3)$$

Damit erhält man vier Gleichungen für die vier Unbekannten u_1, u_2,
v_1, v_2.

Indem man für den Fall, daß der Druckgradient gleich Null ist, die
Integralbeziehungen von v. Kármán benutzt und Pohlhausen befolgt,
wobei man entscheidet

$$u_1 = U \left[3/2 \frac{Y}{\delta_1} - \frac{1}{2} \left(\frac{Y}{\delta_1} \right)^3 \right] \quad 0 \le y \le \delta_1$$

$$u_2 = U \left[3/2 \frac{Y}{\delta_2} - \frac{1}{2} \left(\frac{Y}{\delta_2} \right)^3 \right] \quad 0 \le y \le \delta_2$$

$$u_2 = U \qquad\qquad\qquad\qquad\qquad \delta_2 \le y \le \delta_1$$

erhält man folgende Gleichungen für die Grenzschichtdicken:

$$\frac{d \delta_1}{d x} = \frac{\alpha_1}{\delta_1} - \beta_1 (\delta_1 - \delta_2)$$

$$\frac{d \delta_2}{d x} = \frac{\alpha_2}{\delta_2} + \beta_2 (5 - 6 \lambda + \lambda^3) \delta_2, \qquad (2.4)$$

wobei $\lambda = \delta_2/\delta_1$ ist.

Man kann zeigen, daß, wenn $\alpha_2 < \alpha_1$ ist, so ist $\delta_2 < \delta_1$. In Wirklichkeit
erhält man, wenn man x aus (2.4) eliminiert:

$$\frac{d \delta_2}{d \delta_1} = \frac{\dfrac{\alpha_2}{\delta_2} + \beta_2 \delta_2 (5 - 6 \lambda + \lambda^3)}{\dfrac{\alpha_1}{\delta_1} - \beta_1 (\delta_1 - \delta_2)}. \qquad (2.5)$$

Als Isoklinen dieser Gleichung erweisen sich:

$$\delta_2 = \delta_1, \text{ wobei } \frac{d \delta_2}{d \delta_1} = \frac{\alpha_1}{\alpha_2} < 1$$

$$\delta_2 = \delta_1 - \frac{\alpha_1}{\beta_1 \delta_1} = f (\delta_1), \text{ wobei } \frac{d \delta_2}{d \delta_1} = \infty .$$

Das Isoklinenbild ist in Abb. 2 gezeigt.

Dort ist auch der Charakter der Integralkurve der Gl. (2.5) gezeigt. Wie tatsächliche Berechnungen zeigen, strebt δ_2 sehr schnell nach δ_1 und beide zum Wert der Grenz-schichtdicke der Flüssigkeit mit der Dichte $\varrho_1 + \varrho_2$.

Im Falle der Bewegung mit Druck-gradient werden u_1 und u_2 und folglich auch δ_1, δ_2 sich unbedingt an jeder Stelle voneinander unterscheiden. Hierbei er-gibt die Zweigeschwindigkeits-Theorie unter allen Umständen ein Resultat, das von dem Resultat der die Zwei-komponenten-Flüssigkeit durch eine ho-mogene Flüssigkeit ersetzenden Theorie verschieden ist.

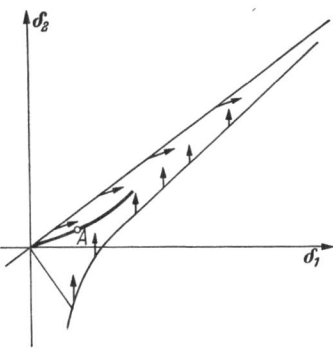

Abb. 2. Das Bild der Isoklinen
in der Ebene δ_1, δ_2

2. Integration der Gleichungen der Grenzschicht einer Zweikompo-nenten-Flüssigkeit

Als Beispiel wird die Grenzschicht einer ebenen Zweikomponenten-Strömung in einem erweiterten Rohr betrachtet.

Zu Beginn findet man die gleichmäßige Strömung einer idealen Zwei-komponenten-Flüssigkeit.

Die Gleichungen der Erhaltung der Masse ergeben:

$$U_1 \varrho_1 \sigma(x) = \text{const} = c_1 \tag{3.1}$$

$$U_2 \varrho_2 \sigma(x) = \text{const} = c_2.$$

Die Gleichungen der Bewegung haben offensichtlich die Form:

$$U_1 \frac{dU_1}{dx} = -\frac{1}{\varrho_{1i}} \cdot \frac{\partial p}{\partial x} + \frac{k}{\varrho_1}(U_2 - U_1)$$

$$U_2 \frac{dU_2}{dx} = -\frac{1}{\varrho_{2i}} \cdot \frac{\partial p}{\partial x} + \frac{k}{\varrho_2}(U_1 - U_2)$$

$$\frac{\varrho_1}{\varrho_{1i}} + \frac{\varrho_2}{\varrho_{2i}} = 1. \tag{3.2}$$

Es wird angenommen, daß $\partial p / \partial x = \text{const} > 0$ und bekannt ist. Im Er-gebnis der Lösung der Aufgabe findet man $\sigma = \sigma(x)$.

Indem man aus (3.1) ϱ_1, ϱ_2 in die letzte Gleichung aus (3.2) einsetzt, erhält man:

$$\frac{c_1}{U_1 \varrho_{1i}} + \frac{c_2}{U_2 \varrho_{2i}} = \sigma(x). \tag{3.3}$$

Indem man ϱ_1 und ϱ_2 aus den beiden ersten Gln. (3.2) eliminiert, erhält man:

$$U_1 \frac{d\,U_1}{d\,x} = -\frac{1}{\varrho_{1i}} \cdot \frac{\partial\,p}{\partial\,x} + \frac{U_1 \cdot \sigma \cdot k}{c_1} \, (U_2 - U_1)$$

$$U_2 \frac{d\,U_2}{d\,x} = -\frac{1}{\varrho_{2i}} \cdot \frac{\partial\,p}{\partial\,x} + \frac{U_2 \cdot \sigma \cdot k}{c_2} \, (U_1 - U_2)\,. \tag{3.4}$$

Aus Gl. (3.3) und (3.4) kann man die Größen U_1, U_2 als Funktionen von $x - \alpha$ bestimmen.

Es wird vorgeschlagen, diese zusammen mit der Gleichung der Grenzschichtdicken zu integrieren.

Bekanntlich kann das System (2.3) in folgende Form umgewandelt werden:

$$\varrho_1 \int_0^{\delta_1} \frac{\partial\,u_1}{\partial\,x} \, (2\,u_1 - U_1)\, d\,y = -f_1 \int_0^{\delta_1} \frac{\partial\,p}{\partial\,x}\, d\,y -$$

$$- f_1 \mu_1 \left(\frac{\partial\,u_1}{\partial\,y}\right)_{y\,=\,0} + k \int_0^{\delta_1} (u_2 - u_1)\, d\,y$$

$$\varrho_2 \int_0^{\delta_2} \frac{\partial\,u_2}{\partial\,x} \, (2\,u_2 - U_2)\, d\,y = -f_2 \int_0^{\delta_2} \frac{\partial\,p}{\partial\,x}\, d\,y -$$

$$- f_2 \mu_2 \left(\frac{\partial\,u_2}{\partial\,y}\right)_{y\,=\,0} + k \int_0^{\delta_2} (u_1 - u_2)\, d\,y\,.$$

Beim Aufbau des Geschwindigkeitsprofils in der Grenzschicht wird folgende Grenzbedingung befriedigt:

bei $y = 0$

$$u_j = v_j = 0$$

und folglich

$$-\frac{1}{\varrho_{ji}} \cdot \frac{\partial\,p}{\partial\,x} + V_{ji}\frac{\partial^2\,u_i}{\partial\,y^2} = 0 \tag{3.6}$$

bei $y = \delta_i$

$$u_j = U_j$$

$$\frac{\partial\,u_j}{\partial\,y} = 0$$

$$\frac{\partial^2\,u_j}{\partial\,y^2} = 0$$

$$j = 1, 2\,. \tag{3.7}$$

Wenn man Pohlhausen befolgt, erhält man:

$$u_j = \bar{a}_j\, y + \bar{b}_j\, y^2 + \bar{c}_j\, y^3 + \bar{d}_j\, y^4\,. \tag{3.8}$$

Aus (3.6) und (3.8) erhält man:

$$\bar{b}_j = \frac{1}{2\,\mu_i} \cdot \frac{\partial\,p}{\partial\,x}$$

$$\bar{a}_j\,\delta_j + \bar{b}_j\,\delta_1{}^2 + \bar{c}_j\,\delta_j{}^3 + \bar{d}_j\,\delta_j{}^4 = U_j \qquad\qquad (3.9)$$

$$\bar{a}_j + 2\,\bar{b}_j\,\delta_j + 3\,\bar{c}_j\,\delta_j{}^2 + 4\,\bar{d}_j\,\delta_j{}^3 = 0$$

$$2\,\bar{b}_j + 6\,\bar{c}_j\,\delta_j + 12\,\bar{d}_j\,\delta_j{}^2 = 0.$$

Aus diesen Gleichungen erhält man:

$$\bar{a}\,\delta \;= 2\,U - 1/3\,\bar{b}\,\delta^2$$

$$\bar{b}\,\delta^2 = \bar{b}\,\delta^2$$

$$\bar{c}\,\delta^3 = -\,2\,U - \bar{b}\,\delta^2$$

$$\bar{d}\,\delta^4 = U + 1/3\,\bar{b}\,\delta^2\,.$$

(In den letzten Formeln ist der Index j weggelassen.)

Nach einfachen, aber umfangreichen Berechnungen bei $b = $ const ergibt sich:

$$\int_0^{\delta} \frac{\partial\,u}{\partial\,x}\,(2\,u - U)\,d\,y = \left[-\frac{53}{252}\,U^2 - \frac{2}{315}\,\bar{b}\,U\,\delta^2 + \frac{863\vdots2}{11\,340}\,\bar{b}^2\,\delta^4\right]\frac{d\,\delta}{d\,x} +$$

$$+ \frac{293}{630}\,U\,U'\,\delta - \frac{71}{3.80}\,\bar{b}\,U'\,\delta^3\,.$$

Gleichfalls ergeben unmittelbare Berechnungen

$$\int_0^{\delta_1} (u_2 - u_1)\,d\,y = \frac{7}{10}\,U_2\,\delta_2 - \frac{1}{60}\,\bar{b}_2\,\delta_2{}^3 - \frac{7}{10}\,U_1\,\delta_1 + \bar{b}_1\,\delta_1{}^3\,\frac{1}{60} + U_2\,(\delta_1 - \delta_2)$$

$$\int_0^{\delta_2} (u_1 - u_2)\,d\,y = U_1\left[\xi - \frac{1}{2}\,\xi^3 + \frac{1}{5}\,\xi^4\right]\delta_2 +$$

$$+ \bar{b}_1\,\delta_1\,\delta_2{}^2\left[-\frac{1}{6} + \frac{1}{3}\,\xi - \frac{1}{4}\,\xi^2 + \frac{1}{15}\,\xi^3\right] -$$

$$- \frac{7}{10}\,U_2\,\delta_2 + \frac{1}{60}\,b_2\,\delta_2{}^3\,;\quad \text{wobei } \xi = \frac{\delta_2}{\delta_1}\ \text{ist.}$$

Demgemäß gelangt man zu folgendem System von Differentialglei-chungen:

$$\varrho_1\left[-\frac{53}{252}\,U_1{}^2 - \frac{2\,\bar{b}_1\,U_1\,\delta_1{}^2}{315} - \frac{135\vdots3}{11\,340}\,\bar{b}_1{}^2\,\delta_1{}^4\right]\frac{d\,\delta_1{}^2}{d\,x} +$$

$$+ \frac{293}{315}\,U_1\,U_1'\,\delta_1{}^2\,\varrho_1 - 71\,\frac{b_1\,\bar{U}_1'\,\delta_1{}^4\,\varrho_1}{1890} = -\,2\,f_1\,\frac{\partial\,p}{\partial\,x}\,\delta_1{}^2 - 2f_1\,\mu_1\,a_1\,\delta_1 +$$

$$+ 2\,k\,\delta_1\left[\frac{7}{10}\,U_2\,\delta_2 - \frac{\bar{b}_2\,\delta_2{}^3}{60} - \frac{7\,U_1\,\delta_1}{10} + \frac{\bar{b}_1\,\delta_1{}^3}{60} + U_2\,(\delta_1 - \delta_2)\right]$$

$$\varrho_2 \left[-\frac{53}{252} U_2{}^2 - \frac{2\bar{b}_1 U_2 \delta_2{}^2}{315} - \frac{13583}{11340} \bar{b}_2{}^2 \delta_2{}^4 \right] \frac{d\,\delta_2{}^2}{d\,x} +$$

$$+ \frac{293}{315} U_2 U_2' \delta_1{}^2 \varrho_1 - 71 \frac{\bar{b}_2 U'_2 \delta_2{}^4 \varrho_2}{1890} = -2 f_2 \frac{\partial p}{\partial x} \delta_2{}^2 -$$

$$-2 f_2 \mu_2 \bar{a}_2 \delta_2 + 2 k \delta_2{}^2 U_1 \left[\xi - \frac{1}{2} \xi^3 + \frac{1}{5} \xi^4 \right] +$$

$$+ 2 \bar{b}_1 \delta_1 \delta_2{}^3 k \left[-\frac{1}{6} + \frac{\xi}{3} - \frac{\xi^2}{4} + \frac{\xi^3}{15} \right] - \frac{7 k U_2 \delta_2{}^2}{5} + \frac{k \bar{b}_2 \delta_2{}^4}{30} \; ;$$

$$\text{wobei } \xi = \delta_2/\delta_1 \,.$$

Parallel ist die Gleichung der Grenzschicht einer homogenen Flüssigkeit zu integrieren, die folgende Form hat:

$$\frac{d\,z}{d\,x} = \frac{0{,}8\,[-9072 + 1670{,}4\,\lambda - (47{,}4 + 4{,}8\,U\,U''/U'^2)\,\lambda^2 - (1 - U\,U''/U'^2)\,\lambda^3]}{U\,[-213{,}12 + 5{,}76\,\lambda + \lambda^2]}$$

wobei

$$z = \frac{\delta^2}{\nu} \; ; \; \lambda = \frac{U\,\delta^2}{\nu} \,.$$

Im vorliegenden Falle werden die Funktionen U, U', U'' nach folgenden Formeln ermittelt:

$$U = \frac{\sigma_0 U_0}{\sigma\,(x)} \; ; \; \sigma\,(x) = \frac{c_1}{\varrho_{1i} U_1} + \frac{c_2}{\varrho_{2i} U_2}$$

$$U' = -\frac{\sigma_0 U_0}{\sigma^2} \cdot \frac{d\,\sigma}{d\,x} \; ; \; U'' = -\frac{\sigma_0 U_0}{\sigma^2} \cdot \frac{\bar{a}^2 \sigma}{d\,x^2} + 2\,\frac{\sigma_0 U_0}{\sigma^3} \left(\frac{d\,\sigma}{d\,x} \right)^2 .$$

Die oben angegebenen Differentialgleichungen können numerisch integriert werden. Die Ergebnisse dieser Berechnungen werden in nächster Zeit veröffentlicht.

Literatur

[1] RACHMATULIN, H. A.: „Grundlagen der Gasdynamik sich durchkreuzender Bewegungen kompressibler Massen" P.M.M. Bd. XX (1956). Eine Zusammenfassung der Arbeit ist in den Werken des Brüsseler Kongresses für angewandte Mechanik veröffentlicht.

Aus der Diskussion

M. REINER (Haifa): I fear, I did not understand you. The equations applied are applicable only to continuous media. One of your two media must be discontinuous.

RACHMATULIN (Moskau): Es liegt die Vorstellung zugrunde, daß für jedes Medium die Teilchen so dicht liegen, daß das Medium als Kontinuum behandelt werden kann. Es werden dabei zwei Arten von Kräften angenommen, und zwar solche, die Teilchen eines Mediums untereinander ausüben, und solche, die die Teilchen eines Mediums auf die des anderen Mediums ausüben. Der Ausdruck „Zweikomponenten-Flüssigkeit" stammt von LANDAU.

Diskussionsveranstaltung zur VII. Sitzung

Contribution to the Subject
Boundary Layer Suction

By **W. Pfenninger,** Northrop Aircraft, Inc., Hawthorne, Cal.

Various boundary layer suction methods were investigated at Northrop for low drag boundary layer control at high Reynolds numbers in the inlet length of laminar flow tubes. Suction was applied in the rear part of the tube through 8 slots, 80 fine slots and 80 rows of closely spaced holes. Laminar flow could be maintained through a considerable pressure rise by means of suction at length Reynolds numbers of $14 \cdot 10^6$ to $20 \cdot 10^6$. The minimum suction quantities for full laminar flow through a moderately strong rear pressure rise were 25% to 30% smaller with suction through 80 slots, as compared with suction through 8 slots. Suction experiments by GOLDSMITH with 80 rows of closely spaced holes showed slightly larger minimum suction rates than with 80 slots.

A film (taken by W. A. MEYER) showed flow observations with smoke in the inlet length of a laminar flow tube with suction through holes. With a single suction hole, trailing vortices were observed downstream of the hole, leading to transition for stronger suction. Similar trailing vortices were found downstream of several suction holes for relatively weak suction, and the flow generally remained laminar. With increasing suction, horseshoe vortices developed periodically between adjacent suction holes, resulting generally in premature transition. With further increasing suction and with closely spaced holes, standing vortices (instead of horseshoe vortices) were observed between adjacent holes, and the flow downstream of the holes became laminar again.

Horseshoe vortices were also noticed downstream of a three-dimensional roughness element or bleed holes.

I. L. VAN INGEN (Amsterdam): At Delft Technical University some preliminary experiments were made on the behaviour of the boundary layer with suction through discrete holes, the holes being positioned just on the boundary between the regions with laminar and turbulent flow if no suction is applied. The china-clay method was used to indicate regions of laminar and turbulent flow.

Although directly downstream of the holes a laminar region is formed, the flow adjacent to both sides of this region is still more "turbulent" than without suction, this being indicated by white streaks in the china-clay. This disturbance seems to be caused by the secondary flow which is induced by suction through discrete sinks.

Beitrag zum Thema
Grenzschicht-Absaugung

Von **F. X. Wortmann**
Institut für Aerodynamik und Gasdynamik der Technischen Hochschule Stuttgart

1. Aus dem Impuls- und Energiesatz gewinnt K. WIEGHARDT [1] für eine einparametrige Profilklasse zwei simultane Differentialgleichungen für den Formparameter $H = \dfrac{\delta_1}{\delta_2}$ und die örtliche Reynoldssche Zahl der Grenzschicht $re = \dfrac{U\delta_2}{\nu}$:

$$\frac{dH}{d\sigma} = -f\,\frac{d\,\dfrac{U}{U_0}}{\dfrac{U}{U_0}\,d\sigma} - g\,\frac{U}{U_0}\,\frac{Re}{re^2} + h\,\frac{Re\,v_0}{re\,U_0}\;;\qquad Re = \frac{U_0\,R}{\nu} \tag{1}$$

$$\frac{dre}{d\sigma} = -\left\{\frac{(H+1)\,d\,\dfrac{U}{U_0}}{\dfrac{U}{U_0}\,d\sigma} + \frac{d\,\dfrac{r_0}{R}}{\dfrac{r_0}{R}\,d\sigma}\right\}re + \varepsilon\,\frac{U}{U_0}\,\frac{Re}{re} + \frac{Re\,v_0}{U_0}\,. \tag{2}$$

Darin bedeuten δ_1 die Verdrängungsdicke, δ_2 die Impulsverlustdicke, U_0 die Anströmgeschwindigkeit, U die Geschwindigkeit außerhalb der Grenzschicht. ν ist die kinematische Zähigkeit, v_0 die Absaugegeschwindigkeit, ε die mit δ_2 dimensionslos gemachte Wandtangente des Grenzschichtprofils. R ist eine charakteristische Länge, z. B. die Profiltiefe oder der größte Radius des Drehkörpers. $\sigma = s/R$ ist die dimensionslose Bogenlänge des umströmten Körpers. r_0 ist der Radius des Drehkörpers. Im ebenen Fall wird $r_0/R = 1$. $f,\, g,\, h$ sind lediglich Funktionen des Formparameters H, die von WIEGHARDT für die von SCHLICHTING angegebenen Absaugeprofile angegeben werden. Das Gleichungssystem (1) (2) benutzt WIEGHARDT zur Behandlung der praktisch besonders interessierenden Frage, wie groß bei vorgegebenen $U\,(\sigma)$, $r_0\,(\sigma)$ die Absaugegeschwindigkeit $v_0\,(\sigma)$ sein muß, wenn die laminare Grenzschicht an jeder Stelle gegen kleine Störungen gerade stabil sein soll. Durch Interpolation theoretischer Stabilitätsgrenzen verschiedener Absaugeprofile ergibt sich für die kritische Reynoldssche Zahl re_{krit} in Abhängigkeit vom Formparameter H in guter Annäherung die Kopplung:

$$re_{\mathrm{krit}} = e^{a-bH}\qquad \text{mit}\qquad \begin{array}{l}a = 26{,}3\\ b = 8\,.\end{array} \tag{3}$$

Damit eliminiert WIEGHARDT aus (1) und (2) $v_0\,(\sigma)$ und $re\,(\sigma)$, um eine Gleichung allein für H zu erhalten:

$$\frac{dH}{d\sigma} = -f_1\,\frac{dU}{U\,d\sigma} + f_2\,\frac{dr_0}{r_0\,d\sigma} - f_3\,\frac{U}{U_0}\,\frac{Re}{re^2}\;;\qquad re = e^{a-bH}. \tag{4}$$

WIEGHARDT löst diese Differentialgleichung durch schrittweise numerische Integration.

An Stelle der numerischen Integration von (4) wird nun folgender Lösungsweg vorgeschlagen:

Aus (1) und (2) läßt sich unter Benutzung von (3) eine zu (4) analoge Gleichung für die Reynoldssche Zahl herleiten:

$$\frac{dre}{d\sigma} = \left(A_1\,\frac{dU}{U\,d\sigma} + A_2\,\frac{dr_0}{r_0\,d\sigma}\right)re + A_3\,\frac{U}{U_0}\,\frac{Re}{re}$$

$$\text{mit}\qquad A_1 = \frac{f-f_1}{h} - (H+1)$$

$$A_2 = 1 + \frac{f_2}{h} \tag{5}$$

$$A_3 = \frac{g-f_3}{h} + \varepsilon,$$

wobei die Funktionen $f_1,\, f_2,\, f_3$ ebenso wie $f,\, g,\, h$ allein von H abhängig sind. Setzt man jetzt $H = \text{const}$, so läßt sich (5) in ähnlicher Weise wie bei Grenzschichten ohne Absaugung geschlossen integrieren:

$$re^2 = re_1{}^2 + 2\,Re\,A_3 \left(\frac{U}{U_0}\right)^{2A_1}\left(\frac{r_0}{R}\right)^{2A_2}\int_{\sigma_1}^{\sigma}\left(\frac{U}{U_0}\right)^{1-2A_1}\left(\frac{r_0}{R}\right)^{-2A_2}d\,\sigma. \tag{6}$$

Die Rechnung beginnt mit der Reynoldsschen Zahl re_1 im Instabilitätspunkt der Grenzschicht, bis zu dem die Grenzschicht gegen kleine Störungen ohnehin stabil ist. Für eine vorgegebene Geschwindigkeits- und Dickenverteilung $U(\sigma)$ und $r_0(\sigma)$ läßt sich (6) leicht für verschiedene Formparameter H_n berechnen. Trägt man die Kurvenscharen $re(\sigma, H_n)$ auf und markiert man die nach (3) jedem H-Wert zugeordnete kritische Reynoldssche Zahl, so erhält man graphisch sofort die gesuchte Beziehung $re_{\mathrm{krit}}(\sigma)$, mit der sich die erforderliche Absauggeschwindigkeit aus

$$\frac{V_0}{U_0} = \frac{(f-f_1)}{h}\frac{d\,U}{U\,d\sigma}\frac{re}{Re} + \frac{(g-f_3)}{h}\frac{U}{U_0\,re} \tag{7}$$

ergibt.

2. Zur Kontrolle seines Verfahrens zieht WIEGHARDT eine Lösung von J. PRETSCH [2] für die längsangeströmte Platte heran, die unter Benutzung der exakten Ge-

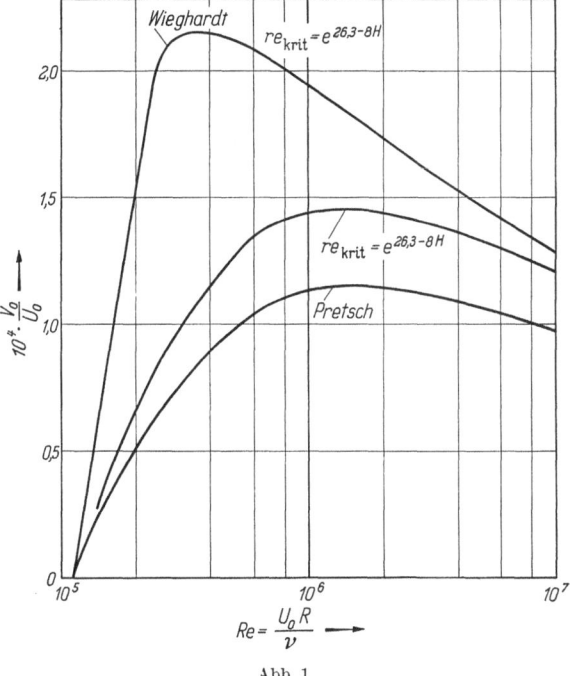

Abb. 1

schwindigkeitsprofile von R. IGLISCH [3] gewonnen wurde. Die zugehörigen kritischen Reynoldszahlen sind von A. ULRICH [4] berechnet worden. Mit den Konstanten $a = 26{,}3$; $b = 8$ für die kritische Reynoldszahl erhält WIEGHARDT die in Abb. 1 dargestellte Verteilung der Absaugemenge. Daß die Absaugemengen bei mäßigen Reynoldszahlen $Re = UR/\nu$ wesentlich zu groß ausfallen, liegt zum Teil daran, daß die Stabilitätsgrenze mit $a = 26{,}3$; $b = 8$ etwas niedriger liegt als die von PRETSCH benutzte Grenze. Unbefriedigend bleibt die verschiedenartige Verteilung der Absaugemengen.

Es wurden deshalb die Funktionen f, g, h und f_1, f_2, f_3 erneut für die Profilklasse

$$F = F_1 + K (F_2 - F_1) ; \quad 0 \leqslant K \leqslant 1 \qquad (8)$$

berechnet, wobei nach dem Vorbild von J. PRESTON [5] für F_1 das exakte Grenzschichtprofil der ebenen Platte (BLASIUS, HOWARTH) benutzt wurde. F_2 ist das asymptotische Absaugeprofil. Die gute Übereinstimmung dieser Profilklasse mit den exakten Profilen von IGLISCH ist bekannt [6]. Das Ergebnis der Rechnung mit den gleichen Konstanten $a = 26{,}3$; $b = 8$ liefert die mittlere Kurve in Abb. 1. Offensichtlich sind also die Abweichungen der WIEGHARDTschen Näherung nicht dem Verfahren, sondern primär den Ungenauigkeiten der benutzten Absaugeprofile (SCHLICHTING) zuzuschreiben. Die verbleibende Abweichung gegenüber der von PRETSCH angegebenen Lösung erklärt sich zwanglos aus den unterschiedlichen Stabilitätsgrenzen, die den Rechnungen jeweils zugrunde gelegt wurden. Die für die Profilklasse (8) berechneten Werte und Funktionen sind mit den WIEGHARDTschen Bezeichnungen in der Tabelle 1 angegeben.

Tabelle 1. *Hilfsfunktionen und Grenzschichtwerte der Absaugeprofile nach Gl. 8.*
Bezeichnungen wie bei WIEGHARDT [1].

H	f	g	h	ε	D	H_{32}	dH_{32}/dH
2,00	6,781	— 1,356	2,712	0,500	0,250	1,667	— 0,246
2,05	7,750	— 1,335	2,890	0,466	0,236	1,655	— 0,224
2,10	8,754	— 1,293	3,118	0,435	0,224	1,644	— 0,207
2,15	9,870	— 1,239	3,331	0,406	0,214	1,634	— 0,190
2,20	11,061	— 1,168	3,546	0,380	0,206	1,625	— 0,176
2,25	12,668	— 1,081	3,774	0,355	0,199	1,617	— 0,163
2,30	13,804	— 0,979	4,018	0,332	0,193	1,609	— 0,151
2,35	15,365	— 0,860	4,274	0,310	0,188	1,601	— 0,141
2,40	17,056	— 0,721	4,543	0,290	0,184	1,595	— 0,131
2,45	18,910	— 0,566	4,831	0,270	0,180	1,588	— 0,122
2,50	20,967	— 0,386	5,146	0,252	0,177	1,582	— 0,113
2,55	23,191	— 0,185	5,474	0,234	0,175	1,577	— 0,105
2,591	25,200	0,000	5,768	0,220	0,173	1,573	— 0,099
2,65	28,354	0,297	6,219	0,202	0,172	1,567	— 0,091
2,70	31,367	0,584	6,645	0,186	0,170	1,563	— 0,085
2,75	34,660	0,903	7,099	0,172	0,169	1,559	— 0,079
2,80	38,288	1,260	7,591	0,158	0,169	1,555	— 0,073

Die Hilfsfunktionen f_1, f_2, f_3 unterscheiden sich praktisch nicht von den WIEGHARDTschen Werten.

Literatur:

[1] WIEGHARDT, K.: Zur Berechnung ebener und drehsymmetrischer Grenzschichten mit kontinuierlicher Absaugung. Ing. Archiv **22**, 368 (1954).
[2] PRETSCH, J.: Die Leistungsersparnis durch Grenzschichtbeeinflussung beim Schleppen einer ebenen Platte. UM Nr. 3048 (1943).
[3] IGLISCH, R.: Exakte Berechnung der laminaren Grenzschicht an der längsangeströmten ebenen Platte mit homogener Absaugung. Schriften d. dt. Akad. d. Lufo, **8 B** Heft 1 (1944), bzw. NACA T.M. 1205 (1949).

[4] ULRICH, A.: Die Stabilität der laminaren Reibungsschicht an der ebenen Platte mit homogener Absaugung. UM Nr. 2033.

[5] PRESTON, J.: The Boundary Layer Flow over a Permeable Surface through which Suction is Applied. R. a. M. 2244 (1946).

[6] LEW, H. G., und R. D. MATHIEU: Boundary Layer Control by Porous Suction. Pennsylvania State Uni. T. R. Nr. 3 (1954).

Contribution to the Subject
Unsteady Boundary Layer Flow

By **I. Tani,** Institute of Science and Technology, University of Tokyo

I would like to supplement Dr. MOORE's lecture by presenting a simple example of unsteady laminar boundary layer problem involving separation. If the outer flow velocity is given in the form

$$U = V - \frac{x}{T - t},$$

where V and T are constants, the three independent variables, x, y, and t, can be grouped together into two variables

$$\xi = \frac{8x}{V(T-t)}, \qquad \eta = \frac{y}{2}\sqrt{\frac{V}{\nu x}},$$

and the solution can be obtained by assuming the stream function in the form

$$\psi = \sqrt{\nu x V} \sum_m \xi^m f_m(\eta),$$

just the same as in HOWARTH's solution for the steady linearly decelerating flow. The solution of this problem can be interpreted to provide some informations for the unsteady flow associated with a diffusor or an airfoil in which angle of divergence or angle of attack varies with time. It is seen that the separation is given by $\xi = 1.20$. The convergency of the series is a little more favorable as compared with HOWARTH's case, but may probably be improved by the new procedure put forward by Prof. GÖRTLER.

Probleme aus der Theorie der dreidimensionalen Grenzschichten

Von

R. Timman

Technische Hogeschool, Delft

1. Einleitung

Im zweidimensionalen Falle genügt die Vereinfachung, die die Grenzschichttheorie der Navier-Stokesschen Gleichungen gibt, gerade um technisch brauchbare Rechenmethoden zu entwickeln. „Exakte" Lösungen sind in einer genügenden Anzahl vorhanden, um Näherungsverfahren, wie das POHLHAUSENsche Verfahren und die Vereinfachung von WALZ und THWAITES, zu prüfen.

Für dreidimensionale laminare Strömungen ist die Lage aber ganz geändert.

Die Grenzschichtgleichungen, die hier nichtlinearen partiellen Differentialgleichungen in drei Variablen sind, sind dermaßen kompliziert, daß man an allgemeine Lösungsverfahren außer mit numerischen Methoden kaum glauben kann, und deshalb sind wir hier gleich von Anfang an genötigt, Näherungsverfahren auszubilden. Als exakte Lösungen sind hier vorhanden die ähnlichen Lösungen, die von GEIS gefunden wurden und in denen die Querströmung, die charakteristisch ist für wesentliche dreidimensionale Effekte, in manchen Fällen auftritt. Hier sucht man Lösungen der Grenzschichtgleichungen zu finden, die invariant sind gegenüber Ähnlichkeitstransformationen in den Geschwindigkeiten, mit Faktoren, die von den Koordinaten abhängen. Eine andere Gruppe von „exakten" Lösungen wird gegeben durch die Strömung um schief angeblasene Zylinder, für die das ganze Strömungsfeld invariant ist gegenüber einer Translation in der Richtung der Zylinderachse.

Wenn man aber um willkürlich geformte Körper die Grenzschichtströmung berechnen will, ist man genötigt, Zuflucht zu nehmen zur Verallgemeinerung des POHLHAUSENschen Verfahrens, die schon durch PRANDTL in einer kleinen Note angedeutet wurde. Während das zweidimensionale Impulsverfahren die partiellen Differentialgleichungen der Grenzschichtströmung überführt in eine gewöhnliche Differentialgleichung in der Variablen x, der Wandbogenlänge, führt hier dasselbe Verfahren die partiellen Differentialgleichungen in drei Variablen x und y entlang der Fläche und z entlang der Normalen über in ein System partieller Differentialgleichungen in den zwei Variablen x und y. Natürlich ist auch hier das Gelingen des Verfahrens abhängig von der Wahl des Geschwindigkeitsprofiles, das hier aber schwieriger nachzuprüfen ist. Das Verfahren ist aber im Prinzip auch geeignet für turbulente Grenzschichten, aber hier sind die Schwierigkeiten der Wahl und der Nachprüfung noch größer. Im folgenden wird eine Ableitung der Differentialgleichungen gegeben werden, die aus dem Impulsverfahren folgen.

2. Impulsgleichungen

Für die allgemeinen Rechnungen ist die Tensordarstellung sehr geeignet.

Wir betrachten die Strömung um eine gekrümmte Fläche und nehmen an, daß die äußere Geschwindigkeit U^i eine bekannte Vektorfunktion der räumlichen charakteristischen Koordinaten y^i ist. An der Fläche soll sie tangentiell sein.

Im Raume führen wir krummlinige Koordinaten ein: x^α ($\alpha = 0, 1, 2$), so daß x^1 und x^2 Gaußsche Koordinaten auf die Fläche F sind und x^0 in der Richtung der Normalen genommen ist.

Wenn wir, wie gewöhnlich, in den Grenzschichttheorien annehmen, daß die Krümmungsradien groß sind im Vergleich zu der Grenzschichtdicke, ist das Linienelement

$$ds^2 = (dy_i)^2 = g_{\alpha\beta}\, dx^\alpha\, dx^\beta + (dx^\circ)^2 \qquad (\alpha, \beta = 1,2).$$

Die Gleichung der Fläche ist $x^\circ = 0$. Die $g_{\alpha\beta}$ sind gegeben durch

$$g_{\alpha\beta} = \frac{\partial y^i}{\partial x^\alpha} \cdot \frac{\partial y^i}{\partial x^\beta}.$$

In den Gaußschen Koordinaten hat die Geschwindigkeit kovariante Komponenten

$$U_\alpha = U_i \frac{\partial y^i}{\partial x^\alpha} \qquad (\alpha = 1,2)$$

$$U_0 = U_i \frac{\partial y^i}{\partial x^0}.$$

Auf der Fläche ist $U_0 = 0$, und auf der Fläche ist ein Vektorfeld U_α gegeben.

Die Grenzschichtgleichungen nehmen dann die Form an

$$u^\beta \bigtriangledown_\beta u_\alpha + u^0 \frac{\partial u_\alpha}{\partial x_0} = U^\beta \bigtriangledown_\beta U_\alpha + \nu \frac{\partial^2 u_\alpha}{\partial (x^0)^2} \qquad (\alpha, \beta = 1,2),$$

und die Kontinuitätsgleichung wird

$$\nabla_\beta u^\beta + \frac{\partial u^0}{\partial x^0} = 0\,.$$

∇_α stellt die kovariante Differentiation nach den Variablen x^α dar

$$\nabla_\alpha u^\beta = \frac{\partial u^\beta}{\partial x^\alpha} + \Gamma^\beta_{\alpha\gamma}\, u^\gamma\,,$$

wobei die $\Gamma^\beta_{\alpha\lambda}$ die Krümmungsgrößen der Fläche sind. Manchmal führt man eine neue Variable

$$\zeta = \frac{x^0}{\sqrt{\nu}}\,, \qquad w = \frac{u_0}{\sqrt{\nu}}$$

ein, und man bekommt

$$u^\beta \nabla_\beta u_\alpha + w\, \frac{\partial u_\alpha}{\partial \zeta} = U^\beta \nabla_\beta U_\alpha + \frac{\partial^2 u_\alpha}{\partial \zeta^2}$$

$$\nabla_\beta u^\beta + \frac{\partial w}{\partial \zeta} = 0\,.$$

Durch Integration nach der Variablen ζ erhält man die Impulsgleichungen

$$\left(\int\limits_0^\infty (U^\beta - u^\beta)\, d\zeta\right) \nabla_\beta U_\alpha + \nabla_\beta \int\limits_0^\infty \{(U_\alpha - u_\alpha)\, u^\beta\}\, d\zeta = \left(\frac{\partial u_\alpha}{\partial \zeta}\right)_0\,.$$

Die Verdrängungsdicke ist jetzt ein Vektor

$$\delta_\alpha = \int\limits_0^\infty (U_\alpha - u_\alpha)\, d\zeta\,,$$

und die Impulsverlustdicke ein Tensor

$$\vartheta_{\alpha\beta} = \int\limits_0^\infty (U_\alpha - u_\alpha)\, u_\beta\, d\zeta\,.$$

Also bekommen wir das Gleichungenpaar

$$(\nabla_\beta U_\alpha)\, \delta^\beta + \nabla_\beta \cdot \vartheta_\alpha^\beta = \left[\frac{\partial u_\alpha}{\partial \zeta}\right]_0 \qquad\qquad \alpha = 1{,}2\,.$$

Dieses Paar von Integralgleichungen ist nicht gleichwertig mit dem ursprünglichen Paar von Differentialgleichungen.

Erst die Randbedingungen bestimmen die Lösung eindeutig. Diese liefern die sogenannten Wandbindungen,

$$\zeta = 0\,; \qquad u_\alpha = u_\alpha''' = 0$$

$$u_\alpha'' = -\, U^\beta \nabla_\beta U_\alpha\,.$$

Die höheren Wandbindungen enthalten Differentiationen nach den Koordinaten auf der Fläche; diese komplizieren die Gleichungen und werden in dem Verfahren nicht berücksichtigt.

3. Die Gleichungen in Stromlinienkoordinaten

Zuerst führen wir Stromlinienkoordinaten ein. Wir bestimmen ein neues orthogonales System von Flächenkoordinaten ξ^λ, so daß in diesen Koordinaten der Geschwindigkeitsvektor die Kennzahlen

$$\overline{U}_1 = 1, \qquad \overline{U}_2 = 0$$

hat. Die Transformationsformel

$$U_\alpha = \frac{\partial \xi^\lambda}{\partial x^\alpha}\, U_\lambda$$

ergibt

$$U_1 = \frac{\partial \xi^1}{\partial x^1}, \qquad U_2 = \frac{\partial \xi^1}{\partial x^2}.$$

Hieraus kann man $\xi^1 = \varphi$ nur bestimmen, wenn

$$\frac{\partial U_1}{\partial x^2} = \frac{\partial U_2}{\partial x^1},$$

wenn also die Komponente des Wirbelvektors in der Normalrichtung verschwindet. Dies ist gewiß der Fall, wenn die Außenströmung wirbelfrei ist. In diesem Falle ist φ das Geschwindigkeitspotential. Die zweite Koordinate ψ muß entlang den Stromlinien eine Konstante sein. Sie ist also eine Flächenstromfunktion (die Normalkomponente des Stromfunktionsvektors).

Auf der Fläche gilt nicht die zweidimensionale Kontinuitätsgleichung, wohl aber

$$\nabla_\alpha U^\alpha + \frac{\partial U^0}{\partial \xi} = 0.$$

Für das Bogenelement einer Stromlinie gilt

$$U\,ds = d\varphi,$$

wo U den Betrag des Geschwindigkeitsvektors darstellt.

Wir setzen jetzt für das Bogenelement einer Äquipotentiallinie an:

$$\sqrt{\varrho}\; U\,ds = d\psi,$$

dann ist das allgemeine Bogenelement in den Stromlinienkoordinaten

$$ds^2 = \frac{1}{U^2}\left[d\varphi^2 + \frac{d\psi^2}{\varrho} \right] = \frac{d\varphi^2}{T} + \frac{d\psi^2}{\varrho T},$$

$$\text{wo } T = U^2 = U_\alpha\, U^\alpha.$$

Die kovariante Ableitung eines Vektors F^β ist gegeben durch

$$\nabla_1 F^\beta = \frac{\partial F^\beta}{\partial \varphi} + \Gamma^\beta_{1\lambda} F^\lambda$$

$$\nabla_2 F^\beta = \frac{\partial F^\beta}{\partial \psi} + \Gamma^\beta_{2\lambda} F^\lambda$$

wobei die Christoffel-Symbole gegeben sind durch:

$$\Gamma^1_{11} = -\frac{T_\varphi}{2\,T}$$

$$\Gamma^1_{12} = \Gamma^1_{21} = -\frac{1}{\varrho}\,\Gamma^2_{11} = -\frac{T_\psi}{2\,T}$$

$$\Gamma^1_{22} = -\frac{1}{\varrho}\,\Gamma^2_{12} = \frac{1}{2\varrho}\left[\frac{T_\psi}{T} + \frac{\varrho_\varphi}{\varrho}\right],$$

$$\Gamma^2_{22} = -\frac{1}{2}\left(\frac{T_\psi}{T} + \frac{\varrho_\psi}{\varrho}\right).$$

$$U_1 = 1, \qquad U_2 = 0, \qquad U^1 = T, \qquad U^2 = 0.$$

Die Funktion ϱ hängt zusammen mit der Kontinuitätsgleichung.

$$\frac{\partial T}{\partial \varphi} + \Gamma^1_{11}\,T + \Gamma^2_{21}\,T + \frac{\partial w}{\partial \zeta} = 0$$

oder

$$\frac{1}{2}\,\frac{\varrho_\varphi}{\varrho} + \frac{\partial w}{\partial \zeta} = 0$$

$$\frac{\partial \log \sqrt{\varrho}}{\partial \varphi} = \frac{\partial w}{\partial \zeta}.$$

Sie ergibt also ein Maß für die Divergenz der Strömung. Für die Stromlinienkoordinaten einer zweidimensionalen Strömung ist $\varrho = $ konst. In diesen Stromlinienkoordinaten wird das Gleichungenpaar:

$$\frac{\partial \vartheta_{11}}{\partial \varphi} + \varrho\,\frac{\partial \vartheta_{12}}{\partial \psi} + \frac{T_\varphi}{2\,T}\,(\delta_1 + \vartheta_{11} + \varrho\,\vartheta_{22}) - \frac{\varrho_\varphi}{2\varrho}\,(\varrho\,\vartheta_{22} - \vartheta_{11}) + \frac{\varrho_\psi}{2}\,\vartheta_{12} =$$

$$= \frac{1}{T}\left[\frac{\partial u_1}{\partial \zeta}\right]_0,$$

$$\frac{\partial \vartheta_{21}}{\partial \varphi} + \varrho\,\frac{\partial \vartheta_{22}}{\partial \psi} + \frac{T_\varphi}{2\,T}\,(\delta_1 + \vartheta_{11} + \varrho\,\vartheta_{22}) - \frac{\varrho_\varphi}{2}\,\vartheta_{21} + \varsigma_\psi\,\vartheta_{22} = \frac{1}{T}\left[\frac{\partial u_2}{\partial \zeta}\right]_0,$$

wobei

$$\delta_1 = \int_0^\infty (1 - u_1)\,d\zeta, \qquad \delta_2 = -\int_0^\infty u_2\,d\zeta,$$

$$\vartheta_{11} = \int_0^\infty (1 - u_1)\,u_1\,d\zeta\,; \qquad \vartheta_{12} = \int_0^\infty (1 - u_1)\,u_2\,d\zeta\,; \qquad \vartheta_{21} = -\int_0^\infty u_1\,u_2\,d\zeta\,;$$

$$\vartheta_{22} = -\int_0^\infty u_2^2\,d\zeta\,.$$

4. Die Verhältnisse im Staupunkt

Die Randbedingungen für die Lösung der Gleichung müssen gefunden werden aus dem Verhalten im Staupunkt.

Setzen wir für das Potential im Staupunkt $x = y = 0$ die Reihenentwicklung

$$\varphi = 1/2\,(ax^2 + by^2) + \cdots \qquad \text{mit } a \leq b$$

an, so sind die Komponenten der Geschwindigkeit

$$u = ax + \cdots$$
$$v = by + \cdots.$$

Die Stromlinien sind die Lösungen der Differentialgleichungen

$$\frac{dx}{ax} = \frac{dy}{by}$$

$$y = \psi \cdot x^{b/a}.$$

Es sind also für $b > a$ Parabeln, die die x-Achse als gemeinsame Tangente haben.

Für die Transformation der x, y auf die φ, ψ Ebene ist also der Ursprung ein singularer Punkt, der auf die ganze Linie $\varphi = 0$ abgebildet wird.

Die Funktion ϱ wird hier bestimmt durch das Bogenelement

$$ds^2 = dx^2 + dy^2 = \frac{d\varphi^2}{T} + \frac{d\psi^2}{\varrho T} = \frac{d\varphi^2 + \dfrac{1}{\varrho}\, d\psi^2}{(a^2 x^2 + b^2 y^2)}.$$

Man findet nach einiger Rechnung

$$\frac{1}{\varrho} = \frac{a^2 x^2 y^2}{\psi^2} = a^2 \cdot x^{2\,(b/a)}.$$

Betrachten wir jetzt die Differentialgleichungen, so sehen wir, daß die Lösung nur regulär sein kann im Ursprung, wenn die Summe der durch T geteilten Glieder verschwindet.

Jede Gleichung liefert eine algebraische Gleichung für die beiden Unbekannten δ und ω, woraus sie numerisch bestimmt werden müssen. Man kann weitergehen und eine Reihenentwicklung ansetzen. Auf diese Weise kann man (aber mit größerem Rechenaufwand) in einer ganzen Umgebung vom Staupunkt die unbekannten Funktionen bestimmen. Da die Differentialgleichungen elliptisch sind, fehlt aber ein allgemeiner Existenz- und Eindeutigkeitsbeweis der auf diese Weise bestimmten Lösung.

Ein rein mathematisches Problem anderer Art, das hiermit eng zusammenhängt, ist das folgende.

Bekanntlich verlieren im Staupunkt die Grenzschichtgleichungen ihre Gültigkeit. Man kann also nicht sagen, daß notwendig die Lösungen dort regulär sein müssen. Bei zweidimensionalen Grenzschichten hat man die durch Erfahrung bestätigte Vermutung, daß die Lösungskurven der bezüglichen gewöhnlichen Differentialgleichungen stabil sein müssen, daß also das Verhalten der Lösung in einigem Abstand vom Staupunkt tatsächlich unabhängig von den Anfangswerten ist. Über das Verhalten der quasilinearen partiellen Differentialgleichungen ist nichts bekannt. Dies ist also ein zweites mathematisches Problem, das seiner Lösung noch weit entfernt ist.

5. Näherungsverfahren

Zur numerischen Lösung des Grenzschichtproblems muß man sich beim Fehlen einer allgemeinen Lösungsmethode geschickter Näherungsverfahren bedienen.

Diese Näherungsverfahren sind hauptsächlich von Zaat entwickelt worden, und sie wurden angewandt auf das Problem der Umströmung eines schräg gestellten Ellipsoids vom Achsenverhältnis $3:1:0,15$, wie das einem Pfeilflügel entspricht.

Bei den Stromlinien muß man zwei Gebiete unterscheiden. Am Äquator entlang fallen die Stromlinien fast gänzlich zusammen, auf der Oberfläche sind es parallele Geraden.

Betrachten wir die Impulsgleichungen, die wir kurz in der Form

$$f^1 \sigma_\varphi + f^2 \sigma_\psi + g^2 \Omega_\psi = d$$
$$F^1 \sigma_\varphi + F^2 \sigma_\psi + G^1 \Omega_\varphi + G^2 \Omega_\psi = D$$

schreiben, wo $\sigma = \delta^2$ und Ω in der Zaatschen Arbeit das Verhältnis der Grenzschichtdicken darstellt, so ist ersichtlich, daß in beiden Fällen die partiellen Ableitungen nach ψ klein sind verglichen mit den Ableitungen nach φ. Also bekommt man für jede Stromlinie gewöhnliche Differentialgleichungen

$$f^1 \frac{d\sigma}{d\varphi} = d$$
$$F^1 \frac{d\sigma}{d\varphi} + G^1 \frac{d\Omega}{d\varphi} = D,$$

die man gesondert löst.

Wenn man jetzt die Ableitungen nach ψ bildet, so kann man verbesserte Werte berechnen, indem man diese Ableitungen als bekannte Funktionen im rechten Gliede einführt. Auch hier ist mathematisch noch nichts bekannt über die Konvergenz des Verfahrens.

Für kleine Querströmungen, wie sie beim Ellipsoid auftreten, hat Zaat noch eine weitere Vereinfachung eingeführt, indem er in der ersten Gleichung von Anfang an die Querstromterme veranlässigt. Also bekommt er längs jeder Stromlinie eine Gleichung vom Typus, wie sie in der zweidimensionalen Theorie auftritt.

Die Lösung kann dann ähnlich dem Walz-Thwaitesschen Verfahren in der Form eines Integrals ganz einfach ermittelt werden. Auch der Querstromparameter ergibt sich dann durch eine einfache Quadratur. Numerisch ergibt dieses vereinfachte Verfahren im Musterbeispiel die gleichen Ergebnisse wie das vorhergehende. Für Fälle mit großen Querströmungen, wie z. B. bei der Umströmung von Ecken, wo der Impulsverlust senkrecht zu den Stromlinien von derselben Ordnung ist wie in der Stromrichtung, ist es noch nicht bekannt, ob diese Verfahren eine brauchbare Lösung ermöglichen.

6. Vereinfachte Methoden

Das allgemeine Verfahren ist sehr kompliziert in der numerischen Auswertung. Deshalb hat ZAAT versucht, es zu vereinfachen, so daß man ein Verfahren bekommt, das es ermöglicht, mit einem erträglichen Rechenaufwand zuverlässige Ergebnisse zu erzielen.

Man setzt voraus, daß die Quergeschwindigkeit U_2 von einer kleinen Ordnung δ ist im Vergleich zu der Hauptgeschwindigkeit. Dann ist

$$\theta_{11} = 0 \ (1) \qquad \theta_{12} = 0 \ (\delta) \qquad \theta_{22} = 0 \ (\delta^2)$$

$$\varDelta_1 = 0 \ (1) \qquad \varDelta_2 = 0 \ (\delta).$$

Die beiden Gleichungen werden dann in erster Ordnung

$$\frac{\partial}{\partial \varphi}\left(\sqrt{\sigma} \cdot \Theta_{11}\right) + \frac{\sqrt{\sigma}}{2\,T}\,\frac{\partial T}{\partial \varphi} \cdot (\Theta_{11} + \varDelta_1) - \frac{\sqrt{\sigma}}{2\varrho}\,\Theta_{11}\,\frac{\partial \varrho}{\partial \varphi} = \frac{1}{T\sqrt{\sigma}}\left(\frac{\partial U_1}{\partial \zeta}\right)_0$$

$$\frac{\partial}{\partial \varphi}\left(\sqrt{\sigma} \cdot \Theta_{21}\right) + \frac{\sqrt{\sigma}\,\varrho}{2\,T}\,\frac{\partial T}{\partial \varphi} \cdot (\Theta_{11} + \varDelta_1) - \frac{\sqrt{\sigma}}{\varrho}\,\Theta_{21}\,\frac{\partial \varrho}{\partial \varphi} = \frac{1}{T\sqrt{\varrho}}\left(\frac{\partial U_2}{\partial \zeta}\right)_0 .$$

Man beachte, daß die erste Gleichung nur von der Hauptgeschwindigkeit abhängt. Man kann sie also für jede Stromlinie gesondert lösen.

Das kann mit genügender Genauigkeit mit Hilfe des WALZ-THWAITESschen Verfahrens geschehen.

$$\varrho\,\frac{\partial}{\partial \varphi}\left(\frac{T\,\sigma}{\varrho}\,\Theta_{11}^2\right) = \frac{4}{3\sqrt{\pi}}\,\Theta_{11}\,(2 + \varLambda + N) - 2\,\varLambda\,\Theta_{11}\,\varDelta_1 .$$

Die rechte Seite dieser Gleichung kann man annähern durch

$$H\,(\varLambda) = 0{,}436 - 2\,a^2\,\varLambda ,$$

wo

$$a = \theta_{11} = 0{,}293 .$$

Das Ergebnis der Integration ist

$$\sigma = 5.08\,\frac{\varrho}{T^2}\left\{c_0\,(\varphi_0)\right\} + \int\limits_{\varphi_0}^{\varphi} \frac{T}{\varrho}\,d\varphi ,$$

wobei φ_0 der Anfangswert von φ ist, und $5{,}08\,c_0\,(\varphi_0)$ den Wert von $\dfrac{T^2\,\sigma}{\varrho}$ für $\varphi = \varphi_0$ bedeutet.

Die Gleichung für die Querströmung kann in der Form

$$\frac{\varrho^2}{2\,\Theta_{21}}\,\frac{\partial}{\partial \varphi}\left(\frac{\sigma\,\Theta_{21}^2}{\varrho^2}\right) = \frac{M}{T}\left(\frac{2}{3\sqrt{\pi}}\,\Omega - a - \varDelta_1\right)$$

geschrieben werden und ergibt nach Integration

$$\Theta_{21}\,(\Omega) = c_1\,(\psi_0) \cdot \frac{\varrho}{\sqrt{\sigma}} \cdot \exp \int\limits_{\varphi_0}^{\varphi} \frac{M}{T\,\sigma\,\Theta_{21}}\left(-\frac{2}{3\sqrt{\pi}}\,\Omega - a\,\varDelta_1\right)d\,\psi$$

wo $c_1\,(\varphi_0)$ den Wert von $\dfrac{\sqrt{\sigma}\,\Theta_{21}}{\varrho}$ für $\varphi = \varphi_0$ darstellt.

23*

Die Funktion ϱ, die nur von der freien Strömung abhängt, spielt in der Theorie eine wesentliche Rolle.

Die Kontinuitätsgleichung ergibt

$$\frac{1}{\varrho}\frac{d\varrho}{d\varphi} = \frac{2}{T}\frac{\partial W}{\partial \zeta},$$

wobei W die Komponente der freien Strömung der Körpernormalen entlang ist.

Um die Anfangswerte $c_0(\varphi_0)$ und $c_1(\varphi_0)$ zu bestimmen, betrachten wir das Strömungsbild um einen dünnen Körper. Im Staupunkt kommen die Stromlinien zusammen und bleiben über einen beträchtlichen Abstand zusammen.

Es gibt eine Trennungsstromlinie, die die Flüssigkeit, die nach oben oder nach unten über den Körper fließt, teilt. Man kann für die Trennungsstromlinie die Größen σ und Ω mittels eines einfachen Iterationsprozesses bequem berechnen.

Man schreibt hier zuerst die Gleichung an in der Form

$$\sigma\frac{d}{d\varphi}\left(\frac{T^2}{\varrho}\right) + \frac{T^2}{\varrho}\frac{\partial \sigma}{\partial \varphi} = \frac{0{,}436}{a^2}\frac{T}{\varrho}.$$

$$\frac{1}{2}\varrho^2\Theta_{21}\frac{\partial}{\partial \varphi}\left(\frac{\sigma}{\varrho^2}\right) + \sigma\frac{\partial \Theta_{21}}{\partial \varphi} = \frac{M}{T}\left(\frac{2}{3\sqrt{\pi}}\Omega - a - \Delta_1\right)$$

und nimmt an, daß sich σ und Ω in der Nähe des Staupunktes nicht viel ändern. Wenn man jetzt $\dfrac{d\sigma}{d\varphi}$ und $\dfrac{d\Omega}{d\varphi}$ vernachlässigt, bekommt man ein System algebraischer Gleichungen für σ und Ω, das man weiterhin korrigiert, indem man die gefundenen Werte numerisch differiert und damit die Gleichungen korrigiert.

Literatur

Timman, R.: A calculation method for threedimensional laminar boundary layers. N.L.L. Report F 66, 1951.

Timman, R., and J. A. Zaat: Eine Rechenmethode für dreidimensionale laminare Grenzschichten. 50 Jahre Grenzschichtforschung 1955. Braunschweig: Vieweg.

Timman, R.: The potential flow about a yawed ellipsoid at zero incidence. N.L.L. Report F 74, 1950.

Zaat, J. A.: A simplified method for the calculation of threedimensional laminar boundary layers. N.L.L. Report F 184, 1956.

Zaat, J. A.: Anwendung der Stromlinienkoordinaten für die Grenzschichtberechnung am schräg angeströmten Zylinder. N.L.L. Report F 194, 1956.

Zaat, J. A., E. van Spiegel and R. Timman: The threedimensional boundary layer flow about a yawed ellipsoid at zero incidence. N.L.L. Report F 165, 1955.

Die Ablösungsbedingung von Grenzschichten

Von

K. Oswatitsch[1]

Deutsche Versuchsanstalt für Luftfahrt, Aachen

1. Einleitung

Die Untersuchung gilt den Bedingungen, unter denen sich eine dreidimensionale Strömung von der Wand ablöst.

Die Voraussetzungen sind möglichst weit gefaßt. Die Wand darf gekrümmt, die Strömung kompressibel, Wärmeleitung vorhanden und die Reynoldszahl beliebig sein, wenn nur die Strömung stationär ist. Es wird lediglich angenommen, daß der Geschwindigkeitsvektor in Wandnähe auch in der Umgebung der Ablösestelle durch den Anfang einer Taylorreihe hinreichend gut approximiert werden kann[2].

Unter Verwendung der Kontinuitätsgleichung, der Navier-Stokesschen Gleichungen (also *nicht* etwa der Grenzschichtgleichungen) und der Haftbedingung stellt sich heraus, daß die Taylor-Koeffizienten bis zu den Gliedern 2. Ordnung einschließlich am Ablösungspunkt durch die Komponenten τ und σ der Wandschubspannung sowie deren Ableitungen und die Ableitungen von p längs der Wand ausgedrückt werden können. Damit erhält man für die Stromlinien einer zähen Strömung in der Nähe der Wand ein System von gewöhnlichen Differentialgleichungen, dessen Lösungen sich vollständig überblicken lassen und zu einer vollständigen Übersicht über das lokale Ablöseverhalten führen. Für den Strömungs*typus* ist nur die Wandschubspannungsverteilung verantwortlich, für Stärke und Richtung der Ablösung jedoch auch der Druckgradient maßgebend.

Die dreidimensionale Ablösung unterscheidet sich insofern wesentlich von der zweidimensionalen, als im ebenen und rotationssymmetrischen Fall die Komponenten der Wandschubspannung entlang der „Ablöselinie", die Anström- und Rückströmgebiet an der Wand trennt, identisch verschwinden, während sie dies bei der allgemeinen dreidimensionalen Strömung nur in einzelnen Punkten tun, falls überhaupt eine Ablöselinie vorhanden ist. Die Ablöselinie ist im dreidimensionalen Fall im

---

[1] Vorgetragen von R. SCHWARZENBERGER.

[2] Ein spezielles singuläres Verhalten wird für den zweidimensionalen Fall in [1] behandelt. (Die Zahlen in eckigen Klammern verweisen auf das Literaturverzeichnis am Ende der Arbeit.)

allgemeinen gleichzeitig Wandstromlinie, im zweidimensionalen Fall
nicht. Überdies ist es zweckmäßig, im dreidimensionalen zwischen
„Ablösung'' und „Abdrängung'' zu unterscheiden. Ablösung liegt vor,
wenn eine Wandstromlinie ihre Fortsetzung in einer von der Wand weg-
führenden Ablösungsstromlinie findet, Abdrängung, wenn sie sich lediglich an der Wand verzweigt und dadurch ein Rückströmungsgebiet
umgeht. Das Verschwinden der Wandschubspannung erweist sich zwar
als *notwendige*, nicht aber als *hinreichende* Voraussetzung für die Ablösung in einem Punkt.

2. Der Geschwindigkeits-, Schubspannungs- und Druckverlauf an der Wand

Im folgenden bedeuten x, y, z kartesische Koordinaten, deren y-Achse
senkrecht auf der als eben vorausgesetzten Wand steht. (Genau entsprechende Überlegungen lassen sich mit denselben Ergebnissen für
gekrümmte Wände durchführen.) Die Richtung von x und z ist nicht an
die Potentialströmung gebunden.

Der Geschwindigkeitsvektor $\mathfrak{w} = \mathfrak{w}\,(u, v, w)$ verschwindet infolge der
Haftbedingung für $y = 0$. Deshalb gilt auch für beliebige m, $n \geqslant 0$

$$\frac{\partial^{m+n}\mathfrak{w}}{\partial x^m \partial z^n} = 0. \tag{1}$$

Durch Hinzunahme der Kontinuitätsbedingung für kompressible Medien
der Dichte ϱ

$$\operatorname{div} \mathfrak{w} + \mathfrak{w}\,\operatorname{grad}\,(\ln \varrho) = 0 \tag{2}$$

folgt

$$\frac{\partial v}{\partial y}\bigg|_{y=0} = \frac{\partial^2 v}{\partial x \partial y}\bigg|_{y=0} = \frac{\partial^2 v}{\partial y \partial z}\bigg|_{y=0} = \cdots = 0. \tag{3}$$

Differentiation von (2) nach y ergibt mit (1) und (3) für $y = 0$:

$$-\frac{\partial^2 v}{\partial y^2} = \frac{\partial^2 u}{\partial x \partial y} + \frac{\partial^2 w}{\partial y \partial z} + \frac{\partial u}{\partial y} \cdot \frac{1}{\varrho} \frac{\partial \varrho}{\partial x} + \frac{\partial w}{\partial y} \cdot \frac{1}{\varrho} \frac{\partial \varrho}{\partial z}. \tag{4}$$

Ist τ die Komponente der Wandschubspannung in x-, σ diejenige in
z-Richtung, μ der Reibungskoeffizient, so gilt wegen (1) auf $y = 0$:

$$\tau = \mu \cdot \frac{\partial u}{\partial y}, \qquad \sigma = \mu \cdot \frac{\partial w}{\partial y}. \tag{5}$$

Nimmt man nun an, daß μ nur von der Temperatur $T\,(x, y, z)$ abhängt,
wie dies bei idealen Gasen und den geläufigen Flüssigkeiten der Fall ist,
so folgt aus (5) für $y = 0$:

$$\left.\begin{aligned}
\tau_x &= \frac{\mu_x}{\mu} \cdot \tau + \mu \cdot \frac{\partial^2 u}{\partial x \partial y}\;; & \tau_z &= \frac{\mu_z}{\mu} \cdot \tau + \mu \cdot \frac{\partial^2 u}{\partial y \partial z}\;; \\
\sigma_x &= \frac{\mu_x}{\mu} \cdot \sigma + \mu \cdot \frac{\partial^2 w}{\partial x \partial y}\;; & \sigma_z &= \frac{\mu_z}{\mu} \cdot \sigma + \mu \cdot \frac{\partial^2 w}{\partial y \partial z}.
\end{aligned}\right\} \tag{6}$$

während aus den vollen Navier-Stokesschen Gleichungen (also nicht etwa durch Grenzschichtannahmen vereinfachten Formen derselben) sich ebenfalls für $y = 0$ ergibt

$$\frac{\partial p}{\partial x} = \frac{\mu_y}{\mu} \cdot \tau + \mu \frac{\partial^2 u}{\partial y^2} \; ; \quad \frac{\partial p}{\partial z} = \frac{\mu_y}{\mu}\sigma + \mu \cdot \frac{\partial^2 w}{\partial y^2} . \tag{7}$$

Damit lassen sich alle 2. Ableitungen von u, v, w geeignet darstellen; (4) kann man nämlich mit (6) und (7) schreiben

$$-\mu \cdot \frac{\partial^2 v}{\partial y^2} = \tau_x + \sigma_z - \frac{\mu_x}{\mu}\tau - \frac{\mu_z}{\mu}\sigma + \frac{\tau}{\varrho}\frac{\partial \varrho}{\partial x} + \frac{\sigma}{\varrho}\frac{\partial \varrho}{\partial z} . \tag{8}$$

Wegen der Haftbedingungen (1) beginnt die Taylorreihe für $\mathfrak{w}$ in der Umgebung des Wandpunktes $x = y = z = 0$

$$\mathfrak{w} = \mathfrak{w}_y \cdot y + \mathfrak{w}_{xy} \cdot x\,y + 1/2\,\mathfrak{w}_{yy} \cdot y^2 + \mathfrak{w}_{yz} \cdot y\,z + \cdots \tag{9}$$

Die in $x = y = z = 0$ zu nehmenden Vektorableitungen haben nach (5) und (8) folgende Komponenten

$\mathfrak{w}$	u	v	w
$\mu\,\mathfrak{w}_y$	τ ;	0	$;\quad \sigma$;
$\mu\,\mathfrak{w}_{xy}$	$\tau_x - \dfrac{\mu_x}{\mu}\tau$;	0	$;\,\sigma_x - \dfrac{\mu_x}{\mu}\sigma$;
$\mu\,\mathfrak{w}_{yy}$	$p_x - \dfrac{\mu_y}{\mu}\tau$;	$-\left[\tau_x + \sigma_z - \dfrac{\mu_x}{\mu}\tau - \dfrac{\mu_z}{\mu}\sigma + \dfrac{\tau}{\varrho}\dfrac{\partial \varrho}{\partial x} + \dfrac{\sigma}{\varrho}\dfrac{\partial \varrho}{\partial z}\right];$	$p_z - \dfrac{\mu_y}{\mu}\sigma$;
$\mu\,\mathfrak{w}_{yz}$	$\tau_z - \dfrac{\mu_z}{\mu}\tau$;	0	$;\,\sigma_z - \dfrac{\mu_z}{\mu}\sigma$.

$$(10)$$

Man bemerkt, daß an einer Stelle $\tau = \sigma = 0$, die sich im folgenden ausschließlich als interessant herausstellt, sowohl der Kompressibilitätseinfluß als auch der Einfluß einer räumlich variierenden Temperatur herausfällt. Damit ist auch der Stromlinienverlauf in der Nähe eines Ablösungspunktes im Rahmen der Approximation (9) auf die Druck- und Schubspannungsverhältnisse an der Wand zurückgeführt.

3. Die Ablösung bei ebener Strömung

Bei ebener bzw. bei achsensymmetrischer (siehe [2]) Strömung läßt sich das Koordinatensystem so legen, daß $w = \sigma = 0$ ist. Gleichzeitig verschwindet die z-Abhängigkeit der übrigen Größen. Man hat dann

$$\left.\begin{aligned}
\mu \cdot u &= \tau \cdot y + \left(\tau_x - \frac{\mu_x}{\mu}\tau\right) x\,y + \frac{1}{2}\left(p_x - \frac{\mu_y}{\mu}\tau\right) y^2 , \\
\mu \cdot v &= \qquad\qquad\quad -\frac{1}{2}\left(\tau_x - \frac{\mu_x}{\mu}\tau + \frac{\tau}{\varrho}\varrho_x\right) y^2
\end{aligned}\right\} \tag{11}$$

und

$$\frac{dy}{dx} = \frac{v}{u} = \frac{-\left(\tau_x - \frac{\mu_x}{\mu}\,\tau + \frac{\tau}{\varrho}\,\varrho_x\right)y}{2\,\tau + 2\left(\tau_x - \frac{\mu_x}{\mu}\cdot\tau\right)x + \left(p_x - \frac{\mu_y}{\mu}\,\tau\right)y}. \qquad (12)$$

Aus (12) ist ersichtlich, daß eine Ablösungsstromlinie (tg $\vartheta \neq 0$) in $x = y = 0$ nur auftreten kann für $\tau = 0$. Damit gilt

$$\frac{dy}{dx} = \frac{-\tau_x \cdot y}{2\,\tau_x \cdot x + p_x \cdot y}. \qquad (13)$$

Diese Gleichung wird erfüllt von der Wandstromlinie $y = 0$ und der Ablösungsstromlinie $y = x \cdot \mathrm{tg}\,\vartheta$;

$$\mathrm{tg}\,\vartheta = -\frac{3\,\tau_x}{p_x}. \qquad (14)$$

ϑ ist der Ablösungswinkel. Er ist um so größer, je schneller die Wandschubspannung abnimmt und je weniger (!) der Druck ansteigt. Trotz $p_x = 0$ kann der Ablösewinkel ϑ gleich 0 sein, wenn auch $\tau_x = 0$ ist. Das Hartree-Ablöseprofil bietet dafür ein Beispiel. Die allgemeine Lösung von (13) lautet

$$y^2\,(x - y \cot \vartheta)\ = \mathrm{const.}. \qquad (15)$$

Abb. 1 zeigt die (14) und (15) entsprechenden Stromlinien und gleichzeitig die Kurve verschwindender u-Komponente $y = 2/3\ \mathrm{tg}\,\vartheta \cdot x$. Sie ist qualitativ bereits in der klassischen Arbeit von Prandtl [3] enthalten.

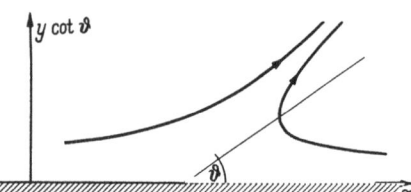

Abb. 1. Ablösungspunkt bei ebener Strömung. Weitere Stromlinien gehen aus den gezeichneten durch Ähnlichkeitstransformation mit dem Ablösungspunkt als Zentrum hervor

Gl. (14) und ihre Verallgemeinerung auf drei Dimensionen (22) finden sich im wesentlichen bereits bei R. Legendre [4].

Offen bleibt die Frage, bis zu welchem Wandabstand die Lösung wirklichkeitstreu ist. Messungen von E. A. Eichelbrenner und H. Werlé [5] zeigen, daß in bestimmten Fällen die Ablösungsstromlinie nach anfänglich starker Neigung rasch in die Wandrichtung abbiegt. Eine nähere Beachtung verdient auch der Fall einer Überschallströmung an einer konkaven Wand, wo sich experimentell trotz stärksten Druckanstiegs keine Ablösung nachweisen läßt, obwohl die bisher bekannten Grenzschichttheorien eine Ablösung bereits im Anfang der Kompression bei durchaus endlichem Ablösungswinkel ergeben.

Die experimentelle Nachprüfung von (14) kommt wegen der schwierigen Messung von τ meist nicht in Frage, während eine theoretische Bestimmung von τ_x Berechnungsmethoden für Grenzschichten voraussetzt, die auch noch am Ablösungspunkt funktionieren. Denn aus der 2. Navier-Stokesschen Gleichung folgt mit (8) im Ablösungspunkt

$$p_y = \mu \cdot \frac{\partial^2 v}{\partial y^2} = -\tau_x = \frac{1}{3}\,\mathrm{tg}\,\vartheta \cdot p_x, \tag{16}$$

so daß bei „starker" Ablösung mit einem Versagen der Grenzschichtnäherung $p_y \ll p_x$ zu rechnen ist.

4. Der symmetrische räumliche Ablösepunkt

Eine einfache Form der Ablösung einer räumlichen Grenzschicht erhält man bei einer zur Ebene $z = 0$ symmetrischen Strömung, wie sie etwa an der Ober- oder Unterseite des Mittelschnitts eines Flügels endlicher Streckung auftritt. In diesem Fall gilt

$$\tau\,(x, y, z) = \tau\,(x,y,\,-z); \qquad p\,(x, y, z) = p\,(x, y,\,-z); \tag{17}$$
$$\sigma\,(x, y, z) = -\,\sigma\,(x, y, z)$$

und folglich

$$\tau_z(x,y,0) = 0; \quad p_z(x,y,0) = 0; \quad \sigma(x,y,0) = 0; \quad \sigma_x(x,y,0) = 0. \tag{18}$$

Die Darstellung der Geschwindigkeitsverteilung reduziert sich dann auf

$$\mu \cdot u = \tau \cdot y + \left(\tau_x - \frac{\mu_x}{\mu}\,\tau\right)x\,y + \frac{1}{2}\left(p_x - \frac{\mu_y}{\mu}\,\tau\right)y^2\,;$$
$$\mu \cdot v = \qquad -\frac{1}{2}\left(\tau_x + \sigma_z - \frac{\mu_x}{\mu}\,\tau + \frac{\tau}{p}\,p_x\right)y^2\,; \tag{19}$$
$$\mu \cdot w = \qquad \sigma_z \cdot y\,z\,.$$

Die Stromlinien sind gegeben durch

$$dx : dy : dz = u : v : w. \tag{20}$$

Durch Bilden von dy/dx stellt man wie bei (12) fest, daß $\tau = 0$ eine *notwendige* Voraussetzung für eine von Null verschiedene Stromlinienneigung ist. Wegen $\sigma = 0$ verschwindet damit wieder in einem Punkt die gesamte Wandschubspannung.

Wählt man y als unabhängige Veränderliche, so ergibt (19) und (20) für die Umgebung des Ablösepunktes

$$\left.\begin{array}{l}\dfrac{u}{v} = \dfrac{dx}{dy} = \dfrac{2\,\tau_x \cdot x + p_x \cdot y}{-(\tau_x + \sigma_z)\,y}\,, \\[3mm] \dfrac{w}{v} = \dfrac{dz}{dy} = \dfrac{2\,\sigma_z \cdot z}{-(\tau_x + \sigma_z)\,y}\,.\end{array}\right\} \tag{21}$$

Als Gleichung der Ablösungsstromlinie findet man in Verallgemeinerung von (14) $y = x \cdot \mathrm{tg}\,\vartheta$, $z = 0$, mit

$$\mathrm{tg}\,\vartheta = -\frac{3\,\tau_x + \sigma_z}{p_x}\,. \tag{22}$$

Durch die Affintransformation

$$\xi = x - y \cot \vartheta, \quad \eta = y, \quad \zeta = z \tag{23}$$

erhält (21) die normierte Form

$$\frac{d\xi}{dy} = -\frac{2\tau_x}{\tau_x + \sigma_z} \cdot \frac{\xi}{y} ; \quad \frac{dz}{dy} = -\frac{2\sigma_z}{\tau_x + \sigma_z} \cdot \frac{z}{y} \tag{24}$$

mit der Lösung

$$\left. \begin{array}{l} y^{\frac{2\tau_x}{\tau_x+\sigma_z}} \cdot \xi \equiv y^{\frac{2\tau_x}{\tau_x+\sigma_z}} \cdot (x - y \cot \vartheta) = C_1 , \\[2mm] y^{\frac{2\sigma_z}{\tau_x+\sigma_z}} \cdot z \qquad\qquad\qquad\quad = C_2 . \end{array} \right\} \tag{25}$$

Für die Wandstromlinien ergibt sich

$$x^{-\frac{\sigma_z}{\tau_x}} \cdot z \qquad\qquad = C . \tag{26}$$

Strömt Materie auf die x-Achse zu bzw. von der x-Achse fort, so spricht man von einem *Konvergenz-* bzw. *Divergenzpunkt*, und es ist $\sigma_z < 0$ bzw. $\sigma_z > 0$. Es ist zweckmäßig, für $0 < \sigma_z/\tau_x < 1$ von schwacher, für $\sigma_z/\tau_x > 1$ von starker Konvergenz zu sprechen. Analog unterscheidet man für $-1/3 < \sigma_z/\tau_x < 0$ und $\sigma_z/\tau_x < -1/3$ schwache und starke Divergenz. Ebene Strömung entspricht dem Wert $\sigma_z/\tau_x = 0$.

Folgende Tabelle gibt eine Aufstellung charakteristischer· symmetrischer Ablösezentren.

Fall	σ_z	τ_x; $-p_x$	σ_z/τ_x	$\dfrac{2\tau_x}{\tau_x+\sigma_z}$	$\dfrac{2\sigma_z}{\tau_x+\sigma_z}$	$1+\dfrac{1}{3}\dfrac{\sigma_z}{\tau_x}$		
1	<0	>0	-4	$-2/3$	$8/3$	$-1/3$	Konvergenz	↑
2	<0	>0	-3	-1	3	0		
3	<0	<0	2	$2/3$	$4/3$	$5/3$		stark
4	<0	<0	1	1	1	$4/3$		—
5	<0	<0	$1/2$	$4/3$	$2/3$	$7/6$		schwach
6	0	<0	0	2	0	1		ebene Strömung
7	>0	<0	$-1/5$	$5/2$	$-1/2$	$14/15$	Divergenz	schwach
8	>0	<0	$-1/3$	3	-1	$8/9$		—
9	>0	<0	$-1/2$	4	-2	$5/6$		stark
10	>0	<0	-1	∞	$-\infty$	$2/3$		↓
11	>0	<0	-2	-2	$+4$	$1/3$		keine Abl.

Die Abb. 2—9 stellen die in den Bildunterschriften erwähnten Fälle obiger Tabelle dar.

Von besonderem Interesse ist Abb. 8 (Fall 11). Hier ist die Divergenz so stark angenommen, daß die Strömung — obwohl von beiden Seiten auf die Trennungsstromfläche $x = y \cot \vartheta$ zufließend — nicht ablöst. Auf

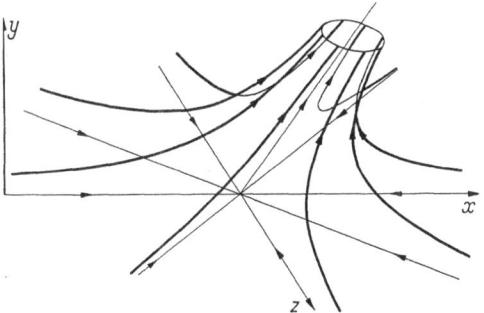

Abb. 2. Konvergenzpunkt mittlerer Konvergenz (Fall 4 der Tabelle). Strömung strebt radial von der Anström- wie von der Rückströmseite auf das Ablösungszentrum zu. Gerade Wandstromlinien, die sich in der Ablösungsstromlinie fortsetzen

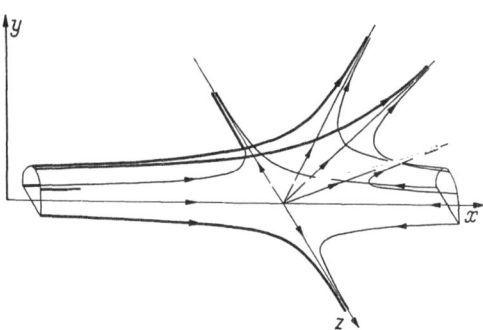

Abb. 3. Divergenzpunkt mittlerer Divergenz (Fall 8 der Tabelle). Zwei Stromflächen halbkreisförmigen Querschnitts, die von Anström- und Rückströmseite auf die Ablösungsebene $x = y \cot \vartheta$ zustreben

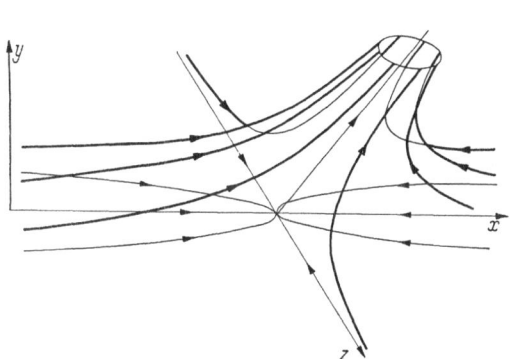

Abb. 4. Ablösezentrum bei schwacher Konvergenz (Fall 5 der Tabelle)

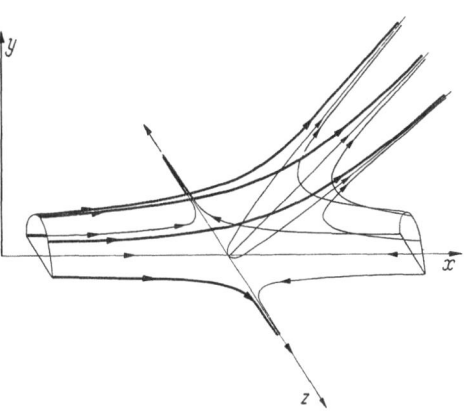

Abb. 5. Divergenzpunkt schwacher Divergenz (Fall 7)

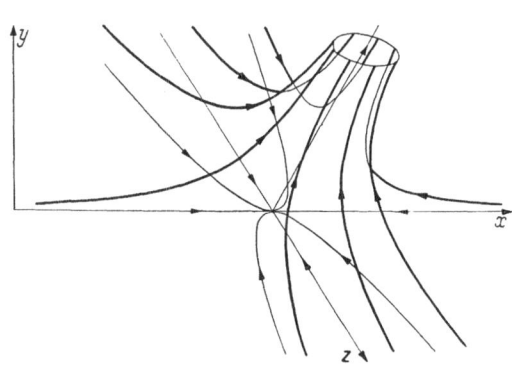

Abb. 6. Ablösezentrum bei starker Konvergenz (Fall 3)

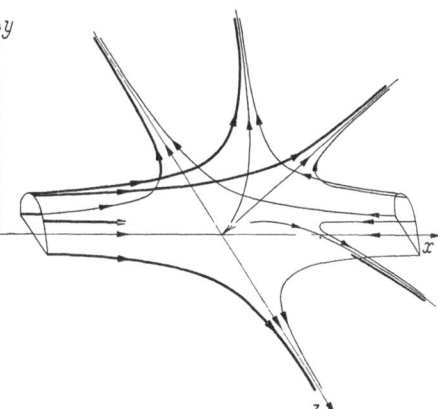

Abb. 7. Divergenzpunkt starker Divergenz (Fall 9)

der „Ablösegeraden" $x = y \cot \vartheta$, $z = 0$ strebt die Strömung, wie in der
ganzen x,y-Ebene, auf das Zentrum zu und setzt sich in der Abströmung
längs der z-Achse fort. Die übrigen Stromlinien nähern sich gleichzeitig
der Trennebene $x = y \cot \vartheta$ und der Wand $y = 0$. Es handelt sich also
nurmehr um eine „Abdrängung" einer Strömung durch eine Rückströmung, nicht mehr aber um eine Ablösung von der Wand. $\tau = \sigma = 0$

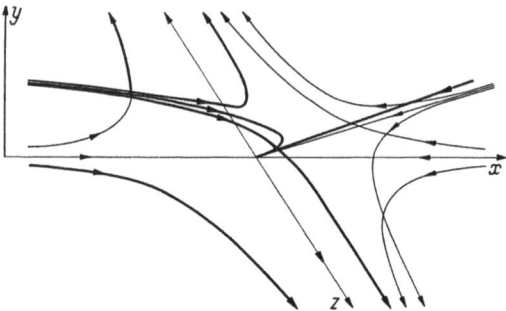

Abb. 8. Divergenzpunkt ohne Ablösung (Fall 11)

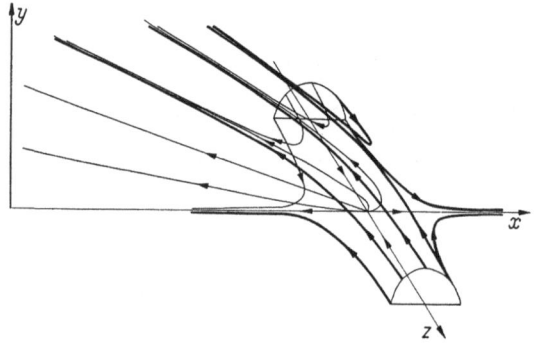

Abb. 9. Konvergenzpunkt mit Ablösung trotz negativem ϑ (Fall 1)

ist deshalb lediglich notwendig, nicht aber hinreichend für Ablösung im
strengen Sinne.

Abb. 9 (Fall 1) nimmt eine Ausnahmestellung ein und bildet gleichzeitig ein Gegenstück zu Abb. 8. Angenommen wird eine besonders
starke Konvergenz bei $p_x < 0$ (Druckabfall!) und $\tau_x > 0$, wie es dem
Wiederanlegen einer ebenen Strömung entsprechen würde. Trotzdem
löst die Strömung wegen der starken Konvergenz zur x,y-Ebene ab.

5. Der asymmetrische, räumliche Ablösepunkt

Auch im allgemeinen Falle ist $\tau = 0$, $\sigma = 0$ die notwendige Voraussetzung für eine Ablösungsstromlinie. Man erhält in Verallgemeinerung
von (21):

$$\frac{dx}{dy} = \frac{2\,\tau_x \cdot x + p_x \cdot y + 2\,\tau_z \cdot z}{-(\tau_x + \sigma_z)\,y}\;;$$

$$\frac{dz}{dy} = \frac{2\,\sigma_x \cdot x + p_z \cdot y + 2\,\sigma_z \cdot z}{-(\tau_x + \sigma_z)\,y}\,. \qquad \left.\begin{matrix}\\[1em]\\\end{matrix}\right\} \quad (27)$$

Die Ablösestromlinie

$$\frac{x}{y} = \cot\vartheta\,, \qquad \frac{z}{y} = \cot\psi\cdot\cot\vartheta \qquad (28)$$

ist gegeben durch

$$\cot\vartheta = \frac{2\,p_z\,\tau_z - p_x\,(\tau_x + 3\,\sigma_z)}{(3\,\tau_x + \sigma_z)(\tau_x + 3\,\sigma_z) - 4\,\sigma_x\,\tau_z}\;;$$

$$\cot\psi\cdot\cot\vartheta = \frac{2\,p_x\,\sigma_x - p_z\,(3\,\tau_x + \sigma_z)}{(3\,\tau_x + \sigma_z)(\tau_x + 3\,\sigma_z) - 4\,\sigma_x\,\tau_z}\,. \qquad \left.\begin{matrix}\\[1em]\\\end{matrix}\right\} \quad (29)$$

Analog zu (23) macht man mittels

$$\xi = x - y\cot\vartheta\;; \quad \eta = y\;; \quad \zeta = z - \cot\psi\cot\vartheta \qquad (30)$$

die Ablösungsstromlinie zur Achse eines Affinkoordinatensystems und erhält statt (27)

$$-(\tau_x + \sigma_z)\,y \cdot \frac{d\xi}{dy} = 2\,\tau_x \cdot \xi + 2\,\tau_z \cdot \zeta\;;$$

$$-(\tau_x + \sigma_z)\,y \cdot \frac{d\zeta}{dy} = 2\,\sigma_x \cdot \xi + 2\,\sigma_z \cdot \zeta\,. \qquad \left.\begin{matrix}\\[1em]\\\end{matrix}\right\} \quad (31)$$

Da der Druck hier nicht mehr explizit auftritt, sind für den Typus des Systems offenbar die Wandschubspannungsableitungen allein maßgebend. Mit $\lambda_{1,2} = \dfrac{-(\tau_x + \sigma_z) \pm \sqrt{(\sigma_z - \tau_x)^2 + 4\,\tau_z\,\sigma_x}}{\tau_x + \sigma_z}$ und 2 willkürlichen Konstanten A_1, B_2 erhält man als Lösung von (31)

$$\xi = A_1\,y^{\lambda_1} + B_2 \cdot \frac{1}{2\,\sigma_x}\,(\tau_x - \sigma_z + \sqrt{(\sigma_z - \tau_x)^2 + 4\,\tau_z\,\sigma_x}\,)\,y^{\lambda_2}\;;$$

$$\zeta = -A_1 \cdot \frac{1}{2\,\tau_z}\,(\tau_x - \sigma_z + \sqrt{(\sigma_z - \tau_x)^2 + 4\,\tau_z\,\sigma_x}\,)\,y^{\lambda_1} + B_2\,y^{\lambda_2}\;; \qquad \left.\begin{matrix}\\[1em]\\\end{matrix}\right\}^{\;\lambda_1 \neq \lambda_2} \quad (32)$$

$$\xi = \frac{A_1}{y} + \frac{2\,\tau_z}{\sigma_z - \tau_x} \cdot \frac{B_2}{y}\,\ln y\;;$$

$$\zeta = -\frac{2\,\sigma_x\,A_1 + (\sigma_z + \tau_x)\cdot B_2}{\sigma_z - \tau_x} \cdot \frac{1}{y} + \frac{B_2}{y}\,\ln y\;; \qquad \left.\begin{matrix}\\[1em]\\\end{matrix}\right\}^{\;\lambda_1 = \lambda_2 = -1} \quad (32')$$

in der im Falle komplexer λ_1, λ_2 von den rechten Seiten der Realteil zu nehmen ist.

Statt (32) läßt sich mit beliebigen A_3, B_3 auch schreiben

$$\xi - \frac{1}{2\,\sigma_x}\,(\tau_x - \sigma_z + \sqrt{(\sigma_z - \tau_x)^2 + 4\,\tau_z\,\sigma_x}\,)\,\zeta = A_3\,y^{\lambda_1}\,,$$

$$\frac{1}{2\,\tau_z}\,(\tau_x - \sigma_z + \sqrt{(\sigma_z - \tau_x)^2 + 4\,\tau_z\,\sigma_x}\,)\cdot\xi + \zeta = B_3\,y^{\lambda_2}\,. \qquad \left.\begin{matrix}\\[1em]\\\end{matrix}\right\} \quad (33)$$

Der Vergleich mit (25) zeigt, daß für

$$(\sigma_z - \tau_x)^2 + 4\,\tau_z\sigma_x > 0 \qquad \text{(also } \lambda_1, \lambda_2 \text{ reell)}, \tag{34}$$

der allgemeine Fall durch eine weitere (reelle) Affintransformation auf den symmetrischen Ablösungspunkt zurückgeführt werden kann. Unter der Voraussetzung (34) erhält man für die Wandstromlinien

$$\left\{2\,\sigma_x\cdot x - \left[\tau_x - \sigma_z + \sqrt{(\sigma_z - \tau_x)^2 + 4\,\tau_z\,\sigma_x}\,\right]\cdot z\right\}\cdot\left\{\left[\tau_x - \sigma_z +\right.\right.$$
$$\left.\left. + \sqrt{(\sigma_z - \tau_x)^2 + 4\,\tau_z\,\sigma_x}\,\right]\cdot x + 2\tau_z\cdot z\right\}^{-\frac{\lambda_1}{\lambda_2}} = \text{const.} \tag{35}$$

Ein *Konvergenzpunkt* liegt vor für $\lambda_1/\lambda_2 > 0$, d.h. $\tau_x\,\sigma_z > \tau_z\,\sigma_x$, ein *Divergenzpunkt* für $\lambda_1/\lambda_2 < 0$, d.h. $\tau_x\,\sigma_z < \tau_z\,\sigma_x$. Wie schon beim symmetrischen Fall erhält man für $\tau_x + \sigma_z < 0$ derart starke Divergenz, daß *keine Ablösung* mehr erfolgt.

Ist jedoch statt (34) $(\sigma_z - \tau_x)^2 + 4\,\tau_z\sigma_x < 0$, so setzt man zweckmäßig

$$(\tau_x + \sigma_z)\,\lambda'' = \sqrt{-(\sigma_z - \tau_x)^2 - 4\,\tau_z\,\sigma_x} \tag{36}$$

und erhält mit a, α als willkürlichen Konstanten die Lösung in der Gestalt

$$\left.\begin{array}{l} \xi + \dfrac{\sigma_z - \tau_x}{2\,\sigma_x}\cdot\zeta = \dfrac{a}{y}\cdot\sin(\lambda''\ln y - \alpha)\,; \\[2ex] \dfrac{\sigma_z + \tau_x}{2\,\sigma_x}\cdot\lambda''\cdot\zeta = \dfrac{a}{y}\cos(\lambda''\ln y - \alpha)\,. \end{array}\right\} \tag{37}$$

Als Gleichung der Wandstromlinien findet man

$$\text{arctg}\,\frac{2\,\sigma_x\cdot x + (\sigma_z - \tau_x)\cdot z}{(\sigma_z + \tau_x)\,\lambda''\cdot z} =$$
$$= -\lambda''\ln\sqrt{\left(x + \frac{\sigma_z - \tau_x}{2\,\sigma_x}\,z\right)^2 + \left(\frac{\sigma_z + \tau_x}{2\,\sigma_x}\,\lambda''\,z\right)^2} + \text{const.} \tag{38}$$

Man hat es im vorliegenden Fall mit einem Strudelpunkt[1] zu tun; λ'' bestimmt die Stärke des Strudels. Abb. 10 und 11 zeigen einen Strudelpunkt mittlerer Konvergenz und mittlerer bzw. schwacher Stärke. Typisch ist, daß eine Ablöselinie als Aufteilung in Anström- und Rückströmgebiet fehlt. Die Wandstromlinien münden erst nach unendlich viel Windungen im Ursprung und finden dort ihre Fortsetzung in der Ablösungsgeraden. Es ist eine Frage der Definition, ob man ein derartiges Verhalten noch als Ablösung bezeichnet.

Eine besondere Gefährdung zur Bildung eines Strudels zeigt der Ablösungspunkt mittlerer Konvergenz (Abb. 2). Es ist zu vermuten, daß die Wirbel, welche am vorderen Ende von stärker angestellten Rotationskörpern und Flügeln kleiner Streckung ihren Ausgang nehmen, in einem Strudelpunkt entspringen.

---

[1] Der Strudelpunkt wurde von R. Legendre [6] vorausgesagt.

Der Ausnahmefall $\lambda_1 = \lambda_2 = -1$ bringt nichts wesentlich Neues.

Für die kürzende Umarbeitung des Manuskriptes bin ich Herrn R. SCHWARZENBERGER zu bestem Dank verpflichtet.

Abb. 10. Strudelpunkt mittlerer Stärke und Konvergenz

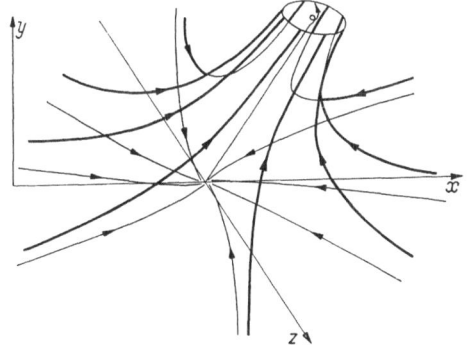

Abb. 11. Strudelpunkt schwacher Stärke und mittlerer Konvergenz

Literatur

[1] GOLDSTEIN, S.: On laminar boundary-layer flow near a position of separation. Quart. J. Mech. Appl. Math. 1, 43—69 (1948).

[2] MANGLER, W.: Zusammenhang zwischen ebenen und rotationssymmetrischen Grenzschichten von kompressiblen Flüssigkeiten. ZAMM **XXVIII**, 97—103 (1948).

[3] PRANDTL, L.: Über Flüssigkeitsbewegung bei sehr kleiner Reibung (Heidelberg, 1904). Vier Abhandlungen zur Hydrodynamik und Aerodynamik, Göttingen 1944.

[4] LEGENDRE, R.: Décollement laminaire régulier. Comptes Rendus **241**, 732—734 (1955).

[5] EICHELBRENNER, E. A., et H. WERLÉ: Décollement laminaire en deux dimensions Rech. Aeron. No. 51, Mai—Juin 1956.

[6] LEGENDRE, R.: Séparation de l'écoulement laminaire — tridimensionel. Rech. Aeron. No. 54, Nov.—Déz. 1956.

Neue Anwendungen des Prinzips der gemittelten Grenzschichtbedingungen nach v. Kármán und Pohlhausen[1]

Von

A. Walz

Emmendingen, Baden

1. Prandtlsche Grenzschichtgleichung als Basis

1.1. Diskussion des vollständigen Systems der gemittelten Bedingungen. Die im Jahre 1921 von v. Kármán und Pohlhausen [1, 2] aufgestellte, über die Grenzschichtdicke gemittelte Impulsbedingung ist — wie zuerst L. S. Leibenson [3] 1935, später unabhängig davon K. Wieghardt [4] 1943 zeigte — ein einzelner Sonderfall unendlich vieler gemittelter Bedingungen in Form von gewöhnlichen Differentialgleichungen 1. Ordnung, in die sich die Prandtlsche Grenzschichtgleichung durch partielle Integrationen überführen läßt.

E. Truckenbrodt [4] fand, daß sich dieses Gleichungssystem (mit u = Geschwindigkeitskomponente in x-Richtung, parallel zur Wand, ϱ = Dichte, Index δ: Werte an der Stelle $y = \delta$) in der allgemeinen Form

$$\frac{1}{\varrho_\delta u_\delta{}^{n+2}} \frac{d}{dx}\left(\varrho_\delta u_\delta{}^{n+2} f_n\right) + \frac{g_n}{u_\delta}\frac{du_\delta}{dx} - e_n = 0 \qquad (n = 0,\, 1,\, 2 \cdots \infty) \tag{1}$$

schreiben läßt[2]. Dabei sind e_n, f_n und g_n bestimmte Integralausdrücke

$$
\begin{aligned}
e_n &= -(1+n)\int_0^\delta \left(\frac{u}{u_\delta}\right)^n \frac{\partial}{\partial y}\left(\frac{\tau}{\varrho_\delta u_\delta{}^2}\right) dy \\
&= -(1+n)\left\{ \left[\left(\frac{u}{u_\delta}\right)^n \frac{\tau}{\varrho_\delta u_\delta{}^2}\right]_0^\delta - \int_0^1 \frac{\tau}{\varrho_\delta u_\delta{}^2}\, d\left(\frac{u}{u_\delta}\right)^n \right\} \\
&= +(1+n)\,n\int_0^1 \frac{\tau}{\varrho_\delta u_\delta{}^2}\left(\frac{u}{u_\delta}\right)^{n-1} d\left(\frac{u}{u_\delta}\right) - (1+n)\underbrace{\left[\left(\frac{u}{u_\delta}\right)^n \frac{\tau}{\varrho_\delta u_\delta{}^2}\right]_0^\delta}_{0 \text{ für } n \neq 0}
\end{aligned}
\tag{2}
$$

---

[1] Dieser Vortrag war, seinem Thema entsprechend, für die II. Sitzung vorgesehen, mußte aber wegen Verhinderung des Vortragenden auf die VIII. Sitzung verlegt werden. — D. Hrsg.

[2] E. Truckenbrodt gibt dieses Gleichungssystem in der noch allgemeineren Form für Grenzschichten an rotationssymmetrischen Körpern an, in der noch der Querschnittsradius des Rotationskörpers vorkommt. Wir wollen uns jedoch hier auf den Fall der ebenen Strömung beschränken.

$$f_n = \int\limits_0^\delta \left[\frac{\varrho\,u}{\varrho_\delta\,u_\delta} - \frac{\varrho}{\varrho_\delta}\left(\frac{u}{u_\delta}\right)^{n+2} \right] dy \tag{3}$$

$$g_n = -(1+n) \int\limits_0^\delta \left[\frac{\varrho\,u}{\varrho_\delta\,u_\delta} - \left(\frac{u}{u_\delta}\right)^n \right] dy \ . \tag{4}$$

Gl. (1) läßt sich durch Ausdifferentiieren und mit Beachtung der am Rande der Grenzschicht (bei $y = \delta$) geltenden Adiabaten-Beziehung

$$\frac{\varrho_\delta{}'}{\varrho} = - \frac{u_\delta{}'}{u_\delta} M_\delta{}^2 \tag{4a}$$

in die übersichtlichere Form

$$\boxed{f_n' + \frac{u_\delta{}'}{u_\delta}\left[f_n\,(2+n-M_\delta{}^2) + g_n\right] - e_n = 0} \tag{5}$$

bringen, wobei $'$ die Ableitung nach x bedeutet.

Von diesem Gleichungssystem hat man bis jetzt nur die Gleichungen für $n = 0$ (Impulssatz) und $n = 1$ (Energiesatz) verwertet in Verbindung mit einem Näherungsansatz für die Geschwindigkeitsprofile u/u_δ und (im kompressiblen Fall) auch für die Dichteprofile ϱ/ϱ_δ, obwohl klar ist, daß erst die Berücksichtigung aller (unendlich vielen) Gleichungen des Systems (5) das volle mathematische Äquivalent für die Prandtlsche Grenzschichtgleichung darstellen würde.

Grund für diese beschränkte Verwertung des Gleichungssystems (5) sind wohl in erster Linie die erstaunlich guten Ergebnisse, die man mit den rasch und bequem arbeitenden „Zwei-Gleichungs-Methoden" mit zwei Unbekannten (meist die Impulsverlustdicke δ_2 und ein Formparameter H des einparametrig angesetzten Geschwindigkeitsprofils) erzielte (vgl. z.B. [5, 6, 7]).

Der vor allem von mathematischer Seite erhobene Einwand, daß dies in unkontrollierbaren Fällen anders sein könnte, also bei den festgestellten guten Ergebnissen in gewisser Hinsicht beispielsbedingte „Zufallstreffer" vorliegen könnten, war bis jetzt nicht stichhaltig zu entkräften.

Es dürfte daher bei der großen Verbreitung, die die „Zwei-Gleichungs-Methoden" in der praktischen Strömungstechnik gefunden haben, eine besonders vordringliche Aufgabe sein, eine Genauigkeitskontrolle oder Fehlerabschätzung für diese Näherungsverfahren zu schaffen.

Eine nähere Betrachtung des Gleichungssystems (5) mit den Beziehungen (2) bis (4) weist hierfür einen Weg.

Zunächst soll eine bis jetzt offenbar nicht beachtete Eigenschaft des Gleichungssystems (2) bis (5) aufgezeigt werden.

Läßt man die Ordnungszahl n des Systems gegen unendlich gehen, so konvergieren die Gln. (5) gegen eine besonders einfache Grenzgleichung

$$(\delta - \delta_1)' + (\delta - \delta_1)(1 - M_\delta^2)\frac{u_\delta'}{u_\delta} - e_\infty = 0. \tag{6}$$

Dies folgt aus

$$\left[\left(\frac{u}{u_\delta}\right)^n\right]_{n \to \infty} = 0 \text{ für } 0 < \frac{u}{u_\delta} < 1, \tag{7}$$

woraus sich dann weiterhin ergibt

$$f_\infty = \int_0^\delta \frac{\varrho\, u}{\varrho_\delta\, u_\delta}\, d\, y = \delta - \delta_1 \tag{8}$$

mit

$$\delta_1 = \int_0^\delta \left(1 - \frac{\varrho\, u}{\varrho_\delta\, u_\delta}\right) d\, y = \text{Verdrängungsdicke}, \tag{9}$$

$$[f_n(n+2) + g_n]_{n \to \infty} = f_\infty, \tag{10}$$

$$e_\infty = -\frac{\tau_w}{\varrho_\delta\, u_\delta^2}\left[\int_0^\delta (1+n)\left(\frac{u}{u_\delta}\right)^n \frac{\partial\,(\tau/\tau_w)}{\partial\, u}\, d\, u\right]_{n \to \infty}. \tag{11}$$

Für $0 < \dfrac{u}{u_\delta} < 1$ gilt

$$\left[(1+n)\left(\frac{u}{u_\delta}\right)^n\right]_{n \to \infty} \to 0. \tag{12}$$

Da außerdem bei $y = \delta$, $u = u_\delta$ die Schubspannung τ und deren Ableitung nach y null werden, folgt für beliebiges Geschwindigkeitsprofil u/u_δ (laminar und turbulent)

$$e_\infty = 0. \tag{11a}$$

Gl. (6) wird dadurch sofort integrierbar. Beachtet man, daß auf Grund der Adiabaten-Beziehung der Gasdynamik

$$(1 - M_\delta^2)\frac{d\, u_\delta/d\, x}{u_\delta} = \frac{d\,(\varrho_\delta\, u_\delta)/d\, x}{\varrho_\delta\, u_\delta} \tag{12a}$$

ist, so lautet die Lösung von (6)

$$(\delta - \delta_1)\,\varrho_\delta\, u_\delta = \text{konst} \tag{12b}$$

oder

$$\delta f\,(H, M_\delta)\,\varrho_\delta\, u_\delta = \text{konst} \tag{12c}$$

wo $f = 1 - \dfrac{\delta_1}{\delta}$ mit der gewählten Klasse von Geschwindigkeitsprofilen und mit M_δ (bei wärmeisolierender Wand) als universelle Funktion festliegt.

Diese Gleichung mit den Unbekannten $\delta(x)$ und $H(x)$ bringt zum Ausdruck, daß der „Massenstrom" $(\delta - \delta_1)\,\varrho_\delta\,u_\delta$ zwischen den durch $\delta(x)$ und $\delta_1(x)$ gebildeten Stromlinien konstant sein muß. Sie folgt auch unmittelbar aus der Kontinuitätsgleichung durch partielle Integration, wenn man berücksichtigt, daß die y-Komponente v_δ der Geschwindigkeit am Rande der Grenzschicht $(y = \delta)$ ausgedrückt werden kann als

$$v_\delta = u_\delta \frac{d\,\delta}{d\,x}. \tag{12d}$$

Die Erfahrung wird zeigen müssen, ob man aus dieser Gleichung für praktische Rechnungen wird Nutzen ziehen können. Es ist anzunehmen, daß die Aussage dieser Gleichung wegen der mehr oder weniger willkürlichen Definition der Grenzschichtdicke δ eine verhältnismäßig geringe Bedeutung haben wird.

Man wird aus ihr nur Hinweise über die Form des Geschwindigkeitsprofils im Bereich des Übergangs der Grenzschichtströmung in die Außenströmung erhalten, die jedoch in bezug auf Wandreibung und Ablösungsverhalten unwesentlich sind.

Es ist nun eine wichtige Frage, wie rasch das Gleichungssystem (5) mit wachsender Ordnungszahl n gegen die Grenzgleichung (6) konvergiert. Man wird offenbar eine Zahl n angeben können, von der ab (zu größeren Werten hin) alle weiteren Gleichungen keine neuen Aussagen mehr bringen, sondern praktisch mit der Grenzgleichung (6) übereinstimmen.

Diese Zahl n hängt, wie aus den Gl. (3) und (4) zu ersehen ist, nur vom Geschwindigkeitsprofil ab. Je weniger völlig das Geschwindigkeitsprofil ist, desto rascher wird das Gleichungssystem (5) gegen die Grenzgleichung (6) konvergieren. Bei laminarer Grenzschicht wird also die Konvergenz besser sein als bei turbulenter Grenzschicht.

Eine zahlenmäßige Untersuchung im Fall der turbulenten Grenzschicht zeigt, daß bei einem Geschwindigkeitsprofil entsprechend dem Potenzgesetz

$$\frac{u}{u_\delta} = \left(\frac{y}{\delta}\right)^\gamma \tag{13}$$

für $\gamma = 1/7$ (etwa der Plattengrenzschicht bei inkompressibler Strömung entsprechend) n den Wert 700 haben müßte, damit die Funktion f_n bis auf etwa 1% mit der Grenzfunktion f_∞ übereinstimmt und damit die Grenzgleichung (6) etwa erreicht wird. Dies bedeutet aber, daß das Gleichungssystem (5) mit wachsender Ordnungszahl n sehr schlecht gegen die Grenzgleichung (6) konvergiert.

Um so mehr ist es zunächst verwunderlich, daß die Näherungsmethoden der Grenzschichtberechnung, die das Gleichungssystem schon (bei $n = 1$) abbrechen, erfahrungsgemäß zu gut brauchbaren Ergebnissen führen.

Eine Erklärung für diesen Sachverhalt kann die sein, daß es offenbar entscheidend auf die Güte der in die Näherungstheorie mit $n = 0$ und $n = 1$ eingeführten Ansätze für das Geschwindigkeitsprofil ankommt. Die bis jetzt vorwiegend benutzten Klassen von Geschwindigkeitsprofilen (Pohlhausen-P4- oder Hartree-Klasse bei laminarer, Potenz-Klasse bei turbulenter Grenzschicht) sind demnach ziemlich „glücklich" gewählt. Man kann aber auch sagen, daß mit wachsender Ordnungszahl n das Reibungsglied e_n, das für das Grenzschicht-Phänomen verantwortlich ist, gegen null geht, die Gleichungen mit niedriger Ordnungszahl n also am meisten über die Grenzschichteigenschaften aussagen können.

Die vorangehenden Ausführungen lassen es jedenfalls ratsam erscheinen, nach einer zumindest rohen Kontrollmethode für die Näherungsrechnungen auf der Zwei-Gleichungs-Basis zu suchen, die man ohne Zweifel nach wie vor wegen ihrer Einfachheit für die meisten strömungstechnischen Berechnungen anwenden wird. Zu einer solchen Kontrollmethode kann man, eventuell mit Verwertung der abgeleiteten Grenzgleichung, folgende Überlegung anstellen:

Es sei angenommen, daß die für ein bestimmtes Problem aus dem Zwei-Gleichungs-System gewonnenen Lösungen $\delta(x)$ und $H(x)$ wegen genau richtiger Vorgabe der Geschwindigkeitsprofile eine exakte Lösung der Prandtlschen Grenzschichtgleichung seien. Dann müßten sich bei Benutzung anderer gemittelter Bedingungen des Systems (5) außer $n = 0$ und 1 die gleichen Lösungen ergeben. Sind nun die Lösungen $\delta(x)$ und $H(x)$ der Zwei-Gleichungs-Methode und die eingeführten Geschwindigkeitsprofile — wie dies normalerweise der Fall sein wird — keine exakten Lösungen der Prandtlschen Grenzschichtgleichung, so wird dies daran zu erkennen sein, daß die mit den Gleichungspaaren $n = 0$ und $n = 1$ einerseits und z.B. den Gleichungspaaren $n = 0$ und $n = 2$ ermittelten Lösungen für $\delta(x)$ und $H(x)$ voneinander abweichen. *Die festzustellende Abweichung wird aber dann ein Maß für die Größenordnung des Fehlers sein, mit dem man bei der Anwendung dieser Näherungsmethode rechnen muß.*

Durch eine Mittelung der Ergebnisse aus den beiden Gleichungspaaren $n = 0$, $n = 1$ und $n = 0$, $n = 2$ kann überdies der Fehler verkleinert werden.

Praktische Erfahrungen mit der hier vorgeschlagenen kontrollierten Zwei-Gleichungs-Methode liegen noch nicht vor.

1.2. Grundgedanke einer Methode zur Bestimmung des Geschwindigkeitsprofils aus gemittelten Grenzschichtbedingungen. Trotz der im vorangegangenen Abschnitt aufgezeigten Möglichkeit einer Genauigkeitskontrolle für die handliche Zwei-Gleichungs-Methode und des damit unternommenen Versuches einer besseren Rechtfertigung dieser Methode bleibt es vom mathematischen Standpunkt aus gesehen ein Nachteil oder zumindest ein Schönheitsfehler dieser Methoden, daß das Geschwindigkeitsprofil in jedem Fall vorgegeben werden muß.

Entschließt man sich nun aber, mehr als zwei oder drei Gleichungen des Systems (5) zu verwerten, so liegt die Frage nahe, ob dann eine Vorgabe des Geschwindigkeitsprofils u/u_δ nicht entbehrlich wird.

Ein möglicher Weg, eine Willkür im Ansatz für das Geschwindigkeitsprofil u/u_δ weitgehend zu vermeiden, wäre natürlich der, eine große Anzahl von Formparametern (z.B. als Koeffizienten eines Reihenansatzes) für u/u_δ vorzusehen und diese Koeffizienten aus einer entsprechenden Anzahl von Gleichungen des Systems (5) zu bestimmen. Bei der Bildung der n^{ten} Potenz von u/u_δ nach den Gln. (2) bis (4) entstehen jedoch recht unübersichtliche Ausdrücke, die die Lösung des simultanen Systems von n gewöhnlichen Differentialgleichungen mit n Unbekannten sehr erschweren.

Folgender Weg scheint, jedenfalls für den inkompressiblen Fall, einfacher zum Ziel zu führen:

Auf einen analytischen Ansatz für u/u_δ wird ganz verzichtet. Die Werte $(u/u_\delta)_\nu$ an bestimmten, am besten äquidistanten Stellen y/δ werden, neben der Grenzschichtdicke δ, als Unbekannte eingeführt. Die Integralausdrücke (2) bis (4) können dann durch Summen mit endlicher Gliederzahl ersetzt werden. Insbesondere kann bei inkompressibler Strömung gesetzt werden

$$f_n = \delta \int_0^1 \left[\frac{u}{u_\delta} - \left(\frac{u}{u_\delta}\right)^{n+2} \right] d\,(y/\delta) = \frac{\delta}{2m} \sum_{\nu=1}^{\nu=m} \left[\frac{u_\nu}{u_\delta} + \frac{u_{\nu-1}}{u_\delta} - \left(\frac{u_\nu}{u_\delta}\right)^{n+2} - \left(\frac{u_{\nu-1}}{u_\delta}\right)^{n+2} \right]$$

(14)

$$g_n = -(1+n)\,\delta \int_0^1 \left[\frac{u}{u_\delta} - \left(\frac{u}{u_\delta}\right)^{n} \right] d\,(y/\delta) = \frac{\delta}{2m} \sum_{\nu=1}^{\nu=m} \left[\frac{u_\nu}{u_\delta} + \frac{u_{\nu-1}}{u_\delta} - \left(\frac{u_\nu}{u_\delta}\right)^{n} - \left(\frac{u_{\nu-1}}{v_\delta}\right)^{n} \right]$$

(15)

Die im Gleichungssystem (5) vorkommende Ableitung f'_n ist dann

$$f'_n = \frac{\delta}{2m} \sum_{\nu=1}^{\nu=m} \left\{ \left(\frac{u_\nu}{u_\delta}\right)' \left[1 - (n+2)\left(\frac{u_\nu}{u_\delta}\right)^{n+1} \right] + \left(\frac{u_{\nu-1}}{u_\delta}\right)' \left[1 - (n+2)\left(\frac{u_{\nu-1}}{u_\delta}\right)^{n+1} \right] \right\}$$
$$+ \frac{\delta'}{2m} \sum_{\nu=1}^{\nu=m} \left[\frac{u_\nu}{u_\delta} + \frac{u_{\nu-1}}{u_\delta} - \left(\frac{u_\nu}{u_\delta}\right)^{n+2} - \left(\frac{u_{\nu-1}}{u_\delta}\right)^{n+2} \right].$$

(16)

Der Ausdruck e_n (Gl. (2)) läßt sich bei inkompressibler laminarer Grenzschicht mit $\tau = \mu\,\partial\,u/\partial\,y$ umwandeln in

$$e_n = n\,\frac{n+1}{2R_\delta} \sum_{\nu=1}^{\nu=m} \left\{ \left[\left(\frac{u_\nu}{u_\delta}\right)^{n-1} + \left(\frac{u_{\nu-1}}{u_\delta}\right)^{n-1} \right] \left[\frac{\dfrac{u_\nu}{u_\delta} - \dfrac{u_{\nu-1}}{u_\delta}}{1/m} \right]^2 \frac{1}{m} \right\} \quad (n \neq 0)$$

(17)

mit

$$R_\delta = \frac{\varrho_\delta\,u_\delta\,\delta}{\mu_w}.$$

(18)

Damit ist das Gleichungssystem (5) für inkompressible laminare
Grenzschichten darstellbar durch ein System gewöhnlicher Differentialgleichungen für die Unbekannten u_ν/u_δ und für δ.

Um aus diesem System m Unbekannte u_ν/u_δ und die Größe δ zu bestimmen, werden allerdings keine $m+1$ Gleichungen dieses Systems benötigt, da noch einige einfache Gleichungen durch die Einführung von Randbedingungen bei $y=0$ (Wandbindungen) und $y=\delta$ gewonnen werden.

Entschließt man sich, das Geschwindigkeitsprofil z.B. mit 7 Punkten
zwischen $y=0$ und $y=\delta$ festzulegen, so wird man mit etwa 4 Gleichungen des Systems (5) auskommen, wobei man zweckmäßig auch die einfache Grenzgleichung (6) mitbenutzt.

Die Lösung des Systems von $m+1$ gewöhnlichen Differentialgleichungen 1. Ordnung mit $m+1$ Unbekannten konnte im einzelnen noch nicht
studiert werden. Sie dürfte jedoch zumindest bei Zuhilfenahme von
Analog-Rechenmaschinen keine grundsätzlichen Schwierigkeiten bieten
und dann auch mit erträglichem Zeitaufwand durchführbar sein.

Bei kompressibler Grenzschicht und auch schon bei inkompressibler
turbulenter Grenzschicht zeichnen sich gewisse Schwierigkeiten für die
Anwendung dieses Verfahrens ab (durch das Auftreten der Dichteprofile,
die ihrerseits wieder von den Geschwindigkeitsprofilen abhängen, bei
turbulenter Grenzschicht durch das Auftreten der nur empirisch gegebenen Schubspannungsprofile). Eine Klärung des Problems in dieser Richtung steht noch aus.

Ob sich solche Verfahren, bei denen man ohne Vorgabe der Geschwindigkeitsprofile auskommt, gegenüber der Zwei-Gleichungs-Methode (ergänzt durch eine Fehlerabschätzung nach Abschnitt 1.1) durchsetzen
können, wird nicht zuletzt von dem dafür nötigen Zeitaufwand abhängen.
Dieser läßt sich jedoch zur Zeit noch nicht einwandfrei übersehen.

2. Navier-Stokessche Gleichungen als Basis

Das Prinzip der gemittelten Grenzschichtbedingungen läßt sich natürlich
grundsätzlich auch auf die Navier-Stokesschen Gleichungen anwenden.

Den Anreiz, so etwas zu tun, geben praktisch wichtige Aufgaben-
stellungen, bei denen die vereinfachenden Voraussetzungen der Prandtlschen Grenzschichttheorie offenkundig nicht mehr erfüllt sind. Dies ist
z.B. bei Grenzschichten im Bereich von Verdichtungsstößen, in der Nähe
von Ablösestellen und bei kleinen Reynoldsschen Zahlen der Fall.

Wenn auf die Prandtlschen Vereinfachungen, insbesondere auf die
Annahme $\partial p/\partial y = 0$ verzichtet werden soll, sind bei zweidimensionaler
Strömung zwei Navier-Stokessche Gleichungen (die eine für die x-Kom-
ponente, die andere für die y-Komponente der Strömung) zu berücksichtigen.

Durch partielle Integration dieser beiden partiellen Differentialgleichungen in y-Richtung erhält man dann zwei Systeme unendlich vieler gewöhnlicher Differentialgleichungen, die beide im wesentlichen in der Form (5) darstellbar sind, wie sie durch partielle Integrationen aus der Prandtlschen Grenzschichtgleichung gewonnen wurde.

Bei dem Gleichungssystem für die x-Richtung läßt sich der über die Grenzschichtdicke δ gemittelte Unterschied zwischen der Prandtlschen und der Navier-Stokesschen Theorie in einem Glied K_n zusammenfassen, so daß also dieses Gleichungssystem die Form

$$f_n' + \frac{u_\delta'}{u_\delta} \left[f_n \left(2 + n - M_\delta^2 \right) + g_n \right] - e_n = K_n \qquad (19)$$

erhält.

Beim ersten Ausbau dieser Theorie wurden von dem Gleichungssystem für die x-Richtung nur die Gleichungen mit der Ordnungszahl $n = o$ und 1 verwertet, von dem Gleichungssystem für die y-Richtung nur die Gleichung für $n = o$ (Impulssatz). Diese Gleichung liefert unmittelbar einen Ausdruck für den Druckunterschied

$$\Delta p \left(x \right) = p \left(x, \delta \right) - p \left(x, 0 \right) \qquad (20)$$

zwischen der Wand ($y = o$) und dem Rande ($y = \delta$) der Grenzschicht. Diese Größe Δp wurde neben der Grenzschichtdicke δ und dem Formparameter H der angenommenen Geschwindigkeitsprofile als dritte Unbekannte in diese Näherungstheorie eingeführt.

Die Größen K_0, K_1 und Δp wurden probeweise als Korrekturgrößen aufgefaßt und in einigen Beispielen eine Lösung des simultanen Systems mit 3 Gleichungen und 3 Unbekannten durch Iteration versucht.

Dieser Lösungsweg erwies sich noch als durchführbar in dem Beispiel einer stark verzögerten Strömung mit Ablösung der turbulenten Grenzschicht (Strömungstyp $\dfrac{u_\delta}{u_\infty} = 1 - 100 \; x/l$).

Dagegen versagte die Iterationsmethode in dem Fall, in dem die Grenzschicht im Bereich eines Verdichtungsstoßes berechnet werden sollte. Die Größen K_0, K_1 und Δp erreichten in diesem Fall die Größenordnung der Funktionen e_0, e_1, konnten dann also nicht mehr als Korrekturgrößen aufgefaßt werden.

Die volle Berücksichtigung der Größen K_0, K_1 und Δp bedingt aber, daß die Gleichungen des Systems (19) von 2. oder 3. Ordnung werden.

Die Lösung des Gleichungssystems (19) auch in diesem Fall dürfte keine grundsätzlichen Schwierigkeiten bereiten. Die hierzu nötigen Untersuchungen sind in Vorbereitung.

Ausführlicher Bericht über den gegenwärtigen Stand der Theorie siehe [8].

Literatur

[1] v. KÁRMÁN, TH.: Über laminare und turbulente Reibung. ZAMM 1, 233 (1921).

[2] POHLHAUSEN, K.: Zur näherungsweisen Integration der Differential-Gleichungen der laminaren Reibungsschicht. ZAMM, 1, 235 (1921).

[3] LEIBENSON, L. S.: Energy form of the integral conditions in the boundary layer theory CAHI Rep. No. 240 (1935), pp. 41—44 (Original-Arbeit russisch).

[4] WIEGHARDT, K.: Über einen Energiesatz zur Berechnung laminarer Grenzschichten. Ing. Arch. 16, 243 (1948) und interner Bericht des KWJ Göttingen (1944).

[5] TRUCKENBRODT, E.: Ein Quadratur-Verfahren zur Berechnung der laminaren und turbulenten Grenzschicht bei ebener und rotationssymmetrischer Strömung. Ing. Arch. 20 (1952), 4. Heft.

[6] WALZ, A.: Anwendung des Energiesatzes von WIEGHARDT auf einparametrige Geschwindigkeitsprofile in laminaren Grenzschichten. Ing.-Arch. 16 (1948), 3. und 4. Heft.

[7] WALZ, A.: Nouvelle méthode approchée de calcul des couches limites laminaires et turbulentes en écoulement compressible. Publications Scientifiques et Techniques du Ministère de l'Air. No. 309 (1956), No. 336 (1957).

[8] WALZ, A.: General approximate theory for compressible laminar and turbulent boundary layers with respect of flow-effects normal to the wall across the boundary layer. ARDC-Report ASTIA Doc. No. AD 86312 (Febr. 1956).

Experimental investigation of flow separation over a step

By

Itiro Tani

University of Tokyo

1. Introduction

Because of the growing importance of the base pressure phenomena at supersonic speeds, much attention has been drawn to the mechanism of flow separation behind a blunt base. In particular, CROCCO and LEES [1] applied to the base pressure problem their general theory based on a flow model exhibiting interaction between the dissipative flow region and the nearly isentropic main flow. Their results could give the qualitative explanation for the observed phenomena. More recently, KORST [2] and CHAPMAN, KUEHN and LARSON [3] independently put forward a simple method of predicting the base pressure by dividing the flow into the dissipative "cavity" flow region wherein the pressure is assumed to be constant and a reattachment zone wherein the compression is assumed to be such that not much total pressure is lost along the dividing streamline.

It appears from these theoretical investigations that the most essential and intriguing part of the problem is concerned with the mixing process between the dissipative cavity flow and the non-dissipative main flow. This kind of interactive mixing occurs equally at subsonic speeds, but no ad hoc measurement seems to have been made at subsonic speeds.

In view of these circumstances, the investigation described in this paper was undertaken by determining at subsonic speeds the distribution of surface pressure and that of mean and fluctuating velocities in the separated flow over a backward-facing step. The work was conducted with the financial support of the Ministry of Education by the Scientific Research Fund. The author wishes to express his indebtedness to Mr. Y. MATSUBARA, Mr. M. IUCHI, Mr. Y. KOBASHI and Mr. H. KOMODA for their efficient cooperation in carrying out the experiments.

2. Experimental procedures

The measurements were made in a closed channel of 1- by 1-meter cross section placed within a 1.5-meter low-speed wind tunnel. A two-dimensional plate 10 centimeters thick and 150 centimeters long was spanned vertically across the test section. A backward-facing step of adjustable height (up to 6 centimeters) was made on one side of the plate at a distance of 80 centimeters from the leading edge.

Most of the measurements were made at the main-flow reference
velocity U_0 of 28 meters per second, which was actually determined by
the static pressure hole located 3 centimeters upstream of the step, the
total pressure being measured outside the boundary layer. Some additional
measurements were made, for comparison, at $U_0 = 22$, 16 and 10 meters
per second. The boundary layer 3 centimeters upstream of the step was
found to be laminar only at the lowest velocity ($U_0 = 10$ meters per
second), having a thickness of 0.6 centimeter. At the higher velocities the
boundary layer was turbulent, and the thickness was 1.1 centimeters for
$U_0 = 28$ meters per second. The boundary layer thickness could be in-

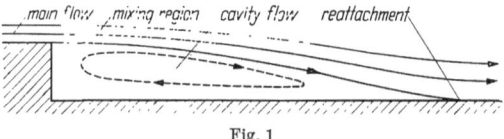

Fig. 1

creased to 3 centimeters at $U_0 = 28$ meters per second by placing a trip
0.5 centimeter high at the distance of 6 centimeters from the leading edge.

The static pressure was measured by the pressure holes on the model
surface and by a small static pressure probe within the mixing region.
It was required to aline the probe approximately with the flow direction
in order to minimize the error. The total pressure and flow direction were
measured by the combined use of two small pitot probes, the one with
normal end and the other with slanted end cut 45 degrees from the probe
axis. The use of static pressure probe and the slanted pitot probe being
questioned in the highly turbulent region, the hot-wire apparatus was
also employed for measuring mean and fluctuating velocities. The
magnitude of mean velocity and the longitudinal component of fluctuating velocity were measured by the usual application of a hot-wire.
The transverse components of mean and fluctuating velocities were
obtained by setting a hot-wire inclined at two different angles with the
flow direction. From these data, determination was made of the distribution of static pressure along the model surface, and the distributions of
longitudinal mean velocity, mean flow direction, turbulence intensity
and turbulent shear stress across the mixing region. It was found that
the slanted pitot probe indicated the mean flow direction agreeing well
with that determined by the hot-wire in the outer part of the mixing
region, where the turbulence intensity was small, while the indication was
somewhat excessive in the highly turbulent central region. Therefore
only the hot-wire results were adopted for the mean flow direction. It is
moreover to be noted that the measurements could not be extended into
the interior of the cavity flow, where the mean velocities were small,
while the fluctuating velocities were comparatively large.

For visualizing the flow pattern, additional observations were made by towing another model in a water tank 1.5 meters wide. The towing velocity was about 15 centimeters per second. The streamlines were made visible by strewing aluminum powders over the water surface, and pictures were taken by a camera moving with the model.

3. Experimental results

Figs. 2 to 5 show the pressure distributions on the step face as well as on the downstream surface. In these results, the distance x measured along the surface is normalized by the step height h, and the pressure p normal-

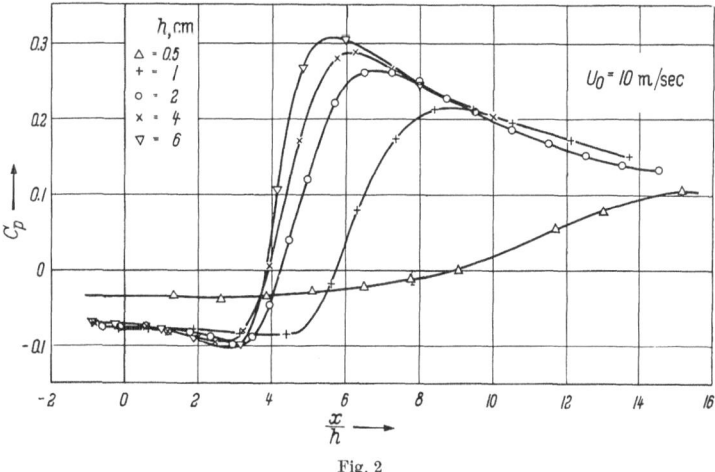

Fig. 2

ized in the form of pressure coefficient, $C_p = (p - p_0)/\frac{1}{2} \varrho U_0^2$, where ϱ is the air density, and p_0 and U_0 are the static pressure and main-flow velocity, respectively, at the location 3 centimeters upstream of the step. It is to be noted that the values of x/h lying between -1 and 0 correspond to the locations on the step face, while the positive values of x/h correspond to those on the downstream surface.

Figs. 2 and 3 are the pressure distributions for several different step heights for $U_0 = 10$ and 28 meters per second, respectively. As mentioned above, the boundary layer approaching the step is laminar at $U_0 = 10$ meters per second and turbulent at $U_0 = 28$ meters per second. The results of similar measurements for $U_0 = 16$ and 22 meters per second are not shown, because they almost agree with those for $U_0 = 28$ meters per second. Fig. 3 also includes measurements for the case of artificially thickened boundary layer (thickness 3 centimeters). In all cases, there is a negative pressure on the step face (base pressure), followed initially by a

slight drop in pressure downstream of the step, and then by a rather
rapid rise of pressure indicating the reattachment of separated flow.
  It is seen from these results that the pressure distribution is rather

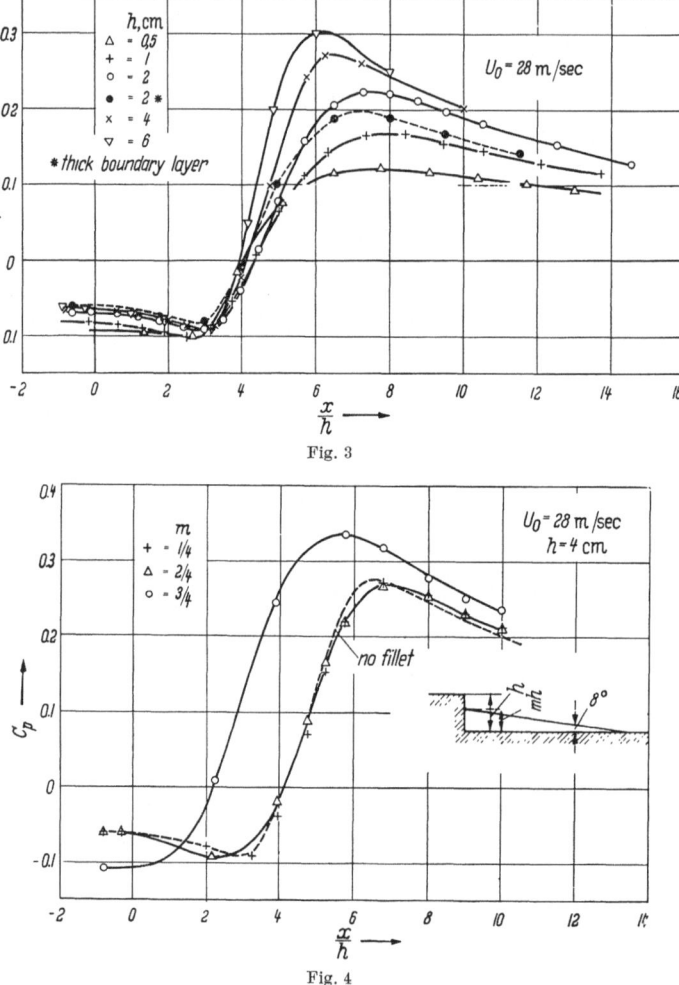

Fig. 3

Fig. 4

insensitive to the changes in the step height as well as the thickness of the
approaching boundary layer. In particular, the base pressure is essentially
the same for different step heights except for very low steps, and the
pressure rise by reattachment increases slightly as the height increases.
  Fig. 4 shows the pressure distribution for the cases when a triangular
fillet of various heights is inserted behind the step. No appreciable change
is observed in the pressure distribution until the fillet height exceeds one

half the step height. Fig. 5 shows the pressure distribution for the cases when a fence of the height of one half the step height is placed at various positions on the downstream surface. The effect of the fence is most appreciable when it is placed at a distance of twice the step height from the step.

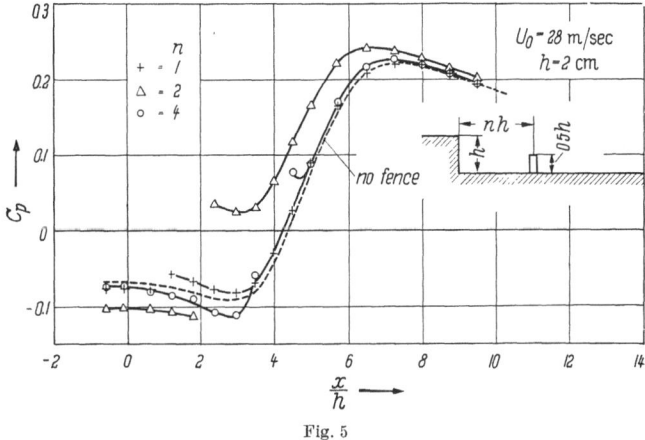

Fig. 5

Fig. 6 shows the typical distribution of longitudinal component of mean velocity across several transverse sections in the mixing region for $h = 2$ centimeters and $U_0 = 28$ meters per second. The figure also includes the distribution of longitudinal mean velocity in the reverse flow region immediately close to the downstream surface. Only a rough estimate can be made for the velocity distribution across the central part of cavity flow by joining the two regions of measurement by dotted lines. From the distribution of mean longitudinal velocity, and also by taking into account of the mean flow direction determined by hot-wire measurements, a mean streamline can be drawn that starts from the step shoulder and approaches the downstream surface in the reattachment region. This line can be considered as that dividing the cavity flow region from the main flow.

Fig. 7 shows the distribution for the same h and U_0 of the turbulence intensity $\overline{u_s^2}$ and turbulent shear stress $- \overline{\varrho\, u_s v_s}$ normalized by U_0^2 and $\varrho\, U_0^2$, respectively, where u_s and v_s are the components of fluctuating velocity in the directions parallel and normal to the mean streamline, respectively. It is seen that both the turbulence and shear increase downstream in the mixing region and that the positions of maximum turbulence and maximum shear approximately coincide at the outset with the mean dividing streamline, but deviate outward as the reattachment is approached.

Figs. 8 and 9 are the aluminum-powder pictures of streamlines for the step of 5-centimeter height, the exposure time being ½ and 5 seconds, respectively. The picture of short exposure reveals the intermingling of distinct eddies in cavity flow region, while that of long exposure indicates

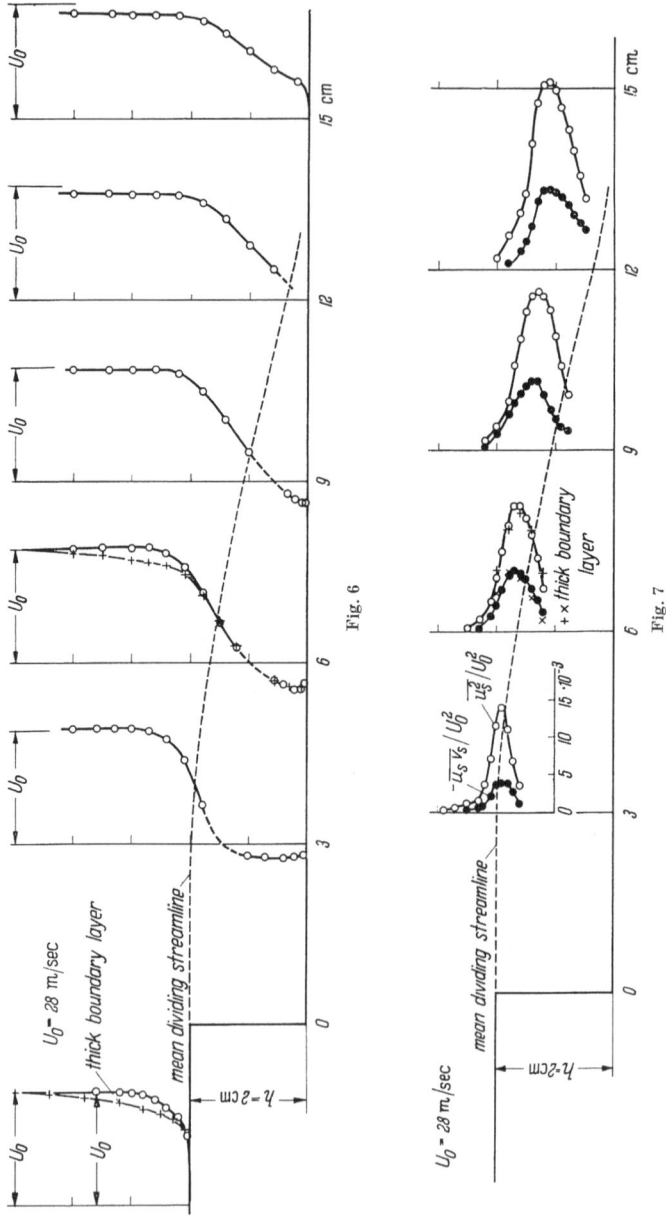

Fig. 6

Fig. 7

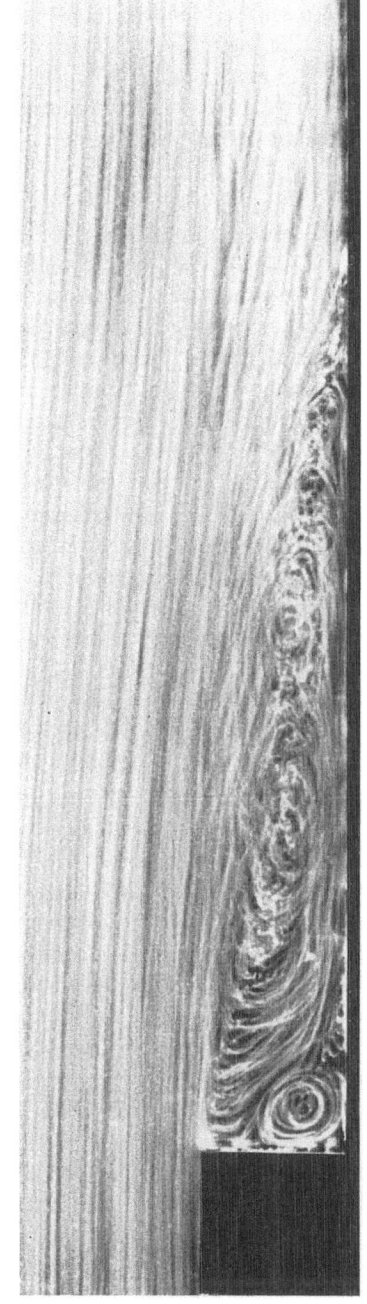

Fig. 8

Fig. 9

something like the mean dividing streamline. The Reynolds numbers based on h and U_0 concerning the water-tank experiments are one order of magnitude lower than those of the wind-tunnel experiments. In view of the above mentioned small effect of Reynolds numbers, however, the flow patterns indicated by aluminum powders are expected to be not much different from those of the wind-tunnel experiments.

4. Discussions

As mentioned above, the base pressure is essentially the same for different values of step height and boundary layer thickness. Even the difference in character of the approaching boundary layer flow, laminar or turbulent, makes no essential difference except for the step of very small height. This is conceivable because the laminar boundary layer when separated becomes turbulent over a very short distance. For $h = 2$ centimeters and $U_0 = 10$ meters per second, for example, the transition to turbulence was observed to occur in the separated layer over a distance of only 2 centimeters from the step shoulder.

The pressure rise by reattachment increases slightly as the step height is increased or the thickness of the approaching boundary layer is reduced. From what mentioned above, it is easily understandable that the effect of boundary layer becomes inappreciable as the step height is increased.

The base pressure, and also the pressure rise by reattachment, are almost unchanged even when a triangular fillet is inserted behind the step. This suggests that the interaction between the cavity flow and the external flow remain unchanged unless the cavity flow region is unduly oppressed. On the other hand, a considerable change in pressure distribution is found when the reverse part of the cavity flow is interrupted by means of a fence. It is to be noted that the maximum velocity of the reverse flow is about one fourth of the main-flow velocity.

The above mentioned approximate similarity and the comparatively steady character of cavity flow seem to be typical of the flow separation associated with a step. This has tempted one to calculate the flow pattern as the irrotational motion of a nonviscous fluid, the cavity flow being replaced by a point vortex captured behind the step. The calculation has been carried out by transforming the physical z-plane conformally into the auxiliary t-plane (fig. 10) and requiring that the vortex is at rest and that the velocity at the visor tip V has a nonzero finite value equal to the undisturbed flow velocity. (If there were no visor, the step shoulder B would become a stagnation point.) It is found that the vortex is stable against small disturbances and that the required length of visor, BV, is only one twentieth of the step height, BC. Yet the result of calculation is at variance with the measurement; the streamline leaving the visor tip rises upward considerably so that the distance to reattachment, CR, is

predicted to be only 1.67 times the step height, while the observed value
is about 7 times the step height.

A correct understanding may probably be obtained by realizing that
the cavity flow is chiefly maintained by the turbulent shear stress, which
is set up in the mixing region approximately independently of the characteristics of the approaching boundary layer. This view seems to be supported by the observed result that the distribution of $-\overline{u_s v_s}/U_0^2$ for
the step of 2-centimeter height is essentially the same both for the normal

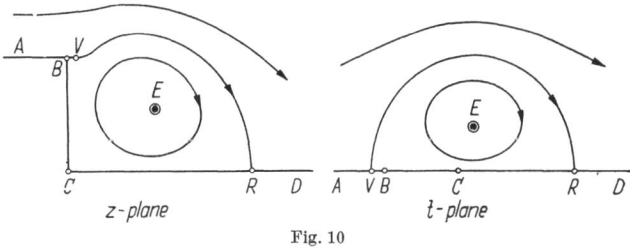

and artificially thickened boundary layer (fig. 7). Even the distribution
of $-\overline{u_s v_s}/U_0^2$ for the step of 4-centimeter height is found not much
different from that for the step of 2-centimeter height.

The model of the flow pattern might be simplified if the cavity flow
were considered to be isolated from the main flow by the above mentioned
mean dividing streamline. Aluminum-powder pictures such as fig. 9
appear more likely to favor the possibility of isolation. If this were the
case, however, the normal and tangential stresses acting on the boundary
of the cavity flow region should be in equilibrium, because otherwise a
certain amount of momentum should be transported into or out of the
region. This reasoning has led ROSHKO [4] to throw doubt on the isola-
bility of cavity flow in a square cutout.

From the results of measurements for $h = 2$ centimeters and $U_0 =$
28 meters per second, a rough examination can be made for the equilibrium
of the stresses acting on the fluid in cavity flow. The pressure force
exerted by the step face is -0.07, when normalized by the area $h \cdot 1$ and
the dynamic pressure $\frac{1}{2} \varrho U_0^2$. Negative sign indicates that the force is
against the direction of undisturbed main flow. Similarly, the pressure
force acting on the dividing streamline is -0.03. The shearing force
exerted by the downstream surface is $+ 0.01$, which is estimated from the
velocity distribution of the reverse flow. Finally, the shearing force
acting on the dividing streamline is $+ 0.07$. The sum of these forces is
not exactly equal to zero, but sufficiently small so that the forces can be
considered to form approximately a system of equilibrium.

According to the experiments of LIEPMANN and LAUFER [5], the
turbulent shear stress in a half jet increases at the outset, but tends to

maintain a constant value after a certain distance is covered. As mentioned above, however, this is not the case for the separated flow over a step, the shear stress continuing to increase downstream. Undoubtedly the development of shear stress is considered to be consequent on the interaction between the dissipative cavity flow and the non-dissipative main flow.

5. Conclusions

The low-speed measurements on the flow separated over a backward-facing step bring out the following features:

1. Except for steps of very small height, the base pressure is essentially the same for different values of step height and boundary layer thickness, and the pressure rise by flow reattachment increases slightly as the step height is increased or the boundary layer thickness is reduced.

2. The base pressure and the pressure rise by reattachment are almost unchanged even when a triangular fillet is inserted behind the step. An appreciable change in pressure distribution is found, however, when the reverse part of cavity flow is interrupted by placing a fence on the downstream surface.

3. The flow pattern calculated by assuming an irrotational motion is at variance with the observation.

4. The insensitivity of base pressure to the step height and boundary layer thickness can be explained by realizing that the cavity flow is chiefly maintained by the turbulent shear stress, which is set up in the mixing region approximately independently of the step height and the approaching boundary layer. The shear stress continues to increase downstream as the result of the interaction between the dissipative cavity flow and the non-dissipative main flow.

5. It seems possible to consider the mean dividing streamline by which the cavity flow region is isolated from the main flow. The normal and tangential stresses acting on the fluid in cavity flow are approximately in equilibrium, so that no momentum seems to be transported into or out of the cavity flow region.

References

[1] CROCCO, L., and L. LEES: A Mixing Theory for the Interaction Between Dissipative Flows and Nearly Isentropic Streams. J. Aero. Sci., **19**, 649—676 (1952).

[2] KORST, H. H.: A Theory for Base Pressures in Transonic and Supersonic Flow. J. App. Mech., **23**, 593—600 (1956).

[3] CHAPMAN, D. R., D. M. KUEHN and H. K. LARSON: Investigation of Separated Flows in Supersonic and Subsonic Streams with Emphasis on the Effect of Transition. NACA Tech. Note no. 3869 (1957).

[4] ROSHKO, A.: Some Measurements of Flow in a Rectangular Cutout. NACA Tech. Note no. 3488 (1955).

[5] LIEPMANN, H. W., and J. LAUFER: Investigations of Free Turbulent Mixing. NACA Tech. Note no. 1257 (1947).

Beitrag zur Phänomenologie des laminar-turbulenten Übergangs im Freistrahl bei kleinen Reynoldszahlen

Von

O. Wehrmann und R. Wille[1]

Technische Universität, Berlin

1. Einleitung

1.1. Abgrenzung des Themas. Als „Freistrahl" sei eine Flüssigkeitsströmung bezeichnet, die sich aus einer Düse heraus in ein Gebiet bewegt, das von der gleichen Flüssigkeit erfüllt ist, die im Unendlichen sich in Ruhe befindet. Im speziellen, und wegen der vielfachen technischen Anwendungen, wird ein rotationssymmetrischer Freistrahl untersucht, dessen Achse die x-Richtung im System der Zylinderkoordinaten x, y (Radius) und ϑ (Winkel in Umfangsrichtung) bildet. Der Begriff „Düse" beinhaltet, daß die Flüssigkeit vor dem Übertritt in die ruhende Umgebung eine konvektive Beschleunigung erfährt, wobei die Geschwindigkeit vor der Düse vernachlässigbar klein gegenüber der Strahlgeschwindigkeit U in der Düsenmündung ist.

Bei den Versuchen wurde Luft aus der Versuchshalle durch die Düse in eine Unterdruckkammer gesaugt, deren Querschnitt und Länge groß gegenüber den Düsenabmessungen sind: Düsendurchmesser $D = 10 - 25 - 50 - 75 - 100 - 150$ mm $\varnothing$, Durchmesser der Kammer $D_K = 2000$ mm $\varnothing$, Länge der Kammer $L = 4000$ mm. Die Kontur der Düsen entspricht den VDI-Meßdüsen, DIN 1952, wie sie zur Mengenmessung in Rohrleitungen verwendet werden.

Die Strahlgeschwindigkeiten liegen im Bereich 1,5 m/s $< U <$ 6,5 m/s; Einflüsse der Kompressibilität der Luft sind hierbei vernachlässigbar. Die Reynoldszahlen liegen im Bereich $2500 < Re_D < 40000$. Die Reynoldssche Zahl ist definiert durch $Re_D = \dfrac{U \cdot D}{\nu}$ (ν = kinematische Zähigkeit der Luft). Einzelheiten der Versuchsanlage können der unter [17] zitierten Arbeit entnommen werden.

Das Ziel der Untersuchung ist, die Strömungsvorgänge in unmittelbarer Nähe stromab von der Düse, etwa im Bereich $0 < x/D < 2$, in Einzelheiten zu beschreiben. Zur Messung der Geschwindigkeiten wurde die bekannte Hitzdrahttechnik nach der Methode des konstanten Stroms

---

[1] Vorgetragen von R. WILLE.

benutzt, die eine Trennung des zeitlichen Mittels $\bar{c}$ und der überlagerten Geschwindigkeitsschwankungen $c' = c'$ (t) $(t = \text{Zeit})$ erlaubt. Im Strömungsfeld nahe der Mündung setzt sich die Wandgrenzschicht der Düse als freie Strahlgrenzschicht mit $dp/dx = 0$ ($p = $ statischer Druck im Strömungsfeld) fort. Der Begriff „Strahlgrenzschicht" wird auf das stromabwärts anwachsende Gebiet angewandt, in dem das zeitliche Mittel der Geschwindigkeit vom Werte Null auf den Wert im Potentialkern des Strahls ansteigt. In dieser freien Strahlgrenzschicht findet ein Übergang von der laminaren in die turbulente Strömungsform statt, die sich weiter stromab auf den gesamten Freistrahl ausdehnt. Das lokale Übergangsgebiet erstreckt sich über eine gewisse Länge stromab von der Düsenmündung, und in diesem Bereich beherrschen Ringwirbel, die aus der Konzentration der mittleren Wirbelstärke periodischer Abschnitte der Düsengrenzschicht entstehen, den Strömungsvorgang. Die meßbaren Phänomene und ihr Ordnungsprinzip werden in den Abschnitten 2 und 3 behandelt[1].

1.2. Ältere Arbeiten. Die turbulente Ausbreitung von Freistrahlen in größerem Abstand von der Düsenmündung ist oft, und besonders im Hinblick auf die Verifizierung der Prandtlschen Mischungswegtheorie, behandelt worden. Die Verteilung der mittleren Geschwindigkeit wurde von Tollmien [1], Görtler [2], Förthmann [3] und Reichardt [4] behandelt. Messungen der mittleren Geschwindigkeit wurden von Zimm [5] und Ruden [6] ausgeführt. Eine zusammenfassende Darstellung findet sich bei Schlichting [7] im Kapitel „Freie Turbulenz". Die Verteilung der mittleren Geschwindigkeit in der Nähe der Düsenmündung ist von Kuethe [8] sowie von Squire und Trouncer [9] untersucht worden.

Der Mechanismus der Geschwindigkeitsschwankungen bei voll ausgebildeter Turbulenz des Freistrahls war Gegenstand der Arbeiten von Corrsin [10], Uberoi [11] und Liepmann und Laufer [12].

Die Untersuchung der Wirbelvorgänge in der Randzone eines Freistrahls wurde durch die Arbeiten von Domm [13] über die Kármánsche Wirbelstraße ausgelöst. Gemeinsame Merkmale der freien Strahlgrenzschicht und der von einem umströmten Körper abgelösten Strömung sind von Wille und Domm [14] beschrieben worden. Theoretische Untersuchungen über die Stabilität abgelöster Grenzschichten stammen von Lessen [15] und Lin [16].

Erste Ergebnisse über Hitzdrahtmessungen im laminar-turbulenten Übergangsgebiet der Freistrahlgrenzschicht sind in zwei Berichten von

[1] Die Durchführung der Arbeiten wurde durch Forschungsbeihilfen des Air Research and Development Command, United States Air Force, ermöglicht. Die Versuchsanlagen wurden aus Mitteln der Deutschen Forschungsgemeinschaft und des Senators für Wirtschaft und Kredit des Landes Berlin (Berliner ERP-Mittel) errichtet. Allen fördernden Stellen sei hiermit gedankt.

DOMM, FABIAN, WEHRMANN und WILLE [*17–18*] enthalten. In diesen Arbeiten findet sich auch eine erste, jetzt revidierte Darstellung des Frequenzgesetzes der Wirbelbildung im Freistrahl, das für die ebene, von der Hinterkante einer Platte abgelöste Grenzschicht seine Parallele in einer Arbeit von SATO [*19*] findet.

Als Konsequenz der in [*17*] und [*18*] mitgeteilten Ergebnisse stellte in einer weiteren Arbeit DOMM [*20*] die Hypothese auf, daß die Konzentration von mittlerer Wirbelstärke zu Wirbelstraßen eine charakteristische Erscheinung im Übergang zur Turbulenz ist und daß der Zerfall der Wirbel nach Erreichung einer kritischen Stärke mit dem Einsatz der Turbulenz identisch ist. WEHRMANN [*21*] entwickelte für Hitzdrahtmessungen ein einfaches Eichverfahren zur numerischen Bestimmung der Geschwindigkeitsschwankungen in periodischen Vorgängen und stellte aus kinematischen Überlegungen und Messungen in einer aufgespaltenen KÁRMÁNschen Wirbelstraße Kriterien für Hitzdrahtsignale auf, die Wirbeln entsprechen.

Außerhalb des Problems Freistrahl wurden periodische Ringwirbel als Übergangsphänomene zur voll ausgebildeten Turbulenz auch bei abgelöster Strömung in Rohreinläufen beobachtet. Hierüber unterrichten die Arbeiten von SCHILLER [*22*], NAUMANN [*23*] und KURZWEG [*24*].

2. Strömungsvorgänge in der Strahlgrenzschicht

2.1. Verteilung der mittleren Geschwindigkeit. Abb. 1 zeigt die mit dem Hitzdraht gemessene Verteilung der mittleren Geschwindigkeit und die Linien gleicher Geschwindigkeit im Freistrahl mit $Re_D = 5000$ und Abb. 2 das gleiche für $Re_D = 20\,000$. Zu dieser Darstellung ist zu bemerken, daß zum Beispiel für den Strahl mit $Re_D = 20\,000$ im Abstand $x/D = 4$ noch ein „Potentialkern" existiert, während, wie später gezeigt werden wird, die Strahlgrenzschicht bereits im Abstand $x/D = 2$ den Einsatz des turbulenten Strömungszustandes erreicht hat.

Abb. 3 zeigt die Geschwindigkeitsverteilung zwischen dem äußeren Strahlrand und dem Potentialkern, also in dem Bereich, der oft als „Mischungszone" bezeichnet wird. Die Darstellung folgt dem Vorschlag von KUETHE [*8*]. Die Ordinate erhält das zeitliche Mittel $\bar{c}$ der Geschwindigkeit im Verhältnis zur Strahlgeschwindigkeit U im Mündungsquerschnitt, während der Abszissenmaßstab sich nach dem Quotienten $\theta = \dfrac{y-a}{b}$ richtet. (y = radialer Abstand des Meßpunktes von der Strahlachse, $a = a(x)$ = Radius des Potentialkerns, $b = b(x)$ radiale Breite der Mischungszone.) Die Kurven zeigen, daß affine Geschwindigkeitsprofile erst in Abständen $x/D > 2$ existieren und dann gut mit der Lösung von TOLLMIEN [*1*] übereinstimmen, die in den rechts liegenden Diagrammen gestrichelt eingezeichnet ist. Der unterschiedliche Kurven-

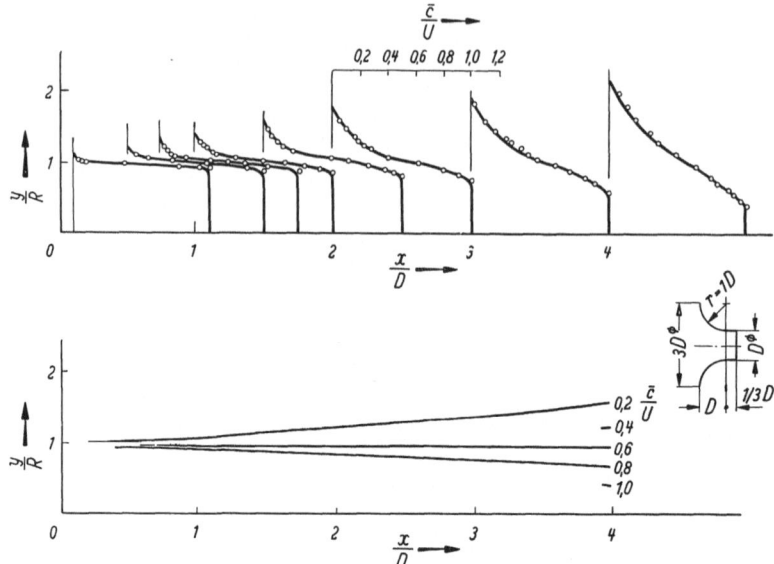

Abb. 1. Verteilung der mittleren Geschwindigkeit im Freistrahl bei $Re_D = 5000$

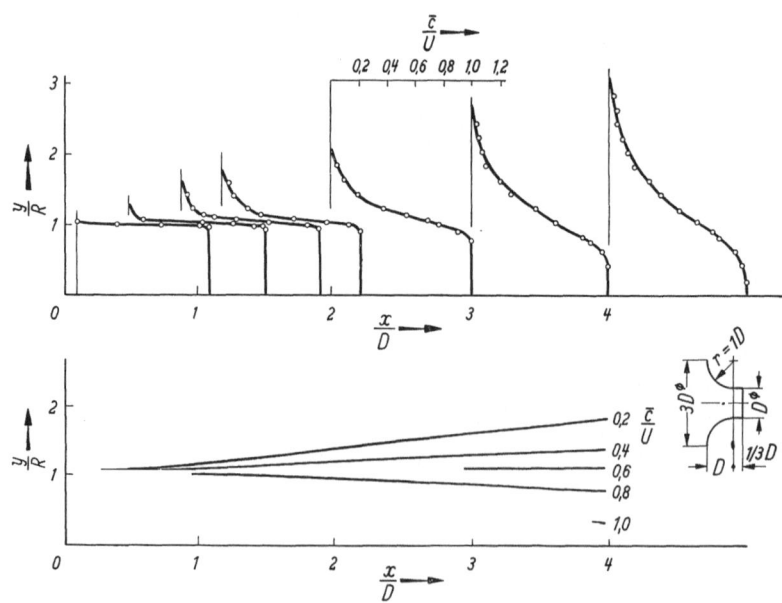

Abb. 2. Verteilung der mittleren Geschwindigkeit im Freistrahl bei $Re_D = 20\,000$

verlauf im Bereich $0 < x/D < 2$ deutet an, daß in unmittelbarer Nähe der Düse die Randzone des Strahls von Strömungsvorgängen beherrscht wird, die nicht durch das zeitliche Mittel der Geschwindigkeit hinreichend beschrieben werden können. Weitere Meßergebnisse dieser Art sind in den unter [17] und [18] zitierten Arbeiten enthalten.

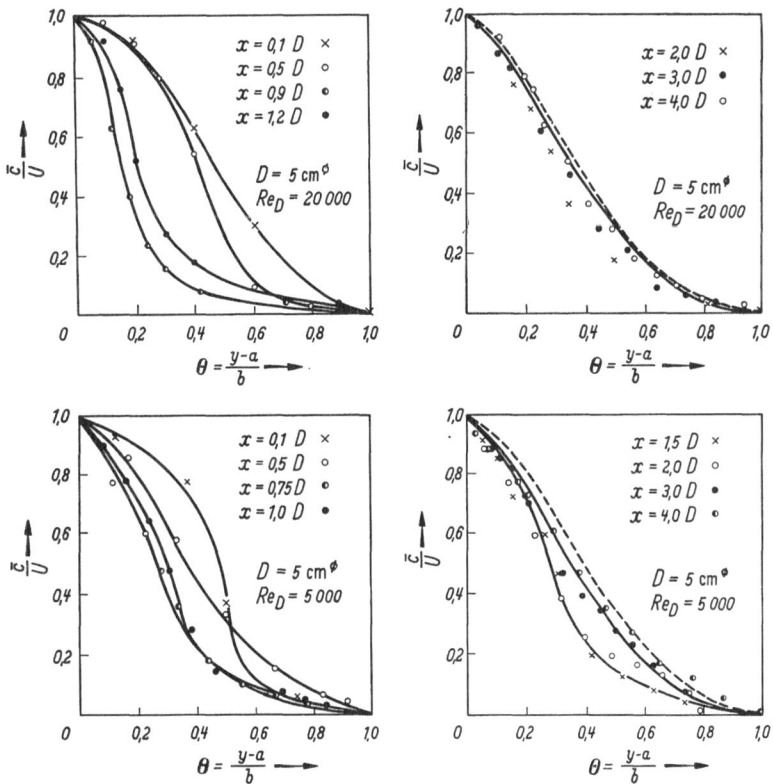

Abb. 3. Verteilung der mittleren Geschwindigkeit in der Randzone des Freistrahls bei $Re_D = 5000$ und $Re_D = 20\,000$

2.2. Ringwirbel der Strahlgrenzschicht. Abb. 4 zeigt die Filmaufnahme eines Freistrahls Wasser in Wasser bei $Re_D = 10\,000$. Filmserien dieser Art und eine Beschreibung der Versuchsanlage sind in der unter [14] zitierten Arbeit zu finden. Bei dieser Filmaufnahme wurde die Randzone des Strahls an zwei gegenüberliegenden Punkten durch Farbzufuhr sichtbar gemacht. Man erkennt, daß die anfänglich ungestörte Farbschicht sich später zu definierten Bereichen zusammenzieht, wobei Wirbelringe entstehen, die das weitere Geschehen beherrschen. Die Ringwirbel-Straße, die den Strahl umgreift, enthält nur wenige voneinander unterscheidbare Wirbelringe. Infolge kleiner Unstabilitäten können die Wirbel Relativbewegungen gegeneinander ausführen, wobei es zu Schlüpfvorgängen mit nachfolgender Vereinigung zweier Wirbel kommen kann. Eine theoretische Betrachtung zu diesem Vorgang ist in einer Arbeit von Domm [20] enthalten. An dieser Stelle sei vermerkt, daß der Schlüpf- und Vereinigungsprozeß vornehmlich bei niedrigen Strahlgeschwindigkeiten beobachtet wird. Der Farbfadenversuch zeigt weiter, daß die Ringwirbel schließlich ihre Gestalt verlieren und in eine allgemeine Mischbewegung übergehen.

Die Filmaufnahme des angefärbten Freistrahls in Wasser zeigt die Gesamtheit der Erscheinungen gleichzeitig an verschiedenen Raum-

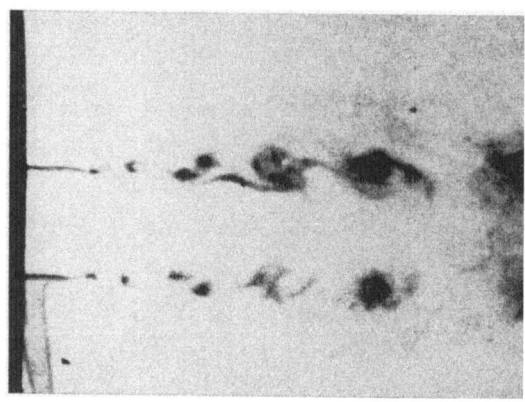

Abb. 4. Filmaufnahme eines Freistrahls Wasser in Wasser bei $Re_D = 10\,000$

punkten. Zur Ergänzung sind in den folgenden Bildern Hitzdrahtsignale wiedergegeben, die in einem Luftfreistrahl aufgenommen wurden. Die Hitzdrahtsignale geben den zeitlichen Verlauf der Geschwindigkeit an verschiedenen Raumpunkten, nämlich am jeweiligen Ort der Meßsonde an. Es handelt sich also um die Darstellung $c' = c'(t)$. Unter jedem Signal sind der axiale Abstand von der Mündungsebene und der radiale Abstand von der Strahlachse vermerkt. Der Hitzdraht befindet sich in jedem Falle in der Strahlgrenzschicht, und der radiale Abstand liegt etwas unterhalb der Bahn der Wirbelkerne.

Abb. 5 gilt für eine Reynoldszahl von $Re_D = 20\,000$. In unmittelbarer Nähe der Düse zeigt das Hitzdrahtsignal nur sehr geringe Schwankungen. Weiter stromab werden die Amplituden größer, und man erkennt, daß es sich um eine periodische Geschwindigkeitsvariation nach Art einer Sinuslinie handelt, deren Frequenz über der Zeit und bei verschiedenen axialen Abständen im Mittel konstant bleibt. Dies gilt bis zum Punkt $x/D = 0{,}9$. Bei größerem Abstand ist die Frequenz des Signals auf die Hälfte gefallen, während die Amplitude weiter gewachsen ist. Dieser

Frequenzabfall ist einfach zu erklären: Im Bereich $0,8 < x/D < 1$ haben sich zwei aufeinanderfolgende Wirbel zu einem vereinigt. Bei noch größeren Abständen, etwa bei $x/D = 1,2$, erfährt das Hitzdrahtsignal eine Veränderung. Dem periodischen Signal sind hochfrequente Schwingungen überlagert, die in zeitlich unregelmäßiger Folge auftreten. Diese

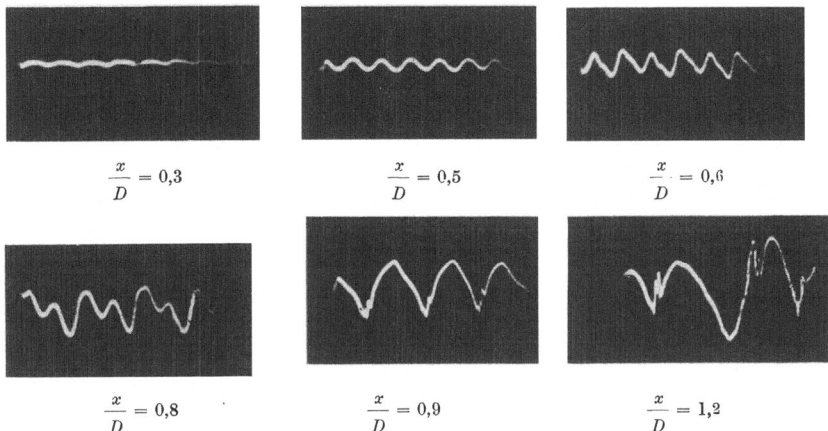

$$\frac{x}{D} = 0,3 \qquad \frac{x}{D} = 0,5 \qquad \frac{x}{D} = 0,6$$

$$\frac{x}{D} = 0,8 \qquad \frac{x}{D} = 0,9 \qquad \frac{x}{D} = 1,2$$

Abb. 5. Hitzdrahtsignale in der Randzone des Freistrahls $Re_D = 20\,000$; $D = 7,5\,\mathrm{cm}$; $y/D = 0,5 = \mathrm{const.}$

,,Turbulenzausbrüche" können, je nachdem, ob man die Erscheinung zeitlich oder räumlich deuten will, als ,,turbulent flashes" (REYNOLDS) oder als ,,turbulent spots" (EMMONS) bezeichnet werden; sie kennzeichnen den intermittierenden Einsatz der Turbulenz, der zuerst von TOWNSEND [25] beschrieben wurde.

Abb. 6 zeigt Hitzdrahtsignale in einem Strahl bei $Re_D = 30\,000$. Der allgemeine Charakter der Signale ist der gleiche wie bei der niedrigen

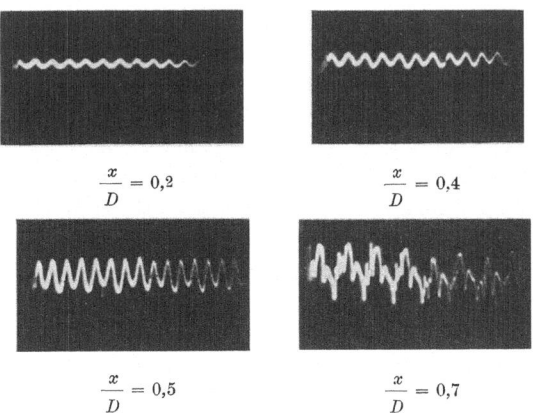

$$\frac{x}{D} = 0,2 \qquad \frac{x}{D} = 0,4$$

$$\frac{x}{D} = 0,5 \qquad \frac{x}{D} = 0,7$$

Abb. 6. Hitzdrahtsignale in der Randzone des Freistrahls $Re_D = 30\,000$; $D = 7,5\,\mathrm{cm}$; $y/D = 0,5 = \mathrm{const.}$

Geschwindigkeit. Unterschiedlich ist jedoch, daß der Frequenzabfall auf die Hälfte fehlt. Bei dieser Geschwindigkeit kommt also eine Wirbelvereinigung nicht zustande.

2.3. Geschwindigkeitsverteilung der Wirbel der Strahlgrenzschicht. Die periodischen Hitzdrahtsignale entstehen dadurch, daß ein Wirbel am Hitzdraht vorbeiläuft. Ein Wirbel besteht aus einem Kern, in dem die Geschwindigkeit aus dem Zentrum heraus anwächst, und einem kreisen-

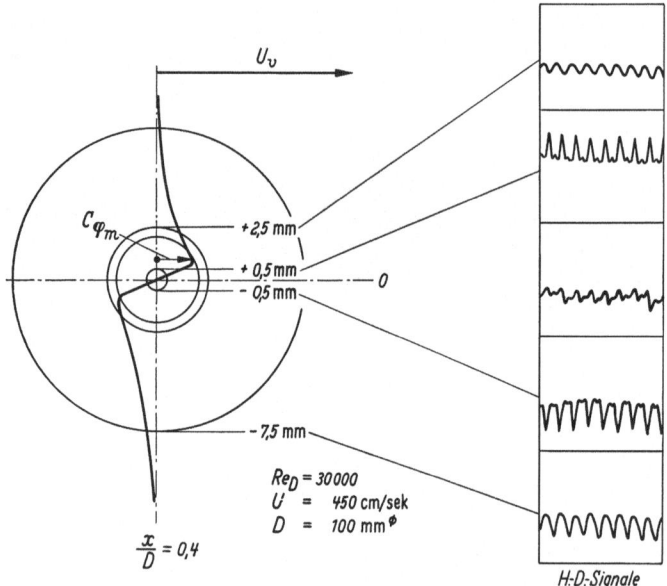

Abb. 7. Hitzdrahtsignale beim Durchgang eines Ringwirbels am Ort der Hitzdrahtsonde. Linke Bildhälfte: Schematische Darstellung der Geschwindigkeitsverteilung eines Wirbels

den Geschwindigkeitsfeld, in dem die Umfangsgeschwindigkeit wieder absinkt. Eine Analyse der Hitzdrahtsignale, die dem Durchgang von Wirbeln entsprechen, wurde von O. WEHRMANN [21] durchgeführt. Abb. 7 zeigt Hitzdrahtsignale im Gebiet der Ringwirbelstraße für verschiedene radiale Abstände von der Strahlachse. Die schematische Darstellung eines Wirbels in der linken Hälfte des Bildes erläutert die Entstehung der verschiedenen Signalformen: der Hitzdraht mißt stets den Betrag der Geschwindigkeit, der am Meßort aus der vektoriellen Addition der Umfangsgeschwindigkeit c_φ und der Transportgeschwindigkeit U_v des Wirbels resultiert. Die Hitzdrahtsignale geben also nicht die zeitliche Variation einer Geschwindigkeitskomponente, etwa u' parallel zu x, wieder. Aus diesem Grunde wurde in der vorliegenden Arbeit für alle Geschwindigkeiten, die in der von Wirbeln beherrschten Strahlzone gemessen wurden, die allgemeine Notierung c bzw. c' gewählt.

Auf jeden Fall ist das Hitzdrahtsignal beim Durchgang eines Wirbels am Ort der Sonde periodisch; seine Form ändert sich, wenn der Wirbelkern erreicht wird. Als ein allgemeines Ergebnis mag hier erwähnt werden, daß bei Hitzdrahtmessungen in Wirbelstraßen die größten Amplituden nicht beim Durchgang des Wirbelzentrums auftreten, sondern dort, wo die Bahn der maximalen Umfangsgeschwindigkeit $c_{\varphi\,\text{max}}$ den Hitzdraht berührt.

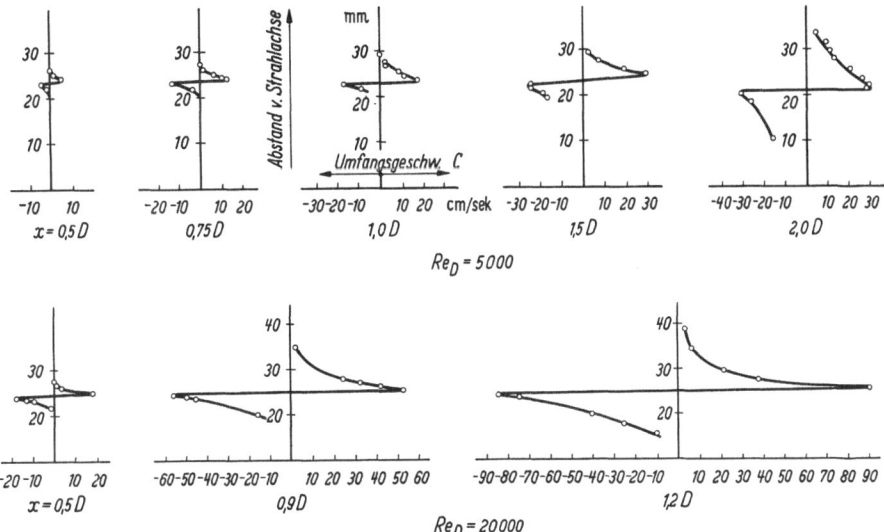

Abb. 8. Geschwindigkeitsverteilung in Wirbeln der Strahlgrenzschicht bei $Re_D = 5000$ und $Re_D = 20\,000$

Für die Mitte einer Signalphase gilt, daß der Vektor der Umfangsgeschwindigkeit c_φ in gleicher Richtung mit dem Vektor der Transportgeschwindigkeit U_v liegt. Für diesen Punkt des periodischen Signals hat O. WEHRMANN [21, 18] eine Eichmethode entwickelt, die es erlaubt, die Umfangsgeschwindigkeiten der Wirbel numerisch zu bestimmen. Die Messungen sind zeitraubend, sie erfordern einen streng stationären Zustand des Strahls und eine periodische Wirbelfolge. Die Meßergebnisse gelten für Ebenen senkrecht zur Strahlachse.

In Abb. 8 sind Geschwindigkeitsverteilungen in Wirbeln der Strahlgrenzschicht als Ergebnisse der Hitzdrahtmessungen dargestellt. Die obere Reihe gilt für einen Strahl mit $Re_D = 5000$, die untere Reihe für einen Strahl mit $Re_D = 20\,000$. Man erkennt, daß mit wachsendem Abstand von der Düse die Umfangsgeschwindigkeiten wachsen. Für die Wirbel, die $Re_D = 5000$ (obere Reihe) zugeordnet sind, gilt, daß im Bereich $1{,}0 < x/D < 1{,}5$ sich zwei Wirbel vereinigt haben. Die Verschmelzung erfolgt sehr schnell; denn die Geschwindigkeitsverteilung für $x/D = 1{,}5$ hat die gleichen allgemeinen Eigenschaften wie die Geschwindigkeitsverteilungen, die vor der Vereinigung liegen.

Würde man die Werte der Umfangsgeschwindigkeit c_φ für verschiedene y längs des Abstands x/D auftragen, so ergäbe sich eine „An-

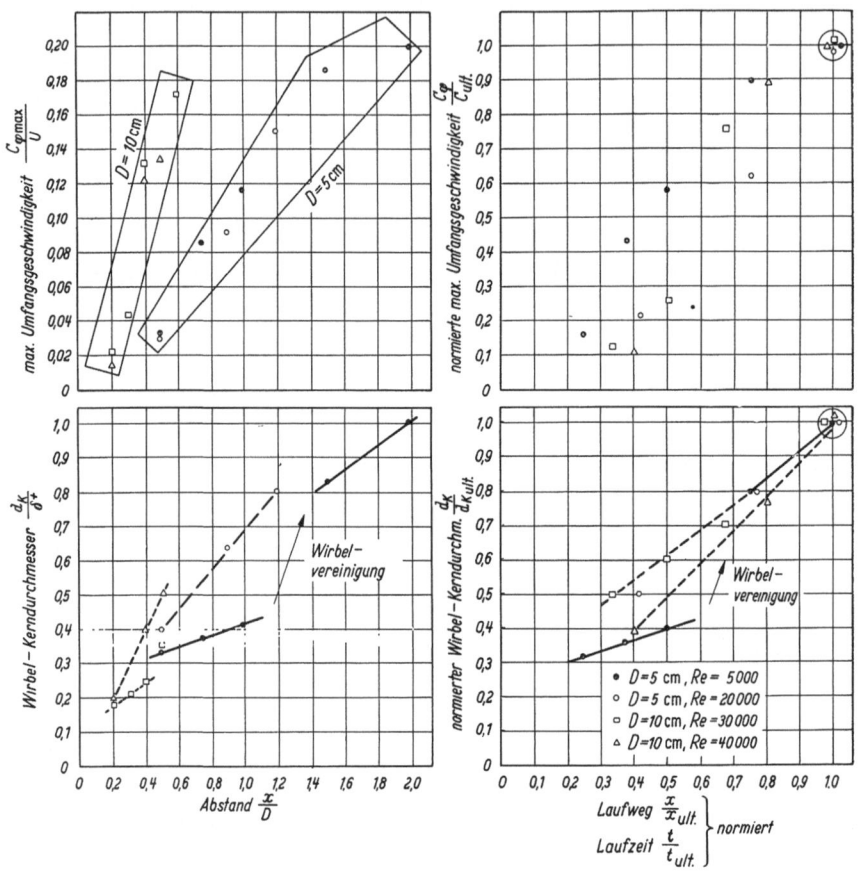

Abb. 9. Charakteristische Daten der Wirbel der Strahlgrenzschicht

fachungskurve" der Geschwindigkeitsschwankungen. In unserem Falle läßt sich die Geschwindigkeitsschwankung phänomenologisch auf kreisende Wirbelbezirke zurückführen.

2.4. Charakteristische Daten der Wirbel der Strahlgrenzschicht. In Abb. 9 sind charakteristische Daten der Grenzschichtwirbel in Diagrammform dargestellt. Es handelt sich hierbei um die Größen: Maximale Umfangsgeschwindigkeit der Wirbel $c_{\varphi\,max}$ und Durchmesser des Wirbelkerns d_k, der hier gleich dem Durchmesser des Kreises der maximalen

Umfangsgeschwindigkeit gesetzt worden ist. In die Darstellung sind Meßwerte für vier Strahl-Reynoldszahlen aufgenommen worden: $Re_D = 5000 - 20000 - 30000 - 40000$. Zwei geometrisch ähnliche Düsen mit $D = 50$ mm $\varnothing$ und $D = 100$ mm $\varnothing$ wurden hierbei verwendet.

In den beiden linken Diagrammen sind die Meßpunkte über dem relativen Düsenabstand x/D aufgetragen, und zwar oben die maximale Umfangsgeschwindigkeit $c_{\varphi\,max}$ als Bruchteil der Strahlgeschwindigkeit U und unten der Kerndurchmesser d_k, mit der Verdrängungsdicke δ^* der Grenzschicht an der Düsenmündung dimensionslos gemacht.

In der rechten Bildhälfte sind die gleichen Meßwerte noch einmal in einer normierten Darstellung aufgetragen. Der Normierung sind „ultimate" Werte, Grenzwerte der Wirbelströmung, zugrunde gelegt worden, die denjenigen Hitzdrahtsignalen entnommen wurden, bei denen sich die ersten in den Abb. 5 und 6 erläuterten hochfrequenten Turbulenzausbrüche zeigten. Die intermittierende Turbulenz kennzeichnet nämlich das Ende der Existenz eines Wirbelrings.

Die Auftragung der Werte $c_{\varphi\,max}$ und d_k läßt kein Ordnungsprinzip erkennen; nur eines tritt klar hervor, nämlich, daß mit wachsendem Laufweg, oder, was dasselbe ist, daß mit wachsendem Lebensalter der Wirbel sowohl die maximale Umfangsgeschwindigkeit als auch der Kerndurchmesser zunehmen. Das Schlüpfen zweier aufeinanderfolgender Wirbel und ihre Vereinigung zeigt sich bei $Re_D = 5000$ besonders deutlich an der Zunahme des Kerndurchmessers.

Es liegt nahe, das Ordnungsprinzip in einer Reynoldszahl der Wirbelströmung zu suchen, wie dies schon früher von uns vorgeschlagen worden ist [17, 18]. Als charakteristische Länge wird hierbei der Wirbelkerndurchmesser d_k, und als charakteristische Geschwindigkeit wird die diesem Kerndurchmesser zugeordnete maximale Umfangsgeschwindigkeit $c_{\varphi\,max}$ gewählt. Die Reynoldssche Zahl der Wirbel lautet demnach $Re_v = \dfrac{c_{\varphi\,max} \cdot d_k}{\nu}$. In Abb. 10 ist diese Größe in normierter Form über dem normierten Laufweg aufgetragen. Die Kennzeichnung der aus den Meßwerten errechneten Punkte erfolgte in gleicher Weise wie in Abb. 9.

Man erkennt, daß Re_v ein Ordnungsprinzip für die wachsenden Wirbel der Strahlgrenzschicht darstellt. Die Streuung der Zahlenwerte ist in der Natur des Vorgangs und in der Schwierigkeit der Messungen begründet.

2.5. Diskussion. Die Bildung der Grenzschicht-Ringwirbel, ihr Anwachsen und ihr dreidimensionaler Zerfall finden in einem Bereich des Freistrahls statt, in dem im Innern des Strahls noch ein Potentialkern existiert.

Die Reynoldszahl der Wirbel Re_v enthält im Zähler die Zirkulation $\Gamma/\pi = c_{\varphi\,max} \cdot d_k$ längs des Kreises mit dem Durchmesser d_k des Wirbelkerns. Für die Ringwirbel der freien Strahlgrenzschicht, die aus der Konzentration der mittleren Wirbelstärke eines Abschnitts der aus der Düse austretenden laminaren Grenzschicht entstehen, ist es charakteristisch, daß dieser Zahlenwert ständig wächst.

Dies steht in Übereinstimmung mit der Auffassung, daß zwischen den TOLLMIENschen Wellen der labilen laminaren Grenzschicht und dem durch „Intermittency"-Signale angezeigten Beginn der Turbulenz ein „Anfachungsvorgang" liegt. Für die freie Strahlgrenzschicht kann gezeigt

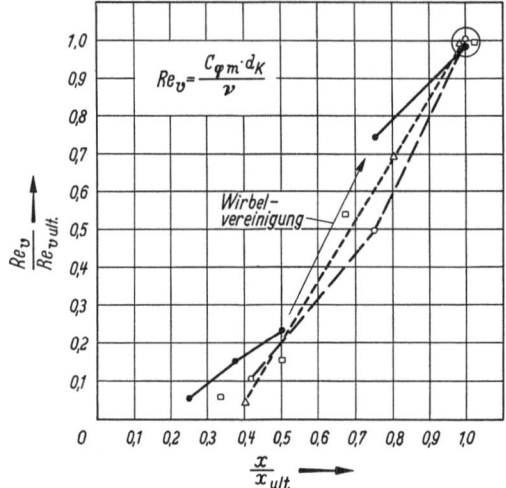

Abb. 10. Anwachsen der Wirbel-Reynoldszahl mit dem Laufweg der Wirbel

werden, daß die „angefachten Störungen" durch wachsende Ringwirbel hervorgerufen werden, deren Eigenschaften definiert werden können.

Der Einsatz der Turbulenz in der Randzone des Strahls ist identisch mit dem Zerfall der Ringwirbel. FABIAN [26] untersucht zur Zeit, ob hierfür eine kritische Größe angegeben werden kann, wie sie von DOMM [20] in seiner Hypothese der Turbulenzentstehung gefordert worden ist.

Die Wirbelstärke des zerfallenen Ringwirbels findet sich in einer Vielzahl kleinerer Wirbel wieder, deren Drehachsen, wie qualitative Messungen mit einem Drallrädchen zeigten, bevorzugt in Richtung der Strahlachse liegen. Die von GÖRTLER [27] berechneten Wirbel der dreidimensionalen Störungen treten offenbar als drittes Stadium des Übergangs laminar-turbulent in der Strahlgrenzschicht auf.

3. Frequenzgesetz der Strahlgrenzschicht

Das „Frequenzgesetz" beschreibt die Abhängigkeit der gemessenen Störungsfrequenzen der Strahlgrenzschicht von der Mündungsgeschwindigkeit U in der Düse. Nach den Ausführungen unter 2 ist die „Störungsfrequenz" identisch mit der Bildungsfrequenz der Ringwirbel.

Bei den Messungen wurden sechs geometrisch ähnliche Düsen mit Mündungsdurchmessern $D = 10 - 25 - 50 - 75 - 100 - 150$ mm verwendet. Als Sonde wurde ein Hitzdraht von $6\,\mu$ Dicke und 2 mm Länge verwendet, der auf den Meßort $x/D = 0,2$, $y/D = 0,45$ eingestellt war. Die genaue Einhaltung des Meßorts ist für das Ergebnis der Frequenzmessungen nicht wichtig, da, wie im Abschnitt 2 ausgeführt wurde, die Ringwirbelstraße von periodischen Vorgängen beherrscht wird.

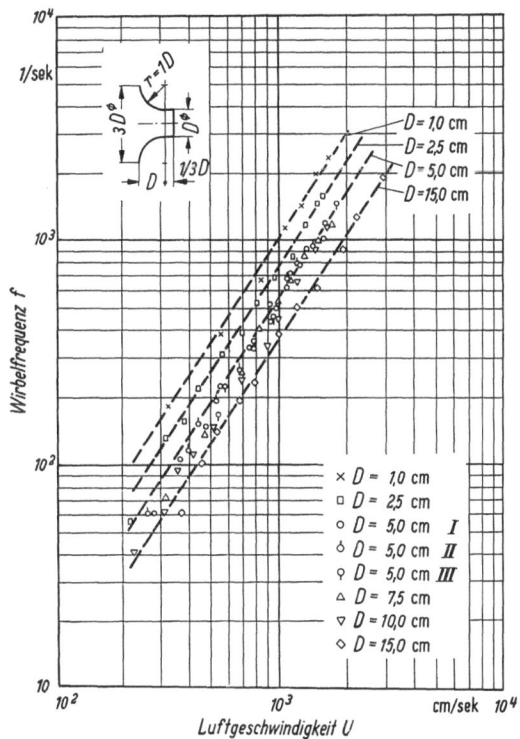

Abb. 11. Wirbelfrequenzen verschiedener Freistrahlen

Die Frequenzmessungen wurden sowohl an einem Freistrahl, der aus der Umgebung in eine Unterdruckkammer gesaugt wurde, als auch an einem Strahl, der aus einer Überdruckkammer in die Umgebung austrat, vorgenommen. Es sei bereits hier erwähnt, daß die Art der Strahlerzeugung keinen Einfluß auf das Frequenzgesetz hat.

Abb. 11 gibt die Meßergebnisse wieder. Die gemessenen Frequenzen f sind über der Mündungsgeschwindigkeit U des Strahls aufgetragen. In der doppelt-logarithmischen Darstellung lassen sich die Meßpunkte auf Geraden mit der Steigung 3/2 anordnen, wobei der Düsendurchmesser D als Parameter eingeht. Die Meßpunkte der Düse $D = 5,0$ III gelten für den Strahl, der aus einer Überdruckkammer austrat, während alle anderen Punkte in Strahlen gemessen wurden, die in eine Unterdruckkammer gesaugt wurden.

In Abb. 12 ist eine andere Auftragung der Meßpunkte gewählt worden. Die Ordinate stellt die „reduzierte" Frequenz $f_{\text{red}} = f \cdot \sqrt[3]{D/D_1}$ dar, wobei die Berechnung der Zahlenwerte für f_{red} auf den Düsendurchmesser

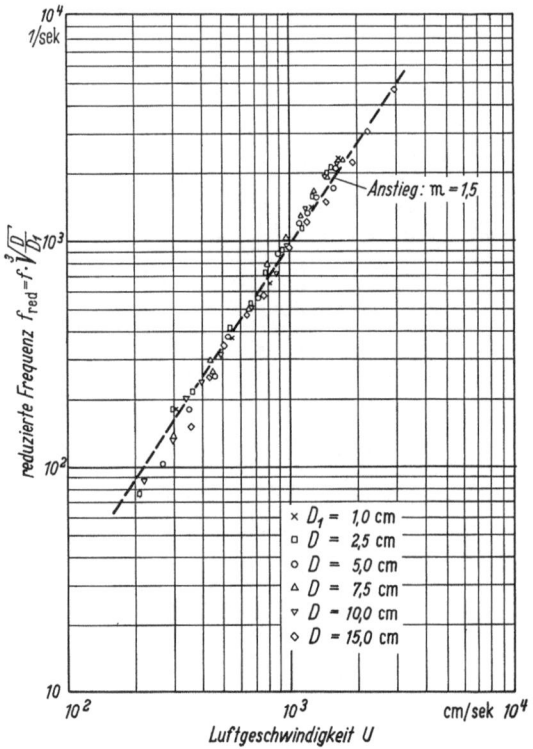

Abb. 12. Reduzierte Wirbelfrequenz verschiedener Freistrahlen

$D_1 = 1$ cm bezogen ist. Die Werte für f_{red} lassen sich gut längs einer einzigen Geraden mit der Steigung 3/2 anordnen.

Für die Gültigkeit des Ordnungsprinzips $\sqrt[3]{D}$ läßt sich zunächst noch keine theoretische Begründung geben. Es läßt sich lediglich folgendes bemerken: Die Wirbelbildung in der freien Strahlgrenzschicht steht mit der Dicke der Wandgrenzschicht in der Düse in Zusammenhang. Diese Grenzschichtdicke wiederum ist, für eine gegebene Mündungsgeschwindigkeit, eine Funktion der Lauflänge der Luft entlang der gekrümmten Düsenkontur. Da bei sämtlichen Messungen geometrisch ähnliche Düsen verwendet wurden, läßt sich D durch eine Lauflänge $L = \text{Zahl} \cdot D$ ersetzen. Es ist also zu vermuten, daß das Ordnungsprinzip $\sqrt[3]{D}$ seine Erklärung im Anwachsen der Düsengrenzschicht in der rotationssymmetrischen Strömung finden wird.

Für die reduzierte Frequenz läßt sich also, nach dem experimentellen Befund, ein Gesetz anschreiben in der Form:

$$f_\text{red} = A \cdot U^{3/2}.$$

Der Exponent 3/2 der Mündungsgeschwindigkeit deutet darauf hin, daß die Grenzschichtdicke an der Düsenmündung sich mit $U^{-1/2}$ ändert.

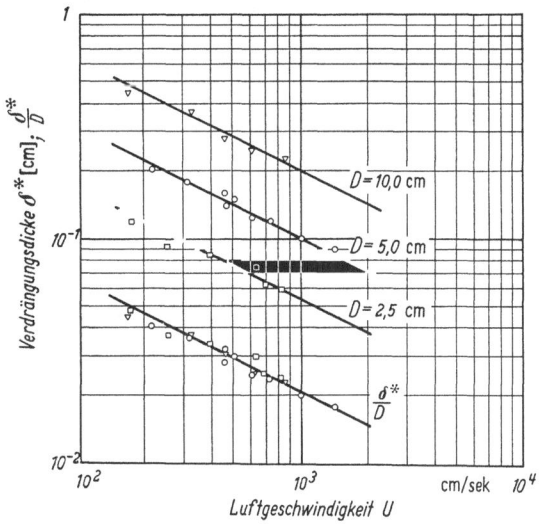

Abb. 13. Grenzschichtdicke an der Düsenmündung

Dieses Gesetz wird durch die Meßergebnisse, wie Abb. 13 zeigt, gut bestätigt. Hierbei ist auf der Ordinate die Verdrängungsdicke δ^* der Düsengrenzschicht in der Definition [17]

$$\delta^* = \int\limits_{0}^{D/2} \frac{y}{D/2}\left(1 - \frac{u}{U}\right) dy$$

aufgetragen.

Setzt man an, daß das Frequenzgesetz nur von den am Vorgang beteiligten Meßgrößen U, δ^* und f abhängt, so ergibt die Dimensionsanalyse: $f = k \cdot U \cdot \delta^{*-1}$. Mit $\delta^* = k_1 \cdot U^{-1/2}$ folgt $f = k_2 \cdot U^{3/2}$. H. Sato [17] fand die gleiche Gesetzmäßigkeit für die Störungsfrequenzen in einer ebenen Wandgrenzschicht, die sich von der Hinterkante einer Platte ablöst.

Die gleichfalls in Abb. 13 aufgetragene Gesetzmäßigkeit $\delta^*/D = B \cdot U^{-1/2}$ findet eine Stütze in der von Schlichting angegebenen Lösung für die

Grenzschichtdicke einer ebenen Senkenströmung längs einer geraden Wand. In unserem Falle handelt es sich zwar um eine räumliche Senkenströmung längs einer gekrümmten Wand, für die keine „ähnliche" Lösung der Grenzschicht-Differentialgleichung existiert, aber es kann offenbar eine Analogiebetrachtung angestellt werden.

Es liegt nahe, die an der Düsenmündung gemessenen Wirbelfrequenzen in gleicher Weise wie die Störungsfrequenzen in der Plattengrenzschicht

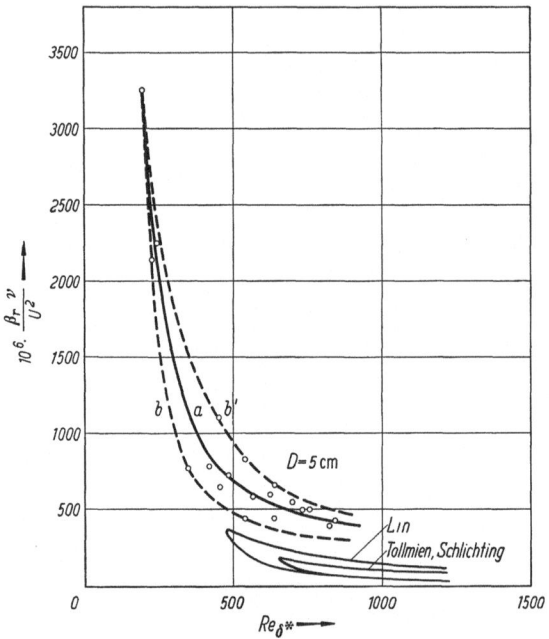

Abb. 14. Bereich der Wirbelfrequenzen im Freistrahl. Zum Vergleich: Neutrale Störungsfrequenzen der Plattengrenzschicht

darzustellen. Dabei ergibt sich Abb. 14. Die Ordinate enthält die dimensionslose Frequenz $\dfrac{\beta\,r\cdot v}{U^2}$ mit $\beta_r = 2\,\pi\,f$, und die Abszisse enthält die mit der Verdrängungsdicke δ^* an der Düsenmündung gebildete Reynoldszahl. Am unteren Bildrand sind, um einen Vergleichsmaßstab zu haben, die Indifferenzkurven für neutrale Störungsfrequenzen bei der längs angeströmten ebenen Platte nach den Rechnungen von TOLLMIEN, SCHLICHTING und LIN eingetragen. Die Meßergebnisse von SCHUBAUER und SKRAMSTAD [28] gruppieren sich längs der Lin-Kurve.

Die in die obere Bildhälfte vorstoßenden Kurvenzüge gelten für die Frequenzmessungen am Freistrahl. Die innere Kurve a gibt den Verlauf der natürlich auftretenden Wirbelfrequenzen an. Die beiden äußeren

Kurven b, b' geben die Grenzen an, innerhalb der durch akustische Beeinflussung Wirbelfrequenzen angeregt werden können. Die hierbei verwendete Versuchstechnik und ältere Ergebnisse sind in der unter [18] zitierten Arbeit zu finden, und weitere Eigenschaften der akustischen Beeinflussung des laminar-turbulenten Übergangs im Freistrahl sind von O. WEHRMANN [29] mitgeteilt worden.

Die experimentell bestimmte Grenzkurve $b - b'$ der Wirbelfrequenzen des Freistrahls hat, strenggenommen, nicht die gleiche Bedeutung wie die ähnlich aussehenden Kurven der Plattengrenzschicht. Für die Plattengrenzschicht geben die Kurven „neutrale", d.h. weder wachsende noch vergehende Störungen an; beim Freistrahl konnten nur wachsende Störungen, d.h. wachsende Ringwirbel, festgestellt werden. Das Innere des von den Kurven eingehüllten Bereichs hat aber beim Freistrahl und bei der Plattengrenzschicht die gleiche Bedeutung: nur im umschlossenen Bereich können Störungen entstehen.

Literatur

[1] TOLLMIEN, W.: ZAMM **6**, 468 (1926).
[2] GÖRTLER, H.: ZAMM **22**, 244 (1942).
[3] FÖRTHMANN, E.: Ing. Arch. **5**, 42 (1934).
[4] REICHARDT, H.: VDI-Forschungsheft **414** (1942).
[5] ZIMM, W.: VDI-Forschungsheft **234** (1921).
[6] RUDEN, P.: Naturwiss. **21** (1933).
[7] SCHLICHTING, H.: Grenzschicht-Theorie (1951), G. Braun, Karlsruhe.
[8] KUETHE, A. M.: Journ. Appl. Mech. **2** (1935), No. 3.
[9] SQUIRE, H. B., and J. TROUNCER: British ARC Rand M 1974 (1944).
[10] CORRSIN, S.: NACA Wartime Rep. W-94 (1943).
[11] UBEROI, M. S., and S. CORRSIN: NACA TN 2124 (1950).
[12] LIEPMANN, H. W., and J. LAUFER: NACA TN 1257 (1947).
[13] DOMM, U.: Ing. Archiv **22** (1954).
[14] WILLE, R., und U. DOMM: Jahrb. S.T.G. **46** (1952).
[15] LESSEN, M.: NACA Techn. Rep. 979 (1950).
[16] LIN, C. C.: NACA Techn. Note 2887 (1953).
[17] DOMM, FABIAN, WEHRMANN und WILLE: AF Techn. Rep. AFOSR-56-9 (1955).
[18] WILLE, R., O. WEHRMANN und H. FABIAN: AFOSR-TR-57-31 (1956).
[19] SATO, H.: Journ. Phys. Soc. Japan, **11**, 6 (1956).
[20] DOMM, U.: DVL-Bericht Nr. 23 (1956).
[21] WEHRMANN, O.: DVL-Bericht Nr. 43 (1957).
[22] SCHILLER, L., A. NAUMANN: Ing. Arch. **11** (1940).
[23] NAUMANN, A.: Forsch. Ing. Wes. **2** (1931).
[24] KURZWEG, H.: Ann. d. Phys. **18**, 5 (1933).
[25] TOWNSEND, A. A.: Proc. Roy. Soc. A, **197** (1949).
[26] FABIAN, H.: Diss. Techn. Universität Berlin (noch nicht veröffentlicht).
[27] GÖRTLER, H.: ZAMM **21**, 250 (1941).
[28] SCHUBAUER, G. B., und H. K. SKRAMSTAD: J. Aero. Sci. **14** (1947).
[29] WEHRMANN, O.: WGL-Vortrag (1957) (noch nicht veröffentlicht).

Aus der Diskussion

I. TANI (Tokyo): In connection with the lecture of Professor WILLE and Dr. WEHRMANN, I want to introduce the work of my collaborator, H. SATO, who has made some measurements on two-dimensional laminar jets. When the initial mean velocity distribution is nearly uniform, he has found that the transition takes place in a similar way as in a half jet, for which his previous investigation indicated the frequency of sinusoidal oscillations to be proportional to the 3/2 power of the mean velocity. When the initial velocity distribution is of the parabolic type, the velocity fluctuation which appears first in the transition region is also sinusoidal, but its frequency approximately coincides with the one for the maximum rate of amplification as predicted by stability calculation for the velocity distribution of hyperbolic secant type. Propagation velocity, rate of amplification and amplitude distribution in transverse direction are very close to those predicted theoretically for infinite Reynolds number.

There seems to be an essential difference in transition mechanism between this kind of free jet and the usual boundary layer flow along a solid wall. In the latter case, the observed sinusoidal velocity fluctuation is very weak and the transition actually takes place by successive intermittent bursts or by turbulent spots as described by Dr. SCHUBAUER and his coworkers. In the case of the free jet, no turbulent spots have been observed so far, except near the outer edge of the jet. The sinusoidal velocity fluctuation seems to grow more and more until the whole flow field becomes turbulent. SATO is now preparing for experiments on the excited transition in two-dimensional laminar jet flow.

Reference:

SATO, H.: Experimental Investigation on the Transition of Laminar Separated Layer. J. Phys. Soc., Japan, **11**, 702-709 (1956).

TANI, I., and H. SATO: Boundary-Layer Transition by Roughness Element. J. Phys. Soc. Japan, **11**, 1284-1291 (1956).

Diskussionsveranstaltung zur VIII. Sitzung

Beitrag zum Thema

Allgemeine Eigenschaften laminarer Grenzschichtströmungen

Von **K. Nickel,** Institut für Angewandte Mathematik der Technischen Hochschule Karlsruhe

Im Verlauf dieses Symposiums wurde mehrfach die Frage gestellt, wie sich ein gegebenes Grenzschicht-Anfangsprofil weiterentwickeln wird bzw. welche geometrische Gestalt ein stromabwärts liegendes Profil sicher *nicht* annehmen kann. Ist es z. B. möglich, daß aus dem in Abb. 1 links dargestellten Profil sich das rechts gezeichnete entwickelt, d. h., kann eine ,,Übergeschwindigkeit" von selbst entstehen ? In einer ganzen Reihe von Fällen kann diese Frage bejaht werden, so z. B. bei dreidimensionalen Grenzschichten mit nichtdrehungsfreier Außenströmung (vgl. etwa H. G. Loos [*1*]). Im folgenden wird über Untersuchungen zu dieser Fragestellung berichtet.

Um einfache Aussagen machen zu können, wird dabei eine zweidimensionale aminare und stationäre Wandgrenzschicht in einem inkompressiblen Medium vorausgesetzt. Die Außenströmung sei drehungsfrei; es seien zwar i. a. beliebige Absauge- und Ausblasegesetze zugelassen, doch möge nur senkrecht zur Wand abgesaugt bzw.

ausgeblasen werden. Mit Hilfe eines von H. Westphal [2] bzw. M. Nagumo —
S. Simoda [3] formulierten Satzes über partielle explizite parabolische Differentialgleichungen, auf den schon früher in diesem Zusammenhang von H. Görtler [4]
hingewiesen worden ist, lassen sich dann folgende Aussagen beweisen:

Übergeschwindigkeiten können a) nicht von selbst entstehen, sind sie jedoch entstanden, so können sie b), c) nicht weiter anwachsen. Genauer gilt:

a) Besitzt das Anfangsprofil keine Übergeschwindigkeiten, so auch kein stromabwärts liegendes Geschwindigkeitsprofil (Abb. 1[1]), und zwar bei beliebiger Außengeschwindigkeit $U(x)$.

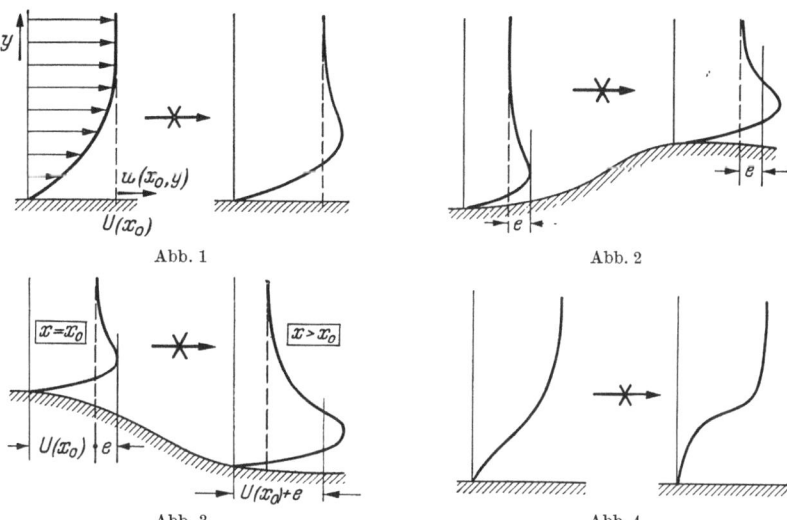

Abb. 1   Abb. 2

Abb. 3   Abb. 4

b) Besitzt das Anfangsprofil eine Übergeschwindigkeit der Größe e, und nimmt die
Außengeschwindigkeit nicht ab (in Abb. 2 durch die Gestalt der Wand angedeutet),
dann kann stromabwärts keine Übergeschwindigkeit größer als e auftreten (Abb. 2).

c) Besitzt das Anfangsprofil eine Übergeschwindigkeit der Größe e, und nimmt
die Außengeschwindigkeit $U(x)$ nicht zu, dann kann stromabwärts nirgends die
Geschwindigkeit $U(x_0) + e$ überschritten werden, wenn $U(x_0)$ die Außengeschwindigkeit im Anfangsprofil ist (Abb. 3).

In einer Arbeit von H. Witting [5] wird für kleine Störungen eines Grenzschichtprofils ein lokales Anwachsen gezeigt. Die vorstehenden Ergebnisse a) bis c) geben
in Ergänzung zur Wittingschen Arbeit eine globale Schranke an.

Die **Schubspannung** nimmt — bei beliebiger Außengeschwindigkeit $U(x)$ — ihre
maximalen Werte im Anfangsprofil und an der Wand an. Das bedeutet also:

a) Ein „Aufsteilen" eines Geschwindigkeitsprofils im Innern der Strömung ist
nicht möglich (Abb. 4), und

---

[1] In den beigefügten Bildern stellt jeweils der linke Teil das Anfangsprofil, der
rechte ein stromabwärts liegendes Geschwindigkeitsprofil dar. Der durchstrichene
Pfeil ⟶✕⟶ zwischen den beiden Teilen soll bedeuten: aus dem dargestellten Anfangsprofil kann *nicht* das dargestellte rechte Geschwindigkeitsprofil entstehen.
Aufgetragen sind jeweils als Abszisse die wandparallele Geschwindigkeitskomponente $u(x, y)$, als Ordinate der Wandabstand y, wie das bei dem Anfangsprofil von
Abb. 1 vermerkt ist.

b) monotone Geschwindigkeitsprofile bleiben auch stromabwärts noch monoton (Abb. 5). Weiter gilt sogar

c) die Anzahl der Extremwerte $E(x)$ eines Geschwindigkeitsprofils kann sich stromabwärts höchstens vermindern (Abb. 6), d. h., es ist stets $E(x) \leqq E(x_0)$ für $x \geqq x_0$.

Rückströmung. Ist $U'(x) \geqq 0$ (nimmt die Außengeschwindigkeit stromabwärts nicht ab) und wird an der Wand höchstens abgesaugt (wobei undurchlässige Wand noch zugelassen ist), so kann

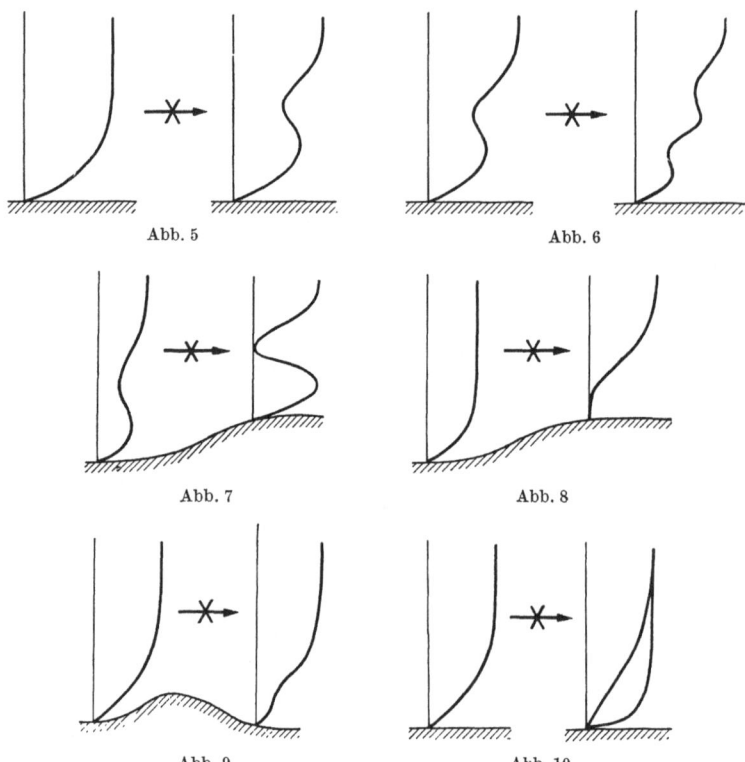

Abb. 5

Abb. 6

Abb. 7

Abb. 8

Abb. 9

Abb. 10

a) im Innern der Strömung die Geschwindigkeit nirgends auf Null absinken, d. h., dort kann keine Rückströmung eintreten (Abb. 7) und

b) auch die Wandschubspannung kann nicht zu Null werden, d. h., auch Strömungsablösung ist ausgeschlossen (Abb. 8).

Für $U'(x) > 0$, wenn also die Außengeschwindigkeit stromabwärts dauernd zunimmt, darf an der Wand sogar noch beliebiges Ausblasen zugelassen werden.

Profilwendepunkte. Undurchlässige Wand! Wechselt der Gradient $U'(x)$ der Außengeschwindigkeit zwischen einem Anfangs- und einem Endprofil G-mal das Vorzeichen, dann hat das Endprofil höchstens um $G + 1$ mehr Wendepunkte als das Anfangsprofil. $G + 1$ kann durch G ersetzt werden, wenn das Anfangsprofil an der Wand konvex ist und dort $U(x)$ ein relatives Minimum besitzt, sowie wenn das Anfangsprofil an der Wand konkav ist und dort $U(x)$ ein relatives Maximum hat (Abb. 9).

Einzigkeit. Bei vorgegebener Außengeschwindigkeit $U(x)$ ist die gesamte Grenzschicht durch das Anfangsprofil eindeutig bestimmt (Abb. 10). Solch ein Einzig-
keitssatz wurde zuerst von H. Görtler [4] angegeben.

Literatur:

[1] Loos, H. G.: A simple laminar boundary layer with secondary flow. Journal
Aer. Sc. **22**, 35—40 (1955).

[2] Westphal, H.: Zur Abschätzung der Lösungen nichtlinearer parabolischer
Differentialgleichungen. Math. Z. **51**, 690—695 (1949).

[3] Nagumo, M., und S. Simoda: Note sur l'inégalité différentielle concernant les
équations du type parabolique. Proceedings of the Japan Academy, **XXVII**,
536—539 (1951).

[4] Görtler, H.: Über die Lösung nichtlinearer partieller Differentialgleichungen
vom Reibungsschichttypus. ZAMM **30**, 265—267 (1950).

[5] Witting, H.: Über die Instabilitäten der Prandtlschen Grenzschichtgleichungen. 50 Jahre Grenzschichtforschung, Braunschweig 1955. S. 334—342.

Contribution to the Subject
Thermal Laminar Boundary Layers

By **N. Frössling,** Royal Institute of Technology, Stockholm

For such small Mach numbers that constant properties can be used (cf. Eckert
and Drewitz, Luftfahrtforschung 1942), solutions have been worked out by series
of Blasius-type for symmetrical two dimensional and for axially symmetrical flow,
thus getting some sorts of exact solutions within the region of convergence. The
cases studied are 1. the temperature distribution at vanishing heat transfer 2. the
heat transfer when the surface temperature varies with a power series.

1. With traditional symbols and with index a referring to quantities just outside
the boundary layer, with velocity distribution $U_a = u_1 x + u_3 x^3 + u_5 x^5 + \cdots$
and (for axial symmetry) axial distance distribution $r = r_1 x + r_3 x^3 + r_5 x^5 + \cdots$,
the temperature factor $\xi = \dfrac{T - T_a}{U_a^2/2\,c_p}$ can be calculated as a function of x and of a

dimensionless quantity η. This quantity is defined by $\eta = y\,\sqrt{u_1/\nu}$ for twodimen-
sional flow and by $\eta = y\,\sqrt{2\,u_1/\nu}$ for axially symmetrical flow. When $\eta = 0$, then
$\xi =$ the local recovery factor ξ_0. The following series are found, similarly to [2]:

$$\xi = 4\,\overline{\overline{F}}_2 + 6\,\frac{u_3}{u_1}\,\overline{\overline{H}}_4\,x^2 + 8\left(\frac{u_5}{u_1}\,\overline{\overline{H}}_5 + \frac{u_3^2}{u_1^2}\,\overline{\overline{K}}_6\right)x^4 + \cdots \text{(for two dimensional flow)}$$

$$\xi = 2\,\overline{\overline{H}}_2 + 3\left(\frac{u_3}{u_1}\,\overline{\overline{J}}_4 + \frac{r_3}{r_1}\,\overline{\overline{Q}}_4\right)x^2 + 4\left(\frac{u_5}{u_1}\,\overline{\overline{L}}_6 + \frac{r_5}{r_1}\,\overline{\overline{M}}_6 + \frac{r_3^2}{r_1^2}\,\overline{\overline{N}}_6 + \right.$$

$$\left. + \frac{u_3^2}{u_1^2}\,\overline{\overline{Q}}_6 + \frac{r_3\,u_3}{r_1\,u_1}\,\overline{\overline{V}}_6\right)x^4 + \cdots \text{(for axial symmetry)}$$

The double-barred quantities are functions of η alone. They are defined by
ordinary linear differential equations of the second order. The boundary conditions
are that the functions vanish at the outer edge of the boundary layer and that their
first derivatives vanish at the surface. Twelve of these functions were calculated
numerically for the Prandtl number 0,7. One result of the calculation is that all the
functions except the first vanish at the surface of the body. This means that the

local recovery factor ξ_0 will be constant over the surface, at least as far as the terms included in the calculation show. This corollary has been confirmed experimentally, e. g. [1].

2. To the solutions for the case with constant surface temperature [2] must be added some new terms, containing universal functions.

References

[1] ECKERT, E., und W. WEISE: Messung der Temperaturverteilung auf der Ober-fläche schnell angeströmter unbeheizter Körper. Forsch. Ing.-Wes. **13**, 246 (1942).

[2] FRÖSSLING, N.: Verdunstung, Wärmeübergang und Geschwindigkeitsverteilung bei zweidimensionaler und rotationssymmetrischer laminarer Grenzschicht-strömung. Lunds Universitets Årsskrift, N. F. Avd. 2, **36**, Nr. 4 (1940). English translation in NACA TM 1432 (1958).

E. WRAGE (Freiburg i. Br.): Nach den Ausführungen von Herrn N. FRÖSSLING scheint es notwendig, auf eine Arbeit von A. N. TIFFORD [1] hinzuweisen, in der eine ähnliche Behandlung der Temperaturgleichung durch eine Reihenentwicklung in den BLASIUSkoordinaten vorgenommen wird.

Im Falle konstanter Wandtemperatur ($T_w = $ const.) wird bei TIFFORD die dimensionslose Temperaturvariable $\Theta = \dfrac{T - T_a}{T_a - T_w}$ in eine Reihe

$$\theta = \frac{1}{U_a} \sum_{k=1}^{\infty} G_{2k-1}(y) \, x^{2k-1} + \frac{1}{c_p (T_a - T_w)} \sum_{k=1}^{\infty} H_{2k}(y) \, x^{2k}$$

entwickelt. Durch den Übergang von den x, y-Koordinaten zu den Blasiuskoordinaten x, η ($\eta = \sqrt{u_1/\nu} \cdot y$) und durch Aufspaltung der G_{2k-1} und H_{2k} in Linearkombinationen erhält TIFFORD universelle Funktionen zur Berechnung des Temperaturfeldes. Die zugehörigen Differentialgleichungen werden aufgestellt für die symmetrische Strömung um einen Körper mit vorderem Staupunkt für die Ordnungen $k = 1, \ldots, 6$. Vertafelt sind aber außer den Funktionen für die Berechnung des Geschwindigkeitsfeldes nur die durch die Aufspaltung der G_{2k-1} entstandenen Funktionen, und zwar für eine PRANDTL-Zahl 1.

Zur Behandlung des Thermometerproblems und des Kühlungsproblems mit variabler Wandtemperatur wird die dimensionslose Temperaturkoordinate $\vartheta = \dfrac{T - T_t}{T_t}$ ($T_t = U_a^2/2 \, c_p + T_a = $ const.) in die folgende Reihe entwickelt:

$$\vartheta = \sum_{K=0}^{\infty} G_k(y) \, x^k + \frac{U_a^2}{2 c_p T_t} \left\{ -1 + \sum_{K=0}^{\infty} H_k(y) \, x^k \right\}.$$

Die Differentialgleichungen für die zugehörigen universellen Funktionen werden aufgestellt für die Ordnungen $k = 0, 1, \ldots, 8$. Schließlich sei noch erwähnt, daß auch die Differentialgleichungen für die universellen Funktionen der Geschwindigkeitsgrenzschicht und der Temperaturgrenzschicht ($T_w = $ const.) mit einer Außengeschwindigkeit der Form

$$U_a = u_0 + u_1 \, x + u_2 \, x^2 + \cdots \qquad (u_0 \neq 0)$$

aufgestellt worden sind.

Literatur:

[1] TIFFORD, A. N.: WADC Techn. Report 53—288 Part 4 (Aug. 1954).

Bemerkung des Herausgebers: Die vorstehende Diskussionsbemerkung von Herrn E. WRAGE ist eine ausführlichere Darstellung seines während des Symposiums gegebenen kurzen mündlichen Hinweises.

Contribution to the Subject

Boundary Layer Separation on Swept Wings and Stall

By **J. E. de Krasinski,** Instituto Aerotecnico, Cordoba

The problem of boundary layer separation on swept wings at high incidences presents a formidable problem when attacked by means of Navier-Stokes equations. The study of this problem however is of great practical importance as it involves the static longitudinal stability of wings near the stall.

The author would like to draw attention to his paper published originally in the Instituto Aerotecnico, Cordoba, Argentine and which was subsequently translated from Spanish into English by the British Ministry of Supply as TIB/T 4295 (Ref. [1]).

The behaviour in the stall of swept wings depends on the 3-dimensional boundary layer flow; it seems that not enough attention has been paid so far whether the stall develops in the direction of the cross flow from the plane of symmetry towards the wing tips or against it. Assuming that the stall must begin somewhere in the vicinity of the maximum local lift coefficient C_l, swept wings were divided into two classes. Class I comprises the wings where the maximum local C_l (estimated by means of Ref. [2]) occurs between the plane of symmetry and the wing tip region. Class II comprises the wings where the maximum local C_l occurs in the wing tip region itself or more exactly between the tip and the 0,92 of the semi-span.

For the wings of Class I an equation of equilibrium for neutral static stability of a wing element was derived in the form:

$$(\eta_{Cl\,max} - \eta_{C.\,P.})\,\mathrm{tg}\,\varLambda - \frac{dx}{d\alpha}\,\frac{C_{l1}}{a_2 - a_1}\,\frac{2}{b} = 0$$

η the non-dimensional fraction of the semi-span.

$\dfrac{dx}{d\alpha}$ a measure of the C. P. movement during the stall.

C_{l1} local lift coefficient of the element.

b span.

$a_1; a_2$ lift slopes before and after the stall of the element.

The first term of this expression can be treated rigorously by means of swept wing theory as a function of the aspect ratio A, taper ratio λ, and sweep angle $\varLambda$; the second term can not be treated in this way. The first term however should change its sign within the range of stable and unstable wings.

An analysis of behaviour in the stall was made of 40 swept wings. This behaviour was plotted against the first term of the equation i. e. against $(\eta_{Cl\,max} - \eta_{C.\,P.})\,\mathrm{tg}\,\varLambda$. The result of this analysis is shown in fig. 1. One can observe that the wings stable and unstable group themselves well around a value of 0,063 of that parameter. It means that for neutral stability $(\eta_{Cl\,max} - \eta_{C.\,P.})\,\mathrm{tg}\,\varLambda = 0,063$ what is equivalent to a statement that $f\,(A;\,\lambda;\,\varLambda) = 0,063$. This condition is the suggested stability criterion for wings of Class I. With the help of WEISSINGER's swept wing theory (Ref. [2]) it was possible to plot that relation as seen on the upper part of the fig. 2.

.

Wings to the left of the curves marking their corresponding aspect ratio are stable, to the right unstable.

This criterion for Class I wings can be compared with the well known SHORTAL's

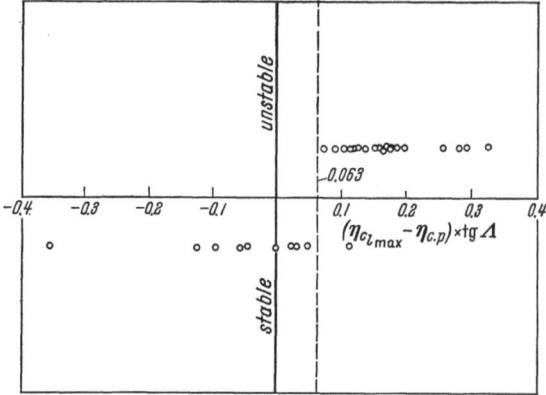

Fig. 1. Stable and unstable wings (experimental) grouped in dependance of the parameter:
$$(\eta_{Cl\,max} - \eta_{C.P.}) \times \mathrm{tg}\,\varLambda$$

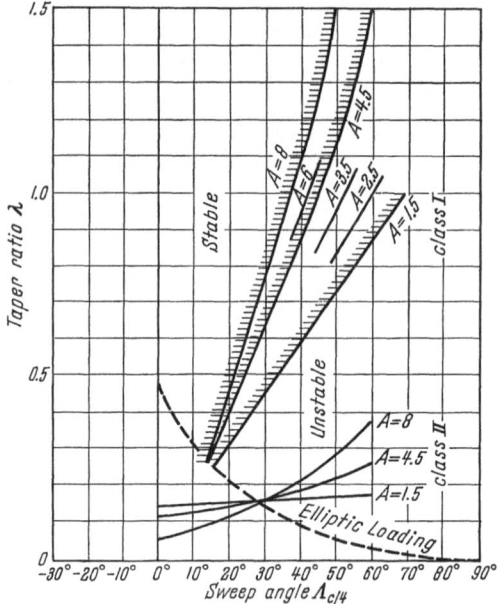

Fig. 2. Longitudinal stability in stall of swept wings (Class I and II) as function of taper, sweep angle, and aspect ratio

empirical curve (Ref. [3]). Neglecting the effects of taper ratio, one can take as a rough approximation that $(\eta_{Cl\,max} - \eta_{C.P.}) \sim A$, hence one obtains $A = \dfrac{K}{\mathrm{tg}\,\varLambda}$. The constant of proportionality K can be shown to be approximately equal some value equal about 2. Curves $\dfrac{2}{\text{tg }\varLambda}$ and $\dfrac{2,72}{\text{tg }\varLambda}$ are plotted against sweep angle $\varLambda$ in fig. 3 together with Shortal's curve, and are in good agreement with each other.

Wings of Class II are in general of a pronounced taper. Their stall begins at the tip; as the surfaces affected are small, the stall moves relatively slowly the cross flow in the boundary layer, and the C. P. of the central part of the wing tends to move slightly backwards at higher incedences, hence stability (delta wings) or very gentle instability seems usually to occur. Wings of this Class plotted by means

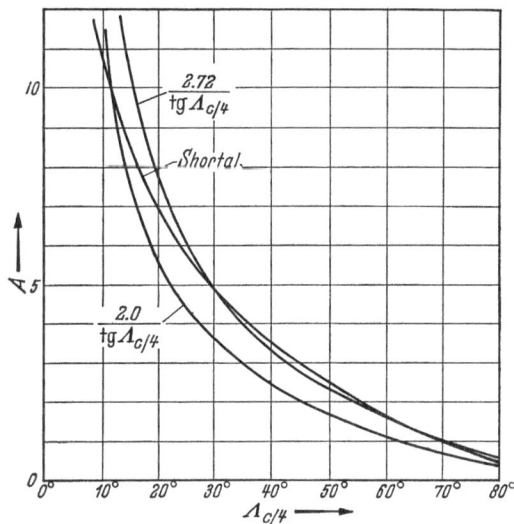

Fig. 3. Comparison of Shortal's empirical curve (Ref. [3]) with simplified premises of the proposed criterion (Ref. [1]).

of Ref. [2] are shown at the lower part of fig. 2. Wings under the curves of their corresponding aspect ratio should possess the above mentioned characteristic. A curve which corresponds to eliptical loading is also shown. Wings of this class can be of particular interest for cargo planes and sailplanes, but more research work on them is needed.

It is not thought that the conclusiones reached for Class I and II of wings would necessarily apply when "vortex flow" establishes itself on the upper surface of the wing due to a very pronounced sweep and small radius of curvature of the leading edge.

References:

[1] DE KRASINSKI, J. E.: Suggested Criterion of Longitudinal Stability in Stalling of Swept-Back Wings. Com. e Informes C. 6. Instituto Aerotecnico. Cordoba. Argentine. Translated Brit. Min. of Supply TIB/T 4295.

[2] YOUNG, J.: Theoretical additional span loading characteristics of wings with arbitrary sweep, aspect and taper ratios. Dec. 1947. NACA T. N. 1491 (WEISSINGER's Theory).

[3] SHORTAL, J.: Effect of sweepback and aspect ratio on longitudinal stab. characteristics of wings at low speed. July 46, NACA T. N. 1093.